AF323677

MODERN

CONCEPTS IN

ANGIOGENESIS

MODERN

CONCEPTS IN

ANGIOGENESIS

edited by

MICHAEL SIMONS
Dartmouth Medical School
Dartmouth–Hitchcock Medical Center, USA

GABOR M RUBANYI
Cardium Therapeutics, Inc., USA

Imperial College Press

Published by

Imperial College Press
57 Shelton Street
Covent Garden
London WC2H 9HE

Distributed by

World Scientific Publishing Co. Pte. Ltd.
5 Toh Tuck Link, Singapore 596224
USA office: 27 Warren Street, Suite 401-402, Hackensack, NJ 07601
UK office: 57 Shelton Street, Covent Garden, London WC2H 9HE

British Library Cataloguing-in-Publication Data
A catalogue record for this book is available from the British Library.

MODERN CONCEPTS IN ANGIOGENESIS

ISBN-13 978-1-86094-763-6
ISBN-10 1-86094-763-8

Typeset by Stallion Press
Email: enquiries@stallionpress.com

Printed in Singapore by Mainland Press

CONTENTS

Contributors   **xiii**

Preface   **xix**

I Components of Angiogenic Cascades

Chapter 1. Semaphorins, Plexins and Neuropilins and Their Role in Vasculogenesis and Angiogenesis   **1**

Gera Neufeld, Niva Shraga-Heled, Tali Lange and Ofra Kessler

1. Introduction and Historical Perspective   1
2. The Semaphorins   3
3. The Plexin Receptor Family   4
4. The Neuropilins   6
5. Vascular Endothelial Growth Factors and Their Receptors   8
6. Signal Transduction by Neuropilins   11
7. The Role of the Neuropilins in the Regulation of Vasculogenesis and Angiogenesis   14
8. Modulation of Angiogenesis by Semaphorins that Bind Directly to Plexins   17
Acknowledgments   17
References   17

Chapter 2.   EPH Receptors and Ephrins   27
Elena B. Pasquale

1. Introduction   27
2. Effects on Vascular Cell Behavior and Signaling Pathways   32
3. Endothelial Cell Fate   42
4. Angiogenic Remodeling of Embryonic Blood Vessels   43
5. Lymphatic Vessels   49
6. Adult Vasculature   50
7. Targeting Eph Receptor-Ephrin Interactions
    to Modulate Angiogenesis   58
8. Perspectives   59
Acknowledgments   60
References   60

Chapter 3.   The FGF Family of Angiogenic Growth Factors   67
Patrick Auguste and Andreas Bikfalvi

1. Introduction   67
2. Molecular Mechanisms   68
3. Role in Vascular Development   73
4. FGFs in Tumor Angiogenesis   75
5. Role of FGFs in Developmental and Tumor
    Lymphangiogenesis   78
6. Role in Repair-Associated Angiogenesis and Ischemia
    Revascularization   78
7. Conclusion   80
Acknowledgments   81
References   81

Chapter 4.   Neuropeptide Y: Neurogenic Mediator of
              Angiogenesis and Arteriogenesis   91
Joanna B. Kitlinska and Zofia Zukowska

1. The NPY System   91
2. NPY as a Growth Factor for Vascular Cells   93

3. DPPIV: A Molecular Switch of the NPY
    Angiogenic System   97
4. Downstream Mediators of NPY Actions   98
5. NPY in Revascularization of Ischemic Tissues   99
6. NPY in Wound Healing   101
7. NPY in Adipose Tissue Growth and Obesity   102
8. NPY in Retinopathy   103
9. NPY-Induced Angiogenesis in Angioplasty-Induced
    Neointima and Atherogenesis   105
10. NPY in Tumor Angiogenesis   105
11. NPY-Mediated Angiogenesis and Neurogenesis   109
References   111

**Chapter 5.   Modulation of Growth Factor Signaling by
               Heparan Sulfate Proteoglycans**   **119**
Nicholas W. Shworak

1. Introduction   119
2. Historical Perspective   121
3. The Structure, Synthesis, and Post-Synthetic Modification
    of HSPGs   123
4. Evolution of HSPGs   129
5. HSPGs in Development   131
6. HSPG Modulation of Ligand-Receptor Interactions   133
7. HSPGs Enable Global Control of EC Phenotype   139
8. Future Therapeutic Directions   139
9. Conclusions   140
References   141

II Angiogenic Regulators

Chapter 6.   Directional Cues in Angiogenesis   **147**
Arie Horowitz

1. Introduction: Blood Vessels and Nerves Use Similar
    Guidance Cues   147
2. Semaphorin Signaling   148

3. Ephrins and Eph Signaling   156
4. Netrin and Slit Signaling   163
5. Open Questions   165
References   166

**Chapter 7.   Regulation of Angiogenesis and Arteriogenesis
             by Hypoxia-Inducible Factor-1**   **175**
Gregg L. Semenza

1. Oxygen Homeostasis: Phylogeny, Ontogeny, Physiology,
    and Pathobiology   175
2. Hypoxia-Inducible Factor 1: Master Regulator of
    O_2 Homeostasis   178
3. Control of Angiogenic Growth Factor and Cytokine
    Production by HIF-1   184
4. Cell-Autonomous Effects of HIF-1 in Vascular
    Endothelial Cells   185
5. Control of Angiogenesis and Arteriogenesis by HIF-1   186
6. Control of Tumor Angiogenesis by HIF-1   198
References   204

**Chapter 8.   Redox State and Regulation of Angiogenic
             Responses**   **217**
Masuko Ushio-Fukai and R. Wayne Alexander

1. Introduction   217
2. Reactive Oxygen Species (ROS) in the Vasculature   218
3. ROS and Angiogenesis   219
4. NAD(P)H Oxidase: A Major Source of ROS in
    the Vasculature   222
5. Role of NAD(P)H Oxidase in Angiogenesis   225
6. ROS as Signaling Molecules in Angiogenesis   229
7. Angiogenesis-Dependent Transcription Factors and
    Genes Regulated by ROS   233

8. Conclusion   235
References   235

**Chapter 9.   Angiogenesis and Arteriogenesis in Cardiac
                Hypertrophy**   **253**
Robert J. Tomanek and Eduard I. Dedkov

 1. Introduction   253
 2. Assessing Coronary Angiogenesis and
     Arteriogenesis   254
 3. Pressure Overload-Induced Hypertrophy   254
 4. Volume Overload-Induced Cardiac Hypertrophy   258
 5. Thyroxine-Induced Hypertrophy   260
 6. Hypoxia-Induced Hypertrophy   261
 7. Exercise-Induced Hypertrophy   263
 8. Myocardial Infarction-Induced Hypertrophy   264
 9. Modulators of Angiogenesis During Hypertrophy   267
10. Stimuli of Angiogenesis During Hypertrophy   269
11. Summary   271
References   272

**Chapter 10.   Regulation of Coronary Vascular Tone and
                Microvascular Physiology**   **281**
Basel Ramlawi, Munir Boodhwani and Frank W. Sellke

1. Introduction   281
2. Coronary Resistance   283
3. Regulation of Coronary Microvascular Tone   285
4. Endothelial Factors in Vascular Growth and Response
   to Injury   300
5. Impact of Disease States on Coronary Circulation   301
6. The Coronary Microcirculation in Hypertophic States   305
7. Summary   306
References   307

III   Clinical Applications

Chapter 11.   Kinase Inhibitors: Cancer Drugs Derived from Mechanistic Considerations   **313**
Karl-Heinz Thierauch and Andreas Chlistalla

1. Kinase Inhibition and Tumor Angiogenesis   313
2. Major Angiogenesis Factors and Receptors   314
3. Further Angiogenesis-Related Signaling   315
4. Need for Selectivity of Anti-Angiogenic Kinase Inhibitors   315
5. Kinase Inhibitors in Clinical Development   316
6. Challenges and Future Directions   332
Acknowledgments   335
References   335

Chapter 12.   Therapeutic Angiogenesis — An Overview   **343**
Masahiro Murakami and Michael Simons

1. Introduction   343
2. Concepts and Rationales   344
3. Strategy   346
4. Clinical Trials   348
5. Issues Regarding Current Strategy   352
6. Emerging Concepts of Therapeutic Angiogenesis   358
7. Future Prospective   361
8. Summary   363
References   363

Chapter 13.   Hepatocyte Growth Factor   **367**
Ryuichi Morishita and Toshio Ogihara

1. Hepatocyte Growth Factor in Cardiovascular System   367
2. HGF Signaling in Endothelial Cells   368
3. Angiogenic Therapy for Ischemic Peripheral Arterial Diseases   370
4. Clinical Trial in PAD   371
5. HGF Gene Therapy for Myocardial Ischemia   375

6. HGF Gene Therapy for Restenosis After Angioplasty   377
7. Next Five Years Perspective — Future Direction
    of HGF Therapy   378
Acknowledgments   379
References   380

**Chapter 14.   Role of Nitric Oxide in Adult Angiogenesis:
                Therapeutic Potential of Endothelial Nitric Oxide
                Synthase Gene Transfer   385**
Gabor M. Rubanyi

1. Endothelial Nitric Oxide in Health and Disease   385
2. Nitric Oxide and Angiogenesis   389
3. NOS Gene Transfer   393
4. Therapeutic Angiogenesis with NOS-III Gene Transfer
    for Critical Limb Ischemia   395
5. Potential Therapeutic Utility of NOS-III Gene Transfer
    in the Heart   407
6. Conclusions   410
Acknowledgments   411
References   411

Index   423

CONTRIBUTORS

R. Wayne Alexander
Division of Cardiology, Department of Medicine
Emory University School of Medicine
1639 Pierce Drive
Atlanta, GA, 30322, USA

Patrick Auguste
Molecular Angiogenesis Laboratory
INSERM EMI 0113
Université Bordeaux I
Avenue des Facultés
33 405 Talence, France

Andreas Bikfalvi
Molecular Angiogenesis Laboratory
INSERM EMI 0113
Université Bordeaux I
Avenue des Facultés
33 405 Talence, France

Munir Boodhwani
Division of Cardiothoracic Surgery, Department of Surgery
Beth Israel-Deaconess Medical Center
Harvard Medical School
110 Francis St., LMOB 2a
Boston, MA 02215, USA

Andreas Chlistalla
Medical Development-Oncology, Schering AG
Muellerstraße 178
D-13342, Berlin, Germany

Eduard I. Dedkov
Department of Anatomy and Cell Biology
    and the Cardiovascular Center
Carver College of Medicine
University of Iowa
Iowa City, IA 52242, USA

Arie Horowitz
Angiogenesis Research Centre and
Section of Cardiology, Department of Medicine
Dartmouth Medical School
Borwell Building 554W, One Medical Center Drive
Lebanon, NH 03756, USA

Ofra Kessler
Cancer and Vascular Biology Research Center
Rappaport Research Institute in the Medical Sciences
The Bruce Rappaport Faculty of Medicine
Technion, Israel Institute of Technology
1 Efron St., P.O. Box 9679
Haifa, 31096, Israel

Joanna B. Kitlinska
Department of Physiology and Biophysics
Georgetown University Medical Center
3900 Reservoir Dr, NW
Washington, DC 20057, USA

Tali Lange
Cancer and Vascular Biology Research Center
Rappaport Research Institute in the Medical Sciences
The Bruce Rappaport Faculty of Medicine
Technion, Israel Institute of Technology
1 Efron St., P.O. Box 9679
Haifa, 31096, Israel

Ryuichi Morishita
Division of Clinical Gene Therapy
Osaka University Graduate School of Medicine
2-2 Yamada-oka, Suita 565-0871, Japan

Masahiro Murakami
Angiogenesis Research Centre and
Section of Cardiology, Department of Medicine
Dartmouth Medical School
Borwell Building 554W, One Medical Center Drive
Lebanon, NH 03756, USA

Gera Neufeld
Cancer and Vascular Biology Research Center
Rappaport Research Institute in the Medical Sciences
The Bruce Rappaport Faculty of Medicine
Technion, Israel Institute of Technology
1 Efron St., P.O. Box 9679
Haifa, 31096, Israel

Toshio Ogihara
Department of Geriatric Medicine
Osaka University Graduate School of Medicine
2-2 Yamada-oka, Suita 565-0871, Japan

Elena B. Pasquale
The Burnham Institute for Medical Research
10901 N. Torrey Pines Rd.
La Jolla, CA 92037, USA
and Pathology Department
University of California San Diego
La Jolla, CA 92093, USA

Basel Ramlawi
Division of Cardiothoracic Surgery, Department of Surgery
Beth Israel-Deaconess Medical Center
Harvard Medical School
110 Francis St., LMOB 2a
Boston, MA 02215, USA

Gabor M. Rubanyi
Chief Scientific Officer
Cardium Therapeutics
3611 Valley Center Drive, Suite 525
San Diego, CA 92130, USA

Frank W. Sellke
Division of Cardiothoracic Surgery, Department of Surgery
Beth Israel-Deaconess Medical Center
Harvard Medical School
110 Francis St., LMOB 2a
Boston, MA 02215, USA

Gregg L. Semenza
Vascular Biology Program, Institute for Cell Engineering
Departments of Pediatrics, Medicine, Oncology,
    and Radiation Oncology
McKusick-Nathans Institute of Genetic Medicine
The Johns Hopkins University School of Medicine
Broadway Research Building, Suite 671

733 North Broadway
Baltimore, MD 21205, USA

Niva Shraga-Heled
Cancer and Vascular Biology Research Center
Rappaport Research Institute in the Medical Sciences
The Bruce Rappaport Faculty of Medicine
Technion, Israel Institute of Technology
1 Efron St., P.O. Box 9679
Haifa, 31096, Israel

Nicholas W. Shworak
Section of Cardiology, Department of Medicine
Dartmouth Medical School and
    Angiogenesis Research Center
HB7504, Dartmouth-Hitchcock Medical Center
Borwell Building 540W, One Medical Center Drive
Lebanon, NH 03756, USA

Michael Simons
Section of Cardiology, Department of Medicine
Dartmouth Medical School and
    Angiogenesis Research Centre
Borwell Building 554W, One Medical Center Drive
Lebanon, NH 03756, USA

Karl-Heinz Thierauch
Global Research Business Area-Oncology, Schering AG
Muellerstraße 178
D-13342, Berlin, Germany

Robert J. Tomanek
Department of Anatomy and Cell Biology
    and the Cardiovascular Center
1-402 BSB, Carver College of Medicine
University of Iowa
Iowa City, IA 52242, USA

Masuko Ushio-Fukai
Division of Cardiology, Department of Medicine
Emory University School of Medicine
1639 Pierce Drive, Rm. 319
Atlanta, GA, 30322, USA

Zofia Zukowska
Department of Physiology and Biophysics
Georgetown University Medical Center
Box 571460
3900 Reservoir Dr, NW
Washington, DC 20057, USA

PREFACE

The field of angiogenesis continues to rapidly evolve in many different directions. Initial discoveries of a few angiogenic growth factors, such as vascular endothelial growth factors (VEGF), fibroblast growth factors (FGF) and angiopoietins, gave an impression of a relatively uncomplicated system with a few straightforward regulatory mechanisms revolving around hypoxia and inflammation. However, failures of therapeutic approaches based on brute force stimulation or inhibition of vessel growth combined with discoveries in fields as diverse as developmental biology, signal transduction and neurosciences began painting a much more nuanced and complex system. The existence of a great number of checks and balances involved in growth and maintenance of the vasculature, a great diversity of angiogenic growth factors and inhibitors, and in particular the concepts of vascular guidance and participation of numerous proteins capable of modifying growth factor activities were some of the more prominent discoveries of the last decade.

It would be utterly impossible to summarize all these developments in one volume. Therefore, we have chosen to focus on a few select developments in the field of therapeutic angiogenesis that have significantly altered our understanding of vascular biology and pathology.

The first part of the book deals with key components of the angiogenic cascade. Semaphorins, plexins and neuropilins have emerged as important regulators of vascular growth by both providing guidance clues and transmitting signaling to the endothelium. The diversity of these families, their structure and interactions, as well as our current understanding of their roles are discussed in Chapter 1 by G. Neufeld and colleagues. Ephrins and their Eph receptors play an important and still poorly understood bidirectional signaling function in various cell-cell interactions that regulates processes as diverse as vascular sprouting, blood and lymphatic vessel morphogenesis and remodeling as well as arterial-venous fate decisions. These and other roles played by numerous members of this family are addressed in Chapter 2 by E. Pasquale.

FGF were the first angiogenic growth factors isolated, yet their function in the vasculature still remains mysterious and poorly appreciated. New insights into FGF biology are presented in Chapter 3 by P. Auguste and A. Bikfalvi. The discovery of various ways in which the nervous system participates in regulation of angiogenesis has been one of the most intriguing recent developments in vascular biology. In Chapter 4, J. Kitlinska and Z. Zukowska address the role of neuropeptide Y (NPY) in regulation of blood vessel growth. The extracellular matrix (ECM) plays a critical role in modulating signaling of various growth factors. One relatively little studied component of the ECM is the heparan sulfate matrix that affects signaling of various heparin-binding growth factors. Recent advances in this field are addressed in Chapter 5 by N. Shworak.

The second part of the book deals with various processes that affect the angiogenic cascade. The concept of arterial guidance is addressed in Chapter 6 by A. Horowitz. The chapter addresses various guidance systems starting with semaphorins, neuropilins and plexins, and progressing to ephrins/Eph receptors, and then to netrins. New development in our understanding of the role of HIF signaling are discussed in Chapter 7 by G. Semenza.

The reactive oxygen species have long been studied in terms of their contribution to vascular wall injury. But new discoveries in this field demonstrate an important role of ROS in regulation of various aspects of endothelial signaling. These and other new ideas about ROS function are examined in Chapter 8 by M. Ushio-Fukai and W. Alexander. The growth of new vessels may affect normal organ function and, conversely, organ growth *per se* can apparently induce an angiogenic program in the absence of hypoxia by using mechanical stretch as a stimulus. These new insights are discussed in the context of myocardial hypertrophy by R. Tomanek and E. Dedkov. The growth of new vessels may affect microvascular milieu in a number of ways, including changes in microvasculature response to various agonists and antagonists. These microvascular physiological aspects of angiogenesis are addressed in Chapter 10 by F. Sellke and colleagues.

The third part of the book deals with new insights into therapeutic applications of discoveries in the field of angiogenic biology. Perhaps one of the most anticipated applications has been the development of various tyrosine kinase inhibitors. This side of therapeutic anti-angiogenesis is discussed in Chapter 11 by K.-H. Thierauch. To date, application of angiogenic therapies to cardiovascular diseases have not fulfilled lofty expectations of the past decade. The reasons for this lack of progress and potential paths forward are addressed in Chapter 12 by M. Murakami and M. Simons. Perhaps some of the more exciting recent therapeutic applications in the angiogenic growth factor field have involved the hepatocyte growth factor (HGF). HGF biology and biological role as well as its therapeutic applications are discussed in Chapter 13 by R. Morishita and T. Ogihara. Nitric oxide, a recent *Science* magazine molecule of the year, has always been at the center of endothelial biology. New advances in the NO field and their potential therapeutic applications are discussed in Chapter 14 by G. Rubanyi.

The composition of a multi-author monograph is never a simple process and many thanks are due. First, we would like to thank all the contributors for a thorough review of their respective areas. The editors at the Imperial College Press and especially Joy Quek have been highly professional and patient in dealing with us and helping us to put together the best book possible. Finally, we would like to thank our colleagues for their support and advice in this project.

M. Simons
G. M. Rubanyi
(May 2007)

1

Semaphorins, Plexins and Neuropilins and Their Role in Vasculogenesis and Angiogenesis

by Gera Neufeld, Niva Shraga-Heled,
Tali Lange and Ofra Kessler

1. Introduction and Historical Perspective

In the late 1980s and early 1990s, the angiogenic factors belonging to the VEGF family were discovered and characterized. It was obvious from the very earliest days that multiple splice forms of VEGF existed. However, these splice forms exhibited relatively similar activities in standard *in vitro* assays although their potencies varied. Nevertheless, it seemed possible that receptors able to differentiate between specific splice forms

---

Note: The names of all semaphorins are abbreviated according to the following rules: s always appears as the first letter denoting a semaphorin. The number after the s designates the semaphorin to a specific class, and the final letter designates its place within the class. Thus, s3f means semaphorin-3F while s4d means semaphorin-4D.

VEGF:Vascular endothelial growth factor, VEGFR-1: VEGF receptor-1, VEGFR-2: VEGF receptor-2, np1: neuropilin-1, np2: neuropilin-2, and HGF: hepatocyte growth factor.

1

of VEGF may exist, and that these receptors should differ from the known tyrosine-kinase VEGF receptors, VEGFR-1 and VEGFR-2, which do not differentiate between VEGF splice forms. An early indication that such splice form-specific VEGF receptors exist was obtained in experiments in which receptors that bind the VEGF splice form $VEGF_{165}$ but not the $VEGF_{121}$ splice form were first identified.[1] However, the molecular identity of these receptors was only revealed two years later when they were found to be the products of the neuropilin-1 (np1) gene.[2] Subsequently, it was realized that the related neuropilin-2 (np2) gene product also functions as a receptor for $VEGF_{165}$ and for $VEGF_{145}$, a VEGF splice form that does not bind to np1.[3] Furthermore, additional VEGF family members such as PlGF-2, VEGF-B, and VEGF-C were also found to interact with np1 or with np2 or with both receptors.[4–6]

Np1 had been identified almost a decade ago as the A5 antigen, a cell surface protein involved in neuronal recognition.[7,8] It was subsequently realized that np1 functions as a receptor for the axon guidance factor semaphorin-3A (s3a).[9,10] S3a belongs to a semaphorin subfamily that includes the seven class-3 semaphorins. S3a was previously known as collapsin-1 because it induces an np1-mediated collapse of the actin cytoskeleton in axon growth cones of responsive nerve cells.[11] At the same time, Np2 was also found to function as a receptor for class-3 semaphorins such as semaphorin-3F (s3f). Following these findings it was realized that np1 is unlikely to be able to transduce semaphorin signals on its own because its intracellular domain is very short. It was postulated therefore that class-3 semaphorin receptors had to contain an independent signal transducing component other than neuropilins. These signaling components turned out to be members of the plexin receptor family.[12,13] Interestingly, plexins serve as direct signal transducing receptors for most of the semaphorins except for some members of the class-3 semaphorin subfamily (S3a–d, and s3f) which do not bind directly to plexins and require neuropilins as obligate semaphorin binding components in semaphorin holo-receptors.

These developments raise several questions with regard to the regulation of angiogenesis. The first is concerned with the elucidation of the mechanisms by which neuropilins modulate the angiogenesis promoting activities of the members of the VEGF family. The second concerns the potential role of the alternative neuropilin ligands, the semaphorins, as potential modulators of angiogenesis. Another, more general question relates to the role of the nervous system in the regulation of developmental angiogenesis. These questions have been addressed to some extent in recent years although our knowledge is still far from complete.

2. The Semaphorins

The semaphorin family consists of more than 30 genes divided into eight classes, of which the first two classes are derived from invertebrates, classes 3–7 are the products of vertebrate semaphorins, and the eighth class contains viral semaphorins (Fig. 1). In the literature, the semaphorins are often referred to by an array of confusing designations.

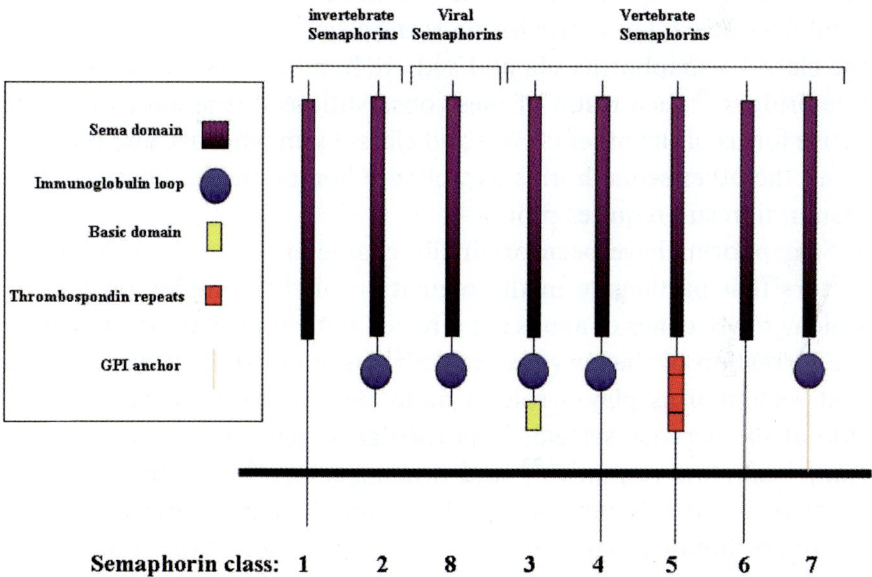

Fig. 1. The semaphorin family. The different semaphorin subclasses are shown. Classes 3–7 contain vertebrate semaphorins. The two main semaphorin subclasses containing members reported to function as angiogenesis regulators are the class-3 and class-4 semaphorins. Class-3 semaphorins are the only secreted vertebrate semaphorins. The subfamily contains seven known members. They are distinguished by a small basic domain and by an Ig-like domain in addition to the sema domain which is present in all semaphorins. Class-4 semaphorins are membrane-anchored semaphorins containing an Ig loop-like domain.

This situation was clarified several years ago by the adoption of a unified nomenclature for the semaphorins.[14] The semaphorins are characterized by the presence of a $\sim$ 500 amino acid-long sema domain that is located close to their N-termini. The sema domain is essential for semaphorin signaling and determines receptor binding specificity.[15] The sema domains of two different semaphorins were recently characterized by X-ray crystallography revealing a beta propeller topology.[16–19] The different semaphorin classes are characterized by class-specific structural motifs. Thus, semaphorins belonging to classes 2–4 and 7 contain immunoglobulin-like domains (Fig. 1), class-5 semaphorins contain thrombospondin repeats and class-3 semaphorins contain a basic domain. Class-3 semaphorins are produced as secreted proteins but other classes of vertebrate semaphorins are produced as membrane anchored or transmembrane proteins that can be further processed into soluble proteins. The active forms of the class-3 semaphorin s3a and of the class-4 semaphorins s4a and s4d are homodimers linked by disulfide bridges.[20] As a result of these observations it is assumed that the active forms of the other class-3 and class-4 semaphorins, and possibly of all the other semaphorins as well, are homodimeric. However, this assumption still requires proof.

Semaphorins have been originally characterized as axon guidance factors that participate in the regulation of the complex process by which growth cones of axons are directed to their proper targets during the formation of the nervous system.[21] In recent years it was realized that semaphorins play a role in many developmental processes outside of the nervous system, in particular as regulators of cell migration,[22] immune responses[23] and organogenesis.[24] It is therefore not surprising that s3b and s3f have been initially characterized as modulators of tumor progression,[25,26] and that these semaphorins as well as other semaphorins have been recently found to function as regulators of angiogenesis and tumor angiogenesis.

3. The Plexin Receptor Family

Most of the semaphorins were found to bind directly to cell surface receptors belonging to the plexin receptor family.[12] The plexins are segregated into four sub-families. There are four A-type plexins, three

B-type plexins, and single C- and D-type plexins.[27,28] Different plexins serve as direct binding receptors for different types of semaphorins. Thus, plexin-B1 is a receptor for s4d,[12] plexin-B3 is a receptor for s5a,[29] plexin-A1 is a s6d receptor,[30] and plexin-D1 is a receptor for s3e,[31] to name but a few examples. The plexins contain a split cytoplasmic SP (sex-plexin) domain (also known as the C1 and C2 domains). The intracellular domain contains putative tyrosine phosphorylation sites but no tyrosine kinase domain. The intracellular part of plexin-B1 contains a domain located between the C1 and C2 domains that functions as a GTPase activating protein (GAP) domain. This GAP-like domain is conserved quite highly throughout the plexin family although it is unclear whether it is functional in all plexins.[32] The extracellular domain of all plexins contains a sema domain which serves as an auto-inhibitory domain in the basal, non-activated state of the receptor.[33] The extracellular domains of the plexins also contain sequences homologous to similar sequences found in the Met subfamily of tyrosine-kinase receptors. These are designated as Met related sequence (MRS) domains and glycine-proline (G-P) rich motifs (Fig. 2).[34]

The intracellular domain of the plexin-A1 receptor contains a binding site for the small GTPase Rac-1 as well as a binding site for the intracellular tyrosine-kinase Fes/Fps, which phosphorylates plexin-A1 in response to s3a.[35] Likewise, it was found that the intracellular tyrosine kinase Fyn binds to the intracellular domain of the plexin-A2 receptor and phosphorylates it in response to s3a.[36] The serine-threonine kinase Cdk-5 also associates with plexin-A2, is activated by s3a and phosphorylates in turn the CRMP2 protein which serves as an important downstream target of plexins in neurons.[36–38] The intracellular domain of the Drosophila homologue of plexin-A1, plexin-A, also contains a binding site that enables association with the flavoprotein oxidoreductase MICAL, which was found to be essential for correct s1a-induced axon repulsion in Drosophila.[39]

Type-B plexins resemble class-A plexins in their primary features, but there are some peculiarities specific to this plexin subclass. Plexin-B1 is the best characterized type-B plexin. It also binds Rac-1 but the purpose of the binding may be sequestration of Rac-1 so as to prevent its interaction with downstream targets.[40,41] Activation of plexin-B1 by s4d sequesters Rac-1 and inhibits Rac1 signaling, while simultaneously

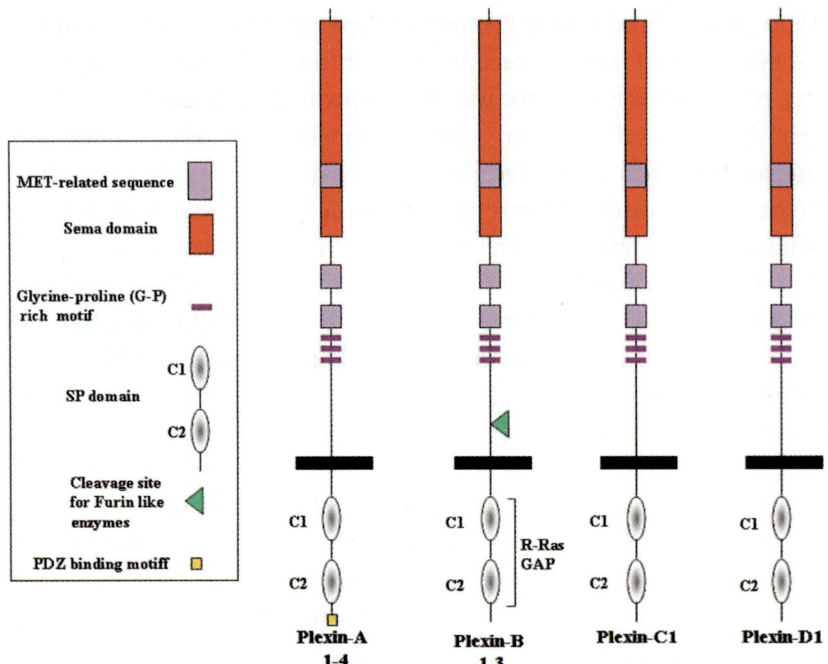

Fig. 2. The plexin receptor family. There are currently nine known mammalian members of this family. They are grouped into four subfamilies. Members of the A, B and D subfamilies have been found to function as modulators of angiogenesis. All plexins contain a sema domain and MET-related sequences. The intracellular part contains tyrosine residues that can be phosphorylated but lack tyrosine-kinase activity.

activating Rho signaling promoting the collapse of growing growth cones.[41,42] Plexin-B1 has a PDZ binding motif at the c-terminal and binds the guanine nucleotide exchange (GEF) factors PDZ_Rho-GEF and LARG.[43,44] The GTPase activating protein (GAP)-like functional domain of plexin-B1 also contains a binding site for the small GTPase Rnd-1, and directly regulates the activity of R-Ras when Rnd-1 and s4d are bound to plexin-B1 via activation of the intrinsic GTPase activity of R-Ras.[32,45]

4. The Neuropilins

Unlike most of the semaphorins which bind directly to plexins, most class-3 semaphorins have been found to bind to the receptors belonging

to the neuropilin family, and to induce signal transduction as a result
of this interaction. The human and mouse neuropilin family consists of
two genes, np1 and np2. The proteins encoded by these two genes are
membrane-bound receptors, although soluble splice forms of np1 and
np2 have been reported.[46,47] The two neuropilins share a very similar
domain structure although the overall homology between np1 and np2
is only 44% at the amino acid level.[48] Both neuropilins contain two complement binding (CUB)-like domains (a1 and a2 domains), two coagulation factor V/VIII homology-like domains (b1 and b2 domains),
and a meprin (MAM) domain thought to be important for neuropilin
dimerization and possibly for the interaction of neuropilins with other
membrane receptors[9,48] (Fig. 3). Various class-3 semaphorins differentiate between the two neuropilins. Thus s3a binds to np1 but not to np2,
while s3f binds to np2 and with a much reduced affinity to np1.[9,10,48]
The binding site of s3a in np1 covers part of the second a-domain and

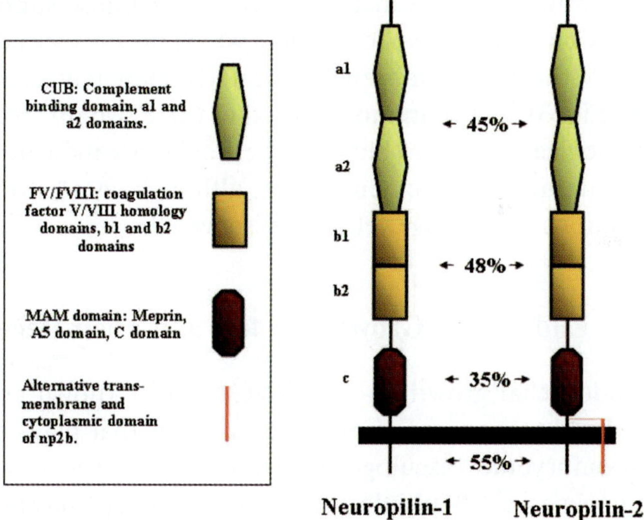

Fig. 3. The neuropilin receptor family. The two members of the neuropilin family are
membrane anchored receptors containing very short intracellular domains. Interestingly, np2 has a splice form in which the transmembrane and intracellular domains are
completely different. The s3a binding domain of np1 is located between the a2 and b1
domains and overlaps partially the VEGF$_{165}$ binding domain. The MAM domain is
required for receptor dimerization and for interaction with other receptors.

part of the first b-domain. The VEGF binding domain of np1 was mapped following the identification of the VEGF binding properties of the neuropilins. It was found that the VEGF binding site partially overlaps the s3a binding site so that VEGF and s3a compete for binding to np1.[49] Interestingly, it was possible to introduce mutations into the ligand binding domain of np1 which resulted in the complete nullification of the VEGF binding ability, but did not compromise the binding of s3a to np1, indicating that the binding domains of VEGF and s3a overlap but are not identical.[49] In contrast, the binding of s3f to np2 is not inhibited by $VEGF_{165}$ indicating that the binding sites of s3f and VEGF on np2 are independent.[50] The last three amino-acids of np1 contain a SEA sequence which functions as a docking site for the PSD-95/Dlg/ZO-1 (PDZ) domain containing protein NIP, also known as RGS–GAIP-interacting protein (GIPC) and synectin.[51] Therefore, although repulsion of np1 expressing growth cones does not require the presence of the intracellular domain of np1,[52] it may yet turn out to be required for additional np1 functions. This notion is also supported by the identification of an np2 splice form in which the c-terminal domain (including also the transmembrane domain) is completely exchanged to yield np2b (Fig. 3).[53] It is unknown whether these two np2 forms have different biological functions, but their mere existence indicates that the intracellular domains of the neuropilins, although short, are likely to possess functions which have yet to be discovered.

5. Vascular Endothelial Growth Factors and Their Receptors

Vascular endothelial growth factor (VEGF) (also known as VEGF-A) is considered to be a major angiogenic factor that plays an essential role in embryonic vasculogenesis and angiogenesis as well as in tumor angiogenesis.[54] Multiple forms of VEGF are produced as a result of alternative splicing (Fig. 4), but three of these forms, $VEGF_{121}$, $VEGF_{165}$, and $VEGF_{189}$ are considered to be the major forms that are most frequently encountered. All the VEGF forms bind and activate the VEGFR-2 tyrosine kinase receptor which seems to be essential for the transmission of VEGF-induced angiogenic signals.[55,56] The VEGFR-1 tyrosine kinase receptor[57] and its soluble form[58] which also binds all

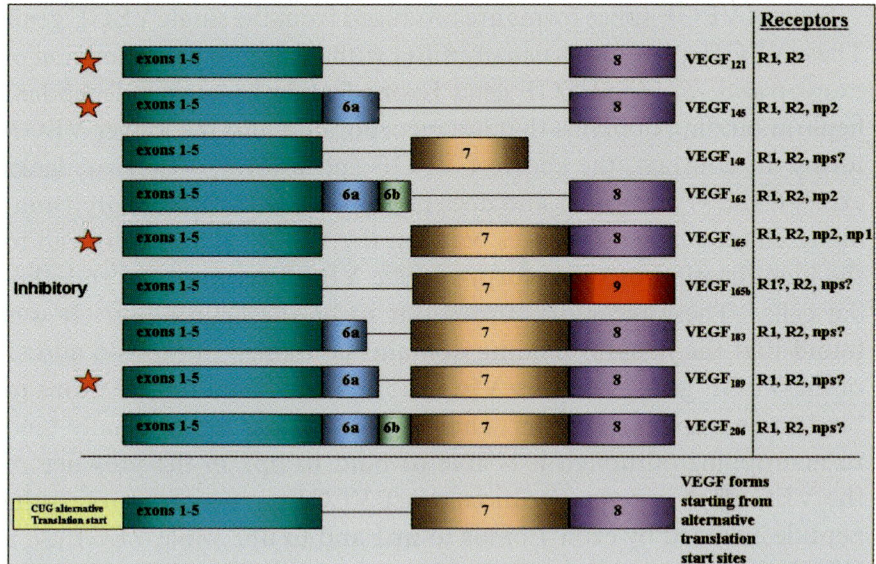

Fig. 4. The VEGF splice forms and their interaction with VEGF receptors. Shown are the various splice forms of VEGF. The active forms of the various VEGF forms are disulfide-linked homodimers. Exons 1–5 are contained in all splice forms and contain the domains that enable VEGF to bind to the VEGFR-1 (R1) and VEGFR-2 (R2) tyrosine-kinase receptors. The presence of the heparin binding peptide encoded by exon-7 enables binding to np1 and np2 while the heparin binding peptide encoded by exon-6 enables binding to np2 but not to np1. In VEGF$_{165b}$ exon-8 is exchanged with exon-9 turning this VEGF form into an inhibitory molecule. There are several longer VEGF forms that are synthesized from alternative CUG translation initiation sites. Their function is unclear at the moment and they encode non-secreted forms. Stars designate the most commonly encountered splice forms.

the VEGF forms are also required for developmental angiogenesis.[59] However, the role of VEGFR-1 in developmental angiogenesis had been considered to be primarily an inhibitory role until recently, because of experiments that demonstrated that the intracellular part of VEGFR-1 is not required for correct vascular development.[60] Nevertheless, recent experiments employing VEGFR-1 function blocking antibodies indicate that contrary to previous assumptions, this receptor also plays an important active role in VEGF-induced angiogenesis, presumably as a recruiter of bone marrow derived precursor cells to sites of active angiogenesis.[61]

Several VEGF splice forms are produced from the single VEGF gene. These VEGF splice forms usually differ with respect to the expression of exons-6 and -7 of the VEGF gene. Exons-6 and -7 encode independent heparin binding domains that are incorporated into the longer VEGF forms. In contrast, the shortest VEGF splice form, $VEGF_{121}$, lacks exons-6 and -7 altogether and does not bind to heparin. Research aiming to characterize differences between the VEGF splice forms lead to the identification of splice form-specific VEGF receptors in endothelial cells.[1] These receptors turned out to be the neuropilins.[2] It was found that the heparin binding domains contained in exons-6 and -7 of the VEGF gene also allow VEGF forms that contain these exons to bind to neuropilin receptors. Thus $VEGF_{121}$ does not normally bind to neuropilins (although it is able to bind to np2 in the presence of the VEGFR-1 receptor[50]). In contrast $VEGF_{165}$, which contains the peptide encoded by exon 7 binds to np1 and to np2 while $VEGF_{145}$, a VEGF form that contains exon-6 but not -7, binds to np2 but not to np1.[2,3,62] Recent evidence indicates that these VEGF splice forms possess somewhat different biological properties. Mice expressing $VEGF_{164}$ exclusively (in mice all VEGF forms are one amino acid shorter than human VEGF) appear to develop normally, while mice that express only $VEGF_{120}$ or only $VEGF_{188}$ do not.[63–65] These differences are attributed in part to differences in heparan-sulfate binding ability which result in different diffusion rates of these VEGF forms in the heparan-sulfate rich extracellular matrix, but may also be attributed in part to the differential neuropilin binding properties of the different VEGF forms.

The VEGF family contains four additional angiogenesis promoting members. These are the angiogenic factors PlGF and VEGF-B and the lymphangiogenesis promoting agents VEGF-C and VEGF-D.[66] VEGF-C and VEGF-D bind to the VEGFR-2 receptor and induce angiogenesis, but in contrast to VEGF both factors do not bind to VEGFR-1.[67] However, both VEGF-C and VEGF-D bind in addition to the third member of the VEGF tyrosine-kinase receptor subfamily, VEGFR-3,[68,69] which is primarily expressed on lymphatic endothelial cells, enabling them to induce proliferation of lymphatic endothelial cells and lymphangiogenesis.[70–72] It was shown that VEGF-C binds to np2, which is highly expressed in lymphatics,[6,73] but the role of np2 in VEGF-C signal transduction is unclear.

PlGF and VEGF-B bind and activate the VEGFR-1 receptor but do not bind to VEGFR-2 or to VEGFR-3.[74] PlGF was reported to potentiate VEGF signaling through activation of VEGFR-1/VEGFR-2 heterodimers.[75] Although PlGF does not seem to affect vasculogenesis and developmental angiogenesis, it does play an important role in pathological angiogenesis, possibly through the recruitment of monocytes.[61,76] VEGF-B was found to bind to np1.[5] while the heparin binding form of PlGF, PlGF-2, binds to both neuropilins.[3,4] The biological function of this binding ability is still unclear. It was recently claimed, in addition, that np1 contains a domain that mimics the structure of the glycosaminoglycan heparin, thus allowing np1 to bind to a wide range of heparin binding growth factors. It was even claimed that neuropilins participate in signal transduction of growth factors such as basic fibroblast growth factor.[77] However, this observation seems to contradict previous observations which indicate that heparin enhances the binding of $VEGF_{165}$ to neuropilins.[1]

6. Signal Transduction by Neuropilins

The intracellular domain of the neuropilins is short, and it was therefore assumed that it does not suffice to for the transduction of biological signals. This view was supported by experiments that have shown that although np1 is required for s3a-induced collapse of axonal growth cones, deletion of the cytoplasmic domain of np1 does not inhibit s3a activity, suggesting the existence of independent signal transducing moieties.[52] These were later found to be the products of genes encoding type-A plexins.[12,13] (Fig. 1). Type-A plexins were found to form complexes with neuropilins, and to serve as the signal transducing components in plexin/neuropilin holo-receptors for various class-3 semaphorins.[12,13,33,78] Plexins belonging to the other three plexin subfamilies may also be able to form functional complexes with neuropilins, as demonstrated in the case of plexin-B1 and np1,[12,79] and recently also in the case of plexin-D1 and np1.[80,81] However, the assumption which predicts that neuropilins cannot transduce biological signals on their own due to their short intracellular domains was recently challenged in experiments in which the extracellular domain of the epidermal growth factor (EGF) receptor was fused to the intracellular and transmembrane domains of np1. These experiments indicate that the chimeric receptor can promote cell migration in response to EGF, suggesting that the intracellular domain of np1 does transduce biological signals independently,[82] perhaps via binding to the PSD-95/Dlg/ZO-1 domain of the NIP protein.[51] These experiments therefore indicate that the short intracellular domain of the neuropilins is not devoid of function.

Nevertheless, it is now widely accepted that to transduce signals of class-3 semaphorins such as s3a, s3f and s3c, neuropilins form complexes with either A-type plexins[12,13] or with plexin-D1.[81] Activation of plexins by semaphorins, either directly or indirectly via neuropilins, leads to diverse biological responses. One of the best studied responses is the s3a- and s3f-induced repulsion of axonal growth cones. The repulsion is apparently triggered by local changes in cell adhesion and actin cytoskeleton organization. Activation of A-type plexins occurs in response to the binding of a class-3 semaphorin to neuropilin, leading to the phosphorylation of tyrosine residues in the cytoplasmic domain, activation of small G proteins and subsequent effects on cell adhesion, cell shape cell migration and cell proliferation. The phosphorylation of plexins is the result of semaphorin-induced recruitment of cytosolic tyrosine kinases. The binding of s3a to np1 induces the association of the tyrosine-kinase Fes/Fps with plexin-A1 leading to plexin-A1 phosphorylation. In growth cones of s3a responsive nerve cells, Fes/Fps forms complexes with brain-specific collapsin response mediator protein-2 (CRMP-2) and with CRMP associated molecule (CRAM). These two proteins are required for s3a signaling to the actin cytoskeleton in responsive nerve cells and are also phosphorylated by Fes/Fps in response to s3a, although their exact role is still unclear.[35] Another cytosolic tyrosine-kinase that was found to associate with the intracellular part of plexin-A1 as well as plexin-A2 is fyn. Fyn phosphorylates plexin-A2 in response to s3a and binds cdk5 kinase, which is also phosphorylated by Fyn. Activation of cdk5 by fyn was found to be essential for s3a-mediated growth cone repulsion.[36]

Another signal transduction mechanism that is activated by class-A plexins involves activation of MICALs (Molecule interacting with CasL). It was found that plexin-A, a Drosophila class-A plexin receptor, associates with MICAL in response to the Drosophila semaphorin s1a. The mammalian homologue, MICAL-1, interacts with intermediate filaments, as well as with actin and small GTPases like Rab1. It also functions as a putative flavoprotein monooxygenase, and its oxidative activity may be important for semaphorin signaling since monooxygenase inhibitors inhibit MICAL-mediated s3a signal transduction.[39,83,84] Another protein, Nervy, a member of the myeloid translocation gene family of A-kinase anchoring proteins (AKAPs), regulates repulsive axon guidance in Drosophila by linking the cyclic adenosine monophosphate (cAMP)-dependent protein kinase (PKA) to the Semaphorin-1a (Sema-1a) receptor Plexin-A (PlexA), and nervey homologues may fulfill similar roles in vertebrates.[85]

Plexins can also affect the organization of the actin cytoskeleton by modulating the activity of small GTPases such as RhoA, Rnd1, Rac1 and CDC42, all of which have been shown to control the organization of the actin cytoskeleton.[86,87] Activation of Rac and CDC42 usually triggers the formation of lamellipodia and filopodia, respectively. In contrast, activation of the Rho family members Rho1 and Rnd1 leads to the formation of stress fibers. S3a induces activation of Rac1 but the mechanism by which it does so is still unclear. This is probably triggered by the binding of Rac1 to plexin-A1 since the interaction is required for the collapsing activity of s3a,[88] although it is unclear why activation of Rac1, usually associated with formation of filopodia and lamellipodia would be associated with the collapse observed in response to s3a. Out of the Rho family GTPases tested only RhoD and Rnd1 were found to bind directly to plexin-A1. Interestingly, Rnd1 binding to plexin-A1 leads to a collapse of the actin cytoskeleton even in the absence of s3a. RhoD on the other hand antagonizes the effects of Rnd1 even though both GTPases belong to the Rho family of GTPases.[89] Various guanine nucleotide exchange factors (GEFs) and GTPase activating proteins (GAPs), which function as regulators of GTPases, also modulate the activity of GTPases that bind to plexins.[43,44] The GTPases in turn regulate the activity of downstream effectors such as the LIM kinase which in turn regulates the phosphorylation state of cofilin, an actin binding/cleaving protein that is required for s3a-induced growth cone collapse.[90] Similar mechanisms are presumably activated in endothelial cells in response to semaphorins such as s3a and s3f. Indeed, s3f and s3a repulse endothelial cells and inhibit angiogenesis, presumably as a result of their effects on the actin cytoskeleton.[91–94]

7. The Role of the Neuropilins in the Regulation of Vasculogenesis and Angiogenesis

Neuropilins as modulators of VEGF function: Since both neuropilins function as splice form Specific VEGF receptors, it was not surprising that they were found to affect VEGF signaling and function in various experimental systems. Initially, the binding of $VEGF_{165}$ to np1 was found to enhance $VEGF_{165}$-induced migration of endothelial cells in cells that express in addition to np1 the VEGF receptor VEGFR-2.[2] It was subsequently observed that soluble dimers of the np1 extracellular domain enhance VEGF-induced vascular development while monomers of the soluble extracellular domain function as $VEGF_{165}$ traps and inhibit VEGF-induced vascular development.[95] The role of np1 in embryonic vascular development was also studied in gene targeting experiments. Mice lacking functional np1 receptors suffer from impaired neural vascularization and from defects in the development of large arteries such as branchial arch arteries. In addition, the development of the heart was strongly impaired in these mice, and failure of heart function was responsible for their premature death.[96] Binding/competition experiments have demonstrated that the $VEGF_{165}$ and s3a binding domains of np1 overlap.[49] Knock-in mice expressing an np1 variant lacking s3a binding ability but retaining VEGF binding displayed normal vascular development but abnormal neural development indicating that the VEGF binding ability of np1 is critical for proper vascular development. In contrast, the s3a binding ability is required in addition to the VEGF binding ability for proper heart development. These results are strengthened by experiments showing that mice in which np1 was targeted in endothelial cells but not in other cell types also suffer from severe vascular abnormalities,[97] and by experiments which show that proper development of the vasculature in zebrafish requires np1.[98]

The mechanism by which np1 enhances $VEGF_{165}$-induced signal transduction via the VEGF receptor VEGFR-2 is unclear. It was suggested that np1 binds $VEGF_{165}$ and presents it to the VEGFR-2 receptor, thereby increasing responsiveness to $VEGF_{165}$. Such a mechanism should function in trans too, and it was indeed found that angiogenesis is enhanced in tumors containing tumor cells expressing high levels of np1.[99] It was recently suggested that np1 contains a heparin-like binding domain that enables np1 to bind a wide variety of heparin binding growth factors such as basic fibroblast growth factor and not just $VEGF_{165}$, and that as a consequence np1 is able to potentiate the activity of a wide variety of heparin binding growth factors.[77] On the other hand, it was also reported that np1 forms complexes with VEGFR-2 directly.[49,100,101] The formation of such complexes may account, at least partially, for the np1-dependent potentiation of $VEGF_{165}$ activity.

The role of np2 in VEGF-induced vasculogenesis and angiogenesis is less clear. Np2 binds $VEGF_{165}$ with a somewhat lower affinity than that of np1. However, the vasculature of mice lacking a functional np2 receptor develops normally except for defects observed at birth in small lymphatic vessels.[73,102] This does not mean that such mice are normal with respect to their responses to VEGF. Indeed, it was recently reported that mice lacking a functional np2 gene do not respond to $VEGF_{165}$ by retinal angiogenesis.[103] The importance of np2 to vascular development is highlighted in experiments in which mice lacking both functional neuropilins were generated. These mice display a total lack of endothelial cells,[104,105] and their phenotype therefore resembles the phenotype of mice lacking functional VEGFR-2 receptors. Furthermore, mice lacking a functional np2 gene and containing only one functional np1 gene also displayed vascular abnormalities that were more severe than those observed in mice that lack both np1 alleles.[104] These experiments therefore indicate that np2 does have an important role in vasculogenesis and developmental angiogenesis as well as in the development of the lymphatic system.

Neuropilins as transducers of semaphorin signals in angiogenesis: Neuropilins bind, in addition to VEGF, several class-3 semaphorins. S3a binds to np1 but not to np2.[9,10] It was therefore natural to assume that s3a may function as an inhibitor of angiogenesis by interfering with np1-mediated VEGF signaling. Indeed, it was found that s3a inhibits the pro-angiogenic effects of $VEGF_{165}$ in *in vitro* angiogenesis experiments, and that it inhibits $VEGF_{165}$ binding to np1.[106] Subsequent experiments demonstrated that s3a can inhibit developmental angiogenesis in chick embryo forelimbs[94] and vascular branching in the developing chick brain.[93] However, no effects of s3a on tumor progression or tumor angiogenesis have been reported to date.

Even though VEGF binds efficiently to np2,[3] it does not inhibit the binding of the np2-specific s3f to np2.[50] Nevertheless, *in vitro* experiments have shown that s3f can inhibit both VEGF-and bFGF-induced proliferation of endothelial cells, and that s3f is also able to inhibit bFGF- and $VEGF_{165}$-induced angiogenesis.[91] S3f also repels endothelial cells in *in vitro* experiments indicating that it could also affect angiogenesis through repulsion of migrating endothelial cells.[92] Both S3f and s3b were identified as tumor suppressor genes whose loss is associated with tumor progression of small cell lung carcinoma.[25,26,107] It is therefore not surprising that both s3f and s3b inhibit tumor formation from small cell lung carcinoma-derived cells.[107,108] Interestingly, it was recently shown that s3f also affects the behavior of tumor cells directly, via np1 and np2 receptors expressed on the tumor cells. S3f inhibits the VEGF-induced spreading, of MCF-7 breast cancer cells by inhibiting np1 mediated signaling,[109] and repulses C100 breast cancer cells as a result of its binding to np2 receptors expressed by this cell type.[110] Similarly, s3b was found to antagonize the anti-apoptotic effects of VEGF in NCI-H1299 lung cancer-derived cells, probably by interfering with neuropilin-mediated VEGF signaling in these cells.[111] These findings indicate that s3f can suppress tumor angiogenesis and that s3b and s3f can directly affect the behavior of tumor cells expressing neuropilins.

S3c is a semaphorin that seems to signal through np2 or through np1/np2 complexes. Like s3f, it can bind to np1 and as such can act as an antagonist of s3a.[112] Recently, it was found that in the presence of plexin-D1, s3c can induce signal transduction via np1 as well as via np2.[81] The heart of mice lacking a functional s3c gene does not develop normally, and mice lacking a functional plexin-D1 gene suffer from heart defects and vascular patterning defects.[24,80,81] It follows that plexin-D1-mediated s3c signaling plays an important role in vascular development, although it is unclear whether these effects are the result of inhibition of endothelial cell migration and proliferation.

8. Modulation of Angiogenesis by Semaphorins that Bind Directly to Plexins

It was recently reported, that contrary to previous assumptions, not all the class-3 semaphorins require neuropilins as binding receptors. It was found that s3e binds directly to plexin-D1 and induces signal transduction without neuropilins. Interestingly, it was found that s3e can disrupt developmental angiogenesis in chick embryos, indicating that s3e may function as a regulator of angiogenesis. This was also verified in gene targeting experiments in which it was shown that disruption of the s3e gene in mice leads to disruption of intersomitic vascular patterning.[31]

It was recently observed that s4d, a membrane-bound semaphorin that binds and activates plexin-B1, is an inducer of angiogenesis. Interestingly, following the binding of s4d to plexin-B1, plexin-B1 forms complexes with the Met tyrosine-kinase receptor. Met serves as a receptor for the angiogenic factor hepatocyte growth factor (HGF) also known as scatter factor. Interestingly, s4d-activated plexin-B1 can activate Met in the absence of HGF resulting in phosphorylation of Met-associated tyrosine residues and induction of angiogenesis.[113,114] Interestingly, all three B-class plexins can form complexes with Met and with the Met-related receptor Ron,[115] indicating that other semaphorins that interact with B-type plexins such as s5a[29] may also be able to induce angiogenesis.

Acknowledgments

This work was supported by grants from the Israel Science Foundation (ISF), German-Israeli Binational Foundation (GIF), International Union against Cancer (AICR) and by the Rappaport Family Institute for Research in the Medical Sciences of the Faculty of Medicine at the Technion, Israel Institute of Technology (to G. Neufeld).

References

1. Gitay-Goren H, Cohen T, Tessler S, Soker S, Gengrinovitch S, Rockwell P, Klagsbrun M, Levi B-Z, Neufeld G (1996). Selective binding of VEGF121 to one of the three VEGF receptors of vascular endothelial cells. *J Biol Chem* **271**: 5519–5523.

2. Soker S, Takashima S, Miao HQ, Neufeld G, Klagsbrun M (1998). Neuropilin-1 is expressed by endothelial and tumor cells as an isoform specific receptor for vascular endothelial growth factor. *Cell* **92**: 735–745.

3. Gluzman-Poltorak Z, Cohen T, Herzog Y, Neufeld G (2000). Neuropilin-2 and Neuropilin-1 are receptors for 165-amino acid long form of vascular endothelial growth factor (VEGF) and of placenta growth factor-2, but only neuropilin-2 functions as a receptor for the 145-amino acid form of VEGF. *J Biol Chem* **275**: 18040–18045.

4. Migdal M, Huppertz B, Tessler S, Comforti A, Shibuya M, Reich R, Baumann H, Neufeld G (1998). Neuropilin-1 is a placenta growth factor-2 receptor. *J Biol Chem* **273**: 22272–22278.

5. Makinen T, Olofsson B, Karpanen T, Hellman U, Soker S, Klagsbrun M, Eriksson U, Alitalo K (1999). Differential binding of vascular endothelial growth factor B splice and proteolytic isoforms to neuropilin-1. *J Biol Chem* **274**: 21217–21222.

6. Karkkainen MJ, Saaristo A, Jussila L, Karila KA, Lawrence EC, Pajusola K, Bueler H, Eichmann A, Kauppinen R, Kettunen MI, Yla-Herttuala S, Finegold DN, Ferrell RE, Alitalo K (2001). A model for gene therapy of human hereditary lymphedema. *Proc Natl Acad Sci USA* **98**: 12677–12682.

7. Takagi S, Tsuji T, Amagai T, Takamatsu T, Fujisawa H (1987). Specific cell surface labels in the visual centers of Xenopus laevis tadpole identified using monoclonal antibodies. *Dev Biol* **122**: 90–100.

8. Takagi S, Hirata T, Agata K, Mochii M, Eguchi G, Fujisawa H (1991). The A5 antigen a candidate for the neuronal recognition molecule has homologies to complement components and coagulation factors. *Neuron* **7**: 295–307.

9. He Z, Tessier-Lavigne M (1997). Neuropilin is a receptor for the axonal chemore-pellent semaphorin III. *Cell* **90**: 739–751.

10. Kolodkin AL, Levengood DV, Rowe EG, Tai YT, Giger RJ, Ginty DD (1997). Neuropilin is a semaphorin III receptor. *Cell* **90**: 753–762.

11. Luo Y, Raible D, Raper JA (1993). Collapsin: a protein in brain that induces the collapse and paralysis of neuronal growth cones. *Cell* **75**: 217–227.

12. Tamagnone L, Artigiani S, Chen H, He Z, Ming GI, Song H, Chedotal A, Winberg ML, Goodman CS, Poo M, Tessier-Lavigne M, Comoglio PM (1999). Plexins are a large family of receptors for transmembrane secreted and GPI-anchored semaphorins in vertebrates. *Cell* **99**: 71–80.

13. Takahashi T, Fournier A, Nakamura F, Wang LH, Murakami Y, Kalb RG, Fujisawa H, Strittmatter SM (1999). Plexin-neuropilin-1 complexes form functional semaphorin-3A receptors. *Cell* **99**: 59–69.

14. Goodman CS, Kolodkin AL, Luo Y, Pueschel AW, Raper JA (1999). Unified nomenclature for the semaphorins collapsins. *Cell* **97**: 551–552.

15. Feiner L, Koppel AM, Kobayashi H, Raper JA (1997). Secreted chick semaphorins bind recombinant neuropilin with similar affinities but bind different subsets of neurons *in situ*. *Neuron* **19**: 539–545.

16. Love CA, Harlos K, Mavaddat N, Davis SJ, Stuart DI, Jones EY, Esnouf RM (2003). The ligand-binding face of the semaphorins revealed by the high-resolution crystal structure of SEMA4D. *Nat Struct Biol* **10**: 843–848.

17. Antipenko A, Himanen JP, van Leyen K, Nardi-Dei V, Lesniak J, Barton WA, Rajashankar KR, Lu M, Hoemme C, Puschel AW, Nikolov DB (2003). Structure of the semaphorin-3A receptor binding module. *Neuron* **39**: 589–598.
18. Koppel AM, Raper JA (1998). Collapsin-1 covalently dimerizes and dimerization is necessary for collapsing activity. *J Biol Chem* **273**: 15708–15713.
19. Gherardi E, Love CA, Esnouf RM, Jones EY (2004). The sema domain. *Curr Opin Struct Biol* **14**: 669–678.
20. Klostermann A, Lohrum M, Adams RH, Puschel AW (1998). The chemorepulsive activity of the axonal guidance signal semaphorin D requires dimerization. *J Biol Chem* **273**: 7326–7331.
21. De Wit J, Verhaagen J (2003). Role of semaphorins in the adult nervous system. *Prog Neurobiol* **71**: 249–267.
22. Tamagnone L, Comoglio PM (2004). To move or not to move? *EMBO Rep* **5**: 356–361.
23. Kumanogoh A, Kikutani H (2003). Roles of the semaphorin family in immune regulation. *Adv Immunol* **81**: 173–198.
24. Feiner L, Webber AL, Brown CB, Lu MM, Jia L, Feinstein P, Mombaerts P, Epstein JA, Raper JA (2001). Targeted disruption of semaphorin 3C leads to persistent truncus arteriosus and aortic arch interruption. *Development* **128**: 3061–3070.
25. Xiang RH, Hensel CH, Garcia DK, Carlson HC, Kok K, Daly MC, Kerbacher K, van den BA, Veldhuis P, Buys CH, Naylor SL (1996). Isolation of the human semaphorin III/F gene (SEMA3F) at chromosome 3p21, a region deleted in lung cancer. *Genomics* **32**: 39–48.
26. Sekido Y, Bader S, Latif F, Chen JY, Duh FM, Wei MH, Albanesi JP, Lee CC, Lerman MI, Minna JD (1996). Human semaphorins A(V) and IV reside in the 3p21.3 small cell lung cancer deletion region and demonstrate distinct expression patterns. *Proc Natl Acad Sci USA* **93**: 4120–4125.
27. Negishi M, Oinuma I, Katoh H (2005). Plexins: axon guidance and signal transduction. *Cell Mol Life Sci* **62**: 1363–1371.
28. Gutmann-Raviv N, Kessler O, Shraga-Heled N, Lange T, Herzog Y, Neufeld G (2005). The neuropilins and their role in tumorigenesis and tumor progression. *Cancer Lett* **231**: 1–11.
29. Artigiani S, Conrotto P, Fazzari P, Gilestro GF, Barberis D, Giordano S, Comoglio PM, Tamagnone L (2004). Plexin-B3 is a functional receptor for semaphorin 5A. *EMBO Rep* **5**: 710–714.
30. Toyofuku T, Zhang H, Kumanogoh A, Takegahara N, Suto F, Kamei J, Aoki K, Yabuki M, Hori M, Fujisawa H, Kikutani H (2004). Dual roles of Sema6D in cardiac morphogenesis through region-specific association of its receptor plexin-A1, with off-tack and vascular endothelial growth factor receptor type 2. *Genes Dev* **18**: 435–447.
31. Gu C, Yoshida Y, Livet J, Reimert DV, Mann F, Merte J, Henderson CE, Jessell TM, Kolodkin AL, Ginty DD (2005). Semaphorin 3E and plexin-D1 control vascular pattern independently of neuropilins. *Science* **307**: 265–268.
32. Oinuma I, Ishikawa Y, Katoh H, Negishi M (2004). The semaphorin 4D receptor plexin-B1 is a GTPase activating protein for R-Ras. *Science* **305**: 862–865.

33. Takahashi T, Strittmatter SM (2001). PlexinA1 autoinhibition by the plexin sema domain. *Neuron* **29**: 429–439.

34. Comoglio PM, Trusolino L (2002). Invasive growth: from development to metastasis. *J Clin Invest* **109**: 857–862.

35. Mitsui N, Inatome R, Takahashi S, Goshima Y, Yamamura H, Yanagi S (2002). Involvement of Fes/Fps tyrosine kinase in semaphorin3A signaling. *EMBO J* **21**: 3274–3285.

36. Sasaki Y, Cheng C, Uchida Y, Nakajima O, Ohshima T, Yagi T, Taniguchi M, Nakayama T, Kishida R, Kudo Y, Ohno S, Nakamura F, Goshima Y (2002). Fyn and Cdk5 mediate semaphorin-3A signaling which is involved in regulation of dendrite orientation in cerebral cortex. *Neuron* **35**: 907.

37. Brown M, Jacobs T, Eickholt B, Ferrari G, Teo M, Monfries C, Qi RZ, Leung T, Lim L, Hall C (2004). Alpha2-chimaerin cyclin-dependent Kinase 5/p35, its target collapsin response mediator protein-2 are essential components in semaphorin 3A-induced growth-cone collapse. *J Neurosci* **24**: 8994–9004.

38. Eickholt BJ, Walsh FS, Doherty P (2002). An inactive pool of GSK-3 at the leading edge of growth cones is implicated in Semaphorin 3A signaling. *J Cell Biol* **157**: 211–217.

39. Terman JR, Mao T, Pasterkamp RJ, Yu HH, Kolodkin AL (2002). MICALs a family of conserved flavoprotein oxidoreductases function in plexin-mediated axonal repulsion. *Cell* **109**: 887–900.

40. Vikis HG, Li W, He Z, Guan KL (2000). The semaphorin receptor plexin-B1 specifically interacts with active Rac in a ligand-dependent manner. *Proc Natl Acad Sci USA* **97**: 12457–12462.

41. Driessens MH, Hu H, Nobes CD, Self A, Jordens I, Goodman CS, Hall A (2001). Plexin-B semaphorin receptors interact directly with active Rac and regulate the actin cytoskeleton by activating Rho. *Curr Biol* **11**: 339–344.

42. Vikis HG, Li W, Guan KL (2002). The plexin-B1/Rac interaction inhibits PAK activation and enhances Sema4D ligand binding. *Genes Dev* **16**: 836–845.

43. Perrot V, Vazquez-Prado J, Gutkind JS (2002). Plexin B regulates Rho through the guanine nucleotide exchange factors Leukemia-associated RhoGEF (LARG) and PDZ-RhoGEF. *J Biol Chem* **278**: 26111–26119.

44. Aurandt J, Vikis HG, Gutkind JS, Ahn N, Guan KL (2002). The semaphorin receptor plexin-B1 signals through a direct interaction with the Rho-specific nucleotide exchange factor LARG. *Proc Natl Acad Sci USA* **99**: 12085–12090.

45. Pasterkamp RJ (2005). R-Ras fills another GAP in semaphorin signalling. *Trends Cell Biol* **15**: 61–64.

46. Gagnon ML, Bielenberg DR, Gechtman Z, Miao HQ, Takashima S, Soker S, Klagsbrun M (2000). Identification of a natural soluble neuropilin-1 that binds vascular endothelial growth factor: *in vivo* expression and antitumor activity. *Proc Natl Acad Sci USA* **97**: 2573–2578.

47. Rossignol M, Gagnon ML, Klagsbrun M (2000). Genomic organization of human neuropilin-1 and neuropilin-2 genes: identification and distribution of splice variants and soluble isoforms. *Genomics* **70**: 211–222.

48. Giger RJ, Urquhart ER, Gillespie SK, Levengood DV, Ginty DD, Kolodkin AL (1998). Neuropilin-2 is a receptor for semaphorin IV: insight into the structural basis of receptor function and specificity. *Neuron* **21**: 1079–1092.
49. Gu C, Limberg BJ, Whitaker GB, Perman B, Leahy DJ, Rosenbaum JS, Ginty DD, Kolodkin AL (2002). Characterization of neuropilin-1 structural features that confer binding to semaphorin 3A and vascular endothelial growth factor 165. *J Biol Chem* **277**: 18069–18076.
50. Gluzman-Poltorak Z, Cohen T, Shibuya M, Neufeld G (2001). Vascular endothelial growth factor receptor-1 and neuropilin-2 form complexes. *J Biol Chem* **276**: 18688–18694.
51. Cai HB, Reed RR (1999). Cloning and characterization of neuropilin-1-interacting protein: A PSD-95/Dlg/ZO-1 domain-containing protein that interacts with the cytoplasmic domain of neuropilin-1. *J Neurosci* **19**: 6519–6527.
52. Nakamura F, Tanaka M, Takahashi T, Kalb RG, Strittmatter SM (1998). Neuropilin-1 extracellular domains mediate semaphorin D/III- induced growth cone collapse. *Neuron* **21**: 1093–1100.
53. Chen H, Chedotal A, He Z, Goodman CS, Tessier-Lavigne M (1997). Neuropilin-2, a novel member of the neuropilin family is a high affinity receptor for the semaphorins sema E and sema IV but not sema III. *Neuron* **19**: 547–559.
54. Neufeld G, Cohen T, Gengrinovitch S, Poltorak Z (1999). Vascular endothelial growth factor (VEGF) and its receptors. *FASEB J* **13**: 9–22.
55. Terman BI, Dougher-Vermazen M, Carrion ME, Dimitrov D, Armellino DC, Gospodarowicz D, Bôhlen P (1992). Identification of the KDR tyrosine kinase as a receptor for vascular endothelial cell growth factor. *Biochem Biophys Res Commun* **187**: 1579–1586.
56. Shibuya M (2003). Vascular endothelial growth factor receptor-2: its unique signaling and specific ligand VEGF-E. *Cancer Sci* **94**: 751–756.
57. Devries C, Escobedo JA, Ueno H, Houck K, Ferrara N, Williams LT (1992). The fms-like tyrosine kinase a receptor for vascular endothelial growth factor. *Science* **255**: 989–991.
58. Hornig C, Weich HA (1999). Soluble VEGF receptors. *Angiogenesis* **3**: 33–39.
59. Fong GH, Rossant J, Gertsenstein M, Breitman ML (1995). Role of the Flt-1 receptor tyrosine kinase in regulating the assembly of vascular endothelium. *Nature* **376**: 66–70.
60. Hiratsuka S, Minowa O, Kuno J, Noda T, Shibuya M (1998). Flt-1 lacking the tyrosine kinase domain is sufficient for normal development and angiogenesis in mice. *Proc Natl Acad Sci USA* **95**: 9349–9354.
61. Luttun A, Tjwa M, Moons L, Wu Y, Angelillo-Scherrer A, Liao F, Nagy JA, Hooper A, Priller J, De Klerck B, Compernolle V, Daci E, Bohlen P, Dewerchin M, Herbert JM, Fava R, Matthys P, Carmeliet G, Collen D, Dvorak HF, Hicklin DJ, Carmeliet P (2002). Revascularization of ischemic tissues by PlGF treatment and inhibition of tumor angiogenesis arthritis and atherosclerosis by anti-Flt1. *Nat Med* **8**: 831–840.

62. Poltorak Z, Cohen T, Sivan R, Kandelis Y, Spira G, Vlodavsky I, Keshet E, Neufeld G (1997). VEGF145: a secreted VEGF form that binds to extracellular matrix. *J Biol Chem* **272**: 7151–7158.

63. Mattot V, Moons L, Lupu F, Chernavvsky D, Gomez RA, Collen D, Carmeliet P (2002). Loss of the VEGF[(164)] and VEGF[(188)] isoforms impairs postnatal glomerular angiogenesis and renal arteriogenesis in mice. *J Am Soc Nephrol* **13**: 1548–1560.

64. Maes C, Stockmans I, Moermans K, Van Looveren R, Smets N, Carmeliet P, Bouillon R, Carmeliet G (2004). Soluble VEGF isoforms are essential for establishing epiphyseal vascularization and regulating chondrocyte development and survival. *J Clin Invest* **113**: 188–199.

65. Carmeliet P, Ng YS, Nuyens D, Theilmeier G, Brusselmans K, Cornelissen I, Ehler E, Kakkar VV, Stalmans I, Mattot V, Perriard JC, Dewerchin M, Flameng W, Nagy A, Lupu F, Moons L, Collen D, D'Amore PA, Shima DT (1999). Impaired myocardial angiogenesis and ischemic cardiomyopathy in mice lacking the vascular endothelial growth factor isoforms VEGF164 and VEGF188. *Nature Med.* **5**: 495–502.

66. Tammela T, Enholm B, Alitalo K, Paavonen K (2005). The biology of vascular endothelial growth factors. *Cardiovasc Res* **65**: 550–563.

67. Cao YH, Linden P, Farnebo J, Cao RH, Eriksson A, Kumar V, Qi JH, Claesson-Welsh L, Alitalo K (1998). Vascular endothelial growth factor C induces angiogenesis *in vivo*. *Proc Natl Acad Sci USA* **95**: 14389–14394.

68. Joukov V, Pajusola K, Kaipainen A, Chilov D, Lahtinen I, Kukk E, Saksela O, Kalkkinen N, Alitalo K (1996). A novel vascular endothelial growth factor VEGF-C is a ligand for the Flt4 (VEGFR-3) and KDR (VEGFR-2) receptor tyrosine kinases. *EMBO J* **15**: 290–298.

69. Achen MG, Jeltsch M, Kukk E, Makinen T, Vitali A, Wilks AF, Alitalo K, Stacker SA (1998). Vascular endothelial growth factor D (VEGF-D) is a ligand for the tyrosine kinases VEGF receptor 2 (Flk1) and VEGF receptor 3 (Flt4). *Proc Natl Acad Sci USA* **95**: 548–553.

70. Veikkola T, Jussila L, Makinen T, Karpanen T, Jeltsch M, Petrova TV, Kubo H, Thurston G, McDonald DM, Achen MG, Stacker SA, Alitalo K (2001). Signalling via vascular endothelial growth factor receptor-3 is sufficient for lymphangiogenesis in transgenic mice. *EMBO J* **20**: 1223–1231.

71. Kaipainen A, Korhonen J, Mustonen T, Vanhinsbergh VWM, Fang GH, Dumont D, Breitman M, Alitalo K (1995). Expression of the fms-like tyrosine kinase 4 gene becomes restricted to lymphatic endothelium during development. *Proc Natl Acad Sci USA* **92**: 3566–3570.

72. Kukk E, Lymboussaki A, Taira S, Kaipainen A, Jeltsch M, Joukov V, Alitalo K (1996). VEGF-C receptor binding and pattern of expression with VEGFR-3 suggests a role in lymphatic vascular development. *Development* **122**: 3829–3837.

73. Yuan L, Moyon D, Pardanaud L, Breant C, Karkkainen MJ, Alitalo K, Eichmann A (2002). Abnormal lymphatic vessel development in neuropilin 2 mutant mice. *Development* **129**: 4797–4806.

74. Terman BI, Khandke L, Doughervermazan M, Maglione D, Lassam NJ, Gospodarowicz D, Persico MG, Bohlen P, Eisinger M (1994). VEGF receptor

subtypes KDR and FLT1 show different sensitivities to heparin and placenta growth factor. *Growth Factors* **11**: 187–195.

75. Autiero M, Waltenberger J, Communi D, Kranz A, Moons L, Lambrechts D, Kroll J, Plaisance S, De Mol M, Bono F, Kliche S, Fellbrich G, Ballmer-Hofer K, Maglione D, Mayr-Beyrle U, Dewerchin M, Dombrowski S, Stanimirovic D, Van Hummelen P, Dehio C, Hicklin DJ, Persico G, Herbert JM, Communi D, Shibuya M, Collen D, Conway EM, Carmeliet P (2003). Role of PlGF in the intra- and intermolecular cross talk between the VEGF receptors Flt1 and Flk1. *Nat Med* **9**: 936–943.

76. Meyer RD, Latz C, Rahimi N (2003). Recruitment and activation of PLC-gamma1 by VEGFR-2 is required for tubulogenesis and differentiation of endothelial cells. *J Biol Chem* **278**: 16347–16355.

77. West DC, Chris RG, Duchesne L, Patey SJ, Terry CJ, Turnbull JE, Delehedde M, Heegaard CW, Allain F, Vanpouille C, Ron D, Fernig DG (2005). Interactions of multiple heparin-binding growth factors with neuropilin-1 and potentiation of the activity of fibroblast growth factor-2. *J Biol Chem* **280**: 13457–13464.

78. Suto F, Murakami Y, Nakamura F, Goshima Y, Fujisawa H (2003). Identification and characterization of a novel mouse plexin plexin-A4. *Mech Dev* **120**: 385–396.

79. Rohm B, Ottemeyer A, Lohrum M, Pueschel AW (2000). Plexin/neuropilin complexes mediate repulsion by the axonal guidance signal semaphorin 3A. *Mech Dev* **93**: 95–104.

80. Torres-Vazquez J, Gitler AD, Fraser SD, Berk JD, Van NP, Fishman MC, Childs S, Epstein JA, Weinstein BM (2004). Semaphorin-plexin signaling guides patterning of the developing vasculature. *Dev Cell* **7**: 117–123.

81. Gitler AD, Lu MM, Epstein JA (2004). PlexinD1 and Semaphorin signaling are required in endothelial cells for cardiovascular development. *Dev Cell* **7**: 107–116.

82. Wang L, Zeng H, Wang P, Soker S, Mukhopadhyay D (2003). Neuropilin-1-mediated vascular permeability factor/vascular endothelial growth factor-dependent endothelial cell migration. *J Biol Chem* **278**: 48848–48860.

83. Suzuki T, Nakamoto T, Ogawa S, Seo S, Matsumura T, Tachibana K, Morimoto C, Hirai H (2002). MICAL a novel CasL interacting molecule associates with vimentin. *J Biol Chem* **277**: 14933–14941.

84. Weide T, Teuber J, Bayer M, Barnekow A (2003). MICAL-1 isoforms novel rab1 interacting proteins. *Biochem Biophys Res Commun* **306**: 79–86.

85. Terman JR, Kolodkin AL (2004). Nervy links protein kinase a to plexin-mediated semaphorin repulsion. *Science* **303**: 1204–1207.

86. Kodama A, Lechler T, Fuchs E (2004). Coordinating cytoskeletal tracks to polarize cellular movements. *J Cell Biol* **167**: 203–207.

87. Fukata M, Nakagawa M, Kaibuchi K (2003). Roles of Rho-family GTPases in cell polarisation and directional migration. *Curr Opin Cell Biol* **15**: 590–597.

88. Turner LJ, Nicholls S, Hall A (2004). The activity of the plexin-A1 receptor is regulated by Rac. *J Biol Chem* **279**: 33199–33205.

89. Zanata SM, Hovatta I, Rohm B, Puschel AW (2002). Antagonistic effects of Rnd1 and RhoD GTPases regulate receptor activity in Semaphorin 3A-induced cytoskeletal collapse. *J Neurosci* **22**: 471–477.

90. Aizawa H, Wakatsuki S, Ishii A, Moriyama K, Sasaki Y, Ohashi K, Sekine-Aizawa Y, Sehara-Fujisawa A, Mizuno K, Goshima Y, Yahara I (2001). Phosphorylation of cofilin by LIM-kinase is necessary for semaphorin 3A-induced growth cone collapse. *Nat Neurosci* **4**: 367–373.

91. Kessler O, Shraga-Heled N, Lange T, Gutmann-Raviv N, Sabo E, Baruch L, Machluf M, Neufeld G (2004). Semaphorin-3F is an inhibitor of tumor angiogenesis. *Cancer Res* **64**: 1008–1015.

92. Bielenberg DR, Hida Y, Shimizu A, Kaipainen A, Kreuter M, Kim CC, Klagsbrun M (2004). Semaphorin 3F a chemorepulsant for endothelial cells induces a poorly vascularized encapsulated nonmetastatic tumor phenotype. *J Clin Invest* **114**: 1260–1271.

93. Serini G, Valdembri D, Zanivan S, Morterra G, Burkhardt C, Caccavari F, Zammataro L, Primo L, Tamagnone L, Logan M, Tessier-Lavigne M, Taniguchi M, Puschel AW, Bussolino F (2003). Class 3 semaphorins control vascular morphogenesis by inhibiting integrin function. *Nature* **424**: 391–397.

94. Bates D, Taylor GI, Minichiello J, Farlie P, Cichowitz A, Watson N, Klagsbrun M, Mamluk R, Newgreen DF (2003). Neurovascular congruence results from a shared patterning mechanism that utilizes semaphorin3A and neuropilin-1. *Dev Biol* **255**: 77–98.

95. Yamada Y, Takakura N, Yasue H, Ogawa H, Fujisawa H, Suda T (2001). Exogenous clustered neuropilin 1 enhances vasculogenesis and angiogenesis. *Blood* **97**: 1671–1678.

96. Kawasaki T, Kitsukawa T, Bekku Y, Matsuda Y, Sanbo M, Yagi T, Fujisawa H (1999). A requirement for neuropilin-1 in embryonic vessel formation. *Development* **126**: 4895–4902.

97. Gu C, Rodriguez ER, Reimert DV, Shu T, Fritzsch B, Richards LJ, Kolodkin AL, Ginty DD (2003). Neuropilin-1 conveys semaphorin and VEGF signaling during neural and cardiovascular development. *Dev Cell* **5**: 45–57.

98. Lee P, Goishi K, Davidson AJ, Mannix R, Zon L, Klagsbrun M (2002). Neuropilin-1 is required for vascular development and is a mediator of VEGF-dependent angiogenesis in zebrafish. *Proc Natl Acad Sci USA* **99**: 10470–10475.

99. Miao HQ, Lee P, Lin H, Soker S, Klagsbrun M (2000). Neuropilin-1 expression by tumor cells promotes tumor angiogenesis and progression. *FASEB J* **14**: 2532–2539.

100. Whitaker GB, Limberg BJ, Rosenbaum JS (2001). Vascular endothelial growth factor receptor-2 and neuropilin-1 form a receptor complex that is responsible for the differential signaling potency of VEGF[165] and VEGF[121]. *J Biol Chem* **276**: 25520–25531.

101. Soker S, Miao HQ, Nomi M, Takashima S, Klagsbrun M (2002). VEGF[165] mediates formation of complexes containing VEGFR-2 and neuropilin-1 that enhance VEGF[165]-receptor binding. *J Cell Biochem* **85**: 357–368.

102. Giger RJ, Cloutier JF, Sahay A, Prinjha RK, Levengood DV, Moore SE, Pickering S, Simmons D, Rastan S, Walsh FS, Kolodkin AL, Ginty DD, Geppert M (2000). Neuropilin-2 is required *in vivo* for selective axon guidance responses to secreted semaphorins. *Neuron* **25**: 29–41.

103. Shen J, Samul R, Zimmer J, Liu H, Liang X, Hackett S, Campochiaro PA (2004). Deficiency of neuropilin 2 suppresses VEGF-induced retinal neovascularization. *Mol Med* **10**: 12–18.

104. Takashima S, Kitakaze M, Asakura M, Asanuma H, Sanada S, Tashiro F, Niwa H, Miyazaki JJ, Hirota S, Kitamura Y, Kitsukawa T, Fujisawa H, Klagsbrun M, Hori M (2002). Targeting of both mouse neuropilin-1 and neuropilin-2 genes severely impairs developmental yolk sac and embryonic angiogenesis. *Proc Natl Acad Sci USA* **99**: 3657–3662.

105. Shalaby F, Rossant J, Yamaguchi TP, Gertsenstein M, Wu XF, Breitman ML, Schuh AC (1995). Failure of blood-island formation and vasculogenesis in Flk-1-deficient mice. *Nature* **376**: 62–66.

106. Miao HQ, Soker S, Feiner L, Alonso JL, Raper JA, Klagsbrun M (1999). Neuropilin-1 mediates collapsin-1/semaphorin III inhibition of endothelial cell motility. Functional competition of collapsin-1 and vascular endothelial growth factor-165. *J Cell Biol* **146**: 233–242.

107. Tomizawa Y, Sekido Y, Kondo M, Gao B, Yokota J, Roche J, Drabkin H, Lerman MI, Gazdar AF, Minna JD (2001). Inhibition of lung cancer cell growth and induction of apoptosis after reexpression of 3p21.3 candidate tumor suppressor gene SEMA3B. *Proc Natl Acad Sci USA* **98**: 13954–13959.

108. Xiang R, Davalos AR, Hensel CH, Zhou XJ, Tse C, Naylor SL (2002). Semaphorin 3F gene from human 3p21.3 suppresses tumor formation in nude mice. *Cancer Res* **62**: 2637–2643.

109. Nasarre P, Constantin B, Rouhaud L, Harnois T, Raymond G, Drabkin HA, Bourmeyster N, Roche J (2003). Semaphorin SEMA3F and VEGF have opposing effects on cell attachment and spreading. *Neoplasia* **5**: 83–92.

110. Nasarre P, Kusy S, Constantin B, Castellani V, Drabkin HA, Bagnard D, Roche J (2005). Semaphorin SEMA3F has a repulsing activity on breast cancer cells and inhibits E-Cadherin-mediated cell adhesion. *Neoplasia* **7**: 180–189.

111. Castro-Rivera E, Ran S, Thorpe P, Minna JD (2004). Semaphorin 3B (SEMA3B) induces apoptosis in lung and breast cancer whereas VEGF165 antagonizes this effect. *Proc Natl Acad Sci USA* **101**: 11432–11437.

112. Takahashi T, Nakamura F, Jin Z, Kalb RG, Strittmatter SM (1998). Semaphorins A and E act as antagonists of neuropilin-1 and agonists of neuropilin-2 receptors. *Nat Neurosci* **1**: 487–493.

113. Conrotto P, Valdembri D, Corso S, Serini G, Tamagnone L, Comoglio PM, Bussolino F, Giordano S (2005). Sema4D induces angiogenesis through Met recruitment by Plexin B1. *Blood* **105**: 4321–4329.

114. Giordano S, Corso S, Conrotto P, Artigiani S, Gilestro G, Barberis D, Tamagnone L, Comoglio PM (2002). The semaphorin 4D receptor controls invasive growth by coupling with Met. *Nat Cell Biol* **4**: 720–724.

115. Conrotto P, Corso S, Gamberini S, Comoglio PM, Giordano S (2004). Interplay between scatter factor receptors and B plexins controls invasive growth. *Oncogene* **23**: 5131–5137.

2

EPH Receptors and Ephrins

by Elena B. Pasquale

1. Introduction

The Eph family comprises ten EphA receptors (EphA1–EphA10) and six EphB receptors (EphB1–EphB6) in vertebrates.[1] EphA9 and EphB5, however, were identified in chicken and do not appear to be present in mammals. The Eph receptors influence the behavior of many cell types, unlike other families of receptor tyrosine kinases such as the vascular endothelial growth factor (VEGF) receptor family and the Tie family of angiopoietin receptors, which function more selectively in blood vessels. The first Eph receptor was identified from a human erythropoietin-producing hepatocellular carcinoma cell line in 1987 and named Eph (and later renamed EphA1).[2,3] Identification of additional Eph receptors and characterization of their expression patterns suggested important roles in the formation of connections between neurons during the wiring of the developing nervous system as well as in the organization of epithelial structures and the vasculature. Functional characterization of the Eph receptors was greatly facilitated by the identification of the ligands that stimulate the signaling activity of these receptors. The

27

first ligand for an Eph receptor to be identified was B61, later renamed ephrin-A1 (<u>Eph</u> <u>r</u>eceptor <u>in</u>teracting protein A1).[4]

Studies with ephrin-A1, and its receptor EphA2, were the first to reveal a role for Eph receptors and ephrins in angiogenesis.[5] These seminal findings were followed by numerous genetic studies in the mouse, which have uncovered the involvement of several Eph receptors and ephrins in the normal development of blood and lymphatic vessels as well as in pathological forms of angiogenesis. The complexities in the expression patterns and signaling mechanisms of Eph receptors and ephrins have so far precluded a detailed understanding of the molecular mechanisms used by these molecules to influence the vasculature *in vivo*. However, *in vitro* studies with vascular cells and *in vivo* angiogenesis assays have shown that ephrin-Eph receptor signaling can regulate many properties of vascular cells — including their shape, adhesion, migration, and proliferation — and promote the assembly of capillary-like structures and capillary sprouting.

1.1. *Eph receptor domain structure*

The Eph receptors have an extracellular region that comprises the ephrin-binding domain at the amino terminus, a cysteine-rich region containing an epidermal growth factor (EGF)-like repeat, and two fibronectin type III repeats (Fig. 1). A single membrane-spanning segment connects the extracellular portion of the Eph receptors to the cytoplasmic portion, which comprises a juxtamembrane segment, the tyrosine kinase domain and a sterile alpha motif (SAM) near the carboxy terminus. Most Eph receptors have a PDZ domain-binding site at their extreme carboxy terminus. Following ephrin binding, the Eph receptors dimerize and further cluster, which leads to phosphorylation of many of their cytoplasmic tyrosines.[1] Several of these phosphorylation sites release the kinase domain from inhibitory interaction with the juxtamembrane segment, thereby promoting kinase activity. Furthermore, the tyrosine phosphorylated motifs recruit cytoplasmic signaling proteins containing SH2 domains, leading to activation of downstream signaling pathways.[1] Two Eph receptors, EphA10 and EphB6, lack residues required for kinase activity and likely can be phosphorylated

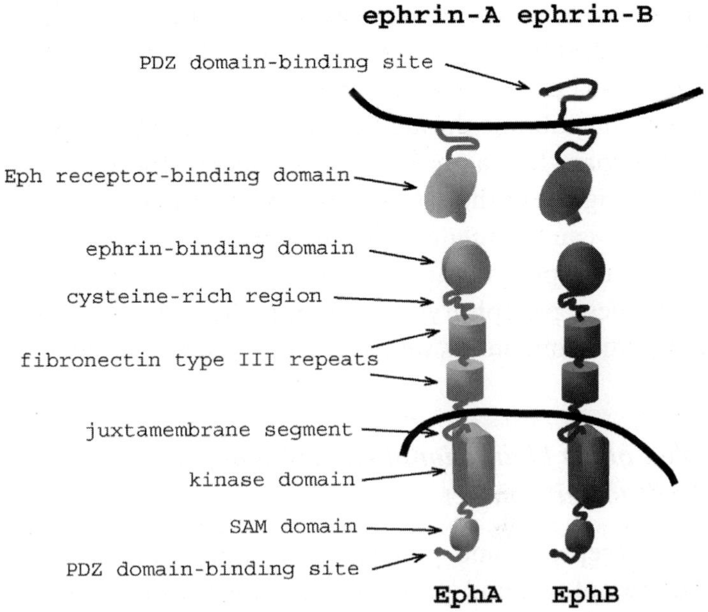

Fig. 1. Prototypical domain structure of Eph receptors and ephrins. The different domains are indicated and the plasma membrane is represented by a thick black line.

only as a result of crosstalk with other Eph receptors or with other signaling molecules. Alternatively spliced forms of the Eph receptors differ from the prototypical Eph receptor described above in their domain structure and, therefore, their function. For example, variant forms of some Eph receptors lack the kinase domain and may be primarily involved in cell adhesion rather than signaling.[6]

1.2. *The ephrin domain structure*

The ephrins are membrane-bound molecules, which is unusual for ligands of receptor tyrosine kinases (Fig. 1). The six vertebrate ephrin-A molecules (ephrin-A1 to ephrin-A6) are associated with the cell surface through a glycosylphosphatidylinositol (GPI) linkage and the three ephrin-B molecules (ephrin-B1 to ephrin-B3) are transmembrane proteins that also contain a cytoplasmic segment. Ephrin-A6, however, was identified in the chicken and does not appear to be present in mammals. The extracellular region of both A and B class ephrins is almost entirely occupied by the receptor-binding domain, which is connected to the plasma membrane through a linker region. The cytoplasmic portion of the ephrin-B ligands is highly conserved and contains a PDZ domain-binding site at the extreme carboxy terminus. Interaction with Eph receptors can stimulate phosphorylation of tyrosine residues in the ephrin-B cytoplasmic domain through the activity of Src family kinases.[1,7] These phosphorylation events affect the conformation of the cytoplasmic domain[8] as well as recruit signaling proteins.[9]

1.3. *Eph-Ephrin bidirectional signaling at sites of cell-to-cell contact*

The EphA receptors bind preferentially ephrins of the A class and the EphB receptors bind preferentially ephrins of the B class, and interactions between receptors and ephrins of the same class are very promiscuous.[10] In addition, EphA4 and EphB2 can bind ligands of the other class. An important exception in the vascular system is the EphB4 receptor, which binds with high affinity only to ephrin-B2. Because both Eph receptors and ephrins are present on the cell surface, their interactions are restricted to sites of cell-to-cell contact.[1] In the vascular system, Eph receptor-ephrin interactions may occur at contact sites between endothelial cells in the same vessel, between venous and arterial endothelial cells and between endothelial and mesenchymal vascular support cells (Fig. 2). Since signals can be generated by both the Eph receptors and the ephrins, bidirectional signals can emanate from sites of cell-to-cell contact and affect both cells. These signals can lead to repulsive effects involving retraction of the cellular processes that were initially engaged in cell-to-cell contact or to attractive effects, such as increased cell-cell and cell-substrate adhesion and forward movement.[1] The outcome of the signals may depend on the degree of Eph receptor activation and clustering, and possibly other as yet unidentified factors. For example, EphA2 signaling mediates positive chemotactic signals and pro-angiogenic effects in endothelial cells, whereas it mediates repulsive effects and apoptosis in cancer cells.[5,11,12]

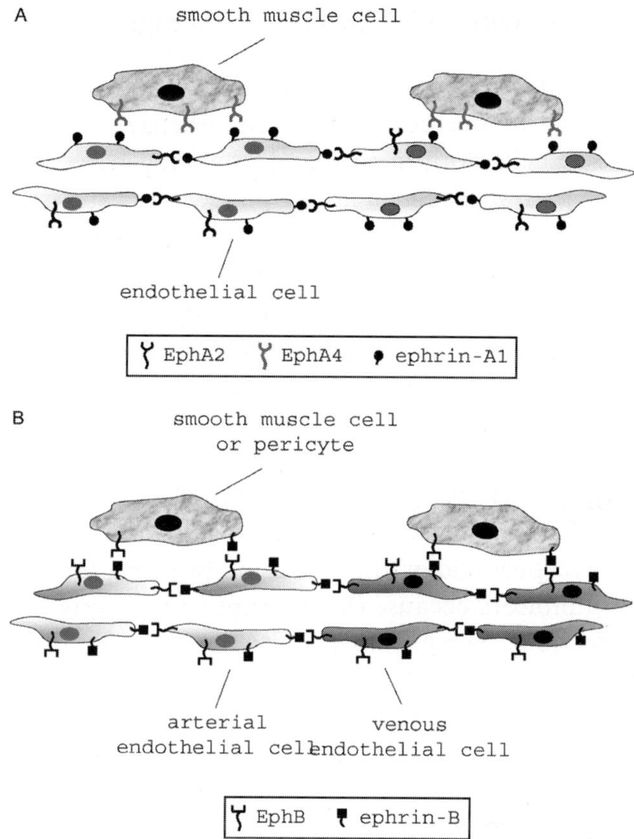

Fig. 2. Eph receptors and ephrins mediate interactions between vascular cells. **(A)** Ephrin-A1 is expressed in embryonic endothelial cells, in postnatal angiogenic endothelial cells, and in endothelial cells in culture. In postnatal and cultured endothelial cells, ephrin-A1 mediates angiogenic effects by activating the EphA2 receptor, which is co-expressed in the endothelial cells. The EphA receptor that is activated by ephrin-A1 in embryonic endothelial cells has not been identified. In culture, endothelial ephrin-A1 also affects the properties of vascular smooth muscle cells, which express the EphA4 receptor, suggesting that a similar interaction may affect vascular smooth muscle cells *in vivo*. **(B)** Ephrin-B2 is expressed in arterial endothelial cells (light gray) and EphB4 is expressed in venous endothelial cells (dark gray). Thus, this B class receptor and ephrin can mediate interactions between arterial and venous vasculature. Ephrin-B1 and other EphB receptors are also expressed in embryonic endothelial cells and in cultured endothelial cells, and could mediate interactions between adjacent endothelial cells of the same type. EphB2 and ephrin-B2 are expressed in vascular smooth muscle cells and pericytes and thus could mediate interactions between these vascular support cells and endothelial cells.

2. Effects on Vascular Cell Behavior and Signaling Pathways

Eph receptor forward signaling and ephrin reverse signaling can dramatically influence the behavior of endothelial cells and vascular smooth muscle cells both *in vitro* and in angiogenesis assays *in vivo*. The signaling pathways regulated by Eph receptors and ephrins in vascular cells are beginning to be elucidated through *in vitro* studies, but the importance of these pathways for physiological angiogenesis *in vivo* remains to be verified. Ephrin extracellular domains fused to the Fc portion of human IgG_1 have been extensively used to activate Eph receptor signaling pathways in angiogenesis assays because they are soluble and dimeric, and can be multimerized by anti-Fc antibodies. Thus, they can be used to induce the Eph receptor dimerization and further clustering that are important for proper signaling.[1,13] Similarly, Eph receptor ectodomains fused to Fc have been used to stimulate ephrin reverse signaling. Eph receptor and ephrin Fc fusion proteins, however, can also function as inhibitors because they disrupt endogenous Eph receptor-ephrin interactions. For example, EphA receptor Fc fusion proteins have been used to inhibit EphA forward signaling, which has established the importance of EphA receptors in postnatal angiogenesis[14,15] (Sec. 6).

2.1. *Ephrin-A1 and EphA2*

Ephrin-A1 and EphA2 are the main ephrin and Eph receptor of the A class that have thus far been implicated in endothelial cell function (Fig. 2A). In a widely used *in vitro* angiogenesis model, endothelial cells plated onto reconstituted basement membrane proteins (Matrigel) respond by forming capillary tube-like structures. Interestingly, one of the consequences of plating human umbilical vein endothelial (HUVE) cells on Matrigel is the upregulation of ephrin-A1.[16] The effect of ephrin-A1 in capillary morphogenesis on Matrigel was confirmed by showing that exogenously added ephrin-A1 Fc promotes the assembly of capillary structures in HUVE cells and mouse pulmonary microvascular endothelial cells, both of which express high levels of the EphA2 receptor.[5,17] Furthermore, reducing expression of the transcription factor Homeobox B3 (HoxB3) with antisense oligonucleotides decreases ephrin-A1 expression and impairs capillary morphogenesis in dermal microvascular endothelial cells.[18] Given that treatment with ephrin-A1 Fc restores capillary-like tube formation in the HoxB3-deficient cells, these data suggest that the HoxB3-dependent expression of ephrin-A1 is important for endothelial capillary morphogenesis driven by extracellular matrix proteins. Not all endothelial cells may respond to ephrin-A1 Fc, however. Human renal microvascular endothelial cells, for example, reportedly do not form capillary-like tubes in response to ephrin-A1 Fc.[17] The reason for this lack of responsiveness to ephrin-A1 Fc remains mysterious, since these cells express EphA2 and can form capillary-like tubes when treated with ephrin-B1 Fc (Sec. 2.3).

Consistent with a role as an angiogenic factor, ephrin-A1 Fc also promotes endothelial cell migration. Ephrin-A1 Fc acts as a chemoattractant for bovine adrenal capillary endothelial cells and microvascular endothelial cells in transwell migration assays and promotes the movement of cells into a "wound" devoid of cells in a confluent endothelial cell monolayer.[5,14,19,20] Ephrin-A1 Fc also induces endothelial cell sprouting in an *in vitro* capillary sprouting assay.[19] In this assay, microvascular endothelial cells are cultured on collagen-coated beads embedded in fibrin gels and form capillary sprouts that extend out from the beads into the fibrin matrix.

Additional studies have shown that ephrin-A1 Fc also promotes the formation of blood vessels in a variety of *in vivo* angiogenesis assays. These assays include: (i) corneal neovascularization assays, where hydron pellets impregnated with ephrin-A1 Fc induce the formation of blood vessels when implanted in a micropocket in the normally avascular rodent cornea;[5,19] (ii) Matrigel assays, where Matrigel injected under the mouse skin forms plugs that promote the assembly of endothelial cells into blood vessels when supplemented with ephrin-A1 Fc; and (iii) assays in which surgical sponges impregnated with ephrin-A1 Fc and implanted in the dorsal flank of mice attract an increased number of host blood vessels compared to control sponges.[20]

Ephrin-A1 appears to have similar pro-angiogenic effects when it is endogenously expressed in endothelial cells and as an exogenous Fc fusion protein. Hence, the angiogenic effects of ephrin-A1 can be mainly attributed to its stimulation of EphA receptor forward signaling because ephrin-A1 Fc lacks the ability to mediate reverse signals. Indeed, mutants of the EphA2 receptor that inhibit EphA receptor forward signaling in a dominant negative manner block the *in vitro* angiogenic effects of ephrin-A1 Fc, while a constitutively active EphA2 mutant enhances angiogenic responses.[20,21] Furthermore, an EphA antagonist such as EphA2 Fc strongly inhibits capillary formation in an *in vitro* rat aortic ring explant assay and in an *in vivo* Matrigel assay.[15] Since EphA2 Fc can also activate ephrin-A1 reverse signaling (Sec. 2), this experiment corroborates the idea that ephrin-A1 reverse signaling may not promote angiogenesis.

The requirement for EphA2 signaling in endothelial cell migration and vascular assembly has been confirmed by the impaired angiogenic responses to ephrin-A1 Fc in microvascular endothelial cells isolated from EphA2 knockout mice as well as in endothelial cells in which EphA2 expression was downregulated with antisense oligonucleotides.[19,20] A signaling pathway involving phosphatidylinositol (PI) 3 kinase and the Rho family GTPase, Rac1, has been implicated in the effects of EphA2 on microvascular endothelial cell migration *in vitro*.[20,22] Although EphA2 is not expressed in the embryonic vasculature, this receptor has been confirmed as a key player in postnatal angiogenesis *in vivo* because EphA2 knockout mice exhibit a diminished angiogenic response to surgical sponges impregnated with ephrin-A1 Fc.[20] Furthermore, microvascular endothelial cells from EphA2 knockout mice fail to elongate and assemble into capillaries in Matrigel plugs implanted into wild-type recipient mice. The EphA2-deficient cells also have impaired survival in the Matrigel plugs, but appear to proliferate normally, which is consistent with the lack of *in vitro* effects of ephrin-A1 Fc on endothelial cell proliferation.[19]

2.2. *Ephrin-A1 and EphA4*

Endothelial ephrin-A1 likely coordinates different aspects of angiogenesis by activating EphA receptors not only in endothelial cells, but also in the surrounding vascular smooth muscle cells (Fig. 2A). In cultured smooth muscle cells, ephrin-A1 Fc causes a repulsive response involving increased assembly and contractility of actin stress fibers and decreased cell substrate adhesion.[23] These effects occur through activation of the EphA4 receptor, which is highly expressed in vascular smooth muscle cells. EphA4 in turn activates an exchange factor of the Ephexin family with selective expression in vascular smooth muscle cells, Vsm-RhoGEF, which increases RhoA activity.[23] Ephrin-A1 Fc treatment also impairs smooth muscle cell spreading on extracellular matrix proteins by inactivating another Rho family GTPase, Rac1, and its downstream effector p21-activated kinase 1 (Pak1).[24] These data support the idea that endothelial ephrin-A1 coordinates the angiogenic responses of both endothelial and vascular support cells by eliciting different responses in these two cell types. By inhibiting the spreading and promoting the contractility of smooth muscle cells, ephrin-A1 may destabilize their interaction with endothelial cells, allowing endothelial cell migration and vascular assembly into new capillary sprouts. These effects all seem to depend on EphA forward signaling. Whether reverse signals mediated by ephrin-A1 may have additional roles in angiogenesis remains to be determined.

2.3. *Ephrin-B and EphB*

2.3.1. *EphB forward signaling*

EphB receptor forward signaling also affects the properties of endothelial cells. Ephrin-B1 Fc stimulates the formation of capillary structures in human renal microvascular endothelial cells, which express the EphB1 and EphB2 receptors.[13,17] Ephrin-B1 Fc and ephrin-B2 Fc also induce capillary sprouting in adrenal cortex-derived microvascular endothelial cells with a potency comparable to that of angiopoietin-1 and VEGF.[25] However, there is some selectivity with regard to endothelial cell type because ephrin-B1 Fc does not induce capillary-like tubes in HUVE cells,[17] which express the EphB1, EphB2, EphB3, and EphB4 receptors.[26] This suggests that EphB forward signaling is not sufficient to mediate capillary assembly in HUVE cells, although it should be noted that ephrin-B1 does not efficiently activate EphB4. Nevertheless, ephrin-B1 Fc reportedly promotes HUVE cell proliferation and migration.[27] Ephrin-B2 Fc, on the other hand, promotes the proliferation and migration of mesenteric microvascular endothelial cells.[28]

In the renal microvascular endothelial cells, ephrin-B1 Fc promotes capillary-like assembly only when clustered with anti-Fc antibodies, whereas dimeric ephrin-B1 Fc is ineffective, even though in both cases EphB receptor tyrosine phosphorylation is similarly induced.[13] In addition, tetrameric ephrin-B1 Fc promotes cell-substrate adhesion and migration but the dimeric ephrin does not.[13,29] EphB-dependent stimulation of endothelial cell attachment is mediated by the $\alpha_v\beta_3$ integrin and depends on the density of ephrin-B1 Fc immobilized on the surface on which the cells grow (and therefore the degree of EphB receptor clustering), with highest adhesion at intermediate levels of clustering.[29] Sprouting angiogenesis of adrenal cortex microvascular endothelial cells induced by ephrin-B2 Fc (but not ephrin-B1 Fc) also requires clustering of the ephrin.[25] In addition, pellets containing ephrin-B2 Fc induce the formation of blood vessels in corneal micropocket assays, although more weakly than pellets containing VEGF.[30] Interestingly, in this assay ephrin-B2 Fc preferentially promotes venous neovascularization, as judged by the few ephrin-B2 positive vessels induced and by the upregulation of EphB4 mRNA.

These results suggest that there are differences in the signaling pathways activated by different EphB receptor oligomeric forms. Indeed, low-molecular-weight phosphotyrosine phosphatase (LMW-PTP) is only recruited to EphB1 and EphB2 when these receptors are activated by clustered ephrin-B1 Fc, and mutants of EphB1 that cannot bind LMW-PTP phosphatase fail to promote microvascular endothelial cell attachment and capillary-like assembly.[13] Another signaling cascade that has been implicated in promoting cell adhesion downstream of EphB1 and EphB2 involves the adaptor Nck, the Nck-interacting serine/threonine kinase (NIK), and activation of the c-Jun N-terminal kinase (JNK).[31] This Nck-NIK-JNK pathway remains to be verified in endothelial cells, however. EphB1-mediated stimulation of renal microvascular endothelial cell migration also requires Nck, in this case to couple EphB1 to paxillin, which is phosphorylated by activated Src and promotes cell migration.[32] EphB1-dependent cell migration also requires phosphorylation of the adaptor Shc by Src, which promotes binding of Shc to another adaptor, Grb2, leading to activation of the Ras-mitogen-activated protein (MAP) kinase pathway.[33] The Crk adaptor protein has also been implicated in endothelial cell spreading and migration downstream of EphB1. When human aortic endothelial cells are stimulated with ephrin-B1 Fc, Crk promotes membrane ruffling through the Rho family GTPase Rac1 and focal complex assembly and cell spreading through the Ras family GTPase Rap1.[34] Other studies have shown the involvement of the phosphatidylinositol-3 (PI-3) kinase-Akt pathway in EphB-dependent endothelial cell migration *in vitro* as well as in corneal and Matrigel neovascularization *in vivo*.[28,35]

In contrast to the attractive effects of EphB receptors described above, other reports have shown that ephrin-B2 Fc stimulation of EphB forward signaling mediates repulsive effects in HUVE cells, in the brain-derived bEnd3 capillary endothelial cell line, and in FACS-sorted EphB4-positive mouse embryonic endothelial cells.[26,36,37] For example, stimulation of EphB4 forward signaling prevents endothelial cell attachment and spreading on immobilized ephrin-B2 Fc as well as inhibits proliferation, capillary-like assembly, and sprouting angiogenesis. Ephrin-B2 Fc also inhibits the migration and proliferation of HUVE cells stimulated with VEGF or angiopoietin-1 (Sec. 2.4), at least in part through the recruitment of the Ras GTPase-activating protein p120RasGAP to activated EphB receptors and subsequent suppression of the Ras-MAP kinase pathway.[26] EphB4 forward signaling also causes anti-adhesive effects that disrupt the integrity of the endothelial monolayer in spheroids of co-cultured endothelial and smooth muscle cells as well as in umbilical cord explants.[37] The repertoire of EphB receptors activated, the presence of co-expressed ephrin-B ligands, or other as yet unknown factors, may influence the type of response mediated by an activated EphB receptor.

2.3.2. *Ephrin-B reverse signaling*

Ephrin-B reverse signaling causes pro-angiogenic effects in a variety of endothelial cell types. In HUVE and bEnd3 cells treated with EphB4 Fc, ephrin-B2 reverse signals enhance endothelial cell attachment and spreading on immobilized EphB4 Fc and increase cell survival, migration, sprouting angiogenesis in a collagen gel, and capillary-like assembly on Matrigel.[36–38] EphB3 Fc and EphB4 Fc also induce angiogenic sprouting in cultured adrenal cortex-derived endothelial cells [39] and

EphB1 Fc promotes attachment and migration of renal microvascular endothelial cells *in vitro* and blood vessel formation *in vivo* in a mouse corneal neovascularization assay.[40] A mixture of EphB1 Fc and EphB3 Fc has also been shown to increase the number and length of microvessel sprouts in a rat aortic ring angiogenesis assay.[15] While EphB1, EphB2, and EphB3 Fc fusion proteins could activate reverse signals through any of the three ephrin-B ligands, EphB4 Fc more selectively activates only ephrin-B2.

Ephrin-B2 reverse signaling may be responsible for some of the pro-angiogenic effects reported for the EphB4 receptor in HUVE cells. EphB4 expression is upregulated by the transcription factor HoxA9 and contributes to the *in vitro* angiogenic effects of HoxA9 in HUVE cells, because reducing EphB4 expression with siRNA or antisense oligonucleotides impairs HoxA9-dependent HUVE cell migration and capillary-like assembly.[41] Consistent with the idea that EphB4 acts by stimulating ephrin-B2 reverse signaling, transfection of a dominant negative form of EphB4 that should block endogenous EphB receptor forward signaling while still mediating reverse signaling fails to inhibit HUVE cell capillary-like assembly on Matrigel.[21]

Src family kinases, which become activated upon ephrin-B stimulation and phosphorylate the ephrin cytoplasmic domain, have been implicated in the angiogenic effects mediated by ephrin-B reverse signaling.[7,42] Ephrin-B reverse signals also activate integrins to promote renal microvascular endothelial cell attachment and migration.[40] These angiogenic effects may involve an as yet unknown protein that binds to the ephrin-B carboxy terminus and mediates activation of JNK, although this signaling pathway remains to be verified in endothelial cells. In addition, the PI3 kinase-Akt and MAP kinase signaling pathways have been implicated in the angiogenic effects of ephrin-B2 reverse signaling in retinal endothelial cells.[42] It is not known whether different ephrin-B molecules activate distinct angiogenic pathways.

Based on the *in vitro* attractive effects of ephrin-B2 and the repulsive effects of EphB4 in endothelial cells, a model has been proposed where ephrin-B2 reverse signals in arterial endothelial cells mediate propulsive effects that act coordinately with the repulsive effects of EphB4 in vein endothelial cells[37] (Sec. 3.1). Consistent with this model, an artery-to-vein direction of sprouting angiogenesis has been observed in the avian yolk sac[43] (Sec. 4.2).

2.4. *Crosstalk with other angiogenic pathways*

Activation of EphA2 by ephrin-A1 has been shown to mediate the angiogenic effects of tumor necrosis factor α (TNFα) both *in vitro* and *in vivo* (Fig. 3A). TNFα and other pro-inflammatory cytokines upregulate ephrin-A1 expression in endothelial cells, which in turn promotes

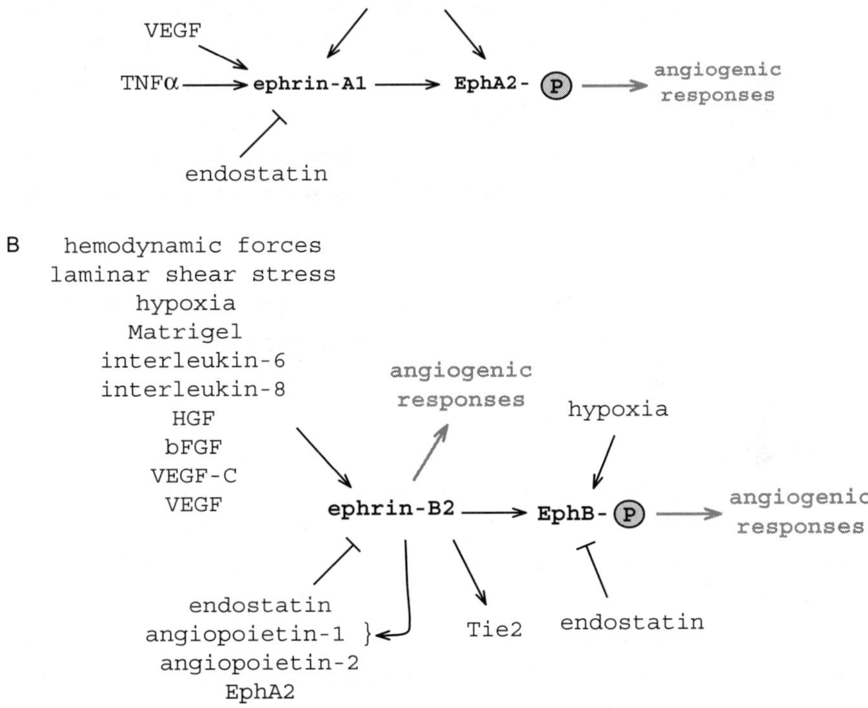

Fig. 3. Various factors that regulate Eph receptor and ephrin expression in endothelial cells. Thin black arrows and bars indicate up- or down-regulation of expression levels, respectively. Thick gray arrows indicate increased angiogenic responses. Activated (tyrosine phosphorylated) EphB receptors mediate angiogenic responses through either attractive or repulsive effects depending on the conditions, the endothelial cell type, and the receptor involved (see text for details). P, tyrosine phosphorylation.

activation of EphA2 (as shown by increased tyrosine phosphorylation of this receptor) and capillary morphogenesis.[5,18,44,45] Furthermore, activation of EphA2 is required for corneal neovascularization induced by TNFα.[5] TNFα regulates ephrin-A1 expression in endothelial cells through the p38 MAP kinase and JNK,[46] and it will be interesting to examine whether these MAP kinases in turn regulate the transcription factor HoxB3 (Sec. 2.1).

Consistent with a role in angiogenesis, ephrin-A1 is downregulated in human microvascular endothelial cells by treatment with the anti-angiogenic factor endostatin.[47] Ephrin-A1 is also an important mediator of the angiogenic effects of VEGF, which instead upregulates ephrin-A1 expression.[19] Studies using EphA2 Fc to block the interaction between endogenous ephrin-A1 and EphA2, or EphA2 antisense oligonucleotides to reduce EphA2 expression, have shown that the ensuing stimulation of EphA forward signaling plays a role in some of the angiogenic activities of VEGF, such as microvascular endothelial cell survival, migration and sprouting *in vitro* as well as the formation of new blood vessels *in vivo* in corneal neovascularization assays and Matrigel assays.[14,15,19,48] In contrast, endothelial cell proliferation induced by VEGF, and the angiogenic effects of basic fibroblast growth factor (FGF2), seem to be independent of ephrin-A1 and EphA2.[5,19]

There is also crosstalk between endostatin and VEGF and the EphB/ephrin-B signaling pathways (Fig. 3B). Endostatin downregulates ephrin-B1 and ephrin-B2 as well as EphB4 in human dermal microvascular endothelial cells.[47] In contrast, VEGF upregulates ephrin-B2 in cultured endothelial cells[49,50] and *in vivo* in arterial endothelial cells of the embryonic skin,[49] in a subset of the blood vessels induced in corneal neovascularization assays,[51] and in capillaries induced by VEGF transgenic expression in the mouse heart.[52] A pathway responsible for ephrin-B2 expression likely involves Notch and TGFβ signaling.[53] Indeed, TGFβx and activin-A can upregulate ephrin-B2 expression in mouse primary embryonic endothelial cells, similar to VEGF.[49] Interestingly, loss-of-function studies in zebrafish embryos have shown that Notch signaling not only upregulates arterial markers like ephrin-B2 but also represses venous markers like EphB4.[54]

In turn, EphB receptor activation by ephrin-B2 Fc has been shown to attenuate VEGF-induced HUVE cell proliferation and migration.[26,37]

Other growth factors in addition to VEGF, as well as plating cells on a Matrigel substrate, have been reported to upregulate ephrin-B2 expression in endothelial cells.[30,50] The other growth factors that have been shown to upregulate ephrin-B2 include VEGF-C, interleukin-6 and interleukin-8 in HUVE cells and hepatocyte growth factor and FGF2 in human aortic and dermal microvascular endothelial cells. Activation of EphB4 by ephrin-B2 Fc in the aortic endothelial cells in turn inhibits the angiogenic effects of FGF2. This effect involves upregulation of syndecan-1 expression and shedding of the ectodomain of this proteoglycan from the cell surface.[55] The overproduced soluble syndecan-1 ectodomain inhibits FGF receptor signaling, likely by sequestering FGF2 away from its receptor. A further twist is that heparitinase, an enzyme that preferentially targets desulphated heparin, converts the soluble syndecan-1 ectodomain from an inhibitor to an activator of FGF2 binding to its receptor. Interestingly, enzymes with activity similar to heparitinase are present in inflamed tissue, where they would be predicted to modify the effects of ephrin-B2 on FGF receptor signaling (Sec. 6.3).

Phorbol myristate acetate (PMA) also promotes the assembly of renal microvascular endothelial cells into capillary-like tubes, and this effect involves activation of EphB1 and EphB2 by endogenously expressed ephrin-B1.[13] Ephrin-B1 levels are not changed by PMA treatment, however, suggesting another form of regulation that may involve ephrin-B1 clustering induced through phosphorylation by protein kinase C (PKC), a serine/threonine kinase that is activated by PMA.

The Tie2 receptor tyrosine kinase has also been shown to phosphorylate tyrosine residues in the cytoplasmic domain of ephrin-B1, at least *in vitro*, which may also modulate ephrin-B angiogenic activities.[25] Ephrin-B2-EphB4 signaling in turn appears to increase the expression of Tie2 and its ligand, angiopoietin-1, because Tie2 and angiopoietin-1 are poorly expressed in ephrin-B2 knockout mice.[39] Interestingly, the phenotype of the angiopoietin-1 and Tie2 knockout mice resembles that of ephrin-B2 and EphB4 knockout mice. This raises the intriguing possibility that ephrin-B2-EphB receptor signaling may mediate blood vessel remodeling at least in part by upregulating the expression of angiopoietin-1 and Tie-2 (Sec. 4.2). In turn, however, angiopoietin-1 has been shown to downregulate ephrin-B2 expression in human dermal microvascular endothelial cells in culture[30] and EphB receptor activation by ephrin-B2 Fc counteracts angiopoietin-1-induced cell migration in HUVE cells.[26] Furthermore, transgenic co-expression of angiopoietin-2 (another member of the angiopoietin family) and VEGF reduces the number of ephrin-B2-positive blood vessels, resulting in a fraction of blood vessels that express neither ephrin-B2 nor EphB4.[52]

Interestingly, pulmonary microvascular endothelial cells isolated from EphA2 knockout mice have normal ephrin-A1 expression but increased expression of EphB4 and ephrin-B2, suggesting compensatory mechanisms and crosstalk between Eph receptors and ephrins of the A and B subclasses.[56] Although all these findings do not yet provide a cohesive picture, they nevertheless show that there are many forms of crosstalk within the Eph family and between the Eph system and other angiogenic signaling pathways, leading to complex networks of positive and negative feedback loops that coordinately regulate the properties of vascular cells.

3. Endothelial Cell Fate

Eph receptors and ephrins are expressed in stem cells and progenitor cells of different lineages, where they regulate the balance of proliferation and self-renewal versus differentiation, cell fate determination, and even cell death.[1,57] Both EphB4 and ephrin-B2 are expressed in mouse undifferentiated embryonic stem cells and in early embryoid bodies derived by culturing those stem cells.[58] Interestingly, EphB4 has been implicated in the differentiation of a common precursor cell, the hemangioblast, towards endothelial and hematopoietic cell lineages.[58,59] When embryoid bodies are used to generate hemangioblast cells *in vitro*, mimicking developmental events occurring in the blood islands of the yolk sac, EphB4-deficient embryoid bodies display defects in hemangioblast development.[58] EphB4 knockout embryoid bodies also produce reduced numbers of endothelial cells and have impaired ability to develop endothelial cells sprouts in the presence of angiogenic growth factors, suggesting a role for EphB4 in the assembly of the primitive vascular network. However, the deficiencies in the production of hemangioblast cells and in the assembly of capillary sprouts are corrected over time, suggesting that EphB4 facilitates these processes, but is not absolutely required. This may explain why major defects in the initial assembly of blood vessels by vasculogenesis have not been noted in the EphB4 knockout mice (Sec. 4.2).

4. Angiogenic Remodeling of Embryonic Blood Vessels

4.1. *Ephrin-A1 and EphA receptors*

Ephrin-A1 is widely expressed in embryonic veins and arteries starting at very early stages of development,[60,61] consistent with its angiogenic role *in vitro* and in angiogenesis assays *in vivo* (Fig. 2A; Sec. 2.1). However, the role of ephrin-A1 in the formation of embryonic blood vessels remains to be determined. It was noted that ephrin-A1 expression is uneven in different blood vessels in the same organ and in different endothelial cells in the same blood vessel and declines as development progresses,[62] but the significance of these observations is unknown. Ephrin-A1 knockout mice will be an important tool to elucidate the function of this ephrin during development of the vascular system, once they become available. The EphA receptors that interact with ephrin-A1 in embryonic blood vessels have not been identified. Despite the expression of EphA2 at sites of adult angiogenesis and the critical role of this receptor in angiogenic responses, EphA2 has not been detected in the embryonic vasculature and EphA2 knockout mice do not exhibit overt defects in vascular development.[20] Other EphA receptors present in the developing vasculature are EphA4 and EphA7, whose expression in the endothelium and mesenchyme of umbilical arteries is regulated by the Homeobox A13 (HoxA13) transcription factor.[63] Deficient expression of EphA4 and EphA7 in HoxA13 knockout mice may contribute to the observed narrowing of the umbilical arteries and loss of stratification of vascular mesenchyme and endothelium in these vessels.

4.2. *EphB4 and Ephrin-B2*

Ephrin-B2 was the first molecular marker of arterial endothelial cells to be identified and EphB4 the first marker of venous endothelial cells[64,65] (Fig. 2B). These expression patterns of ephrin-B2 and EphB4, which are observed from the earliest stages of angiogenesis, revealed for the first time that the separate identity of venous and arterial endothelial cells is specified before blood flow is established. However, recent studies support the idea that the arterial or venous identity of endothelial cells remains plastic and can be regulated by local cues such as hemodynamic forces after onset of the embryonic circulation and laminar shear stress in cultured endothelial cells.[43,66,67] Although ephrin-B2 and EphB4 have been increasingly used as markers of arterial versus venous vessel identity, recent data suggest that additional criteria should also be considered. For example, moderate hypoxia causes loss of arterial expression of ephrin-B2 and its possible upstream regulator, Delta-like 4, in the developing mouse retinal vasculature.[68] However, other markers of arterial identity remain present suggesting that vessels can lose ephrin-B2 expression but still maintain arterial characteristics. There are also some exceptions to the arterial and venous segregation of ephrin-B2 and EphB4. For example, in human embryonic lung vasculature, EphB4 and ephrin-B2 are not segregated in veins and arteries as they are in the mouse.[69]

Interestingly, VEGF secreted locally by peripheral nerves in the skin is responsible for the alignment of blood vessels with the nerves and for the arterial differentiation of these nerve-associated blood vessels, which involves upregulation of ephrin-B2[49] (Sec. 2.3). Less is known about the regulation of EphB4 expression in the vasculature, although the HoxA9 transcription factor has been reported to upregulate EphB4 in endothelial cells, at least *in vitro*[41] (Sec. 2.3). EphB4 may also positively regulate its own expression because homozygous knockout mice in which the β-galactosidase reporter gene is expressed in place of EphB4 show lower levels of β-galactosidase compared to heterozygous mice, which have one functional EphB4 allele.[65]

The phenotype of ephrin-B2 and EphB4 knockout mice does not reveal prominent defects in the initial assembly of blood vessels during vasculogenesis.[39,64,65] The networks of evenly sized capillaries of the primary vascular plexus in the yolk sac and in the head of these mice appear to form normally, although it is not known whether subtle delays may occur in the generation of endothelial cells (Sec. 3). Ephrin-B2 and EphB4 are, however, required for normal development of the dorsal aorta and cardinal veins, which are assembled from coalescing endothelial precursor cells by vasculogenesis.[25,64,65]

The most notable early angiogenic defect in the ephrin-B2 and EphB4 knockout mice is a failure to remodel the primary vascular plexus into a mature, functional vasculature consisting of large and small interconnected branches.[64,65] For example, the vasculature in the yolk sac and in the head persists as an immature network of evenly sized capillaries in the absence of ephrin-B2 or EphB4, and fails to undergo reorganization. This reorganization normally involves sprouting of new blood vessels and pruning of existing vessels as well as fusion and splitting of vessels and, as recently shown by time-lapse imaging, disconnection of small arterial vessels and their reconnection to form new junctions with the venous network.[43,70] EphB-ephrin-B bidirectional signaling could in principle regulate all of these processes (Sec. 2.3). Remarkably, despite the endothelial expression of ephrin-B2 only in arteries and EphB4 mainly in veins, the vascular defects in the ephrin-B2 and EphB4 knockout mice are quite similar and affect both arteries and veins. In addition, decreased capillary formation in mice overexpressing ephrin-B2 in endothelial cells suggests that a precise level of ephrin-B2 expression is essential for arterial-venous capillary boundary formation.[71] Interestingly, exposure of the developing mouse retinal vasculature to decreased (10%) oxygen levels causes loss of ephrin-B2 and inappropriate separation of the arterial and venous networks, supporting a role for ephrin-B2-EphB4-mediated cell repulsion in the normal segregation of arteries and veins.[68] Ephrin-B2 expressed in the intersomitic arteries and EphB4 expressed in the intersomitic veins are also required for the subsequent angiogenic remodeling of the intersomitic vasculature.[65,72]

The similar phenotypes of the ephrin-B2 and EphB4 knockout mice led to the proposal that bidirectional signals mediated by these molecules between arterial and venous endothelial cells are critical for angiogenic remodeling.[64,65] For example, repulsive interactions could prevent arteries and veins from fusing during angiogenic remodeling, while allowing the fusion of vessels of the same type. Repulsive signals could also be important in establishing the proper balance of arteries and veins in capillary beds as well as in the formation and maintenance of the arteriovenous boundary. Although only limited interfaces between the arterial and venous sides exist in the mature vasculature, there is some evidence that extensive transient connections form during angiogenic vascular remodeling in the embryo.[72] Thus, proper signaling at the arterial-venous boundaries may be essential for remodeling of the entire network to occur.

In vitro studies suggest that unidirectional EphB4 forward signaling is sufficient to segregate EphB4-expressing endothelial cells from ephrin-B2-expressing cells by restricting cell intermingling[37] and mouse knock-in studies have shown that ephrin-B2 reverse signaling is not required for angiogenic vascular remodeling in the early embryo. Replacement of wild-type ephrin-B2 with an ephrin-B2 mutant lacking all tyrosine phosphorylation sites in the cytoplasmic domain or with a mutant lacking the carboxy-terminal PDZ domain binding site rescues the early vascular defects observed in ephrin-B2 knockout mice.[73] Furthermore, a form of ephrin-B2 in which the entire cytoplasmic domain is replaced by β-galactosidase supports normal angiogenesis in the early embryo, similar to wild-type ephrin-B2.[74] These data indicate that the function of ephrin-B2 in the early arteries is to stimulate EphB forward signaling, whereas reverse signaling appears to be dispensable. Different results obtained with a mutant ephrin-B2 in which most of the cytoplasmic domain was replaced by the short HA tag may be explained by a partially defective ability of this mutant to elicit EphB reverse signaling.[39,73] These data indicate that the main function of ephrin-B2 in early embryonic angiogenesis is as a ligand that stimulates EphB receptor forward signaling, while its intrinsic reverse signaling function is dispensable. Although in the embryo ephrin-B2 is expressed not only in endothelial cells but also in adjacent mesenchymal cells, the phenotype of conditional knockout mice that lack ephrin-B2 only in endothelial and endocardial cells (due to Cre expression driven by the Tie2 promoter) is indistinguishable from the phenotype of mice lacking ephrin-B2 in all cells.[72] Taken together, these data indicate that the essential function of ephrin-B2 in arterial endothelial cells is to stimulate EphB forward signaling, which is sufficient to mediate remodeling of both arteries and veins.

Although repulsive effects mediated by unidirectional EphB4 signaling in veins at the boundaries with arteries likely play a critical role in blood vessel maturation in the embryo, additional mechanisms may contribute to the observed arterial defects of the mutant mice. For example, ephrin-B2 could affect arterial endothelial cells by stimulating forward signaling by the EphB4 receptor present at low levels in scattered arterial endothelial cells.[65] Arterial defects in the ephrin-B2 knockout mice may also be a secondary consequence of defective blood flow due to an abnormal heart and/or abnormal veins.[65] Although defects in vascular remodeling begin in the ephrin-B2 and EphB4 knockout mice before the blood circulation is established and continue to be present at later stages in a subset of embryos that maintain blood flow,[64,65] perfusion could nevertheless be perturbed in these embryos. Interestingly, recent experiments have shown that hemodynamic forces and laminar shear stress can modify the expression of arterial markers such as ephrin-B2.[43,67] Another intriguing possibility is that ephrin-B2 stimulates EphB4 forward signaling pathways that regulate expression of other molecules that are important for angiogenic remodeling of the embryonic vasculature, such as angiopoietin-1 and its receptor Tie2[39] (Sec. 2.4). However, given the effects of ephrin-B2 and EphB4 on endothelial cell shape, attachment, migration and proliferation *in vitro* (Sec. 2.3), it seems probable that these molecules must exert some direct effects on the behavior of endothelial cells *in vivo* independently of the angiopoietin-1/Tie2 system.

Whether ephrin-B2 also plays an essential role in vascular support cells, in addition to endothelial cells, awaits examination of conditional knockout mice lacking ephrin-B2 only in these cells. A clue as to the function of ephrin-B2 in vascular smooth muscle cells may be the finding that ephrin-B2-expressing stromal cells promote growth and angiogenic sprouting of ephrin-B2-expressing endothelial cells in a co-culture explant system.[75] A functional role for non-endothelial ephrin-B2 in the ascending aorta is also suggested by the abnormalities in the recruitment of smooth muscle cells that have been observed in mice ubiquitously overexpressing ephrin-B2 but not in mice overexpressing ephrin-B2 only in endothelial cells.[71]

4.3. *Other EphB receptors and Ephrin-Bs*

In situ hybridization studies have revealed the presence of additional EphB and ephrin-B molecules in the developing vasculature.[25] Ephrin-B1 is expressed in both arterial and venous endothelial cells. However, it cannot compensate for the loss of ephrin-B2, perhaps due to its poor ability to activate EphB4. The EphB3 receptor is expressed in veins and in the aortic arches, while EphB2 is not expressed in endothelial cells but it is present in surrounding mesenchymal cells. These expression patterns indicate that ephrin-B-EphB interactions are not restricted to the arterial-venous boundaries but can also occur between endothelial cells in the same vessel and at endothelial-mesenchymal interfaces and involve more than just ephrin-B2 and EphB4 (Fig. 2B). For example, endothelial ephrin-B2 could also function by stimulating EphB2 expressed in vascular support cells surrounding the endothelial cells.

Genetic evidence indicates that these additional interactions are functionally important. Although mice lacking either EphB2 or EphB3 do not exhibit overt vascular defects, some of the double knockout mice lacking both receptors have vascular defects similar to those of the ephrin-B2 (and EphB4) knockout mice.[25] Thus, EphB2 and EphB3 also play a role in the maturation of the primitive vascular plexus and the formation of major vessel primordia, presumably through interactions with ephrin-B2 (Sec. 4.2) and possibly ephrin-B1. Defects in the association of arterial endothelium with support cells in the yolk sac of ephrin-B2 knockout embryos suggest a role for ephrin-B2 in the interaction of endothelial cells with surrounding cells.[64] A role for endothelial ephrin-B2 in mediating interactions with EphB2-positive surrounding cells in the neural tube has been suggested to promote the angiogenic sprouting of the blood vessels that penetrate into the neural tube, based on the finding that the neural tube remains avascular in ephrin-B2 knockout mice.[64] However, in EphB2/EphB3 double knockout mice the neural tube is vascularized, suggesting either that another receptor for endothelial ephrin-B2, such as EphB1 or EphA4, is the counterpart for ephrin-B2 or that a different mechanism requiring endothelial ephrin-B2 mediates sprouting angiogenesis in the neural tube.[25]

EphB2/EphB3 double knockout mice, and ephrin-B2 knockout mice in at least some genetic backgrounds, also exhibit aberrant sprouting of intersomitic vessels into the adjacent somitic tissue.[25] This evidence suggests that forward signaling by EphB3 and EphB4, which are expressed in the mouse intersomitic veins, mediates repulsive signals in response to ephrin-B2 expressed in the posterior portion of the somites and the intersomitic arteries, two structures that flank an ephrin-B2-free path where the intersomitic veins extend.[25,71] Consistent with this, widespread expression of ephrin-B2 in the mouse embryo under the control of a ubiquitous promoter, which abolishes discontinuous presentation of endogenous ephrin-B2, causes abnormalities in the projection of intersomitic veins.[71] A similar phenotype is caused in Xenopus embryos by ectopic expression of ephrin-B ligands, or overexpression of a dominant negative form of EphB4 that impairs the ability of endogenous endothelial EphB4 to signal.[76] Taken together, these data support a model where EphB receptor forward signaling in intersomitic venous endothelial cells inhibits the formation of vascular sprouts extending into ephrin-B territories.

In summary, it appears that a balance of several EphB receptors and ephrin-B ligands expressed in endothelial and vascular support cells is required to achieve proper blood vessel sprouting and remodeling during embryonic development.

5. Lymphatic Vessels

Recent findings have shown that ephrin-B2 and EphB4 are also expressed in lymphatic blood vessels, where they play a critical role in the formation of a functional vascular tree.[73] Analysis of LacZ reporter mice revealed that ephrin-B2 is expressed in the endothelial cells of collecting lymphatic vessels, which have smooth muscle cell coverage and contain valves, and EphB4 is widely expressed throughout the lymphatic networks, including capillaries. Interestingly, ephrin-B2 reverse signaling is important in many of the lymphatic vessels. Knock-in mice engineered to express a mutated ephrin-B2 lacking the PDZ domain-binding site have accumulation of chylous lymphatic fluid in body cavities and exhibit major lymphatic defects. For example, the primary lymphatic plexus in the skin (which expresses both ephrin-B2 and EphB4) forms normally but there are defects in subsequent sprouting of new capillaries (expressing only EphB4) and in vascular remodeling. In addition, collecting lymphatic vessels are hyperplastic and lack the luminal valves that allow proper lymphatic drainage. In contrast, knock-in of a mutated form of ephrin-B2 lacking all the cytoplasmic tyrosine phosphorylation sites almost fully compensates for the lack of wild type ephrin-B2. These data indicate that ephrin-B2 interaction with PDZ domain-containing proteins is required for normal development of the lymphatic vasculature, whereas ephrin-B2 tyrosine phosphorylation and interaction with SH2 domain-containing proteins are dispensable. Consistent with this idea, several known ephrin-B2 binding proteins that contain PDZ domains were found to have altered subcellular distribution in lymphatic vessels expressing the mutant ephrin-B2 without the PDZ domain-binding site. Ephrin-B2 reverse signaling therefore seems to play a more important role in lymphatic vascular morphogenesis than in blood vessel morphogenesis.

6. Adult Vasculature

Given the importance of Eph receptors and ephrins in the formation of the embryonic vasculature, it is not surprising that these molecules have also been implicated in physiological and pathological forms of postnatal angiogenesis.

6.1. *Quiescent vasculature*

Ephrin-A1 is downregulated during embryonic development and is not detectable in adult quiescent vasculature,[60] but some EphA receptors are expressed in normal adult blood vessels. For example, EphA7 has been detected in the blood vessels of the liver septa and in blood vessels of the renal parenchyma.[77] However, the function of these receptors in the adult vasculature and the identity of their ephrin-A ligand counterpart are not known.

A stable molecular difference between arteries and veins persists in the adult quiescent vasculature, where ephrin-B2 remains expressed in arterial endothelial cells.[51,78] EphB4 remains expressed in small diameter venous microvessels and capillaries as well as in large veins such as the vena cava, where this receptor exhibits a patchy heterogeneous expression. In addition, however, low levels of ephrin-B2 have been detected in some cells of adult veins, such as the vena cava, and EphB4 has been detected in some arteries. Another receptor for ephrin-B2 in the adult vasculature is EphB1, which in adult kidney glomeruli is expressed at higher levels than during development.[17] Thus, interactions between ephrin-B2 and EphB receptors likely continue to play a role in the maintenance of a mature vessel configuration.

An increasingly important role of ephrin-B2 at later stages of vascular development may be to regulate vascular smooth muscle cells. The endothelial expression domain of ephrin-B2 expands as development progresses to include the vascular smooth muscle cells of many arteries where this ligand continues to be expressed in the adult.[51,78] Interestingly, ephrin-B2 is upregulated in vascular smooth muscle cells only after they have already lined blood vessels and expression starts from the smooth muscle cells directly in contact with endothelial cells and gradually progresses towards more external regions.[51]

6.2. *Physiological angiogenesis*

EphA2 is one of the key Eph receptors that play a role in postnatal angiogenesis, even though it does not seem to be involved in angiogenesis during embryonic development (Sec. 4.1). With regard to the B class, ephrin-B2 continues to be expressed in the arterial vasculature at sites of secondary angiogenesis in the embryo, such as the heart, neural tube, kidney and lung. Ephrin-B2 expression also persists at sites of adult angiogenesis, such as the ovarian follicles and the corpus luteum in the female reproductive system.[78] Here, ephrin-B2 presumably plays an important role in the continuous vascular remodeling that occurs during the estrous cycle. It has also been recently proposed that the ephrin-B/EphB system plays a role in connecting the blood vessels of the human placenta to the maternal circulation, a process mediated by fetal cytotrophoblast cells that invade the uterine wall to reach arterioles and remodel them.[79] During their differentiation, the cytotrophoblast cells lose EphB4 expression and acquire ephrin-B1 and ephrin-B2 expression, which results in repulsive signals and reduced responsiveness to attractive cytokines in response to EphB4. These repulsive signals likely limit interaction of the cytotrophoblast cells with uterine veins, which express EphB4, and promote selective invasion and remodeling of the uterine arterioles, which express ephrin-B2.

6.3. *Inflammation and wound healing*

Ephrin-A1 and EphA2 are involved in angiogenesis in response to inflammatory cytokines such as tumor necrosis factor α[5] (Sec. 2.4). Ephrin-B2 and EphB4 may also play a role in adult inflammatory neovascularization because according to recent data they are upregulated in pyogenic granuloma of human gingiva, which is a benign inflammatory lesion.[80] Furthermore, ephrin-B2 expression is upregulated in HUVE cells not only by VEGF but also by the inflammatory cytokines interleukin-6 and interleukin-8.[50] Ephrin-B2 also becomes expressed in a subset of blood vessels in the skin during wound healing[51] and, presumably, plays a general role in restoring blood vessels at sites of tissue injury. Interestingly, one of the effects of EphB4 activation by ephrin-B2 in a murine would healing model is upregulation of syndecan-1. This likely promotes angiogenesis *in vivo* if it is accompanied by increased secretion of pro-inflammatory enzymes such heparanases, which could switch syndecan 1 from an inhibitor to a stimulator of FGF2 binding to its receptor[55] (Sec. 2.4).

Hypoxia has recently been shown to upregulate expression of ephrin-B2 and its receptor EphB4, as well as ephrin-A1 and EphA2, in the blood vessels of the mouse skin[81] (Fig. 3). Ephrin-B2 is also upregulated in the new arterial vessels that grow to restore circulation after tissue ischemia in the limb.[30] Furthermore, hypoxia upregulates ephrin-B2 in human umbilical arterial endothelial cells *in vitro*.[82] Therefore, hypoxia may link Eph receptor and ephrin expression to adult neovascularization under both physiological and pathological conditions.

6.4. *Tumor angiogenesis*

The ephrins and Eph receptors that have been most prominently detected in tumor blood vessels are ephrin-A1, ephrin-B2 and EphA2 (Fig. 4). An emerging theme is that the interplay between Eph receptors

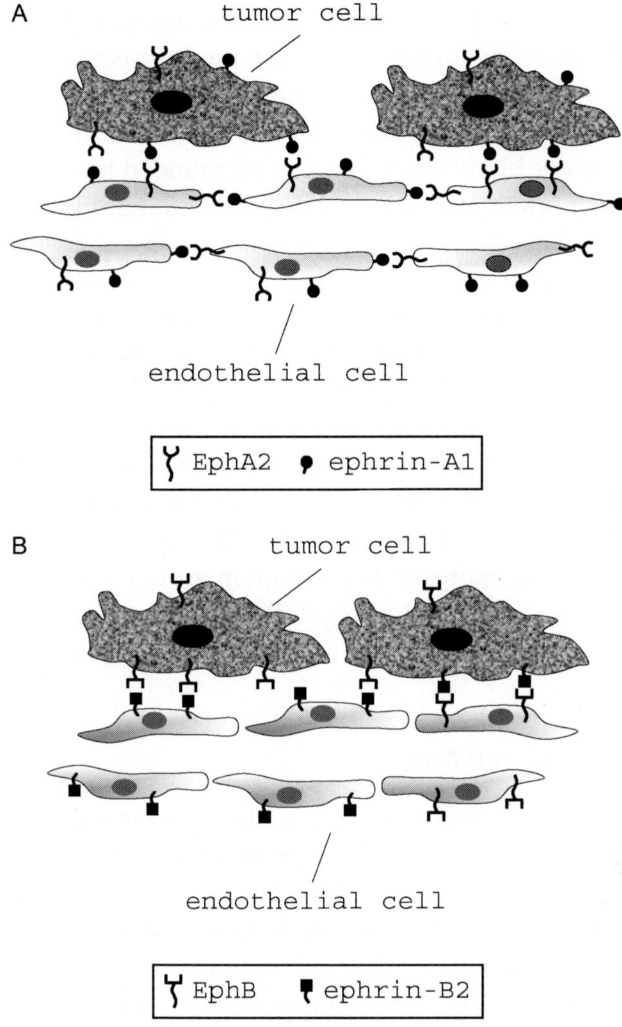

Fig. 4. Eph receptors and ephrins mediate interactions between tumor cells and endothelial cells. **(A)** Tumor endothelial cells as well as many tumor cells express both ephrin-A1 and EphA2. EphA2 signaling in tumor endothelial cells, which could be elicited by ephrin-A1 expressed in either the endothelial cells or the tumor cells, has been shown to promote tumor angiogenesis. **(B)** Ephrin-B2 is expressed in tumor endothelial cells and mediates pro-angiogenic reverse signals when interacting with EphB receptors expressed in tumor cells. Ephrin-B1 on tumor cells has also been shown to promote tumor angiogenesis, presumably by interacting with an EphB receptor expressed in tumor vasculature, which remains to be identified.

and ephrins expressed in endothelial and tumor cells plays an important role in tumor angiogenesis.[38,48] Interactions between tumor cells and endothelial cells may occur particularly during the sprouting and remodeling of new blood vessels. In addition, however, the endothelial cells in tumor blood vessels can be surrounded by a discontinuous basement membrane or lack supporting smooth muscle cells.[83] Hence, tumor endothelial cells have the opportunity to interact with the tumor cells (Fig. 4). In addition, tumor cells are sometimes interspersed among endothelial cells and participate in lining the blood vessel wall and can even form blood vessel-like channels.[83,84] Interestingly, hypoxia has been shown to upregulate ephrin-A1, EphA2, ephrin-B2 and EphB4 in both endothelial and tumor cells.[81] In at least some tumor cells, this effect is mediated by the transcription factor Hypoxia-inducible factor-1α (HIF-1α), which is also known to upregulate VEGF expression. It is therefore possible that VEGF secreted by the tumor cells in turn stimulates the endothelial expression of ephrins in the tumor vasculature, thus resulting in the coordinated upregulation of Eph receptor and ephrin expression in both tumor cells and the tumor vasculature in hypoxic regions.

6.4.1. *Ephrin-A1 and EphA2*

Both ephrin-A1 and EphA2 are upregulated in the vasculature of different types of tumors, while neither protein has been detected in quiescent adult vasculature[14,21] (Fig. 4A). The factors upregulating EphA2 expression in the tumor vasculature have not been identified. On the other hand, there is evidence that inflammatory cytokines and hypoxia contribute to upregulate ephrin-A1 expression in the blood vessels of a tumor and the surrounding tissue leading to EphA2 activation[21,48,81] (Sec. 2.1). Furthermore, VEGF produced by tumor cells likely plays an important role in the upregulation of ephrin-A1 in tumor endothelial cells.[19] Indeed, conditioned medium from islet carcinoma cells, which are known to produce VEGF, promotes EphA2 tyrosine phosphorylation and the migration of HUVE endothelial cells in transwell migration assays.[48] This pro-migratory effect is blocked by VEGF-neutralizing antibodies and also by EphA2 Fc, a soluble EphA2 antagonist. These data implicate ephrin-A1 upregulation and the consequent EphA2 activation in mediating some of the effects of VEGF on tumor angiogenesis (Fig. 3A).

In many tumors, ephrin-A1 and EphA2 are also expressed in the tumor cells (Fig. 4A), and therefore a complex interplay between EphA2 and ephrin-A1 expressed in tumor cells and tumor blood vessels likely takes place, with positive effects on angiogenesis. Supporting this idea, EphA2 overexpression in cancer cells has been recently reported to correlate with high microvessel counts in human colorectal cancer specimens.[85] In addition, mammary and pancreatic tumor cell lines that express ephrin-A1 attract endothelial cells in co-culture transwell migration assays.[14,48] A gain-of-function EphA2 mutant enhances the tumor cell-induced endothelial cell migration, while a dominant negative mutant and soluble EphA2 Fc inhibit migration.[14,48] Furthermore, microvascular pulmonary endothelial cells from EphA2 knockout mice migrate less efficiently in response to 4T1 mammary tumor cells and endothelial cells from wild type mice migrate less efficiently in response to tumor cells with decreased ephrin-A1 expression.[56] These data suggest that VEGF and ephrin-A1/EphA2 cooperate in promoting blood vessel recruitment by the tumor. A possible scenario is that VEGF provides a long-range signal that upregulates ephrin-A1 in endothelial cells leading to activation of endothelial EphA2, whereas ephrin-A1 on tumor cells provides a contact-dependent signal by interacting with endothelial EphA2.[14]

Soluble EphA2 antagonists — such as EphA2 Fc and EphA3 Fc — have been shown to inhibit tumor angiogenesis and progression in mouse xenograft models of breast and pancreatic cancer when administered either systemically or subcutaneously in the vicinity of the tumor.[14,15,19,48] Among the effects documented are a decrease in cell proliferation and an increase in apoptosis of both the tumor cells and the endothelial cells.[14,15] The effects on the tumor cells, however, may be secondary to decreased vascularization because EphA2 Fc has no direct effect on the survival of mammary and pancreatic tumor cells in culture.[14,15] Consistent with effects on blood vessels, EphA2 Fc inhibits mammary and pancreatic tumor-induced angiogenesis in an *in vivo* cutaneous window angiogenesis assay, in which a small tumor placed in a subcutaneous chamber becomes vascularized by host blood vessels.[14] EphA Fc fusion proteins also inhibit the formation of pre-malignant lesions and reduce tumor volume in the RIP-Tag transgenic mouse pancreatic cancer model.[48] Furthermore, tumor growth is impaired and vascular density is decreased when mouse 4T1 mammary adenocarcinoma cells are implanted in EphA2 knockout mice, demonstrating the importance of EphA2 in tumor blood vessels and the tumor microenvironment.[56] Taken together, the available evidence suggests that EphA2 forward signaling in tumor blood vessels is critical for tumor growth and that inhibition of endothelial EphA2 forward signaling is the main mechanism underlying the anticancer effects of EphA receptor Fc fusion proteins.

6.4.2. *Ephrin-B2 and EphB4*

Ephrin-B2 has been detected in the vasculature of a variety of tumor types, where it is expressed in a proportion of the endothelial cells[38,51,78,86] (Fig. 4B). This suggests that tumor capillaries can have arterial or venous identity, contrary to previous beliefs based on morphological criteria. EphB4 expression in tumor endothelial cells remains to be characterized, however, and it is not known whether the ephrin-B2-negative tumor blood vessels express EphB4 and whether a complementary expression of endothelial ephrin-B2 and EphB4 is important for the formation of new vascular networks in tumors. Ephrin-B2 expression in tumor blood vessels is likely upregulated by VEGF (Sec. 2.4) and hypoxia.[81,82] However, expression of ephrin-B2 was not detected in the smooth muscle cells associated with the vasculature of Lewis lung carcinomas grown subcutaneously in mice.[78]

Enhanced expression of the EphB4 receptor ectodomain on the tumor cell surface has been shown to promote tumor growth by promoting ephrin-B2 reverse signaling in the tumor blood vessels.[38] Interestingly, increased EphB4 ectodomain expression in the tumor cells increases the size of the blood vessels and the blood content of the tumors, consistent with the enlarging effects of EphB4 Fc on the chicken allantoic arteries and the enlarged ear skin blood vessels in transgenic mice overexpressing ephrin-B2 in endothelial cells.[43,71] Conversely, A375 melanoma cells secreting soluble monomeric EphB4 ectodomain form much smaller tumors when injected subcutaneously in nude mice, presumably because the EphB4 ectodomain blocks the interaction between EphB2 on the surface of the tumor cells and endothelial ephrin-B2.[87] Indeed, tumor regions where the EphB4 ectodomain was present at highest levels had low microvessel densities and more collapsed vessels lacking a lumen.

Although ephrin-B2 reverse signaling does not appear to be required during early embryonic angiogenesis, angiogenic effects of ephrin-B2 in tumors are consistent with the pro-angiogenic effects of ephrin-B2 reverse signaling in cultured endothelial cells.[36–38] Ephrin-B2 reverse signaling can also regulate the vasculature *in vivo*. For example, defects in the postnatal lung vasculature have been reported in mice expressing mutated ephrin-B2 lacking the PDZ domain-binding site[73] and EphB4 Fc treatment of the allantoic membrane blood vessels of the E4 chicken embryo, which have already undergone remodeling into a network of large and small vessels, causes retraction or regression of venules, enlargement of arteries and formation of arterial-venous shunts.[43] Furthermore, defects in capillary architecture in the kidney glomeruli and the mammary gland of mice overexpressing EphB4 in epithelial cells support a role for ephrin-B2 reverse signaling in endothelial cells stimulated by EphB4 expression in surrounding cells.[88] Interestingly, ephrin-B2 is upregulated in Kaposi's sarcoma, an angioproliferative tumor derived from endothelial cells, and is required for the viability of the tumor cells.[50]

Additional evidence suggests that endothelial EphB receptors also play an important role in tumor angiogenesis. For example, ephrin-B1 is upregulated in hepatocellular carcinomas and promotes hepatocellular tumor growth in a mouse xenograft model without affecting the growth of the tumor cells in culture.[27] Thus, ephrin-B1 on the surface of the tumor cells may be involved in tumor progression *in vivo* by promoting tumor angiogenesis through endothelial EphB receptors (Fig. 4B). Consistent with this hypothesis, the number of blood vessels in tumor xenografts is increased by ephrin-B1 expression in the tumor cells and ephrin-B1 Fc enhances HUVE cell migration and proliferation *in vitro*.[27] Interestingly, treatment of the E4 chicken embryo allantois with ephrin-B2 Fc causes morphological effects in the vasculature.[43]

These morphological changes, presumably mediated by EphB4, include increased branching of veins, dramatic enlargement of both veins and arteries and formation of arterious-venous shunts. However, despite increasing blood vessel density, ephrin-B2 overexpression in human colorectal cancer cells decreases tumor growth and blood perfusion in a mouse xenograft model.[89] Thus, additional studies are required in order to fully elucidate the molecular mechanisms by which Eph receptors and ephrins affect tumor progression by contributing to tumor vascularization.

7. Targeting Eph Receptor-Ephrin Interactions to Modulate Angiogenesis

In conditions where angiogenesis is part of the disease pathology, it is therapeutically desirable to inhibit it. A number of receptor tyrosine kinases expressed in endothelial cells are validated targets for anti-angiogenic therapies. Eph receptors and ephrins are also emerging as new attractive targets. EphA receptor Fc fusion protein and soluble monomeric forms of EphB4 have been successfully used as antagonists in animal tumor models.[14,19,48,87] Furthermore, intravitreal injection of ephrin-B2 Fc or EphB4 Fc, which presumably perturb the function of the corresponding endogenously expressed proteins, reduce the pathological neovascularization occurring in a mouse model of oxygen-induced retinopathy.[90] Thus, ephrin-B2 and EphB4 might be useful targets for therapies to treat retinopathy of prematurity and the abnormal retinal vascularization characterizing macular degeneration and diabetic retinopathy. Although Fc fusion proteins are quite stable when administered systemically *in vivo*, Eph receptor- and ephrin-based agents lack selectivity because of the promiscuity of Eph receptor-ephrin binding (Sec. 1.3). More selective Eph receptor-targeting reagents that have been developed include antagonistic peptides that target the ephrin-binding site of individual Eph receptors,[91,92] siRNAs and antisense oligodeoxynucleotides.[93-95] Peptides as well as antibodies that target extracellular regions of Eph receptors and ephrins could also be used to deliver anti-angiogenic drugs.

In other conditions that are characterized by ischemia, such as heart disease, stroke and wound healing, it may be desirable to use reagents that enhance the pro-angiogenic activities of Eph receptors and ephrins, such as activating peptides or antibodies.[96,97] Ephrins or Eph receptor ectodomains incorporated into fibrin matrices could also be useful to promote angiogenesis, as shown for ephrin-B2.[98] Targeting Eph receptors and ephrins to stimulate or inhibit angiogenesis is an area just beginning to be explored and where rapid developments are expected.

8. Perspectives

The last few years have seen the discovery of Eph receptors and ephrins as new families of angiogenic factors that can discriminate between arteries and veins and that play a role not only in endothelial cells but also in the surrounding support cells. EphB4 and ephrin-B2 are expressed in angiogenic vasculature during embryonic development, at sites of adult neovascularization, and also in the mature quiescent vasculature. These expression patterns suggest diverse roles in neovascularization and the maintenance of mature blood vessels. The EphA2 receptor is also expressed at sites of adult neovascularization but, surprisingly, it has not been detected in normal developing vasculature or quiescent vasculature. This expression pattern makes EphA2 a particularly attractive target for selective anti-angiogenic therapies. On the other hand, ephrin-A1 is in developing but not adult vasculature. Therefore, additional A class Eph receptors and ephrins with angiogenic activities likely remain to be discovered. Antibodies with well-defined specificities will be critical in order to accurately map the expression patterns of different Eph receptors and ephrins in developing and postnatal vasculature. Despite the remarkable progress made so far, much remains to be learned about the mechanisms and signaling pathways used by Eph receptors and ephrins to regulate blood and lymphatic vessels. Areas of particular interest for the future will be to better characterize the roles of Eph receptors and ephrins in the expansion versus the differentiation of vascular stem cells and the role of different levels of hypoxia in regulating the expression of Eph and ephrin genes in endothelial cells and the surrounding tissue. It will also be important to better understand the complex interplay between Eph receptors and other families of angiogenic factors. New therapeutic approaches for targeting Eph receptors and ephrins to promote or inhibit angiogenesis are undoubtedly also forthcoming.

Acknowledgments

The author thanks R. Bayer and N. Noren for helpful comments on the manuscript. The work in the author's laboratory is supported by grants from the NIH and the Department of Defense.

References

1. Pasquale EB (2005). Eph receptor signalling casts a wide net on cell behaviour. *Nat Rev Mol Cell Biol* **6**: 462–475.
2. Hirai H, Maru Y, Hagiwara K, Nishida J, Takaku F (1987). A novel putative tyrosine kinase receptor encoded by the eph gene. *Science* **238**: 1717–1720.
3. Eph-Nomenclature-Committee (1997). Unified nomenclature for Eph family receptors and their ligands, the ephrins. *Cell* **90**: 403–404.
4. Bartley TD, Hunt RW, Welcher AA, *et al.* (1994). B61 is a ligand for the ECK receptor protein-tyrosine kinase. *Nature* **368**: 558–560.
5. Pandey A, Shao H, Marks RM, Polverini PJ, Dixit VM (1995). Role of B61, the ligand for the Eck receptor tyrosine kinase, in TNF-alpha-induced angiogenesis. *Science* **268**: 567–569.
6. Holmberg J, Clarke DL, Frisen J (2000). Regulation of repulsion versus adhesion by different splice forms of an Eph receptor. *Nature* **408**: 203–206.
7. Palmer A, Zimmer M, Erdmann KS, *et al.* (2002). EphrinB phosphorylation and reverse signaling: regulation by Src kinases and PTP-BL phosphatase. *Mol Cell* **9**: 725–737.
8. Song J (2003). Tyrosine phosphorylation of the well packed ephrinB cytoplasmic beta-hairpin for reverse signaling. Structural consequences and binding properties. *J Biol Chem* **278**: 24714–24720.
9. Cowan CA, Henkemeyer M (2001). The SH2/SH3 adaptor Grb4 transduces B-ephrin reverse signals. *Nature* **413**: 174–179.
10. Pasquale EB (2004). Eph-ephrin promiscuity is now crystal clear. *Nat Neurosci* **7**: 417–418.
11. Miao H, Burnett E, Kinch M, Simon E, Wang B (2000). Activation of EphA2 kinase suppresses integrin function and causes focal-adhesion-kinase dephosphorylation. *Nat Cell Biol* **2**: 62–69.
12. Dohn M, Jiang JY, Chen XB (2001). Receptor tyrosine kinase EphA2 is regulated by p53-family proteins and induces apoptosis. *Oncogene* **20**: 6503–6515.
13. Stein E, Lane AA, Cerretti DP, *et al.* (1998). Eph receptors discriminate specific ligand oligomers to determine alternative signaling complexes, attachment, assembly responses. *Genes Dev* **12**: 667–678.

14. Brantley DM, Cheng N, Thompson EJ, *et al.* (2002). Soluble Eph A, receptors inhibit tumor angiogenesis and progression *in vivo*. *Oncogene* **21**: 7011–7026.

15. Dobrzanski P, Hunter K, Jones-Bolin S, *et al.* (2004). Antiangiogenic and antitumor efficacy of EphA2 receptor antagonist. *Cancer Res* **64**: 910–919.

16. Sarma V, Wolf FW, Marks RM, Shows TB, Dixit VM (1992). Cloning of a novel tumor necrosis factor-alpha-inducible primary response gene that is differentially expressed in development and capillary tube-like formation *in vitro*. *J Immunol* **148**: 3302–3312.

17. Daniel TO, Stein E, Cerretti DP, *et al.* (1996). ELK and LERK-2 in developing kidney and microvascular endothelial assembly. *Kidney Int — Suppl* **57**: S73–81.

18. Myers C, Charboneau A, Boudreau N (2000). Homeobox B3 promotes capillary morphogenesis and angiogenesis. *J Cell Biol* **148**: 343–351.

19. Cheng N, Brantley DM, Liu H, *et al.* (2002). Blockade of EphA receptor tyrosine kinase activation inhibits vascular endothelial cell growth factor-induced angiogenesis. *Mol Cancer Res* **1**: 2–11.

20. Brantley-Sieders DM, Caughron J, Hicks D, *et al.* (2004). EphA2 receptor tyrosine kinase regulates endothelial cell migration and vascular assembly through phosphoinositide 3-kinase-mediated Rac1 GTPase activation. *J Cell Sci* **117**: 2037–2049.

21. Ogawa K, Pasqualini R, Lindberg RA, *et al.* (2000). The ephrin-A1 ligand and its receptor, EphA2, are expressed during tumor neovascularization. *Oncogene* **19**: 6043–6052.

22. Pandey A, Lazar DF, Saltiel AR, Dixit VM (1994). Activation of the Eck receptor protein tyrosine kinase stimulates phosphatidylinositol 3-kinase activity. *J Biol Chem* **269**: 30154–30157.

23. Ogita H, Kunimoto S, Kamioka Y, *et al.* (2003). EphA4-mediated Rho activation via Vsm-RhoGEF expressed specifically in vascular smooth muscle cells. *Circ Res* **93**: 23–31.

24. Deroanne C, Vouret-Craviari V, Wang B, Pouyssegur J (2003). EphrinA1 inactivates integrin-mediated vascular smooth muscle cell spreading via the Rac/PAK pathway. *J Cell Science* **116**: 1367–1376.

25. Adams RH, Wilkinson GA, Weiss C, *et al.* (1999). Roles of ephrinB ligands and EphB receptors in cardiovascular development: demarcation of arterial/venous domains, vascular morphogenesis, sprouting angiogenesis. *Genes Dev* **13**: 295–306.

26. Kim I, Ryu YS, Kwak HJ, *et al.* (2002). EphB ligand, ephrinB2, suppresses the VEGF- and angiopoietin 1-induced Ras/mitogen-activated protein kinase pathway in venous endothelial cells. *FASEB J* **16**: 1126–1128.

27. Sawai Y, Tamura S, Fukui K, *et al.* (2003). Expression of ephrin-B1 in hepatocellular carcinoma: possible involvement in neovascularization. *J Hepatol* **39**: 991–996.

28. Steinle JJ, Meininger CJ, Forough R, *et al.* (2002). Eph B4 receptor signaling mediates endothelial cell migration and proliferation via the phosphatidylinositol 3-kinase pathway. *J Biol Chem* **277**: 43830–43835.

29. Huynh-Do U, Stein E, Lane AA, *et al.* (1999). Surface densities of ephrin-B1 determine EphB1-coupled activation of cell attachment through alphavbeta3 and alpha5beta1 integrins. *EMBO J* **18**: 2165–2173.

30. Hayashi S, Asahara T, Masuda H, Isner JM, Losordo DW (2005). Functional ephrin-B2 expression for promotive interaction between arterial and venous vessels in postnatal neovascularization. *Circulation* **111**: 2210–2218.

31. Becker E, Huynh-Do U, Holland S, *et al.* (2000). Nck-interacting Ste20 kinase couples Eph receptors to c-Jun N-terminal kinase and integrin activation. *Mol Cell Biol* **20**: 1537–1545.

32. Vindis C, Teli T, Cerretti DP, Turner CE, Huynh-Do U (2004). EphB1-mediated cell migration requires the phosphorylation of paxillin at Tyr-31/Tyr-118. *J Biol Chem* **279**: 27965–27970.

33. Vindis C, Cerretti DP, Daniel TO, Huynh-Do U (2003). EphB1 recruits c-Src and p52Shc to activate MAPK/ERK and promote chemotaxis. *J Cell Biol* **162**: 661–671.

34. Nagashima K, Endo A, Ogita H, *et al.* (2002). Adaptor protein crk is required for Ephrin-B1-induced membrane ruffling and focal complex assembly of human aortic endothelial cells. *Mol Biol Cell* **13**: 4231–4242.

35. Maekawa H, Oike Y, Kanda S, *et al.* (2003). Ephrin-B2 induces migration of endothelial cells through the phosphatidylinositol-3 kinase pathway and promotes angiogenesis in adult vasculature. *Arterioscler Thromb Vasc Biol* **23**: 2008–2014.

36. Hamada K, Oike Y, Ito Y, *et al.* (2003). Distinct roles of ephrin-B2 forward and EphB4 reverse signaling in endothelial cells. *Arterioscler Thromb Vasc Biol* **23**: 190–197.

37. Fuller T, Korff T, Kilian A, Dandekar G, Augustin HG (2003). Forward EphB4 signaling in endothelial cells controls cellular repulsion and segregation from ephrinB2 positive cells. *J Cell Sci* **116**: 2461–2470.

38. Noren NK, Lu M, Freeman AL, Koolpe M, Pasquale EB (2004). Interplay between EphB4 on tumor cells and vascular ephrin-B2 regulates tumor growth. *Proc Natl Acad Sci USA* **101**: 5583–5588.

39. Adams RH, Diella F, Hennig S, *et al.* (2001). The cytoplasmic domain of the ligand ephrinB2 is required for vascular morphogenesis but not cranial neural crest migration. *Cell* **104**: 57–69.

40. Huynh-Do U, Vindis C, Liu H, *et al.* (2002). Ephrin-B1 transduces signals to activate integrin-mediated migration, attachment and angiogenesis. *J Cell Sci* **115**: 3073–3081.

41. Bruhl T, Urbich C, Aicher D, *et al.* (2004). Homeobox A9 transcriptionally regulates the EphB4 receptor to modulate endothelial cell migration and tube formation. *Circ Res* **94**: 743–751.

42. Steinle JJ, Meininger CJ, Chowdhury U, Wu G, Granger HJ (2003). Role of ephrin B2 in human retinal endothelial cell proliferation and migration. *Cell Signal* **15**: 1011–1017.

43. le Noble F, Moyon D, Pardanaud L, *et al.* (2004). Flow regulates arterial-venous differentiation in the chick embryo yolk sac. *Development* **131**: 361–375.

44. Dixit VM, Green S, Sarma V, *et al.* (1990). Tumor necrosis factor-alpha induction of novel gene products in human endothelial cells including a macrophage-specific chemotaxin. *J Biol Chem* **265**: 2973–2978.

45. Holzman LB, Marks RM, Dixit VM (1990). A novel immediate-early response gene of endothelium is induced by cytokines and encodes a secreted protein. *Mol Cell Biol* **10**: 5830–5838.

46. Cheng N, Chen J (2001). Tumor necrosis factor-alpha induction of endothelial ephrin A1 expression is mediated by a p38 MAPK– and SAPK/JNK-dependent but nuclear factor-kappa B-independent mechanism. *J Biol Chem* **276**: 13771–13777.

47. Abdollahi A, Hahnfeldt P, Maercker C, *et al.* (2004). Endostatin's antiangiogenic signaling network. *Mol Cell* **13**: 649–663.

48. Cheng N, Brantley D, Fang WB, *et al.* (2003). Inhibition of VEGF-dependent multistage carcinogenesis by soluble EphA receptors. *Neoplasia* **5**: 445–456.

49. Mukouyama YS, Shin D, Britsch S, Taniguchi M, Anderson DJ (2002). Sensory nerves determine the pattern of arterial differentiation and blood vessel branching in the skin. *Cell* **109**: 693–705.

50. Masood R, Xia G, Smith DL, *et al.* (2005). Ephrin B2 expression in Kaposi sarcoma is induced by human herpesvirus type 8: phenotype switch from venous to arterial endothelium. *Blood* **105**: 1310–1318.

51. Shin D, Garcia-Cardena G, Hayashi SI, *et al.* (2001). Expression of ephrinB2 identifies a stable genetic difference between arterial and venous vascular smooth muscle as well as endothelial cells, marks subsets of microvessels at sites of adult neovascularization. *Dev Biol* **230**: 139–150.

52. Visconti RP, Richardson CD, Sato TN (2002). Orchestration of angiogenesis and arteriovenous contribution by angiopoietins and vascular endothelial growth factor (VEGF). *PNAS* **99**: 8219–8224.

53. Rossant J, Howard L (2002). Signaling pathways in vascular development. *Annu Rev Cell Dev Biol* **18**: 541–573.

54. Zhong TP, Childs S, Leu JP, Fishman MC (2001). Gridlock signalling pathway fashions the first embryonic artery. *Nature* **414**: 216–220.

55. Yuan K, Hong TM, Chen JJ, Tsai WH, Lin MT (2004). Syndecan-1 up-regulated by ephrinB2/EphB4 plays dual roles in inflammatory angiogenesis. *Blood* **104**: 1025–1033.

56. Brantley-Sieders DM, Fang WB, Hicks DJ, *et al.* (2005). Impaired tumor microenvironment in EphA2-deficient mice inhibits tumor angiogenesis and metastatic progression. *FASEB J* **19**: 1884–1886.

57. Depaepe V, Suarez-Gonzalez N, Dufour A, *et al.* (2005). Ephrin signalling controls brain size by regulating apoptosis of neural progenitors. *Nature* **435**: 1244–1250.

58. Wang Z, Cohen K, Shao Y, *et al.* (2004). Ephrin receptor, EphB4, regulates ES cell differentiation of primitive mammalian hemangioblasts, blood, cardiomyocytes, blood vessels. *Blood* **103**: 100–109.

59. Wang Z, Miura N, Bonelli A, *et al.* (2002). Receptor tyrosine kinase, EphB4 (HTK), accelerates differentiation of select human hematopoietic cells. *Blood* **99**: 2740–2747.

60. McBride JL, Ruiz JC (1998). Ephrin-A1 is expressed at sites of vascular development in the mouse. *Mech Dev* **77**: 201–204.

61. Flenniken AM, Gale NW, Yancopoulos GD, Wilkinson DG (1996). Distinct and overlapping expression patterns of ligands for Eph-related receptor tyrosine kinases during mouse embryogenesis. *Dev Biol* **179**: 382–401.

62. Takahashi H, Ikeda T (1995). Mol cloning and expression of rat and mouse B61 gene: implications on organogenesis. *Oncogene* **11**: 879–883.

63. Stadler HS, Higgins KM, Capecchi MR (2001). Loss of Eph-receptor expression correlates with loss of cell adhesion and chondrogenic capacity in Hoxa13 mutant limbs. *Development* **128**: 4177–4188.

64. Wang HU, Chen ZF, Anderson DJ (1998). Molecular distinction and angiogenic interaction between embryonic arteries and veins revealed by ephrin-B2 and its receptor Eph-B4. *Cell* **93**: 741–753.

65. Gerety SS, Wang HU, Chen ZF, Anderson DJ (1999). Symmetrical mutant phenotypes of the receptor EphB4 and its specific transmembrane ligand ephrin-B2 in cardiovascular development. *Mol Cell* **4**: 403–414.

66. Othman-Hassan K, Patel K, Papoutsi M, *et al.* (2001). Arterial identity of endothelial cells is controlled by local cues. *Dev Biol* **237**: 398–409.

67. Goettsch W, Augustin HG, Morawietz H (2004). Down-regulation of endothelial ephrinB2 expression by laminar shear stress. *Endothelium* **11**: 259–265.

68. Claxton S, Fruttiger M (2005). Oxygen modifies artery differentiation and network morphogenesis in the retinal vasculature. *Dev Dyn* **233**: 822–828.

69. Hall SM, Hislop AA, Haworth SG (2002). Origin, differentiation, maturation of human pulmonary veins. *Am J Respir Cell Mol Biol* **26**: 333–340.

70. Risau W (1997). Mechanisms of angiogenesis. *Nature* **386**: 671–674.

71. Oike Y, Ito Y, Hamada K, *et al.* (2002). Regulation of vasculogenesis and angiogenesis by EphB/ephrin-B2 signaling between endothelial cells and surrounding mesenchymal cells. *Blood* **100**: 1326–1333.

72. Gerety SS, Anderson DJ (2002). Cardiovascular ephrinB2 function is essential for embryonic angiogenesis. *Development* **129**: 1397–1410.

73. Makinen T, Adams RH, Bailey J, *et al.* (2005). PDZ interaction site in ephrinB2 is required for the remodeling of lymphatic vasculature. *Genes Dev* **19**: 397–410.

74. Cowan CA, Yokoyama N, Saxena A, *et al.* (2004). Ephrin-B2 reverse signaling is required for axon pathfinding and cardiac valve formation but not early vascular development. *Dev Biol* **271**: 263–271.

75. Zhang XQ, Takakura N, Oike Y, *et al.* (2001). Stromal cells expressing ephrin-B2 promote the growth and sprouting of ephrin-B2(+) endothelial cells. *Blood* **98**: 1028–1037.

76. Helbling PM, Saulnier DM, Brandli AW (2000). The receptor tyrosine kinase EphB4 and ephrin-B ligands restrict angiogenic growth of embryonic veins in Xenopus laevis. *Development* **127**: 269–278.

77. Hafner C, Schmitz G, Meyer S, *et al.* (2004). Differential gene expression of Eph receptors and Ephrins in benign human tissues and cancers. *Clin Chem* **50**: 490–499.

78. Gale NW, Baluk P, Pan L, *et al.* (2001). Ephrin-B2 selectively marks arterial vessels and neovascularization sites in the adult, with expression in both endothelial and smooth-muscle cells. *Dev Biol* **230**: 151–160.

79. Red-Horse K, Kapidzic M, Zhou Y, *et al.* (2005). EPHB4 regulates chemokine-evoked trophoblast responses: a mechanism for incorporating the human placenta into the maternal circulation. *Development* **132**: 4097–4106.

80. Yuan K, Jin YT, Lin MT (2000). Expression of Tie-2, angiopoietin-1, angiopoietin-2, ephrinB2 and EphB4 in pyogenic granuloma of human gingiva implicates their roles in inflammatory angiogenesis. *J Periodontal Res* **35**: 165–171.

81. Vihanto MM, Plock J, Erni D, *et al.* (2005). Hypoxia up-regulates expression of Eph receptors and ephrins in mouse skin. *FASEB J* **19**: 1689–1691.

82. Suenobu S, Takakura N, Inada T, *et al.* (2002). A role of EphB4 receptor and its ligand, ephrin-B2, in erythropoiesis. *Biochem Biophys Res Commun* **293**: 1124–1131.

83. Carmeliet P, Jain RK (2000). Angiogenesis in cancer and other diseases. *Nature* **407**: 249–257.

84. Hess AR, Seftor EA, Gardner LM, *et al.* (2001). Molecular regulation of tumor cell vasculogenic mimicry by tyrosine phosphorylation: role of epithelial cell kinase (Eck/EphA2). *Cancer Res* **61**: 3250–3255.

85. Kataoka H, Igarashi H, Kanamori M, *et al.* (2004). Correlation of EPHA2 overexpression with high microvessel count in human primary colorectal cancer. *Cancer Sci* **95**: 136–141.

86. Papoutsi M, Othman-Hassan K, Christ B, Patel K, Wilting J (2002). Development of an arterial tree in C6 gliomas but not in A375 melanomas. *Histochem Cell Biol* **118**: 241–249.

87. Martiny-Baron G, Korff T, Schaffner F, *et al.* (2004). Inhibition of tumor growth and angiogenesis by soluble EphB4. *Neoplasia* **6**: 248–257.

88. Andres AC, Munarini N, Djonov V, *et al.* (2003). EphB4 receptor tyrosine kinase transgenic mice develop glomerulopathies reminiscent of aglomerular vascular shunts. *Mech Dev* **120**: 511–516.

89. Liu W, Jung YD, Ahmad SA, *et al.* (2004). Effects of overexpression of ephrin-B2 on tumour growth in human colorectal cancer. *British J Cancer* **90**: 1620–1626.

90. Zamora DO, Davies MH, Planck SR, Rosenbaum JT, Powers MR (2005). Soluble forms of EphrinB2 and EphB4 reduce retinal neovascularization in a model of proliferative retinopathy. *Invest Ophthalmol Vis Sci* **46**: 2175–2182.

91. Murai KK, Nguyen LN, Koolpe M, *et al.* (2003). Targeting the EphA4 receptor in the nervous system with biologically active peptides. *Mol Cell Neurosci* **24**: 1000–1011.

92. Koolpe M, Burgess R, Dail M, Pasquale EB (2005). EphB receptor-binding peptides identified by phage display enable design of an antagonist with ephrin-like affinity. *J Biol Chem* **280**: 17301–17311.

93. Xia G, Kumar SR, Masood R, *et al.* (2005). Up-Regulation of EphB4 in Mesothelioma and its biological significance. *Clin Cancer Res* **11**: 4305–4315.

94. Xia G, Kumar SR, Masood R, *et al.* (2005). EphB4 expression and biological significance in prostate cancer. *Cancer Res* **65**: 4623–4632.

95. Duxbury MS, Ito H, Zinner MJ, Ashley SW, Whang EE (2004). EphA2: a determinant of malignant cellular behavior and a potential therapeutic target in pancreatic adenocarcinoma. *Oncogene* **23**: 1448–1456.

96. Koolpe M, Dail M, Pasquale EB (2002). An ephrin mimetic peptide that selectively targets the EphA2 receptor. *J Biol Chem* **277**: 46974–46979.

97. Carles-Kinch K, Kilpatrick KE, Stewart JC, Kinch MS (2002). Antibody targeting of the EphA2 tyrosine kinase inhibits malignant cell behavior. *Cancer Res* **62**: 2840–2847.
98. Zisch AH, Zeisberger SM, Ehrbar M, *et al.* (2004). Engineered fibrin matrices for functional display of cell membrane-bound growth factor–like activities: study of angiogenic signaling by ephrin-B2. *Biomaterials* **25**: 3245–3257.

3

The FGF Family of Angiogenic Growth Factors

by Patrick Auguste and Andreas Bikfalvi

1. Introduction

Fibroblast growth factors (FGFs) are potent multifunctional growth factors that play significant roles in early and late embryonic development such as mesoderm induction or brain or lung development. In the adult, FGFs are thought to be implicated in tissue repair, wound healing, or neuronal stem cell proliferation and neuron migration (for review, see Refs 1 and 2).

FGFs were amongst the earliest angiogenesis molecules identified. FGFs were found to stimulate endothelial cell proliferation, migration and differentiation *in vitro* and *in vivo*.[1] In addition, tumors were found to produce significant amounts of FGFs which suggested a role in tumor angiogenesis. However, the most important FGF prototypes such as FGF-1 and -2 lack a classical signal sequence that allows efficient export from cells. Thus, the role of endogenous FGFs in developmental or pathological angiogenesis processes remained uncertain.

Some of these unresolved issues were recently taken up by several investigators, thus FGFs becoming, once again the focus of angiogenesis research.

2. Molecular Mechanisms

The FGF/FGF receptor system comprises, to date, 23 FGF family members and four tyrosine kinase receptor prototypes (FGFRs). This repertoire is additionally increased by the presence of a number of isoforms and proteolytic processed derivatives within FGF family members and of spliced variants within FGFRs.[3]

FGF receptor activation requires heparan sulfate proteoglycans (HSPGs), such as syndecans or glypicans as co-receptors (Fig. 1). In one model, HSPGs induces ligand dimerization which in turn leads to FGFR dimerization and activation.[4] Furthermore, a small sequence within FGF-2 spanning between amino acid 48–58 has been recently identified to participate in ligand dimerization.[5] This domain seems to be functionally important since a peptide containing this sequence is able to inhibit ligand dimerization and biological activity. However, in another proposed model derived from crystallographic studies at 3A resolution of the FGF/FGFR complex with the heparan sulfate analogue heparin,[6] heparin makes multiple contacts with FGF and FGFR and promotes receptor dimerization but not ligand dimerization. The reason for these differences is at present not known.

HSPGs may regulate FGF/FGFR interactions in a membrane-bound form or in a soluble form after shedding of heparan sulfates or HSPG fragments by heparanase or proteolytic enzymes. These different HSPGs forms may have distinct regulatory functions in FGF receptor activation. Indeed, it has been reported that membrane-bound or heparinolytically-shedded HSPGs enhance receptor activation whereas proteolytically-shedded HSPG are inhibitory.[7] In perlecan heparan sulphate-deficient mice, FGF-2-induced corneal angiogenesis is severely impaired but mice develop normally.[8] Membrane-associated gangliosides (GM1) were also added to the list of molecules able to regulate FGF/FGFR interactions.[9] Treatment of endothelial cells with ganglioside biosynthesis inhibitors impairs FGF-2-induced endothelial cell proliferation. ^{125}I-FGF2 binds to cell membrane GM1 with high

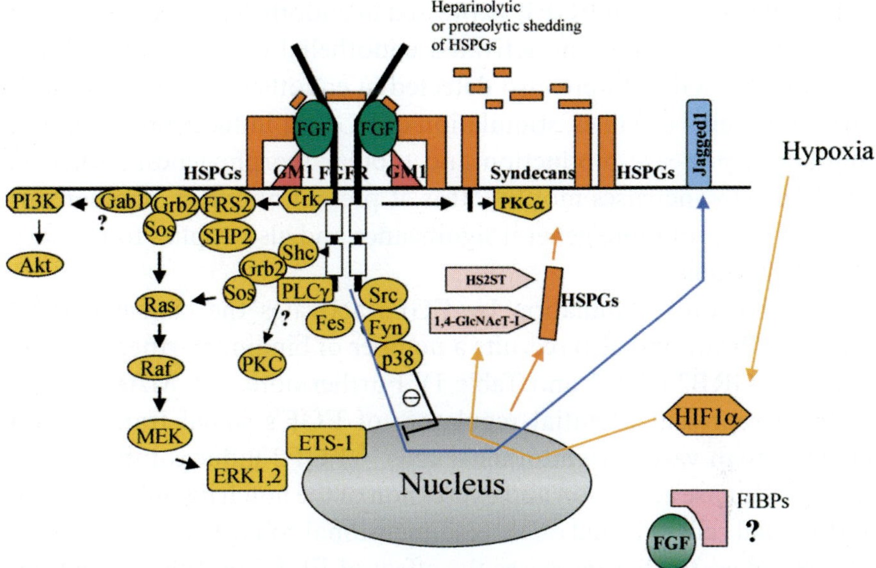

Fig. 1. Signaling mechanisms regulated by FGFs. Fibroblast growth factors (FGFs) bind to FGF receptors (FGFR) and induce receptor dimerization. Heparan sulfate proteoglycans (HSPGs) and gangliosides (GM1) actively participate in the extracellular activation mechanism. Whether, ligand dimerization is involved is not clear. Furthermore, gangliosides (GM1) function also as co-receptors for FGF. FGF induces the mitogen-associated protein (MAP) kinase pathway, src and p38 MAP kinases. P38 kinase negatively regulates Jagged1. FGF activates PI3K/Akt pathway, but in endothelial cells it is not clear if FRS2/Grb2/Grab1 is implicated. Phospholipase C-γ (PLCγp in FGF signaling in vascular cells is not clear. Sef, a transmembrane protein associates with FGFR and inhibits its activation (not represented). The HSPG syndecan-4 also participates in intracellular signaling and induces protein kinase C-α (PKCαp activation. Hypoxia induces hypoxia inducible factor-1α (HIF1α), transcriptionally activates the expression of 1,4-GlcNAc transferase (GlcNAcT-I) and of heparan sulfate 2-O sulfotransferase (HS2ST), the enzyme responsible for sulfation of iduronic acid (IdoA), ghich increases the expression of HSPGs at the cell surface. Proteolytic or heparinolytic cleavage of HSPGs generates modulators of FGF/FGFR interactions. Intracellular fibroblast growth factor-binding proteins (FIBPs) that interact with FGF have also been described. However, their role in the vasculature is not yet established.

affinity (Kd of 3 nM). FGF-2 binding to GM1 and its mitogenic activity are abrogated by cholera toxin-B, a ligand for gangliosides. This indicates that cell-associated gangliosides may act as functional FGF co-receptors.[9]

FGFR1 is the main FGFR expressed in endothelial cells *in vitro* and has also been detected in activated endothelial cells *in vivo*.[10] Small amounts of FGFR2 were also detected in endothelial cells.[11] In capillary endothelial cell lines, stimulation of FGFR1 induces proliferation, migration, protease production and tubular morphogenesis, whereas FGFR2 only increases motility. It is at present not clear whether this observation is of more general significance and also applies to primary endothelial cells.

FGF receptor stimulation by FGFs activates the classical MAP kinase pathway and also recruits a number of binding partners such as FRS2 or GRB2 (Fig. 1 and Table 1). Furthermore, p38 MAP kinases seem to represent essential regulators of FGF's signal transduction machinery in vascular endothelial cells.[12] FGF2 induces tube formation, p38 kinase activation and expression of the notch ligand jagged1 in endothelial cells cultured on three-dimensional collagen gels. Inhibition of p38 kinase further increases the effect of FGF2 on tube formation, decreases apoptosis, stimulates DNA synthesis and further increases jagged1 expression. This correlates with the *in vivo* results in the chicken chorioallantoid membrane (CAM), where FGF-2 and SB202190, an inhibitor of p38 kinase, induce together endothelial cell hyperplasia and aberrant blood vessels.[12] Thus, p38 may be part of a negative feedback loop providing an auto-inhibitory mechanism for FGF effects on the vasculature, or as a component of the vessel maturation pathway. A role for p38 in the vessel maturation pathway is also supported by gene inactivation studies in mice.[13,14]

Besides MAPK activation, a long-lasting PLCγ or PKC (downstream of PLCγ) activation is required for FGF-2's full mitogenic activity in some endothelial cell lines.[15] Nevertheless, FGF-2 is unable to activate PLCγ in HUVEC.[16] In non-endothelial cells, PLCγ is required for FGFR internalization.[17] This observation needs to be confirmed in vascular cells.

In non-vascular cells, it has been reported that FGFs induce phosphatidyl inositol (PI)-3 kinase activation via fibroblast growth factor receptor substrate-2 (FRS2)/Gab1.[18] In culture, FGF-2 induces endothelial cell Akt phosphorylation resulting in an anti-apoptotic effect and increased cell motility. This is confirmed *in vivo*

Table 1. Molecular players involved in FGF-mediated effects on the vasculature.

Molecules	Effects in vascular cells	Ref(s).
FGFR1	Upregulated in angiogenic endothelium; induces proliferation, migration, tubulogenesis through receptor dimerization and autophosphorylation	2, 4, 6, 10, 21
HSPGs	Co-receptors for FGFRs, also implicated as transduction module in FGFR signaling	4, 6, 27
Gangliosides (GM1)	Co-receptors for FGFRs	9
Src	Tubulogenesis	21, 23, 26
Fyn	Tubulogenesis	25
ERK1 ERK2	Migration, proliferation	21
P38 MAP kinase	Negatively regulates migration, tubulogenesis and/or vessel maturation	12
Shb	Tubular morphogenesis	24
c-fes	Migration, tubular morphogenesis	22
PLCγg PKC	Long lasting activation induces full mitogenic activity in some endothelial cells	
PI-3 kinase	Survival, motility. Connected via Gab1, Grb2 and FRS2 to the FGFR?	19
ETS-1	Transcription factor mediating signaling to the nucleus	20
SEF	Transmembrane protein providing a negative control of FGFR activity	29, 30
FIF Translokin P34	Intracellular binding proteins for endogenous FGF: role in the vasculature?	33–35
FGF-BP	Extracellular binding protein for FGF	39, 40

where inhibition of the PI3K/Akt pathway inhibits FGF-2-induced angiogenesis.[19]

It has been recently reported that the transcription factor ETS-1 is a downstream effector of FGF-2 signaling in FGF-2-induced angiogenesis *in vivo* in the mouse ear and in tumor angiogenesis[20] (Table 1). Retroviral expression of a dominant-negative form of ETS-1 lacking the transactivation domain abrogates the *in vivo* effects of FGF-2.

Other signaling modules involved in FGF-driven angiogenesis are Fyn, Src, Fes or Shb[21-26] (Table 1).

Recently, HSPGs have also been reported to play an active role in the FGF-induced signaling.[27] (Table 1). Phosphatidyl-inositol-4,5-bisphosphate (PIP2) or postsynaptic density-95 (PDZ)-binding domain-mutated syndecan-4 overexpressed in vascular endothelial cells, inhibits protein kinase C-α (PKCα) activation. Furthermore, expression levels of HSPGs are modulated in endothelial cells during angiogenesis. Hypoxia increases the ratio of heparan sulfate to chondroitin sulfate at the endothelial cell surface and the binding of[125]I-FGF-2. Hypoxia upregulates heparan sulfate 2-O sulfotransferase (HS2ST) and 1,4-N-acetyl glucosamine (GlcNAc) transferase (GlcNAcT-I) through hypoxia inducible factor-1α (HIF1α)[28] (Fig. 1). This increases the effect of FGF on the vasculature.

Several feedback inhibitors of FGFR activity have been identified by genetic screening in Drosophila or Zebrafish. One of them is Sef, a transmembrane protein that associates with FGFR1 at the level of the cytoplasmic domain.[29] The human Sef homologue is expressed in HUVEC, interacts with FGFR1 and inhibits FGF-2-induced Erk activation[30] (Table 1).

In addition to the paracrine effects of FGF, vascular cells also express different FGF forms which may act in an autocrine or intracellular manner. For example, FGF-2 exists as a cytoplasmic 18 kDa isoform and as four nuclear high molecular weight (HMW) isoforms. HMW isoforms but not 18 kDa FGF-2 has an N-terminal sequence responsible for nuclear targeting/retention signal. Dominant-negative strategies in cultured cells have demonstrated that HMW FGF-2 acts on cell growth by a cell surface receptor-independent mechanism.[31,32] Intracellular FGFs may bind a number of molecules such as p34 for FGF-1,[33] FGF interacting factor (FIF) or Translokin for FGF-2[34,35] may participate in intracellular effects (Table 1). The studies mentioned above have conducted in non vascular cells and it is, therefore, not firmly established whether intracellular FGFs play a role in vascular development. By contrast, released FGF isoforms clearly stimulate endothelial cell migration. This contention is supported by studies in FGF-2 $(-/-)$ mice because endothelial cells from these mice have a defect in cell migration which can be compensated by the addition of exogenous FGF-2.[36] In endothelial cells from $(-/-)$ mice, MAP kinase activity after wounding is also diminished compared to MAP kinase of endothelial cells from wild type (wt) mice. This is in frame with earlier studies, in which an inhibition of migration is observed in cultured fibroblasts expressing dominant-negative FGF receptors.[31]

The mechanisms by which FGFs lacking a classical signal sequence are released from the cells are not well understood. These FGF prototypes (such as FGF-1 or FGF-2) may be released after cell injury or actively exported by Golgi-independent mechanisms. It has been reported that FGF-1 is exported as a multiprotein aggregate that contains copper, the p40 extravesicular domain of synaptotagmin and S100A13.[37] In contrast, FGF-2 exocytosis seems to require a Na/K ATPase subunit.[38]

FGFs molecules are typically stored on heparin sulphate proteoglycans in the extracellular matrix limiting growth factor bio-availability and requiring release from this storage to trigger a distant response. An FGF-binding protein (FGF-BP) binds FGF molecules (FGF-1, -2, -7, -10 and -22), releases growth factors from extracellular matrix and increases FGF-dependent angiogenesis as shown in the CAM assay or in chicken embryos.[39,40]

FGFs may also indirectly regulate vascular morphogenesis by inducing secondary angiogenesis regulators such as VEGF-A or VEGF-C.[41–43]

3. Role in Vascular Development

Contrary to FGFs, the function of ligand/receptor pairs such as VEGF/VEGFR, Eph/EphR, Ang/Tie,[44] and more recently Delta/Notch,[45] has been better assigned and characterized. *In vitro* culture experiments indicate that FGFs regulate vessel formation in the developing mammals.[46] FGFs promote the formation of Flk1+ hemangioblastic precursor cells from mesoderm,[47] sprouting angiogenesis[43,48] or endothelial cell assembly during vasculogenesis.[49] Nevertheless, these culture systems do not reflect completely the embryonic or organ-specific multicellular environment and the precise state of differentiation of endothelial cells.

Disruptions of FGF/FGFR genes have not been very informative with regard to their role in vessel formation *in vivo*. FGF-1 $(-/-)$ or FGF-2 $(-/-)$ mice (single or double knockouts) do not present a vascular phenotype during development.[50] This suggests functional redundancy or a non-essential role of FGFs in developmental vasculogenesis or angiogenesis. Gene knockouts in mice of FGFR1 and FGFR2, the main FGF receptors, yield to embryos that are arrested in their development before the onset of vascularization because they are lacking mesoderm-inducing signals.[51,52]

Studies with embryonic or tissue explant cultures have given better insights into the role of endogenous FGFs in vascular development. Injection of an adenoviral construct encoding a dominant-negative FGF receptor, which blocks FGF signaling, into cultured day 9 mouse embryos demonstrates incomplete branching of the yolk sac vasculature (although endothelial cells are detected) and intersomitic vessels, heart septation defects and angiogenesis defects in organs such as the brain.[53] Furthermore, endogenous FGF also induces vessel outgrowth from embryonic heart explants.[54] Vessel outgrowth induced by VEGF in these explant cultures is completely inhibited by neutralizing anti-FGF-2 antibodies. This demonstrates that endogenous FGF is required for VEGF's effect in this model. Explant cultures are closer to the physiological situation than *in vitro* cell culture because paracrine signaling and cell-to-cell interactions are preserved. However, explant cultures may introduce biases because of misregulated gene expression or exogenous factors added to the cultures.

Gene expression can be manipulated *in vivo* in a tissue-specific manner by transgenesis. A latent pro-angiogenic phenotype is observed in mice that overexpress FGF-2, either ubiquitously[55] or in the retina.[56] Overexpression of FGF-2[57] or FGF1[58] in the heart shows an increase in vessel density and arborescence. These observations suggest that endogenous FGFs may have similar functions in the developing embryo as well.

To solve this issue, we have generated several lines of transgenic mice that overexpress a dominant-negative FGFR1 truncated at the c-terminal tyrosine-kinase domain, specifically in the retinal pigmented epithelium (RPE) of the developing eye.[59] We observed a poorly branched vascular bed in the choroid and an avascular neonatal retina.[46] This, to our knowledge, is the first observation that demonstrates that FGFs play a role in developmental angiogenesis. This is in agreement with branching defects dependent on FGF signaling described in other model systems. In the mouse ureteric bud, FGF-1, FGF-2, FGF-7 and FGF-10 all affect branching morphogenesis.[60] In addition, FGF signaling is also critical for branching morphogenesis in the mouse lung.[61] In the *Drosophila* tracheal system, Branchless, a homologue of mammalian FGF that is hypoxia-regulated, controls the formation of the terminal tracheal ramifications.[62] Taken together, these observations suggest a general scheme for a role of FGFs in branching morphogenesis that may also apply to the vasculature as well. This contention is supported by the observation that the mouse homologue of *Drosophila* Sprouty-4, an inhibitor of tracheal branching by antagonizing Branchless activity, also inhibits angiogenesis in the mouse embryo, by possibly interfering with FGF signaling.[63]

4. FGFs in Tumor Angiogenesis

FGFs have complex roles in tumor development. They can be oncogenic proteins, induce proliferation of the tumor cells, modify the angiogenic phenotypes of the tumor cell or directly act on the vascular and non-vascular stroma.

Many tumor cell lines, such as pancreatic carcinoma, lung carcinoma or glioma, synthesize FGF prototypes, particularly FGF-2 and FGF-1. In fibrosarcoma cells, FGF-2 secretion is correlated with tumor progression.[64] Moreover, FGF-2 urine or serum levels are correlated with the degree of malignancy and tumor progression.[65,66]

A number of observations are in favor of a decisive role of FGFs in tumor angiogenesis. In the TRAMP mice where SV40T is specifically expressed in the prostate, angiogenesis develops in two steps.[67] The "initial switch" induces interductal angiogenic vessels and the second ("progression switch") leads to intraductal vessels. During the first step, 25 kDa HMW FGF-2 is found to be upregulated in the nuclei of tumor cells which is probably responsible for the induction of secondary angiogenesis factors such as VEGF. Furthermore, VEGFR2 is found upregulated in intraductal vessels. During the "progression switch," 18 kDa FGF-2 is upregulated in the tumor and, with FGFR1 IIIb, in the endothelial cells. This suggests the existence of paracrine and autocrine FGF loops to control tumor angiogenesis.[68]

Inhibition of the biological activity by antisense molecules, neutralizing antibodies, dominant-negative FGFRs or soluble receptors can provide functional proof for the role of FGF prototypes in tumor angiogenesis. Blockade of the FGF pathway in mice by FGF-2 or FGFR1 antisense molecules or by FGF-2 neutralizing antibody inhibits tumor angiogenesis and growth.[69,70] Using a dominant-negative FGFR strategy, it was demonstrated that FGFs have a dual action on glioma development.[41,71] C6 glioma cells transfected with dominant negative FGFR1 or FGFR2, are growth inhibited *in vitro*. This indicates a direct autocrine action on tumor cells. When implanted *in vivo* in immunodeficient mice, tumor angiogenesis and tumor development are both impaired and accompanied a downregulation of VEGF expression in tumors. This is in agreement with the induction of VEGF by exogenous FGFs in NIH 3T3 cells and the mouse corneal pocket assay.[43] Similar results are found after transplantation of dominant-negative FGFR expressing C6 cells into the rat and mouse brain.[41,72] The mechanism for inhibition of the tumor vasculature in the brain is different since cooption plays a significant role in the development of brain malignancies. During cooption, tumor cells first grow around normal brain vessels, synthesize Angiopoietin-2, which leads to vessel apoptosis and hypoxia. Hypoxia increases VEGF synthesis and angiogenesis.[73,74] Thus, when FGF activity is blocked, both cooption and reactive neoangiogenesis are inhibited.

Another strategy to inhibit FGF in tumor cells is to express soluble FGFRs. Ogawa *et al.* have shown that inhibition of tumor growth in mice by adenoviral expression of either a soluble form of FGFR1 or VEGFR1 is dependent on the presence of the respective growth factor in tumor cells.[75] Soluble FGFR1 does not inhibit tumor growth *in vitro* but decreases the number of capillaries in tumors *in vivo* which suggests that *in vivo* tumor growth is impaired through an angiogenesis-dependent mechanism. If FGF and VEGF are present together in tumor cells, inhibition of tumor growth and angiogenesis by the respective soluble receptors is additive. Similar results are obtained in the RIP Tag transgenic mice model of multistage pancreatic carcinogenesis where a synergistic effect of both soluble receptors on tumor development was observed.[76]

Another approach to study the role of FGFs in tumor angiogenesis is to overexpress FGF prototypes in tumor cells that do not make FGF themselves.[77-79] After subcutaneous injection in mice of bladder carcinoma cells (NBT-II) expressing the 18 kDa FGF-2 form only, angiogenesis is increased, but not tumor growth. Tumor growth is only increased when cells express HMW FGF-2 and 18 kDa FGF-2 simultaneously. Similar results are found with human endometrial cancer cells (HEC-1-B). If FGF-2 expression is turned off in 20 days after tumor cell injection by using the tet-off expression system, no difference in tumor growth is observed but the number of vessels is significantly decreased when compared to control cells in which FGF-2 expression is maintained. This indicates that the induction of tumor angiogenesis is primarily dependent on 18 kDa FGF-2.

Pepper *et al.* in early studies showed a synergism between FGFs and VEGF to induce *in vitro* angiogenesis.[80] This was confirmed *in vivo* where FGF-2 and VEGF stimulate tumor vascularization synergistically but with different effects — FGF-2 increases vessel size whereas VEGF increases vessel maturation and permeability.[81] Furthermore, FGFs are implicated in escape mechanisms that may explain resistance to anti-angiogenesis treatment. RipTag that become resistant to anti-VEGF treatment, undergo a molecular switch and become dependent on FGFs that are upregulated in escaping tumors.[82] Inhibition of FGFs in escaping tumors leads to inhibition of tumor angiogenesis and tumor burden.

A ribozyme-mediated FGF-BP depletion in squaemous cell carcinoma and in colon carcinoma decreases the release of biologically active FGF and tumor growth and angiogenesis.[83] Fibstatin is a newly characterized extracellular inhibitor of FGF-2 derived from fibronectin. Expression of fibstatin in tumors inhibits tumor growth and angiogenesis.[84]

5. Role of FGFs in Developmental and Tumor Lymphangiogenesis

FGF-2 induces corneal lymphangiogenesis.[42,85] This effect seems to be mediated through VEGF-C and possibly VEGF-D. Recently a direct mechanism of induction of lymphangiogenesis has been evidenced.[86] In fact, FGF-2 binds to high- and low-affinity receptors on lymphatic endothelial cells. Furthermore, Prox-1, a molecule responsible for the lymphangiogenic switch, induces expression of FGFR3 in lymphatic endothelial cells. Budding lymphatic vessels in the mouse embryo also expresses FGFR3. Finally, inhibition of FGFR3 by siRNA inhibits FGF-induced effects on lymphangiogenesis. These results indicate that FGFR3 is a crucial component for lymphangiogenic signaling. However, in tumor lymphangiogenesis, FGF signaling seems to be mainly dependent on the upregulation of VEGF-C (Larrieu Lahargue *et al.*, unpublished results).

6. Role in Repair-Associated Angiogenesis and Ischemia Revascularization

FGF-2 has been implicated in repair-associated angiogenesis. As indicated above, FGF-2 null mice have a delay in wound healing.[87] The delay in wound healing may be explained by inhibition of endothelial cell migration as observed *in vitro* with FGF-2 null endothelial cells[36] or by inhibition of the activation of another angiogenic factor such as VEGF by FGF-2.[43]

FGFs are able to promote neoangiogenesis and the formation of the collateral circulation in cardiac or hindlimb models of ischemia. Several studies have shown that the administration of FGF-2 protein, naked DNA encoding FGF-1, adenovirus encoding FGF-2 in collagen-based matrices or FGF-2 bioreactors increases the formation of neovessels and collaterals.[88–92]

FGF-2 and FGF-1 were successfully used in the salvage of ischemic damage in different animal models such as dog,[88,93,94] pig,[95] rabbit[89] and rat.[96,97] In these models, FGFs increase the vascularization of the ischemia zone by increasing the number of arterioles and capillaries.

The route of FGF administration seems to be important. Intravenous administration of the FGF-2 protein is less effective than intra-arterial injection (intracoronary), which suggests a stability or delivery problem of the protein. These results explain, why, in some studies, FGF-2 is less effective in improving ischemic revascularization.[95,98,99] However FGF-2 intrapericardial administration provides higher myocardial deposition and retention and lower systemic recirculation than intracoronary or intravenous administration.[100] Alternatively, gene transfer techniques may be used instead of protein injection; naked FGF-2 plasmid DNA injected in myocardium or intracoronary adenoviral-containing FGF genes were successfully used in myocardial ischemia.[101,102]

In humans, FGF-2 is in therapeutic trials for coronary disease and for intermittent claudication. Intracoronary injection of FGF-2 protein in patients with ischemic heart disease not subjected to treatment with coronary arterial bypass grafting (CABG) or percutaneous transluminal angioplasty (PTAC) is well tolerated. Only a minor hypotensive effect appears at high doses after intracoronary injection.[103,104] Moreover, a diminution of the ischemic zone and an improvement of myocardial perfusion are observed.[104,105] However, in a second clinical study with more patients, a single bolus intracoronary injection of FGF-2 had no effect in myocardial perfusion.[106] Thus, it is at present unclear whether FGF-2 protein alone is of any benefit, at least in long term, in patients with coronary artery disease. Nevertheless, FGF-2 may be efficient in combination with surgery. Indeed, FGF-2 administration in the ungraftable myocardial territory of a patient concomitantly with CABG improves myocardial revascularization.[107,108]

In peripheral arterial ischemic disease, intra-arterial administration of FGF-2 can improve peak walking but only if patients receive a single FGF-2 dose. The effect is greater among smokers. Nevertheless, during this phase II trial, vascularization of the leg was not studied.[109]

Other FGF prototypes are perhaps more efficient than FGF-2 in stimulating coronary or peripheral artery angiogenesis. This is supported by the absence of a role of endogenous FGF-2 in ischemic hindlimb revascularization in FGF-2 $-/-$ mice.[110] Indeed, FGF-4, another FGF prototype, can induce angiogenesis and arteriogenesis in a rabbit hindlimb ischemia model, and its effect is mediated by

VEGF.[111] In a human study, FGF-4 was delivered in myocardium by intracoronary infusion in an adenoviral vector. This allows a sustained production of the FGF-4 protein. A clear benefit was observed in treated patients.[112] However, few patients (3%) developed brain and colon tumors several months after the administration, indicating a possible risk of malignant disease in patients when FGF-4 is administered via adenoviral vectors.[113]

The combination of FGF with other factors such as PDGF may be more efficient than the administration of single FGF protypes alone. In a recent study, Cao *et al.*[114] have demonstrated that the administration of FGF-2 together with PDGF-BB induces synergistically the formation of vascular networks in rat and rabbit hindlimb ischemic models that remained functional after one year even after depletion of the growth factors. This constitutes a promising new venue for the treatment of ischemic disease.

7. Conclusion

Recent observations indicate that FGF prototypes certainly participate in normal and pathological vascular development. FGFs also promote the formation of larger and better organized blood vessels and can synergize with PDGFs to induce a fully mature and functional vascular network. A role in lymphangiogenesis has been also recently attributed to FGFs that is most likely dependent on the induction of VEGF-C. The respective FGF prototypes involved in vascular development are not yet fully known. Among these prototypes, FGF-2 most likely plays a role in wound healing and tumor angiogenesis.

The signal transduction mechanisms induced by FGFs in endothelial cells are now better known. However, the specific molecular events that occur during tubulogenesis, remodeling and branching remain to be elucidated. The mechanism of action of intracellular FGFs in angiogenesis is also unclear. A number of molecules that bind intracellular FGF have been discovered, however their functional role remains to be determined. The ultimate challenge is to unravel how FGFs control vascular development in an integrated fashion together with other angiogenesis regulators.

Acknowledgments

The work from our laboratory described here was supported by grants from the Association de la Recherche sur le Cancer, the Ligue Nationale contre le Cancer, Rétina France, the Institut National de la Santé et de la Recherche Médicale (INSERM), and the Ministère de la Recherche. Due to space constraints, we are not able to quote all the relevant literature and would like to apologize to those authors whose work is not mentioned.

References

1. Bikfalvi A, Klein S, Pintucci G, Rifkin DB (1997) Biological roles of fibroblast growth factor-2. *Endocr Rev* **18**: 26–45.
2. Javerzat S, Auguste P, Bikfalvi A (2002) The role of fibroblast growth factors in vascular development. *Trends Mol Med* **8**: 483–489.
3. Ornitz DM, Itoh N (2001) Fibroblast growth factors. *Genome Biol* **2**: Reviews 3005.
4. Kwan CP, Venkataraman G, Shriver Z, Raman R, Liu D, Qi Y, Varticovski L, Sasisekharan R (2001). Probing fibroblast growth factor dimerization and role of heparin-like glycosaminoglycans in modulating dimerization and signaling. *J Biol Chem* **276**: 23421–23429.
5. Facchiano A, Russo K, Facchiano AM, De Marchis F, Facchiano F, Ribatti D, Aguzzi MS, Capogrossi MC (2003). Identification of a novel domain *of fibroblast growth factor 2* controlling its angiogenic properties. *J Biol Chem* **278**: 8751–8760.
6. Schlessinger J, Plotnikov AN, Ibrahimi OA, Eliseenkova AV, Yeh BK, Yayon A, Linhardt RJ, Mohammadi, M (2000). Crystal structure of a ternary FGF-FGFR-heparin complex reveals a dual role for heparin in FGFR binding and dimerization. *Mol Cell* **6**: 743–750.
7. Zhang Z, Coomans C, David G (2001). Membrane heparan sulfate proteoglycan-supported FGF2-FGFR1 signaling: evidence in support of the "cooperative end structures" model. *J Biol Chem* **276**: 41921–41929.
8. Zhou Z, Wang J, Cao R, Morita H, Soininen R, Chan KM, Liu B, Cao Y, Tryggvason K (2004). Impaired angiogenesis, delayed wound healing and retarded tumor growth in perlecan heparan sulfate-deficient mice. *Cancer Res* **64**: 4699–4702.
9. Rusnati M, Urbinati C, Tanghetti E, Dell'Era P, Lortat-Jacob H, Presta M (2002). Cell membrane GM1 ganglioside is a functional coreceptor for fibroblast growth factor 2. *Proc Natl Acad Sci USA* **99**: 4367–4372.
10. Arbeit JM, Olson DC, Hanahan D (1996). Upregulation of fibroblast growth factors and their receptors during multi-stage epidermal carcinogenesis in K14-HPV16 transgenic mice. *Oncogene* **13**: 1847–1857.
11. Nakamura T, Mochizuki Y, Kanetake H, Kanda S (2001) Signals via FGF receptor 2 regulate migration of endothelial cells. *Biochem Biophys Res Commun* **289**: 801–806.

12. Matsumoto T, Turesson I, Book M, Gerwins P, Claesson-Welsh L (2002). p38 MAP kinase negatively regulates endothelial cell survival, proliferation, and differentiation in FGF-2-stimulated angiogenesis. *J Cell Biol* **156**: 149–160.

13. Lin Q, Schwarz J, Bucana C, Olson EN (1997). Control of mouse cardiac morphogenesis and myogenesis by transcription factor MEF2C. *Science* **276**: 1404–1407.

14. Lin Q, Lu J, Yanagisawa H, Webb R, Lyons GE, Richardson JA, Olson EN (1998). Requirement of the *MADS*-box transcription factor MEF2C for vascular development. *Development* **125**: 4565–4574.

15. Presta M, Tiberio L, Rusnati M, Dell'Era P, Ragnotti G (1991). Basic fibroblast growth factor requires a long-lasting activation of protein kinase C to induce cell proliferation in transformed fetal bovine aortic endothelial cells. *Cell Regul* **2**: 719–726.

16. Wu LW, Mayo LD, Dunbar JD, Kessler KM, Baerwald MR, Jaffe EA, Wang D, Warren RS, Donner D B (2002). Utilization of distinct signaling pathways by receptors for vascular endothelial cell growth factor and *other* mitogens in the induction of endothelial cell proliferation. *J Biol Chem* **275**: 5096–5103.

17. Sorokin A, Mohammadi M, Huang J, Schlessinger J (1994). Internalization of fibroblast *growth* factor receptor is inhibited by a point mutation at tyrosine. *J Biol Chem* **269**: 17056–17061.

18. Ong SH, Hadari YR, Gotoh N, Guy GR, Schlessinger J, Lax I (2001). Stimulation of phosphatidylinositol 3-kinase by fibroblast growth factor receptors is mediated by coordinated recruitment of multiple docking proteins. *Proc Natl Acad Sci USA* **98**: 6074–6079.

19. Maffucci T, Piccolo E, Cumashi A, Iezzi M, Riley AM, Saiardi A, Godage HY, Rossi C, Broggini M, Iacobelli S, Potter BV, Innocenti P, Falasca M (2005). Inhibition of the phosphatidylinositol 3-kinase/Akt pathway by inositol pentakisphosphate results in antiangiogenic and antitumor effects. *Cancer Res* **65**: 8339–8349.

20. Pourtier-Manzanedo A, Vercamer C, Van Belle E, Mattot V, Mouquet F, Vandenbunder B (2003). Expression of an Ets-1 dominant-negative mutant perturbs normal and tumor angiogenesis in a mouse ear model. *Oncogene* **22**: 1795–1806.

21. Cross MJ, Claesson-Welsh L (2001). FGF and VEGF function in angiogenesis: signaling *pathways*, biological responses and therapeutic inhibition. *Trends Pharmacol Sci* **22**: 201–207.

22. Kanda S, Lerner EC, Tsuda S, Shono T, Kanetake H, Smithgall TE (2000). The nonreceptor protein-tyrosine kinase c-Fes is involved in fibroblast growth factor-2-induced chemotaxis of murine brain capillary endothelial cells. *J Biol Chem* **275**: 10105–10111.

23. LaVallee TM, Prudovsky IA, McMahon GA, Hu X, Maciag T (1998). Activation of the MAP kinase pathway by FGF-1 correlates with cell proliferation induction while activation of the Src pathway correlates with migration. *J Cell Biol* **141**: 1647–1658.

24. Lu L, Holmqvist K, Cross M, Welsh M (2002). Role of the Src homology 2 Domain-containing protein Shb in murine brain endothelial cell proliferation and differentiation. *Cell Growth Differ* **13**: 141–148.
25. Tsuda S, Ohtsuru A, Yamashita S, Kanetake H, Kanda S (2002). Role of c-Fyn in FGF-2-mediated tube-like structure formation by murine brain capillary endothelial cells. *Biochem Biophys Res Commun* **290**: 1354–1360, 2002.
26. Zhan X, Plourde C, Hu X, Friesel R, Maciag T (1994). Association of fibroblast growth factor receptor-1 with c-Src correlates with association between c-Src and cortactin. *J Biol Chem* **269**: 20221–20224,.
27. Horowitz A, Tkachenko E, Simons M (2002). Fibroblast growth factor-specific modulation of cellular response by syndecan-4. *J Cell Biol* **157**: 715–725.
28. Li J, Shworak NW, Simons M (2002). Increased responsiveness of hypoxic endothelial cells to FGF2 is mediated by HIF-1alpha-dependent regulation of enzymes involved in synthesis of heparan sulfate FGF2-binding sites. *J Cell Sci* **115**: 1951–1959.
29. Kovalenko D, Yang X, Nadeau RJ, Harkins LK, Friesel R (2003). Sef inhibits fibroblast growth factor signaling by inhibiting FGFR1 tyrosine phosphorylation and subsequent ERK activation. *J Biol Chem* **24**: 24.
30. Yang RB, Ng CK, Wasserman SM, Komuves LG, Gerritsen ME, Topper JN (2003). A novel interleukin-17 receptor-like protein identified in human umbilical vein endothelial cells antagonizes basic fibroblast growth factor-induced signaling. *J Biol Chem* **278**: 33232–33238.
31. Bikfalvi A, Klein S, Pintucci G, Quarto N, Mignatti P, Rifkin DB (1995). Differential modulation of cell phenotype by different molecular weight forms of basic fibroblast growth factor: possible intracellular signaling by the high molecular weight forms. *J Cell Biol* **129**: 233–243.
32. Arese M, Chen Y, Florkiewicz RZ, Gualandris A, Shen B, Rifkin DB (1999). Nuclear activities of basic fibroblast growth factor: potentiation of low-serum growth mediated by natural or chimeric nuclear localization signals. *Mol Biol Cell* **10**: 1429–1444.
33. Skjerpen CS, Wesche J, Olsnes S (2002). Identification of ribosome-binding protein p34 as an intracellular protein that binds acidic fibroblast growth factor. *J Biol Chem* **8**: 6.
34. Bossard C, Laurell H, Van den Berghe L, Meunier S, Zanibellato C, Prats H (2003). Translokin is an intracellular mediator of FGF-2 trafficking. *Nat Cell Biol* **5**: 433–439.
35. Van den Berghe L, Laurell H, Huez I, Zanibellato C, Prats H, Bugler B (2000). FIF [fibroblast growth factor-2 (FGF-2)-interacting-factor], a nuclear putatively antiapoptotic factor, interacts specifically with FGF-2. *Mol Endocrinol* **14**: 1709–1724.
36. Pintucci G, Moscatelli D, Saponara F, Biernacki PR, Baumann FG, Bizekis C, Galloway AC, Basilico C, Mignatti P (2002). Lack of ERK activation and cell migration in FGF-2–deficient endothelial cells. *FASEB J* **25**: 25.
37. Landriscina M, Bagala C, Mandinova A, Soldi R, Micucci I, Bellum S, Prudovsky I, Maciag T (2001). Copper induces the assembly of a multiprotein aggregate

implicated in the release of fibroblast growth factor 1 in response to stress. *J Biol Chem* **276**: 25549–25557.

38. Florkiewicz RZ, Anchin J, Baird A (1998). The inhibition of fibroblast growth factor-2 export by cardenolides implies a novel function for the catalytic subunit of Na+,K+-ATPase. *J Biol Chem* **273**: 544–551.

39. McDonnell K, Bowden ET, Cabal-Manzano R, Hoxter B, Riegel AT, Wellstein A (2005). Vascular leakage in chick embryos after expression of a secreted binding protein for fibroblast growth factors. *Lab Invest* **85**: 747–755.

40. Tassi E, Al-Attar A, Aigner A, Swift MR, McDonnell K, Karavanov A, Wellstein A (2001). Enhancement of fibroblast growth factor (FGF) activity by an FGF-binding protein. *J Biol Chem* **276**: 40247–40253.

41. Auguste P, Gursel DB, Lemiere S, Reimers D, Cuevas P, Carceller F, Di Santo JP, Bikfalvi A (2001). Inhibition of fibroblast growth factor/fibroblast growth factor receptor activity in glioma cells impedes tumor growth by both angiogenesis-dependent and -independent mechanisms. *Cancer Res* **61**: 1717–1726.

42. Kubo H, Cao R, Brakenhielm E, Makinen T, Cao Y, Alitalo K (2002). Blockade of vascular endothelial growth factor receptor-3 signaling inhibits fibroblast growth factor-2-induced lymphangiogenesis in mouse cornea. *Proc Natl Acad Sci USA* **17**: 17.

43. Seghezzi G, Patel S, Ren CJ, Gualandris A, Pintucci G, Robbins ES, Shapiro RL, Galloway AC, Rifkin DB, Mignatti P (1998). Fibroblast growth factor-2 (FGF-2) induces vascular endothelial growth factor (VEGF) expression in the endothelial cells of forming capillaries: an autocrine mechanism contributing to angiogenesis. *J Cell Biol* **141**: 1659–1673.

44. Carmeliet P (2000). Mechanisms of angiogenesis and arteriogenesis. *Nat Med* **6**: 389–395.

45. Mailhos C, Modlich U, Lewis J, Harris A, Bicknell R, Ish-Horowicz D (2001). Delta4, an endothelial specific notch ligand expressed at sites of physiological and tumor angiogenesis. *Differentiation* **69**: 135–144.

46. Rousseau B, Larrieu-Lahargue F, Bikfalvi A, Javerzat S (2003). Involvement of fibroblast growth factors in choroidal angiogenesis and retinal vascularization. *Exp Eye Res* **77**: 147–156.

47. Poole TJ, Finkelstein EB, Cox CM (2001). The role of FGF and VEGF in angioblast induction and migration during vascular development. *Dev Dyn* **220**: 1–17.

48. Feraud O, Cao Y, Vittet D (2001). Embryonic stem cell-derived embryoid bodies development in collagen gels recapitulates sprouting angiogenesis. *Lab Invest* **81**: 1669–1681.

49. Flamme I, Risau W (1992) Induction of vasculogenesis and hematopoiesis *in vitro*. *Development* **116**: 435–439, 1992.

50. Miller DL, Ortega S, Bashayan O, Basch R, Basilico C (2002). Compensation by fibroblast growth factor 1 (FGF1) does not account for the mild phenotypic defects observed in FGF2 null mice. *Mol Cell Biol* **20**: 2260–2268.

51. Deng CX, Wynshaw-Boris A, Shen MM, Daugherty C, Ornitz DM, Leder P (1994). Murine FGFR-1 is required for early postimplantation growth and axial organization. *Genes Dev* **8**: 3045–3057.

52. Arman E, Haffner-Krausz R, Chen Y, Heath JK, Lonai P (1998). Targeted disruption of fibroblast growth factor (FGF) receptor 2 suggests a role for FGF signaling in pregastrulation mammalian development. *Proc Natl Acad Sci USA* **95**: 5082–5087.

53. Lee SH, Schloss DJ, Swain JL (2002). Maintenance of vascular integrity in the embryo requires signaling through the fibroblast growth factor receptor. *J Biol Chem* **275**: 33679–33687.

54. Tomanek RJ, Sandra A, Zheng W, Brock T, Bjercke RJ, Holifield JS (2001). Vascular endothelial growth factor and basic fibroblast growth factor differentially modulate early postnatal coronary angiogenesis. *Circ Res* **88**: 1135–1141.

55. Fulgham DL, Widhalm SR, Martin S, Coffin JD (1999). FGF-2 dependent angiogenesis is a latent phenotype in basic fibroblast growth factor transgenic mice. *Endothelium* **6**: 185–195.

56. Yamada H, Yamada E, Kwak N, Ando A, Suzuki A, Esumi N, Zack DJ, Campochiaro PA (2000). Cell injury unmasks a latent proangiogenic phenotype in mice with increased expression of FGF2 in the retina. *J Cell Physiol* **185**: 135–142.

57. Sheikh F, Sontag DP, Fandrich RR, Kardami E, Cattini PA (2001). Overexpression of FGF-2 increases cardiac myocyte viability after injury in isolated mouse hearts. *Am J Physiol Heart Circ Physiol* **280**: H1039–H1050.

58. Fernandez B, Buehler A, Wolfram S, Kostin S, Espanion G, Franz WM, Niemann H, Doevendans PA, Schaper W, Zimmermann R (2000). Transgenic myocardial overexpression of fibroblast growth factor-1 increases coronary artery density and branching. *Circ Res* **87**: 207–213.

59. Rousseau B, Dubayle D, Sennlaub F, Jeanny JC, Costet P, Bikfalvi A, Javerzat S (2000). Neural and angiogenic defects in eyes of transgenic mice expressing a dominant-negative FGF receptor in the pigmented cells. *Exp Eye Res* **71**: 395–404.

60. Qiao J, Bush KT, Steer DL, Stuart RO, Sakurai H, Wachsman W, Nigam SK (2001). Multiple fibroblast growth factors support growth of the ureteric bud but have different effects on branching morphogenesis. *Mech Dev* **109**: 123–135.

61. Peters K, Werner S, Liao X, Wert S, Whitsett J, Williams L (1994). Targeted expression of a dominant negative FGF receptor blocks branching morphogenesis and epithelial differentiation of the mouse lung. *EMBO J* **13**: 3296–3301.

62. Jarecki J, Johnson E, Krasnow MA (1999). Oxygen regulation of airway branching in Drosophila is mediated by branchless FGF. *Cell* **99**: 211–220.

63. Lee SH, Schloss DJ, Jarvis L, Krasnow MA, Swain JL (2000). Inhibition of angiogenesis by a mouse sprouty protein. *J Biol Chem* **26**: 26.

64. Kandel J, Bossy-Wetzel E, Radvanyi F, Klagsbrun M, Folkman J, Hanahan D (1991). Neovascularization is associated with a switch to the export of bFGF in the multistep development of fibrosarcoma. *Cell* **66**: 1095–1104.

65. Nguyen M, Watanabe H, Budson AE, Richie JP, Folkman J (1993). Elevated levels of the angiogenic peptide basic fibroblast growth factor in urine of bladder cancer patients. *J Natl Cancer Inst* **85**: 241–242.

66. Nguyen M, Watanabe H, Budson AE, Richie JP, Hayes DF, Folkman J (1994). Elevated levels of an angiogenic peptide, basic fibroblast growth factor, in

the urine of patients with a wide spectrum of cancers. *J Natl Cancer Inst* **86**: 356–361.

67. Huss WJ, Hanrahan CF, Barrios RJ, Simons JW, Greenberg NM (2001). Angiogenesis and prostate cancer: identification of a molecular progression switch. *Cancer Res* **61**: 2736–2743.

68. Huss WJ, Barrios RJ, Foster BA, Greenberg NM (2003). Differential expression of specific FGF ligand and receptor isoforms during angiogenesis associated with prostate cancer progression. *Prostate* **54**: 8–16.

69. Rofstad EK, Halsor EF (2000). Vascular endothelial growth factor, interleukin 8, platelet-derived endothelial cell growth factor, basic fibroblast growth factor promote angiogenesis and metastasis in human melanoma xenografts. *Cancer Res* **60**: 4932–4938.

70. Wang Y, Becker D (1997). Antisense targeting of basic fibroblast growth factor and fibroblast growth factor receptor-1 in human melanomas blocks intratumoral angiogenesis and tumor growth. *Nat Med* **3**: 887–893.

71. Stan AC, Nemati MN, Pietsch T, Walter GF, Dietz H (1995). *In vivo* inhibition of angiogenesis and growth of the human U-87 malignant glial tumor by treatment with an antibody against basic fibroblast growth factor. *J Neurosurg* **82**: 1044–1052.

72. Miraux S, Lemiere S, Pineau R, Pluderi M, Canioni P, Franconi JM, Thiaudiere E, Bello L, Bikfalvi A, Auguste P (2004). Inhibition of FGF receptor activity in glioma implanted into the mouse brain using the tetracyclin-regulated expression system. *Angiogenesis* **7**: 105–113.

73. Holash J, Maisonpierre PC, Compton D, Boland P, Alexander CR, Zagzag D, Yancopoulos GD, Wiegand SJ (1999). Vessel cooption, regression, growth in tumors mediated by angiopoietins and VEGF. *Science* **284**: 1994–1998.

74. Zagzag D, Amirnovin R, Greco MA, Yee H, Holash J, Wiegand SJ, Zabski S, Yancopoulos GD, Grumet M (2000). Vascular apoptosis and involution in gliomas precede neovascularization: a novel concept for glioma growth and angiogenesis. *Lab Invest* **80**: 837–849.

75. Ogawa T, Takayama K, Takakura N, Kitano S, Ueno H (2002). Antitumor angiogenesis therapy using soluble receptors: enhanced inhibition of tumor growth when soluble fibroblast growth factor receptor-1 is used with soluble vascular endothelial growth factor receptor. *Cancer Gene Ther* **9**: 633–640.

76. Compagni A, Wilgenbus P, Impagnatiello MA, Cotten M, Christofori G (2000). Fibroblast growth factors are required for efficient tumor angiogenesis. *Cancer Res* **60**: 7163–7169.

77. Billottet C, Janji B, Thiery JP, Jouanneau J (2002). Rapid tumor development and potent vascularization are independent events in carcinoma producing FGF-1 or FGF-2. *Oncogene* **21**: 8128–8139.

78. Jouanneau J, Plouet J, Moens G, Thiery JP (1997). FGF-2 and FGF-1 expressed in rat bladder carcinoma cells have similar angiogenic potential but different tumorigenic properties *in vivo*. *Oncogene* **14**: 671–676.

79. Giavazzi R, Giuliani R, Coltrini D, Bani MR, Ferri C, Sennino B, Tosatti MP, Stoppacciaro A, Presta M (2001). Modulation of tumor angiogenesis by

conditional expression of fibroblast growth factor-2 affects early but not established tumors. *Cancer Res* **61**:309–317.

80. Pepper MS, Ferrara N, Orci L, Montesano R (1992). Potent synergism between vascular endothelial growth factor and basic fibroblast growth factor in the induction of angiogenesis *in vitro*. *Biochem Biophys Res Commun* **189**: 824–831.

81. Giavazzi R, Sennino B, Coltrini D, Garofalo A, Dossi R, Ronca R, Tosatti MP, Presta M (2003). Distinct role of fibroblast growth factor-2 and vascular endothelial growth factor on tumor growth and angiogenesis. *Am J Pathol* **162**: 1913–1926.

82. Casanovas O, Hicklin DJ, Bergers G, Hanahan D (2005). Drug resistance by evasion of antiangiogenic targeting of VEGF signaling in late-stage pancreatic islet tumors. *Cancer Cell* **8**: 299–309.

83. Czubayko F, Liaudet-Coopman ED, Aigner A, Tuveson AT, Berchem GJ, Wellstein AA (1997). A secreted FGF-binding protein can serve as the angiogenic switch in human cancer. *Nat Med* **3**: 1137–1140.

84. Bossard C, Van den Berghe L, Laurell H, Castano C, Cerutti M, Prats AC, Prats H (2004). Antiangiogenic properties of fibstatin, an extracellular FGF-2-binding polypeptide. *Cancer Res* **64**: 7507–7512.

85. Chang LK, Garcia-Cardena G, Farnebo F, Fannon M, Chen EJ, Butterfield C, Moses MA, Mulligan RC, Folkman J, Kaipainen A (2004). Dose-dependent response of FGF-2 for lymphangiogenesis. *Proc Natl Acad Sci USA* **101**: 11658–11663.

86. Shin JW, Min M, Larrieu-Lahargue F, Canron X, Kunstfeld R, Nguyen L, Henderson JE, Bikfalvi A, Detmar M, Hong YK (2006). Prox1 promotes lineage-specific expression of FGF receptor-3 in lymphatic endothelium: a role for FGF signaling in lymphangiogenesis. *Mol Biol Cell* **17**: 576–584.

87. Ortega S, Ittmann M, Tsang SH, Ehrlich M, Basilico C (1998). Neuronal defects and delayed wound healing in mice lacking fibroblast growth factor 2. *Proc Natl Acad Sci USA* **95**: 5672–5677.

88. Yanagisawa-Miwa A, Uchida Y, Nakamura F, Tomaru T, Kido H, Kamijo T, Sugimoto T, Kaji K, Utsuyama M, Kurashima C, *et al.* (1992). Salvage of infarcted myocardium by angiogenic action of basic fibroblast growth factor. *Science* **257**: 1401–1403.

89. Tabata H, Silver M, Isner JM (1997). Arterial gene transfer of acidic fibroblast growth factor for therapeutic angiogenesis *in vivo*: critical role of secretion signal in use of naked DNA. *Cardiovasc Res* **35**: 470–479.

90. Horvath KA, Doukas J, Lu CY, Belkind N, Greene R, Pierce GF, Fullerton DA (2002). Myocardial functional recovery after fibroblast growth factor 2 gene therapy as assessed by echocardiography and magnetic resonance imaging. *Ann Thorac Surg* **74**: 481–486; discussion 487.

91. Rinsch C, Quinodoz P, Pittet B, Alizadeh N, Baetens D, Montandon D, Aebischer P, Pepper MS (2001). Delivery of FGF-2 but not VEGF by encapsulated genetically engineered myoblasts improves survival and vascularization in a model of acute skin flap ischemia. *Gene Ther* **8**: 523–533.

92. Yamamoto M, Sakakibara Y, Nishimura K, Komeda M, Tabata Y (2003). Improved therapeutic efficacy in cardiomyocyte transplantation for myocardial

infarction with release system of basic fibroblast growth factor. *Artif Organs* **27**: 181–184.

93. Lazarous DF, Scheinowitz M, Shou M, Hodge E, Rajanayagam S, Hunsberger S, Robison WG, Jr, Stiber JA, Correa R, Epstein SE, *et al.* (1995). Effects of chronic systemic administration of basic fibroblast growth factor on collateral development in the canine heart. *Circulation* **91**: 145–153.

94. Lazarous DF, Shou M, Scheinowitz M, Hodge E, Thirumurti V, Kitsiou AN, Stiber JA, Lobo AD, Hunsberger S, Guetta E, Epstein SE, Unger EF (1996). Comparative effects of basic fibroblast growth factor and vascular endothelial growth factor on coronary collateral development and the arterial response to injury. *Circulation* **94**: 1074–1082.

95. Sato K, Laham RJ, Pearlman JD, Novicki D, Sellke FW, Simons M, Post MJ (2000). Efficacy of intracoronary versus intravenous FGF-2 in a pig model of chronic myocardial ischemia. *Ann Thorac Surg* **70**: 2113–2118.

96. Yang HT, Feng Y (2000). bFGF increases collateral blood flow in aged rats with femoral artery ligation. *Am J Physiol Heart Circ Physiol* **278**: H85–H93.

97. Yang HT, Yan Z, Abraham JA, Terjung RL (2001). VEGF (121)- and bFGF-induced increase in collateral blood flow requires normal nitric oxide production. *Am J Physiol Heart Circ Physiol* **280**: H1097–H1104.

98. Lazarous DF, Shou M, Stiber JA, Dadhania DM, Thirumurti V, Hodge E, Unger EF (1997). Pharmacodynamics of basic fibroblast growth factor: route of administration determines myocardial and systemic distribution. *Cardiovasc Res* **36**: 78–85.

99. Rajanayagam MA, Shou M, Thirumurti V, Lazarous DF, Quyyumi AA, Goncalves L, Stiber J, Epstein SE, Unger EF (2000). Intracoronary basic fibroblast growth factor enhances myocardial collateral perfusion in dogs. *J Am Coll Cardiol* **35**: 519–526.

100. Laham RJ, Rezaee M, Post M, Xu X, Sellke FW (2003). Intrapericardial administration of basic fibroblast growth factor: myocardial and tissue distribution and comparison with intracoronary and intravenous administration. *Catheter Cardiovasc Interv* **58**: 375–381.

101. Gao MH, Lai NC, Hammond HK (2005). Signal Peptide Increases the Efficacy of Angiogenic Gene Transfer for Treatment of Myocardial Ischemia. *Hum Gene Ther* **16**: 1058–1064.

102. Heilmann CA, Attmann T, Thiem A, Haffner E, Beyersdorf F, Lutter G (2003). Gene therapy in cardiac surgery: intramyocardial injection of naked plasmid DNA for chronic myocardial ischemia. *Eur J Cardiothorac Surg* **24**: 785–793.

103. Lazarous DF, Unger EF, Epstein SE, Stine A, Arevalo JL, Chew EY, Quyyumi AA (2000). Basic fibroblast growth factor in patients with intermittent claudication: results of a phase I trial. *J Am Coll Cardiol* **36**: 1239–1244.

104. Laham RJ, Chronos NA, Pike M, Leimbach ME, Udelson JE, Pearlman JD, Pettigrew RI, Whitehouse MJ, Yoshizawa C, Simons M (2000). Intracoronary basic fibroblast growth factor (FGF-2) in patients with severe ischemic heart disease: results of a phase I open-label dose escalation study. *J Am Coll Cardiol* **36**: 2132–2139.

105. Udelson JE, Dilsizian V, Laham RJ, Chronos N, Vansant J, Blais M, Galt JR, Pike M, Yoshizawa C, Simons M (2000). Therapeutic angiogenesis with recombinant fibroblast growth factor-2 improves stress and rest myocardial perfusion abnormalities in patients with severe symptomatic chronic coronary artery disease. *Circulation* **102**: 1605–1610.

106. Simons M, Annex BH, Laham RJ, Kleiman N, Henry T, Dauerman H, Udelson JE, Gervino EV, Pike M, Whitehouse MJ, Moon T, Chronos NA (2002). Pharmacological treatment of coronary artery disease with recombinant fibroblast growth factor-2: double-blind, randomized, controlled clinical trial. *Circulation* **105**: 788–793.

107. Pecher P, Schumacher BA (2000). Angiogenesis in ischemic human myocardium: clinical results after 3 years. *Ann Thorac Surg* **69**: 1414–1419.

108. Ruel M, Laham RJ, Parker JA, Post MJ, Ware JA, Simons M, Sellke FW (2002). Long-term effects of surgical angiogenic therapy with fibroblast growth factor 2 protein. *J Thorac Cardiovasc Surg* **124**: 28–34.

109. Lederman RJ, Mendelsohn FO, Anderson RD, Saucedo JF, Tenaglia AN, Hermiller JB, Hillegass WB, Rocha-Singh K, Moon TE, Whitehouse MJ, Annex BH (2002). Therapeutic angiogenesis with recombinant fibroblast growth factor-2 for intermittent claudication (the TRAFFIC study): a randomised trial. *Lancet* **359**: 2053–2058.

110. Sullivan CJ, Doetschman T, Hoying JB (2002). Targeted disruption of the Fgf2 gene does not affect vascular growth in the mouse ischemic hindlimb. *J Appl Physiol* **93**: 2009–2017.

111. Rissanen TT, Markkanen JE, Arve K, Rutanen J, Kettunen MI, Vajanto I, Jauhiainen S, Cashion L, Gruchala M, Narvanen O, Taipale P, Kauppinen RA, Rubanyi GM, Yla-Herttuala S (2003). Fibroblast growth factor 4 induces vascular permeability, angiogenesis and arteriogenesis in a rabbit hindlimb ischemia model. *FASEB J* **17**: 100–102.

112. Grines CL, Watkins MW, Helmer G, Penny W, Brinker J, Marmur JD, West A, Rade JJ, Marrott P, Hammond HK, Engler RL (2002). Angiogenic gene therapy (AGENT) trial in patients with stable angina pectoris. *Circulation* **105**: 1291–1297.

113. Bliznakov EG (2002). Therapeutic angiogenesis: hope or hype. *Circulation* **106**: e220–e221; author reply e220–e221.

114. Cao R, Brakenhielm E, Pawliuk R, Wariaro D, Post MJ, Wahlberg E, Leboulch P, Cao Y (2003). Angiogenic synergism, vascular stability and improvement of hind-limb ischemia by a combination of PDGF-BB and FGF-2. *Nat Med* **31**: 31.

4

Neuropeptide Y: Neurogenic Mediator of Angiogenesis and Arteriogenesis

by Joanna B. Kitlinska and Zofia Zukowska

1. The NPY System

Neuropeptide Y (NPY) is a 36-amino acid peptide originally isolated from porcine brains by the method detecting C-terminal amidation, characteristic of this and many other biologically active peptides.[1,2] Subsequently, NPY was found to be one of the most abundant peptides in the central nervous system. In the periphery, the peptide is present in the sympathetic nerves, where it co-localizes with norepinephrine, and in the adrenal medulla, in epinephrine-containing chromaffin cells.[3] During sympathetic activation, NPY is released into the bloodstream, often together with catecholamines. Thus, circulating levels of the peptide are increased by various types of stress, such as exposure to cold and strenuous exercise, particularly when combined with hypoxic conditions.[4,5] In addition, expression of the peptide has been detected in non-neuronal cells, such as megakaryocytes and platelets, and endothelial and immune cells.[3,6-8] Extraneuronal expression is not always constitutive and depends on animal species, ethnicity and patho-logical conditions. For example, rats and some murine strains contain high platelet-derived NPY.[6] On the other hand, platelet NPY is not detected in healthy humans, but has been found in chronically depressed patients.[9]

NPY is a pleiotropic factor (Fig. 1). In the brain, the peptide exerts a variety of functions, but is most known for its ability to stimulate food intake, inhibit anxiety and regulate energy balance and pituitary secretion. The best known peripheral actions of NPY include vaso-constriction, potentiation of other vasoconstrictors' actions, inhibition of neurotransmitter release and regulation of immune responses.[3] The peptide mediates stress-induced, slow-onset, prolonged vasoconstric-tion of small resistance vessels, and hence it has been suggested to be involved in coronary vasospasm and ischemic heart disease.[10] Recently, however, new growth-regulatory activities of NPY have emerged, which occur at concentrations lower than the vasoconstrictive ones, indicating that NPY's primary physiological role may not be as a vasoconstrictor but that of a neurogenic trophic factor. The peptide exerts proliferative effects in a variety of cells, including neuronal precursors, lymphocytes and tumor cells, such as neuroblastoma and prostate cancer cells.[11–16] On the other hand, NPY can also inhibit cell growth, as seen in Ewing's sarcoma and retinal glial cells.[13,17] However, some of the most potent

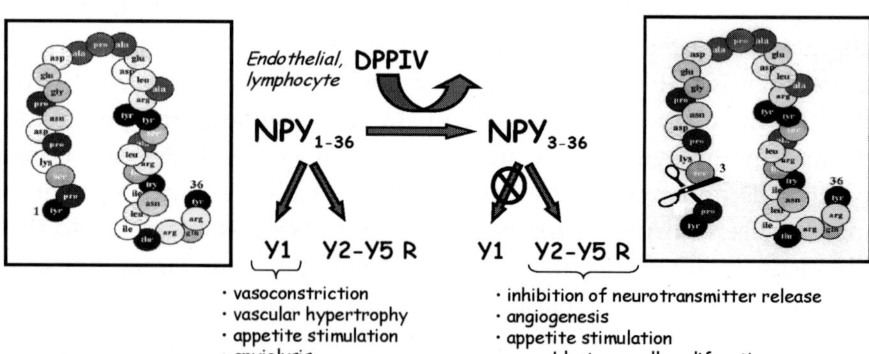

Fig. 1. Structure and functions of NPY peptides, enzymatic cleavage and receptor subtypes.

activities of NPY involve its vascular growth promoting and angiogenic actions, and this chapter will focus on these effects.

Diverse actions of NPY are mediated by five Gi/o-protein-coupled receptors designated Y1–Y5. The Y1 receptor is the predominant vascular receptor mediating vasoconstriction and the major brain receptor involved in anxiety and, together with Y5, in regulation of food intake The Y2 receptor, on the other hand, is the primary receptor responsible for the presynaptic neuro-inhibitory actions of NPY in the central and peripheral nervous system.[3,18] Receptor expression pattern is also an important factor determining effects of NPY on cell growth. For example, Y1 receptor mediates NPY-induced proliferation of neuronal precursors, while Y2 is the main receptor responsible for its mitogenic effect in neuroblastoma cells.[12,13] On the other hand, activation of both Y1 and Y5 receptors in Ewing's sarcoma cells leads to cell death.[13] Hence, growth-regulatory effects of NPY are cell- and receptor-specific.

In addition, actions of NPY are modified by a serine protease, dipeptidyl peptidase IV (DPPIV), which functions in the NPY system as a "converting enzyme" and a "receptor switch." The protease converts the full length NPY_{1-36} to a shorter form, NPY_{3-36}, which is no longer able to bind to the Y1 receptor but retains affinity for all other receptors.[19] Hence, DPPIV is an important regulatory molecule in the NPY system shifting the actions of the peptide from Y1- to non-Y1 receptor mediated (Fig. 1).

2. NPY as a Growth Factor for Vascular Cells

Zukowska's group was the first to discover that NPY, at sub-picomolar concentrations, exerts growth-promoting effects on both endothelial and vascular smooth muscle cells (VSMCs), and is potently angiogenic (Fig. 2).[8,20,21] The angiogenic properties of the peptide have been established using a variety of models, such as aortic sprouting, rodent hindlimb ischemia, retinopathy and wound healing.[20,22–25] The role of NPY in vascularization has been further supported by hypervascularization of skeletal muscles of NPY overexpressing rats[20] and reduced basal angiogenic activity in $NPY^{-}/_{-}$ mice, which exogenous peptide restores.

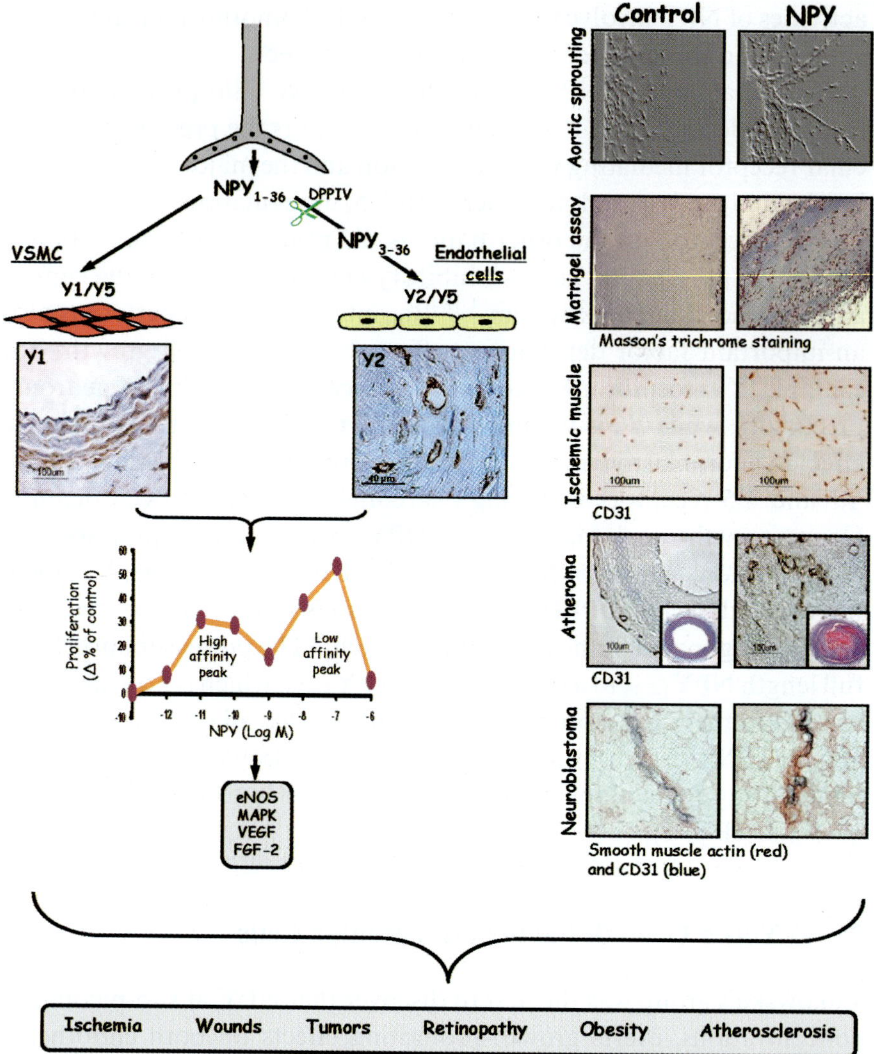

Fig. 2. NPY and its receptor-specific vascular growth-promoting and angiogenic actions: role in neovascularization in various pathological conditions.

Human and rodent endothelial cells constitutively express NPY and its Y1 receptors, whereas Y2 and Y5 receptors are not always expressed but are upregulated/induced during ischemia and after stimulation with NPY.[20,21] The simultaneous expression of the ligand and its receptors in the endothelium suggests possible autocrine effects of the peptide. Indeed, NPY has been found to be a potent growth factor for endothelial cells. The peptide stimulates not only their proliferation, but also other processes involved in angiogenesis, such as cell adhesion, migration and capillary tube formation.[21]

Despite the presence of Y1 receptors, NPY growth-promoting actions in endothelial cells seem to be mediated mainly by Y2, with some contribution of Y5 receptors. The Y2 receptors are abundantly expressed in newly formed microvessels in murine corneas implanted with NPY, as well as bFGF and VEGF pellets.[22] The Y2 is also a predominant NPY receptor in growing vasculature of neuroblastoma tumors (unpublished data) and in neointimal atherosclerotic-like lesions induced by NPY administered at the time of arterial balloon angioplasty.[26] The effect of NPY on endothelial cell proliferation and migration can be mimicked by Y2/Y5 receptor agonist, NPY$_{3-36}$.[21] Moreover, both Y2 and Y5 receptors are upregulated in ischemia, while age-dependent impairment of angiogenesis is associated with Y2 receptor downregulation.[20,27] The angiogenic activities of NPY, as measured by its effect on aortic sprouting, revascularization of ischemic tissues and vascularization of mouse corneas, are either completely abolished or significantly reduced in Y2$^{-}/_{-}$ mice.[22,28] Moreover, deletion of Y2 receptor in mice results in severe impairment of angiogenesis-related processes, such as oxygen-induced retinopathy and wound healing but also reduced vascularization of non-ischemic limbs.[22,24] Hence, Y2 seems to be the main angiogenic receptor in the NPY system, active during physiological developmental angiogenesis as well as in pathological conditions. The presence of the Y5 receptor, on the other hand, appears to provide an additional amplification of the angiogenic signal (see below: "mechanisms").

In addition to its growth-promoting effects on endothelial cells, NPY is also a growth factor for VSMCs. However, in contrast to Y2/Y5 receptor-dependent proliferation of endothelial cells, the mitogenic effect of the peptide in VSMCs is mediated primarily by Y1, and again, amplified by Y5 receptors.[8,29] Physiological consequences of NPY-induced VSMCs proliferation include its role in vascular remodeling and atherosclerosis.[26] This process is also important in neovascularization. Even though the first steps of angiogenesis involve mainly activation of endothelial cells and formation of capillaries, development of the mature blood vessels additionally requires co-option of VSMCs. Thus, by combining proliferative effects on both endothelial and VSMCs (Fig. 1), NPY is able to stimulate formation of functional, mature arteries. Such simultaneous activation of angio- and arteriogenesis upon NPY treatment has been observed, e.g. in rat hindlimb ischemia model and in neuroblastoma xenografts (see below).[13,20]

Although mitogenic actions of NPY in endothelial and VSMCs are mediated by different receptors, in both types of cells the peptide exerts its proliferative effects at sub-picomolar and sub-nanomolar concentrations, significantly lower than those necessary for vasoconstriction.[8,21,27,29] Moreover, in both endothelial and VSMCs, NPY stimulates proliferation in a bi-modal fashion, with two peaks of activity — a high affinity peak at picomolar concentrations and a low affinity peak at nanomolar concentrations of the peptide.[8,21,27,29] Interestingly, this bi-modality seems to be universal for a variety of NPY-induced processes, such as neuroblastoma cell proliferation, endothelial cell migration and differentiation of adipocytes.[21,27] The mechanism of this phenomenon has not been elucidated yet. However, some experimental data indicate that it may involve interaction between multiple NPY receptors, since the bi-modal effect occurs only in cells expressing more than one receptor subtype. In endothelial cells expressing only Y2 receptors, as well as in CHO-K1 cells stably transfected with single Y1 or Y2 receptors, NPY stimulates proliferation only at high concentrations, corresponding to the low affinity peak.[30] Interestingly, the high affinity peak occurs at concentrations below Kd values for any known single NPY receptors. Thus, this first peak of activity seems to be dependent on the presence of multiple NPY receptors and their interactions leading to amplification of NPY's growth-promoting actions.

Such effects have previously been reported to occur as a result of receptor oligomerization, described for other G protein-coupled receptors, and recently also found to be the case for NPY receptors.[31,32] NPY receptor homodimerization has been shown for Y1, Y2 and Y4 receptors, whereas heterodimerization so far only for Y1 and Y5 receptors.[33–35] These interactions are believed to change receptor affinity, trafficking and signaling pathways. Further studies are necessary to determine the role of NPY receptor oligomerization in its growth-promoting activity and its possible physiological, clinical and therapeutic implications. However, the ability of NPY to exert potent angiogenic activity at sub-picomolar concentrations makes the peptide an attractive molecule for either angiogenic or anti-angiogenic therapies

3. DPPIV: A Molecular Switch of the NPY Angiogenic System

An important feature of NPY's potent growth promoting activity in VSM and endothelial cells is that they are mediated by different receptors (Fig. 2). The mitogenic effect of the peptide in endothelial cells is mediated by Y2/Y5 receptors, while Y1/Y5 receptors are involved in VSMC proliferation.[8,20,21,28,29] Additionally, Y1 receptors present on VSMCs are also responsible for NPY's vasoconstrictive effects. These divergent receptor-specific actions of NPY in vascular cells implicate a "converting enzyme," DPPIV, as an important receptor switch and a regulator of angiogenesis. This membrane protease is abundantly expressed in endothelium and often considered a marker of microvascular endothelial cells.[23] As mentioned before, the enzyme cleaves the full length NPY_{1-36} to NPY_{3-36}, a fragment which is not able to bind to the Y1 receptors but is a specific Y2/Y5 agonist.[19] In this way, the protease acts as an endogenous antagonist of Y1 receptor-mediated vasoconstriction and VSMC proliferation and an enhancer/amplifier of the Y2/Y5 receptor-mediated growth-promoting effects of NPY in endothelial cells. The protease also controls the balance between peptide's effects is VSMC and endothelial cells. Such a balance is an important factor regulating NPY actions at different stages of vascularization. The initial steps of vascularization involve mainly changes in endothelium and require Y2 receptor-dependent activity of the peptide. However, maturation of the vessels and formation of collateral vessels involves also Y1 receptor-mediated VSMC proliferation. Hence, changes in DPPIV proteolytic activity may be an important mechanism controlling NPY-induced angio- and arteriogenesis.

A crucial role of DPPIV in NPY-driven angiogenesis has been confirmed using a variety of models. As described above, DPPIV is co-expressed with NPY and its receptors in endothelial cells and, along with them, upregulated in tissues undergoing active angiogenesis, e.g. during ischemia and wound healing (see below).[20,23] On the other hand, age-dependent impairment of spontaneous and NPY-driven angiogenesis is associated with downregulation of the Y2 as well as DPPIV expression in old animals.[27]

Taken together, DPPIV is an integral element of NPY's angiogenic system, augmenting its growth-promoting effects in endothelial cells. The protease also regulates the balance between Y1 receptor-mediated effects of NPY on VSMCs and its Y2 receptor-dependent actions in endothelial cells.

4. Downstream Mediators of NPY Actions

NPY is not only a potent but also a multifunctional peptide and involved in several steps of angiogenesis: endothelial cell adhesion, migration, proliferation and differentiation, as well as growth of VSMCs.[8,21,29] NPY treatment leads to the formation of what appears to be fully functional vessels (Fig. 2), unlike those formed by vascular endothelial growth factor (VEGF), which are often poorly matured and leaky.[13,21,22] This phenomenon may be associated with the fact that NPY is positioned upstream from other angiogenic factors, and acts not only directly, but also indirectly, via stimulation of various angiogenic pathways. Several lines of evidence support this notion. For example, NPY-induced aortic sprouting is completely abolished in mice deficient in endothelial nitric oxide synthase (eNOS), indicating that nitric oxide is a critical mediator of the NPY angiogenic effect.[20] Blockade of VEGF also inhibits NPY-driven formation of capillaries, however, the inhibition is incomplete and does not affect peptide-induced endothelial cell migration.[20] Hence, VEGF is involved in some, but not all, steps of the NPY angiogenic activity. Recent unpublished data from our laboratory indicate that NPY-induced angiogenesis may be also partially mediated by fibroblast growth factor-2 (FGF-2). NPY upregulates expression of FGF-2, its receptors, as well as VEGF receptor-2 in endothelial cells. Consequently, NPY-induced endothelial cell proliferation is blocked by both anti-FGF-2 neutralizing antibody and the VEGF antagonist, VEGF R2/Fc chimera.

In contrast, NPY receptor antagonists do not reduce FGF-2- and VEGF-stimulated endothelial cell proliferation. Moreover, both bFGF and VEGF induce full sprouting response in $Y2^-/\_$ aortic rings, where NPY's angiogenesis is reduced. Therefore, NPY is an upstream angiogenic factor acting, at least partially, via bFGF and VEGF pathways. Interactions of NPY with other angiogenic factors also cannot be excluded. Such activation of multiple downstream pathways by NPY leads to a complete, tightly controlled process of angiogenesis and to formation of mature, functional blood vessels. As a result, in mouse corneas, for example, NPY induces formation of microvessels that are well separated and organized, as opposed to the leaky and hemorrhagic vasculature stimulated by VEGF.[22] Similar effects can also be observed in aortic sprouts.[21]

5. NPY in Revascularization of Ischemic Tissues

Ischemia is one of the most important stimuli activating angiogenic processes. A number of angiogenic factors, such as VEGF, are known to be directly upregulated by hypoxia-inducible factor (HIF-1α).[36] However, hypoxia is also a powerful activator of perivascular nerves. All blood vessels, except the thoracic aorta, are innervated by sensory and sympathetic nerves, which alert the body of vascular injury or tissue ischemia. In consequence, sympathetic activity is augmented in ischemic conditions, which is associated with increased release of norepinephrine and its co-transmitters — NPY and purines.[5,37–39] For years, sympathetic activity was believed to be deleterious for ischemic tissues by impairing tissue blood flow, due to their vasoconstrictive effects. However, discovery of angiogenic and vascular growth-promoting activity of NPY has put this notion in doubt, as new trophic effects of sympathetic nerves have emerged.

The role of NPY in revascularization of ischemic tissues was established using the rodent hindlimb model, where femoral artery occlusion results in calf muscle ischemia and capillary angiogenesis (Fig. 2). At the same time, in the area of vessel occlusion, in the thigh, shear stress above the occluded artery activates arteriogenesis and formation of collaterals.[20] Both processes are associated with upregulation of the NPY system. Venous plasma levels of the peptide are elevated in ischemic leg, as compared to the contralateral non-ischemic limb.[20] This increase is reduced by ipsilateral lumbar sympathectomy, indicating it is, at least in part, associated with hyperactivity of sympathetic nerves, while the remaining, sympathectomy-resistant NPY release may be due to non-neuronal sources, such as platelets, immune or endothelial cells.[20]

Hindlimb ischemia is also associated with changes in NPY receptor expression. In the gastrocnemius muscle, which is located below the occlusion and represents the "capillary angiogenesis zone," ischemia upregulates expression of Y2 receptors and DPPIV, which converts NPY to the Y2/Y5 agonist.[20] In contrast, in the adductor muscle, above the occlusion, Y2 receptor mRNA levels are only slightly increased, whereas the expression of Y1 receptors is markedly upregulated.[40] Such differential regulation of NPY receptor expression may reflect divergent functions of the peptide in the particular areas of the injured limb. In the muscles below the occlusion, ischemia upregulates NPY angiogenic system, Y2 and DPPIV, mediating growth-promoting effects of the peptide in endothelial cells and leading to capillary angiogenesis. On the other hand, in the areas above the occlusion, which correspond to the "collateralization zone," Y1 receptors responsible for NPY mitogenic effect in VSMCs are upregulated, supporting the role of these receptors in arteriogenesis.

Despite activation of various angiogenic mechanisms, spontaneous revascularization does not fully restore blood flow in the ischemic muscle. This effect, however, can be achieved by treatment with exogenous NPY. The peptide, administered to the ischemic leg as a slow release pellet induces additional expression of Y5 receptors, which are the amplifiers of angiogenesis, and this leads to stimulation of capillary formation in the gastrocnemius muscle.[20] Such amplification may be due to dimerization of the Y2 and Y5 receptors. In addition to its angiogenic effect, the exogenous NPY also stimulates formation of collateral vessels above the occlusion, which, along with its angiogenic effect, leads to the complete restoration of the blood flow and improvement of contractile functions of the ischemic muscle.[20]

Taken together, due to its growth-promoting effects on both endothelial and VSMCs NPY stimulates both capillary angiogenesis and arteriogenesis. Moreover, as described above, the peptide also activates other angiogenic pathways, such as VEGF and FGF-2. Thus, by combining its direct effects on vascular cells and actions of other NPY-induced growth factors, the peptide initiates a complex, multi-step cascade of events leading to revascularization of the ischemic tissues and restoration of its functions.

6. NPY in Wound Healing

Wound repair is another example of angiogenesis which occurs throughout the adult life. It is a complicated process involving inflammation, changes in extracellular matrix, proliferation and migration of cells, as well as angiogenesis. After injury, formation of new blood vessels is necessary to restore oxygen and nutrient supply to the healing tissue.[41] Thus, wound healing is associated with induction of various angiogenic pathways, including NPY.

It has been shown that after wounding of confluent human umbilical vein-derived endothelial cells, NPY stimulates their migration and wound closure. This effect is mimicked by Y2/Y5 agonist, NPY_{3-36}, and accompanied by upregulation of DPPIV expression in migrating cells and its localization to the edge of wound.[23] Additionally, DPPIV neutralizing antibody completely blocks endothelial cell proliferation induced by NPY_{1-36}, whereas it has no effect on the same actions of NPY_{3-36}.[23] These findings once again support the crucial role of Y2 receptor and DPPIV in NPY-induced angiogenesis. The fact that DPPIV is necessary for NPY-mediated chemotaxis may be associated with high constitutive expression of Y1 receptors observed in endothelial cells.[21] Since full length NPY can bind both Y1 and Y2 receptors, and the Y1 receptors are normally more abundant, the selectivity for the Y2 activation is provided by DPPIV-mediated cleavage to NPY_{3-36}, which has no affinity for the Y1 and is selective for Y2/Y5 receptors Thus, DPPIV acts here as an "angiogenic switch."

The role of NPY in physiological wound healing has been further confirmed *in vivo*. In mice, NPY accelerates the healing of excisional full skin wounds and this is associated with increased vascularization of the regenerating tissue. Consistently, wound repair is significantly delayed in $Y^-/\_$ mice, which is accompanied by decreased vessel density of the regenerating skin.[22] Exogenous NPY does not improve the healing process in $Y2^-/\_$ mice, whereas other growth factors, such as FGF-2, maintain their wound healing abilities[22] supporting the notion that NPY is upstream from them. Interactions with the immune cells are also possible but not supported by the initial data which indicate no differences in the number of immune cells in the wounds of $Y2^-/\_$ and wild-type mice.[22]

7. NPY in Adipose Tissue Growth and Obesity

One tissue which undergoes a continued growth and remodeling during an adult life is the white adipose tissue (WAT). Since the seminal observation by Rupnick *et al.*,[42] who reported that anti-angiogenic agents lead to reduction of adipose tissue mass and weight loss by blocking vessel development and inducing their apoptosis, the interplay of adipogenesis with angiogenesis has become a hot topic. WAT is also well innervated by autonomic nerves, primarily sympathetic, which, in addition to their vasoconstrictive properties, have long been known to influence lipolysis and cell proliferation.[43,44] These effects were primarily ascribed to catecholamines and their β-adrenergic receptors on preadipocytes and adipocytes, whose activation leads to lipolysis and reduced WAT mass.[43,44] Hence, sympathetic activity is considered the body's major weight loss mechanism.

The role of NPY in this process was unknown until Turtzo *et al.*[45,46] reported that this sympathetic co-transmitter actually opposes β-adrenergic activity in WAT and leads to lipogenesis. They showed that co-culture of 3T3-L1 preadipocytes with rat sympathetic neurons leads to upregulation of NPY in the neuronal cells, and released peptide almost completely prevents isoproterenol-induced lipolysis. Our recent studies extended these observations to NPY-mediated control of adipogenesis. In addition to direct stimulation of preadipocyte proliferation, NPY is potently angiogenic in normal and obese mice. As in other tissues where NPY stimulated angiogenesis, this effect is mediated primarily by Y2 receptors in WAT. Inhibition by either local administration of Y2 antagonist or by receptor knockout reduces adipose tissue mass and vascularization in obese mice, and to a lesser degree, also in lean mice. This suggests that NPY is physiologically active during adipose tissue development and is an important nerve-derived regulator of tissue remodeling, in large part due to its angiogenic activity. While NPY has long been known for its pro-obesity actions, those actions were solely attributed to its centrally mediated stimulation of food intake and thermogenesis. Our study is the first to demonstrate that sympathetic nerves, via NPY, have powerful local effects on tissue growth, via angiogenesis and adipogenesis. Blockade of NPY-Y2 receptors directly in the adipose tissue presents a new avenue for treatment of obesity and site-specific reduction of adiposity.

8. NPY in Retinopathy

Angiogenesis is also a basis of several pathological conditions. One of them is retinopathy, a common and severe complication of diabetes mellitus and one of the leading causes of blindness. It is also a major complication of oxygen therapy of premature babies. In both forms, retinopathy occurs as a result of pathological neovascularization of the retina with abnormal leaky vessels, and is triggered by tissue ischemia.[47] In retinopathy of prematurity, relative retinal hyperoxia caused by exposure of the preterm newborns to hyperbaric oxygen results in excessive vasoconstriction and subsequent decrease of retinal blood flow.[25] On the other hand, in diabetic retinopathy damage of the vessels and ischemia are believed to be triggered by hyperglycemia.[48,49] However, in both cases the initial vascular insult is of the ischemic nature and, consequently, leads to upregulation of angiogenic factors and induction of retinal neovascularization. Although, like in tumor angiogenesis, VEGF has been considered the main molecule implicated in this process, involvement of other factors has not been excluded.

NPY is also present in fetal and adult retinas, and its angiogenic activity implicated in the pathogenesis if retinopathy.[50,51] In the mouse model of oxygen-induced retinopathy, both NPY and its Y2 receptors are significantly upregulated in the retinas of newborn mice subjected to hyperoxia In contrast, the retinal Y1 receptors, which are constitutively expressed, are unchanged in spite of increased oxygen level.[25] Oxygen-induced retinopathy is also significantly reduced in $Y2^-/_-$ mice, as well as in newborn rats treated systemically with Y2 receptor antisense (where scrambled or sense oligonucleotides are ineffective).[24] The above data strongly suggest that NPY, acting via its Y2 receptors, is an important factor involved in pathological vascularization of the retina. Interestingly, in the mouse model of oxygen-induced retinopathy, NPY mRNA levels are elevated starting with the initial vasoconstrictive phase-induced oxygen exposure[25] suggesting that vasoconstrictive activities of the peptide may also play a role, in spite of unchanged Y1 receptor expression. Thus, receptor-specific activities of NPY may contribute to retinopathy at both stages of its development.

The role of NPY in retinopathy is further supported by the clinical data. A functional polymorphism in NPY signal peptide, Leu[7] Pro[7], has been positively correlated with the occurrence of retinopathy in patients with type II diabetes, with 20% of patients who developed severe retinopathy having that mutation as compared to its frequency of 6–8% in the healthy population.[24] This Leucine to Proline substitution appears to result in increased release of NPY in response to stress, and has been already linked to impaired metabolism and elevated levels of blood lipids, decreased levels of insulin and accelerated atherosclerosis.[24,52–55] The exact mechanism by which this polymorphism may contribute to diabetic retinopathy has not been elucidated. However, the enhanced angiogenic activity of NPY associated with its increased secretion may be at least partially responsible for this phenomenon. Interestingly, no correlation between NPY polymorphism and retinopathy has been found in patients with type I diabetes.[24] Thus, although in both types of diabetes retinopathy is believed to be triggered by hyperglycemia, the mechanisms underlying this disorder and, consequently, role of NPY in its development, are different.

Taken together, NPY angiogenic activity mediated by its Y2 receptors plays an important role in development of certain types of retinopathy. Hence, blocking NPY-induced angiogenesis may be an effective therapy in treatment of this disease.

9. NPY-Induced Angiogenesis in Angioplasty-Induced Neointima and Atherogenesis

Development of atherosclerotic plaques does not require activation of angiogenesis but when it occurs, it can dramatically influence the outcome of these vascular lesions, leading to their instability, thrombogenicity and accelerated progression. Paradoxically, the two processes, atherogenesis and neovascularization not only co-exist in the same tissue but are often triggered by the same growth factors, for example, VEGF and bFGF, suggesting that they are the two sides of the same phenomenon, called by some, as the "Janus phenomenon."[56] The situation appears to be somewhat different with NPY.

NPY is a vascular mitogen and as such, potently stimulates neointima formation by activating a set of receptors which are of a different type than those involved in angiogenesis, i.e. Y1/Y5 receptors.[29] Interestingly, however, this neointima, unlike that induced by the balloon angioplasty alone, is also vascularized (Fig. 2). Zukowska's group has recently demonstrated that in rodents undergoing balloon angioplasty, NPY either administered exogenously[26] or released endogenously by chronic cold stress,[57] leads to development of vascularized lesions which also contain macrophages, lipids, thrombus and matrix, components of advanced atherosclerotic plaques. While Y2 receptors appear to be upregulated in these new vessels, the NPY-induced lesions are completely prevented by Y1 or Y5 receptor antagonists,[26,57] suggesting that these receptors are up-stream from the angiogenic ones. Hence, these two activities of the peptide, angiogenic and atherogenic, appear to be mediated by a different set of NPY receptors, unlike in the case of other angiogenic factors, whose pro-angiogenic revascularizing and pro-atherogenic effects seem to be mediated by similar pathways. Such a differential activity of Y1 and Y2 receptors offers an interesting possibility of reducing atherogenesis without impairing revascularization therapy required for treatment of ischemic vascular diseases.

10. NPY in Tumor Angiogenesis

Angiogenesis is also crucial for development of solid tumors where it determines their growth and metastases as blood vessels are necessary to provide oxygen and nutrients to rapidly growing tumor tissues. Thus, only those tumors, which are able to induce formation of new blood vessels by releasing angiogenic stimulators and downregulating angiogenic inhibitors, can advance to a phase of exponential growth. This phenomenon is often termed the "angiogenic switch."[58–60] The most ubiquitous angiogenic factor involved in tumor vascularization is VEGF, which is upregulated in most of the known solid tumors.[58–60] Our recent studies on NPY have revealed that this peptide also promotes tumor angiogenesis and growth.

As a neuronal peptide, NPY is often expressed in neural crest-derived tumors. The peptide is particularly abundant in tumors originating in the autonomic nervous system, such as sympathetic neuroblastomas and pheochromocytomas, as well as parasympathetic Ewing's sarcoma family of tumors (ESFT).[13,61–66] High expression of NPY is often associated with its increased release from the tumors leading to elevated plasma levels in patients with neuroblastoma and pheochromocytoma.[61,62,64–69] Neuroblastoma cell lines release high amounts of NPY to the culture media[13] whereas the peptide is not detectable in conditioned media obtained from rat pheochromocytoma cells, PC12, despite its high mRNA expression,[13] unless additional stimulants are present.[70] In ESFTs, extracellular release of NPY into the culture media is significantly lower than that of neuroblastomas, in spite of the intracellular NPY being comparable.[13] Consequently, no elevated NPY plasma levels in ESFT patients have been reported. Therefore, release of NPY from the tumor cells is an active and tightly regulated process and likely to be an important factor determining functions of the peptide in these tumors.[71,72]

Since NPY is potently angiogenic, its release from neural crest-derived tumors suggests a potential role for the peptide in regulation of their growth. Indeed, endothelial cell proliferation stimulated by neuroblastoma-conditioned media is *completely* blocked by NPY receptor antagonists, suggesting that the peptide is *crucial* for vascularization of these tumors.[13] In ESFT cells, which release less NPY than neuroblastomas, NPY receptor antagonists only partially reduce angiogenic activity of the conditioned media.[13] Thus, in ESFTs, NPY is important, but not the only factor responsible for their vascularization.

The critical role of NPY in tumor angiogenesis is further confirmed by its stimulatory effect on vascularization of neuroblastoma and ESFT xenografts. In both types of tumors, the exogenous peptide significantly increases vessel density within tumor tissues.[13] Interestingly, this effect is accompanied by an increased content of pericytes/vascular smooth muscle cells in the vasculature of NPY-treated tumors, indicating its more mature character.[13] This effect may be due to the mitogenic activity of the peptide on vascular smooth muscle cells. Thus, as seen in other models, NPY not only stimulates formation of the tumor blood vessels, but also facilitates their maturation, which can improve delivery of oxygen and nutrients to the tumor tissue, and further promote tumor growth.

NPY promotes growth not only of vascular smooth muscle and endothelial cells, but also of a variety of other cells, including neuronal precursors.[8,12,21,29] Thus, its angiogenesis-dependent stimulation of tumor growth is additionally modified by direct autocrine effects, which, depending on the type of NPY receptors the tumor cells express, may lead to their proliferation or apoptosis. Neuroblastomas, which are derived from sympathetic neurons, express Y2 receptors and, in some cell lines, also Y5 receptors,[13,73,74] and their activation leads to tumor cell proliferation.[13] Y2 and Y5 receptor antagonists significantly decrease basal proliferation of neuroblastomas, which further confirms the autocrine growth-promoting effect of endogenous NPY.[13] Thus, NPY stimulates growth of neuroblastomas directly, by activation of tumor cell proliferation and indirectly, by its angiogenic effect.

The role of NPY as a critical neuroblastoma growth factor is further supported by clinical data. Elevated plasma levels of NPY have been found mainly in patients with advanced disease, in association with poor clinical outcome.[62,67,69] Moreover, expression of Y2 receptors in human neuroblastoma tissues has been detected in both tumor[74] and endothelial cells (unpublished data). Since both NPY-induced neuroblastoma cell proliferation and angiogenesis are Y2 receptor-dependent, blocking Y2-NPY pathway may be an effective, bidirectional therapy for these tumors. Such therapies combining angiostatics with low doses of chemotherapeutic agents have been already tested and proven successful in several models, including neuroblastomas.[75–80]

Unlike Y2 receptor-bearing neuroblastomas, ESFT cells express Y1 and Y5 receptors, which, interestingly, is associated with an opposite effect of NPY on their growth.[13,81] In contrast to its proliferative actions in neuroblastomas, in ESFT cells, endogenous NPY stimulates apoptosis and this requires activation of both Y1 and Y5 receptors.[13,82] Therefore, in ESFT the peptide exerts two opposite effects — inhibitory, via Y1/Y5-dependent apoptosis and stimulatory, via Y2 receptor-mediated angiogenesis. The overall effect of NPY on ESFT growth, therefore, depends on balance between these two processes. The activity of the peptide is additionally modified by DPPIV, which is expressed along with NPY and its receptors in ESFT cells and is abundant in tumor vasculature.[13] By converting NPY to the non-Y1 receptor agonist, the protease prevents Y1/Y5 receptor-mediated apoptosis of ESFT cells, and shifts activity of the peptide toward Y2 receptor-mediated angiogenesis. Blocking the DPPIV activity with specific inhibitors enhances NPY-induced apoptosis of the ESFT cells.[13] Hence, DPPIV acts as a "survival factor" for ESFT by augmenting angiogenesis-dependent, growth-promoting effects of NPY. These activities make the protease a potential target for therapy of these tumors.

Direct effects of NPY on the growth of pheochromocytomas has not been established yet. Clinically, elevated plasma levels of the peptide have been associated with the malignant phenotype of the disease.[63] Hence, even if NPY does not exert direct proliferative effects on pheochromocytoma cells, the peptide may facilitate tumor growth and spread via its angiogenic activity. This notion is supported by the fact that nerve growth factor (NGF), which is known to upregulate NPY in PC12 cells, stimulates vascularization of PC12 xenografts in a VEGF-dependent manner.[83] Since NPY-induced angiogenesis is also partially driven by VEGF,[20] NPY may be a mediator of NGF's angiogenic actions in pheochromocytomas. However, further studies would be needed to fully elucidate peptide's functions in these tumors.

The NPY effect on tumor vascularization and growth seems to be most relevant to neural crest-derived tumors, which express their own NPY. However, all tissues of the body have blood vessels, and all of them, except the aorta, are, to a greater or lesser extent, innervated by the sympathetic nerves. Hence, the role of nerve-derived peptide cannot be excluded in tumors of any origin. Furthermore, NPY release from the sympathetic nerves is increased in conditions of tissue ischemia, which occurs during tumor growth, as well as in several systemic diseases such as hypertension or renal failure and during stress.[4,5,84,85] Although the role of stress in tumor growth is still very controversial, there are animal and human studies linking it with cancer development and progression. For example, stressful life events have been associated with increased risk of breast and colon cancer.[86–90] The potential effect of stress on tumor growth is usually explained by changes in the immune system.[91] However, given NPY's powerful angiogenic and growth-promoting activities and its release during stress, a potential role of NPY as a stress-induced neurogenic regulator of tumor growth demands investigation.

Taken together, NPY is an important factor in growth and development of tumors, particularly, but not solely, those of neural crest origin. The angiogenesis-dependent growth-stimulatory activities of NPY are modified by its tumor- and receptor-specific direct effects on tumor cell proliferation. These bidirectional actions make NPY, its receptors and converting enzyme, DPPIV, attractive new targets in tumor therapies directed against both tumor cell proliferation and vascularization.

11. NPY-Mediated Angiogenesis and Neurogenesis

Angiogenesis and arteriogenesis are essential not only for tissue repair in ischemic and degenerative diseases but also during organ development. While the last two decades of angiogenesis research have identified many factors, which *in vitro* and/or *in vivo* are able to stimulate growth of new capillaries, stimulation of the formation of new fully matured arteries has remained a challenge. One of the reasons for our poor understanding of mechanisms of arteriogenesis may be attributed to the fact that most of our knowledge came out of tumor biology. Tumor angiogenesis, however, may not represent physiological but rather a thwarted process of vessel formation. During organogenesis and normal tissue growth, blood vessels develop in a tissue-specific well-organized pattern, and lay alongside of peripheral autonomic and sensory nerves.[92,93] This association, however, had long been neglected until recently when molecular signals governing angiogenesis and neurogenesis have begun to emerge. Growth factors known to be angiogenic and released by innervated tissues, such as VEGF or bFGF, were also found to possess neurogenic activity.[92] For example, VEGF, derived from the Schwann cells, was shown to signal sprouting of vessels alongside of nerves[93] while neurotrophins secreted by vascular smooth muscle and endothelium are both angiogenic and neurogenic.[94] Coordination of vessel and nerve guidance is provided by endothelial and neuronal expression of specific attracting or repelling signaling molecules such as semaphorins and their receptors, neuropilins,[95] or Eph/ephrin family of proteins.[96]

NPY appears to be one of the signals which are used during sprouting of both vascular as well as neuronal cells, particularly from arteries and sympathetic nerves. Recent studies have shown that the peptide, in addition to being angiogenic and arteriogenic,[97] also stimulates neurogenesis. Through a series of elegant studies, Gray and Scharfman have demonstrated that the peptide increases proliferation and sprouting of neuronal stem cells in the dentate gyrus of the adult hippocampus[98] and implicated NPY in recovery from epileptic seizures, depression after electroconvulsive shock therapy and some cognitive disorders.[99] The neurogenic activity of NPY, however, appears to be mediated by Y1 receptors expressed by neuronal progenitor cells,[12] the type of receptors different from those which induce angiogenesis (Y2). Thus, NPY's angiogenic and neurogenic activities can be differentially regulated by expression of specific receptors on respective cells. In addition, endothelial expression of DPPIV, an enzyme which forms Y2/Y5 agonist and which is not expressed by neuronal cells, may serve as a "stop-and-go" signal for neurogenesis and angiogenesis, respectively.

Whether or not NPY-mediated angiogenesis and neurogenesis, are temporally related and how, has not been determined yet. However, one can speculate that they are interdependent, and one process can facilitate the other. For example, increased neuronal sprouting may be started by Y1 receptor activation but subsequently amplified and guided by formation of vessels, arteries in particular, which are themselves stimulated by NPY and endothelial activation of DPPIV and Y2 receptors. How these processes are executed remains to be determined, but even at the present time, one thing is certain, that proper neurogenesis is required for proper, physiological arteriogenesis, and *vice versa*.

Thus, the challenge of inducing formation of a new fully functional circulation for ischemic tissues will depend on our better understanding of how these two processes, neurogenesis and arteriogenesis, can be reproduced and coordinated.

References

1. Tatemoto K (1982). Neuropeptide Y: complete amino acid sequence of the brain peptide. *Proc Natl Acad Sci USA* **79**: 5485–5489.
2. Tatemoto K, Carlquist M, Mutt V (1982). Neuropeptide Y–a novel brain peptide with structural similarities to peptide YY and pancreatic polypeptide. *Nature* **296**: 659–660.
3. Colmers WF, Wahlestedt C (1993). *The Biology of Neuropeptide and Related Peptides* (Humana Press, Totowa, N.J.), p. xvi, 564.
4. Pernow J, Ohlen A, Hokfelt T, Nilsson O, Lundberg JM (1987). Neuropeptide Y: presence in perivascular noradrenergic neurons and vasoconstrictor effects on skeletal muscle blood vessels in experimental animals and man. *Regul Pept* **19**: 313–324.
5. Zukowska-Grojec Z (1993). Origin and actions of neuropeptide Y in the cardio-vascular system. In: Wahlestedt C (ed.), *The Biology of Neuropeptide Y and Related Peptides* (Humana Press, Totowa, N.J.), pp. 315–388.
6. Myers AK, Farhat MY, Vaz CA, Keiser HR, Zukowska-Grojec Z (1988). Release of immunoreactive-neuropeptide by rat platelets. *Biochem Biophys Res Commun* **155**: 118–122.
7. Schwarz H, Villiger PM, von Kempis J, Lotz M (1994). Neuropeptide Y is an inducible gene in the human immune system. *J Neuroimmunol* **51**: 53–61.
8. Zukowska-Grojec Z, Karwatowska-Prokopczuk E, Fisher TA, Ji H (1998). Mechanisms of vascular growth-promoting effects of neuropeptide Y: role of its inducible receptors. *Regul Pept* **75–76**: 231–238.
9. Nilsson C, Karlsson G, Blennow K, Heilig M, Ekman R (1996). Differences in the neuropeptide Y-like immunoreactivity of the plasma and platelets of human volunteers and depressed patients. *Peptides* **17**: 359–362.
10. Ullman B, Hulting J, Lundberg JM (1994). Prognostic value of plasma neuropeptide-Y in coronary care unit patients with and without acute myocardial infarction. *Eur Heart J* **15**: 454–461.
11. Elitsur Y, Luk GD, Colberg M, Gesell MS, Dosescu J, Moshier JA (1994). Neuropeptide Y (NPY) enhances proliferation of human colonic lamina propria lymphocytes. *Neuropeptides* **26**: 289–295.
12. Hansel DE, Eipper BA, Ronnett GV (2001). Neuropeptide Y functions as a neuroproliferative factor. *Nature* **410**: 940–944.
13. Kitlinska J, Abe K, Kuo L, Pons J, Yu M, Li L, Tilan J, Everhart L, Lee EW, Zukowska Z, Toretsky JA (2005) Differential effects of neuropeptide Y on the growth and vascularization of neural crest-derived tumors. *Cancer Res* **65**: 1719–1728.

14. Magni P, Motta M (2001). Expression of neuropeptide Y receptors in human prostate cancer cells. *Ann Oncol* (Suppl 12) **2**: S27–29.

15. Medina S, Del Rio M, Hernanz A, De la Fuente M (2000). Age-related changes in the neuropeptide Y effects on murine lymphoproliferation and interleukin-2 production. *Peptides* **21**: 1403–1409.

16. Medina S, Rio MD, Cuadra BD, Guayerbas N, Fuente MD (1999). Age-related changes in the modulatory action of gastrin-releasing peptide, neuropeptide Y and sulfated cholecystokinin octapeptide in the proliferation of murine lymphocytes. *Neuropeptides* **33**: 173–179.

17. Milenkovic I, Weick M, Wiedemann P, Reichenbach A, Bringmann A (2004). Neuropeptide Y-evoked proliferation of retinal glial (Muller) cells. *Graefes Arch Clin Exp Ophthalmol* **242**: 944–950.

18. Grundemar SR (1997). *Neuropeptide Y and Drug Development* (Academic Press, San Diego, London).

19. Mentlein R (1999). Dipeptidyl-peptidase IV (CD26) — role in the inactivation of regulatory peptides. *Regul Pept* **85**: 9–24.

20. Lee EW, Michalkiewicz M, Kitlinska J, Kalezic I, Switalska H, Yoo P, Sangkharat A, Ji H, Li L, Michalkiewicz T, Ljubisavljevic M, Johansson H, Grant DS, Zukowska Z (2003). Neuropeptide Y induces ischemic angiogenesis and restores function of ischemic skeletal muscles. *J Clin Invest* **111**: 1853–1862.

21. Zukowska-Grojec Z, Karwatowska-Prokopczuk E, Rose W, Rone J, Movafagh S, Ji H, Yeh Y, Chen WT, Kleinman HK, Grouzmann E, Grant DS (1998). Neuropeptide Y: a novel angiogenic factor from the sympathetic nerves and endothelium. *Circ Res* **83**: 187–195.

22. Ekstrand AJ, Cao R, Bjorndahl M, Nystrom S, Jonsson-Rylander AC, Hassani H, Hallberg B, Nordlander M, Cao Y (2003). Deletion of neuropeptide Y (NPY) 2 receptor in mice results in blockage of NPY-induced angiogenesis and delayed wound healing. *Proc Natl Acad Sci USA* **100**: 6033–6038.

23. Ghersi G, Chen W, Lee EW, Zukowska Z (2001). Critical role of dipeptidyl peptidase IV in neuropeptide Y-mediated endothelial cell migration in response to wounding. *Peptides* **22**: 453–458.

24. Koulu M, Movafagh S, Tuohimaa J, Jaakkola U, Kallio J, Pesonen U, Geng Y, Karvonen MK, Vainio-Jylha E, Pollonen M, Kaipio-Salmi K, Seppala H, Lee EW, Higgins RD, Zukowska Z (2004). Neuropeptide Y and Y2–receptor are involved in development of diabetic retinopathy and retinal neovascularization. *Ann Med* **36**: 232–240.

25. Yoon HZ, Yan Y, Geng Y, Higgins RD (2002). Neuropeptide Y expression in a mouse model of oxygen-induced retinopathy. *Clin Exp Ophthalmol* **30**: 424–429.

26. Li L, Lee EW, Ji H, Zukowska Z (2003). Neuropeptide Y-induced acceleration of postangioplasty occlusion of rat carotid artery. *Arterioscler Thromb Vasc Biol* **23**: 1204–1210.

27. Kitlinska J, Lee EW, Movafagh S, Pons J, Zukowska Z (2002). Neuropeptide Y-induced angiogenesis in aging. *Peptides* **23**: 71–77.

28. Lee EW, Grant DS, Movafagh S, Zukowska Z (2003). Impaired angiogenesis in neuropeptide Y (NPY)-Y2 receptor knockout mice. *Peptides* **24**: 99–106.

29. Pons J, Kitlinska J, Ji H, Lee EW, Zukowska Z (2003). Mitogenic actions of neuropeptide Y in vascular smooth muscle cells: synergetic interactions with the beta-adrenergic system. *Can J Physiol Pharmacol* **81**: 177–185.

30. Kitlinska J, Pons J, Jacques D, Bkaily G, Sader S, Kurban G, Abdel-Malak N, Perreault C, Zukowska Z (2002). Neuropeptide Y-Y1 and -Y2 receptors: mitogenic activities and intracellular trafficking. *Mol Biol Cell* **13**: 233a.

31. Breitwieser GE (2004). G protein-coupled receptor oligomerization: implications for G protein activation and cell signaling. *Circ Res* **94**: 17–27.

32. Rios CD, Jordan BA, Gomes I, Devi LA (2001). G-protein-coupled receptor dimerization: modulation of receptor function. *Pharmacol Ther* **92**: 71–87.

33. Berglund MM, Schober DA, Esterman MA, Gehlert DR (2003) Neuropeptide Y Y4 receptor homodimers dissociate upon agonist stimulation. *J Pharmacol Exp Ther* **307**: 1120–1126.

34. Dinger MC, Bader JE, Kobor AD, Kretzschmar AK, Beck-Sickinger AG (2003). Homodimerization of neuropeptide y receptors investigated by fluorescence resonance energy transfer in living cells. *J Biol Chem* **278**: 10562–10571.

35. Schober DA, Berglund MM, Gehlert DR (2003). Neuropeptide Y (NPY) Y1 and Y5 receptors form constitutive dimers that elicit an enhanced response to Y5 agonist. In: *Society for Neuroscience Annual Conference, Washington, DC*, pp. 615.

36. Kelly BD, Hackett SF, Hirota K, Oshima Y, Cai Z, Berg-Dixon S, Rowan A, Yan Z, Campochiaro PA, Semenza GL (2003). Cell type-specific regulation of angiogenic growth factor gene expression and induction of angiogenesis in nonischemic tissue by a constitutively active form of hypoxia-inducible factor 1. *Circ Res* **93**: 1074–1081.

37. Burnstock G (1987). Local control of blood pressure by purines. *Blood Vessels* **24**: 156–160.

38. Ganguly PK, Dhalla KS, Shao Q, Beamish RE, Dhalla NS (1997). Differential changes in sympathetic activity in left and right ventricles in congestive heart failure after myocardial infarction. *Am Heart J* **133**: 340–345.

39. Takatsu H, Duncker CM, Arai M, Becker LC (1997). Cardiac sympathetic nerve function assessed by [131I]metaiodobenzylguanidine after ischemia and reperfusion in anesthetized dogs. *J Nucl Cardiol* **4**: 35–41.

40. Tilan J, Karlsson J, Abe K, Kuo L, Lee EW, Everhart L, Li L, Kitlinska J, Fricke S, Thornell L, Zukowska Z (2005). Rapid activation of the neuropeptide Y (NPY) receptor system during ischemic angiogenesis in rat and human limb muscles. *FASEB J* **19**: A1663.1931.1616.

41. Arnold F, West DC (1991) Angiogenesis in wound healing. *Pharmacol Ther* **52**: 407–422.

42. Rupnick MA, Panigrahy D, Zhang CY, Dallabrida SM, Lowell BB, Langer R, Folkman MJ (2002) Adipose tissue mass can be regulated through the vasculature. *Proc Natl Acad Sci USA* **99**: 10730–10735.

43. Bartness TJ, Kay Song C, Shi H, Bowers RR, Foster MT (2005). Brain-adipose tissue cross talk. *Proc Nutr Soc* **64**: 53–64.

44. Dodt C, Lonnroth P, Wellhoner JP, Fehm HL, Elam M (2003). Sympathetic control of white adipose tissue in lean and obese humans. *Acta Physiol Scand* **177**: 351–357.

45. Turtzo LC, Lane MD (2006). NPY and neuron-adipocyte interactions in the regulation of metabolism. *EXS* **95**: 133–141.
46. Turtzo LC, Marx R, Lane MD (2001). Cross-talk between sympathetic neurons and adipocytes in coculture. *Proc Natl Acad Sci USA* **98**: 12385–12390.
47. Timar J, Dome B, Fazekas K, Janovics A, Paku S (2001). Angiogenesis-dependent diseases and angiogenesis therapy. *Pathol Oncol Res* **7**: 85–94.
48. Porta M, Bandello F (2002). Diabetic retinopathyA clinical update. *Diabetologia* **45**: 1617–1634.
49. Stitt AW (2003). The role of advanced glycation in the pathogenesis of diabetic retinopathy. *Exp Mol Pathol* **75**: 95–108.
50. Jen PY, Li WW, Yew DT (1994). Immunohistochemical localization of neuropeptide Y and somatostatin in human fetal retina. *Neuroscience* **60**: 727–735.
51. Sinclair JR, Nirenberg S (2001). Characterization of neuropeptide Y-expressing cells in the mouse retina using immunohistochemical and transgenic techniques. *J Comp Neurol* **432**: 296–306.
52. Kallio J, Pesonen U, Kaipio K, Karvonen MK, Jaakkola U, Heinonen OJ, Uusitupa MI, Koulu M (2001). Altered intracellular processing and release of neuropeptide Y due to leucine 7 to proline 7 polymorphism in the signal peptide of preproneuropeptide Y in humans. *FASEB J* **15**: 1242–1244.
53. Karvonen MK, Pesonen U, Koulu M, Niskanen L, Laakso M, Rissanen A, Dekker JM, Hart LM, Valve R, Uusitupa MI (1998). Association of a leucine(7)-to-proline(7) polymorphism in the signal peptide of neuropeptide Y with high serum cholesterol and LDL cholesterol levels. *Nat Med* **4**: 1434–1437.
54. Karvonen MK, Valkonen VP, Lakka TA, Salonen R, Koulu M, Pesonen U, Tuomainen TP, Kauhanen J, Nyyssonen K, Lakka HM, Uusitupa MI, Salonen JT (2001). Leucine7 to proline7 polymorphism in the preproneuropeptide Y is associated with the progression of carotid atherosclerosis, blood pressure and serum lipids in Finnish men. *Atherosclerosis* **159**: 145–151.
55. Niskanen L, Voutilainen-Kaunisto R, Terasvirta M, Karvonen MK, Valve R, Pesonen U, Laakso M, Uusitupa MI, Koulu M (2000). Leucine 7 to proline 7 polymorphism in the neuropeptide y gene is associated with retinopathy in type 2 diabetes. *Exp Clin Endocrinol Diab* **108**: 235–236.
56. Burnett Jr, JC (2005). Urocortin: advancing the neurohumoral hypothesis of heart failure. *Circulation* **112**: 3544–3546.
57. Li L, Jonsson-Rylander AC, Abe K, Zukowska Z (2005). Chronic stress induces rapid occlusion of angioplasty-injured rat carotid artery by activating neuropeptide Y and its Y1 receptors. *Arterioscler Thromb Vasc Biol* **25**: 2075–2080.
58. Bergers G, Benjamin LE (2003). Tumorigenesis and the angiogenic switch. *Nat Rev Cancer* **3**: 401–410.
59. Folkman J (2002). Role of angiogenesis in tumor growth and metastasis. *Semin Oncol* **29**: 15–18.
60. Tonini T, Rossi F, Claudio PP (2003). Molecular basis of angiogenesis and cancer. *Oncogene* **22**: 6549–6556.
61. Biedler JL, Roffler-Tarlov S, Schachner M, Freedman LS (1978). Multiple neurotransmitter synthesis by human neuroblastoma cell lines and clones. *Cancer Res* **38**: 3751–3757.

62. Dotsch J, Christiansen H, Hanze J, Lampert F, Rascher W (1998). Plasma neuropeptide Y of children with neuroblastoma in relation to stage, age and prognosis, tissue neuropeptide Y. *Regul Pept* **75–76**: 185–190.
63. Grouzmann E, Comoy E, Bohuon C (1989). Plasma neuropeptide Y concentrations in patients with neuroendocrine tumors. *J Clin Endocrinol Metab* **68**: 808–813.
64. Helman LJ, Cohen PS, Averbuch SD, Cooper MJ, Keiser HR, Israel MA (1989). Neuropeptide Y expression distinguishes malignant from benign pheochromocytoma. *J Clin Oncol* **7**: 1720–1725.
65. O'Hare MM, Schwartz TW (1989). Expression and precursor processing of neuropeptide Y in human and murine neuroblastoma and pheochromocytoma cell lines. *Cancer Res* **49**: 7015–7019.
66. Pruszczyk P, Wocial B, Ignatowska-Switalska H, Feltynowski T, Ellafi M, Januszewicz A, Lapinski M, Zukowska-Grojec Z, Januszewicz W (1995). Does plasma neuropeptide-Y immunoreactivity in patients with pheochromocytoma depend on hormonal activity of the tumor? *Clin Chim Acta* **243**: 205–212.
67. Cohen PS, Cooper MJ, Helman LJ, Thiele CJ, Seeger RC, Israel MA (1990). Neuropeptide Y expression in the developing adrenal gland and in childhood neuroblastoma tumors. *Cancer Res* **50**: 6055–6061.
68. de SS P, Denker J, Bravo EL, Graham RM (1995). Production, characterization, expression of neuropeptide Y by human pheochromocytoma. *J Clin Invest* **96**: 2503–2509.
69. Kogner P, Bjork O, Theodorsson E (1994). Plasma neuropeptide Y in healthy children: influence of age, anaesthesia and the establishment of an age-adjusted reference interval. *Acta Paediatr* **83**: 423–427.
70. Rajakumar PA, Westfall TC, Devaskar SU (1998). Neuropeptide Y gene expression in immortalized rat hippocampal and pheochromocytoma-12 cell lines. *Regul Pept* **73**: 123–131.
71. Chen X, DiMaggio DA, Han SP, Westfall TC (1997). Autoreceptor-induced inhibition of neuropeptide Y release from PC-12 cells is mediated by Y2 receptors. *Am J Physiol* **273**: H1737–1744.
72. Mori Y, Higuchi M, Masuyama N, Gotoh Y (2004). Adenosine A2A receptor facilitates calcium-dependent protein secretion through the activation of protein kinase A and phosphatidylinositol-3 kinase in PC12 cells. *Cell Struct Funct* **29**: 101–110.
73. Hofliger MM, Castejon GL, Kiess W, Beck Sickinger AG (2003). Novel cell line selectively expressing neuropeptide Y-Y2 receptors. *J Recept Signal Transduct Res* **23**: 351–360.
74. Korner M, Waser B, Reubi JC (2004). High expression of neuropeptide y receptors in tumors of the human adrenal gland and extra-adrenal paraganglia. *Clin Cancer Res* **10**: 8426–8433.
75. Katzenstein HM, Rademaker AW, Senger C, Salwen HR, Nguyen NN, Thorner PS, Litsas L, Cohn SL (1999). Effectiveness of the angiogenesis inhibitor TNP-470 in reducing the growth of human neuroblastoma in nude mice inversely correlates with tumor burden. *Clin Cancer Res* **5**: 4273–4278.
76. Klement G, Baruchel S, Rak J, Man S, Clark K, Hicklin DJ, Bohlen P, Kerbel RS (2000). Continuous low-dose therapy with vinblastine and VEGF receptor-2

antibody induces sustained tumor regression without overt toxicity. *J Clin Invest* **105**: R15–24.

77. Ribatti D, Ponzoni M (2005). Antiangiogenic strategies in neuroblastoma. *Cancer Treat Rev* **31**: 27–34.

78. Ribatti D, Raffaghello L, Marimpietri D, Cosimo E, Montaldo PG, Nico B, Vacca A, Ponzoni M (2003). Fenretinide as an anti-angiogenic agent in neuroblastoma. *Cancer Lett* **197**: 181–184.

79. Shusterman S, Maris JM (2005). Prospects for therapeutic inhibition of neuroblastoma angiogenesis. *Cancer Lett* **228**: 171–179.

80. Wassberg E, Hedborg F, Skoldenberg E, Stridsberg M, Christofferson R (1999). Inhibition of angiogenesis induces chromaffin differentiation and apoptosis in neuroblastoma. *Am J Pathol* **154**: 395–403.

81. van Valen F, Winkelmann W, Jurgens H (1992). Expression of functional Y1 receptors for neuropeptide Y in human Ewing's sarcoma cell lines. *J Cancer Res Clin Oncol* **118**: 529–536.

82. Reubi JC, Gugger M, Waser B, Schaer JC (2001). Y(1)-mediated effect of neuropeptide Y in cancer: breast carcinomas as targets. *Cancer Res* **61**: 4636–4641.

83. Middeke M, Hoffmann S, Hassan I, Wunderlich A, Hofbauer LC, Zielke A (2002). *In vitro* and *in vivo* angiogenesis in PC12 pheochromocytoma cells is mediated by vascular endothelial growth factor. *Exp Clin Endocrinol Diab* **110**: 386–392.

84. Zukowska Z, Pons J, Lee EW, Li L (2003). Neuropeptide Y: a new mediator linking sympathetic nerves, blood vessels and immune system? *Can J Physiol Pharmacol* **81**: 89–94.

85. Zukowska-Grojec Z (1995). Neuropeptide Y. A novel sympathetic stress hormone and more. *Ann NY Acad Sci* **771**: 219–233.

86. Chorot P, Sandin B (1994). Life events and stress reactivity as predictors of cancer, coronary heart disease and anxiety disorders. *Int J Psychosom* **41**: 34–40.

87. Ginsberg A, Price S, Ingram D, Nottage E (1996). Life events and the risk of breast cancer: a case-control study. *Eur J Cancer* **32A**: 2049–2052.

88. Kune S, Kune GA, Watson LF, Rahe RH (1991). Recent life change and large bowel cancer. Data from the Melbourne Colorectal Cancer Study. *J Clin Epidemiol* **44**: 57–68.

89. Levav I, Kohn R, Iscovich J, Abramson JH, Tsai WY, Vigdorovich D (2000). Cancer incidence and survival following bereavement. *Am J Public Health* **90**: 1601–1607.

90. Lillberg K, Verkasalo PK, Kaprio J, Teppo L, Helenius H, Koskenvuo M (2003). Stressful life events and risk of breast cancer in 10,808 women: a cohort study. *Am J Epidemiol* **157**: 415–423.

91. Levy SM, Herberman RB, Maluish AM, Schlien B, Lippman M (1985). Prognostic risk assessment in primary breast cancer by behavioral and immunological parameters. *Health Psychol* **4**: 99–113.

92. Carmeliet P (2003). Blood vessels and nerves: common signals, pathways and diseases. *Nat Rev Genet* **4**: 710–720.

93. Mukouyama YS, Shin D, Britsch S, Taniguchi M, Anderson DJ (2002). Sensory nerves determine the pattern of arterial differentiation and blood vessel branching in the skin. *Cell* **109**: 693–705.

94. Young HM, Anderson RB, Anderson CR (2004). Guidance cues involved in the development of the peripheral autonomic nervous system. *Auton Neurosci* **112**: 1–14.
95. Bagri A, Tessier-Lavigne M (2002). Neuropilins as Semaphorin receptors: *in vivo* functions in neuronal cell migration and axon guidance. *Adv Exp Med Biol* **515**: 13–31.
96. Coulthard MG, Duffy S, Down M, Evans B, Power M, Smith F, Stylianou C, Kleikamp S, Oates A, Lackmann M, Burns GF, Boyd AW (2002). The role of the Eph-ephrin signalling system in the regulation of developmental patterning. *Int J Dev Biol* **46**: 375–384.
97. Zukowska Z, Grant DS, Lee EW (2003). Neuropeptide Y: a novel mechanism for ischemic angiogenesis. *Trends Cardiovasc Med* **13**: 86–92.
98. Howell OW, Doyle K, Goodman JH, Scharfman HE, Herzog H, Pringle A, Beck-Sickinger AG, Gray WP (2005). Neuropeptide Y stimulates neuronal precursor proliferation in the post-natal and adult dentate gyrus. *J Neurochem* **93**: 560–570.
99. Gray WP, Scharfman HE (2005). NPY in hippocampal neurogenesis. In: Zukowska Z, Feuerstein GZ (eds.), *The NPY Family of Peptides in Immune Disorders, Inflammation, Angiogenesis and Cancer* (Birkhauser Verlag, Basal, Switzerland), pp. 201–222.

5

Modulation of Growth Factor Signaling by Heparan Sulfate Proteoglycans

by Nicholas W. Shworak

1. Introduction

Heparan sulfate proteoglycans (HSPGs) are produced by virtually all cell types and regulate a multitude of biological processes. HSPGs are hybrid molecules composed of protein cores to which are attached one or more long chains of heparan sulfate (HS). There are multiple HSPG core proteins, each of which engenders unique biologic properties. Functional diversity is further enhanced by the profound structural complexity of the attached HS chains. This polysaccharide is a form of glycosaminoglycan (GAG) — a long unbranched copolymer comprised of alternating acid and amino sugars. The HS sugar residues are decorated at defined positions with sulfate groups. These modifications create a large array of short sequence motifs that bind and thereby modulate the functional properties of numerous regulatory molecules including signaling ligands/receptors, proteases, enzymes and lipoproteins (Table 1). In turn, these HS:protein interactions control numerous cellular processes such as signaling, vesicular trafficking, migration,

Table 1. Representative heparan sulfate-binding proteins. Most of these ligands can be defined as "heparin-binding growth factors" (derived from Refs. 1 and 2).

Growth factors	Growth factor-binding proteins	Morphogens	ECM/plasma components
Angiogenin	Follistatin	Activin	Fibrin
Amphiregulin	IGF BP-3, -5	BMP-2, -4	Fibronectin
Betacellulin	TGF-β BP	Chordin	Interstitial collagens
Most FGFs	FGF receptors	Frizzled-type peptides	Laminins
Heparin-binding EGF		Sonic hedgehog	Pleiotropin (HB-GAM)
HGF	**Anti-angiogenic**	Sprouty peptides	Tenascin
IGF-II	Angiostatin	Wnts (1–13)	Thrombospondin
Midkine	Endostatin		Vitronectin
Neuregulin	Tgf-β	**Coagulation**	
Pleiotrophin	Interferon-γ	Antithrombin	
PDGF-AA	GCP-2	Heparin co-factor II	**Cell adhesion**
TGF-β	IP-10	Leuserpin	L-selectin
VEGF-165, 189	PF-4	Plasminogen activator inhibitor	MAC-1
		Tissue factor pathway inhibitor	N-CAM
		Tissue plasminogen activator	PECAM-1
Chemokines	**Cytokines**	Thrombin	
GRO-α	IL-2, -3, -4, -5, -7, -12		
GRP-β	Gm-CSF	**Proteases**	**Energy metabolism**
IL-8	Interferon-γ	Cathepsin G	Agouti-related protein
GCP-2	TNF-α	Neutrophil elastase	ApoB, ApoE
IP-10		Protease Nexin I	Lipoprotein lipase
PF-4			Triglyceride lipases

and adhesion. Such cellular actions of HSPGs regulate many biological events including angiogenesis, inflammation, hemostasis, lipoprotein metabolism, axonal guidance, and developmental inductions.

Although HSPGs control a myriad of functions, this chapter selectively focuses on the role of the HS chain in regulating endothelial cell (EC) signaling by "heparin-binding growth factors." For ease of discussion, this term is defined as any signaling ligand that exhibits high affinity to heparin, a particular flavor of highly sulfated HS. This definition encompasses classic growth factors [such as vascular endothelial growth factors (VEGFs), fibroblast growth factors (FGFs), heparin-binding epidermal growth factor, and Wnts], cytokines (such as GM-CSF, interleukin-3, and interferon-γ), most chemokines and even growth inhibitors such as endostatin. Consequently, this term includes the majority of HS-binding ligands (Table 1). "Heparin-binding" is a historical term, which derives from many of these factors being originally purified by heparin chromatography. However, heparin only occurs within mast cell granules, so it is unlikely to contribute to the *in vivo* roles of most of these ligands. The physiological activities of heparin-binding growth factors instead predominantly involve their interaction with the HS chains of extracellular matrix and cell surface HSPGs, which are produced by almost all cell types.

2. Historical Perspective

Although heparin is only found in mast cells, it was the first form of HS discovered and consequently dominated the early history of the field. In 1916, Jay McLean, a second-year medical student, found that an extract from dog liver exhibited strong anticoagulant activity.[1] This activity was designated as heparin to denote its isolation from liver. By 1935, clinical trials were started, which heralded the long-standing use of heparin as an efficacious anticoagulant (reviewed in Ref. 2). The acceptance of heparin as a therapeutic agent motivated studies to discover its structure and mechanism of action.

The paradigm for how HS motifs convey biological activity derives from the original elucidation of heparin's mechanism of action. McLean's mentor, William Henry Howell, determined in 1925 that heparin was a form of polysaccharide and proposed that its anticoagulant activity required a plasma cofactor.[3] Potentially, this cofactor might be "antithrombin" — an activity in defibrinated plasma, discovered at the end of the 19th century, known to slowly neutralize thrombin.[4] However, almost six decades of research were required to unravel how heparin's intricate structure conveys its mechanism of action (reviewed in Ref. 2). In 1968, Abildgaard ultimately determined that Howell's hypothetical plasma cofactor was indeed antithrombin.[5] Shortly after, Rosenberg and Damus found that the binding of heparin to antithrombin dramatically catalyzed antithrombin's ability to neutralize coagulation proteases.[6] Initially, such binding was largely considered to occur by non-specific ionic interactions. This notion was discounted in 1976, when the three independent groups of Lindahl, Rosenberg and Sims showed that only one-third of heparin molecules could bind antithrombin and only this population of molecules exhibited anticoagulant activity.[7–9] This landmark observation suggested structural specificity must exist. In the early 1980s, the groups of Choay, Lindahl and Rosenberg demonstrated that the active component of heparin was a pentasaccharide motif with a specific arrangement of sulfate groups (reviewed in Ref. 2). It is now appreciated that many of the distinct activities of HS are conveyed by specific motifs, with a given motif bound by a distinct effector protein.

Although the initial HS landscape was dominated by heparin, it was appreciated as early as 1937 that heparin must play only a limited role in the body, as it is only present in the basophilic granules of mast cells.[10] The rich heparin content of liver derives from high levels of mast cells in the liver capsule. The ubiquitous nature of HS was eventually realized in the early 1970s, when it was determined that virtually all cell types produce HSPGs (reviewed in Ref. 2). In the 1980s, the various core proteins were identified, and the 1990s heralded the cloning of the HS biosynthetic enzymes. The same family of enzymes were found to generate both the ubiquitous HS and heparin, which led to the appreciation that heparin is simply one type of HS (reviewed in Refs. 11 to 13). Indeed, certain cell types can produce an HS subpopulation that exhibits a high sulfate content indistinguishable from that of mast cell heparin.

The functional diversity of HS began to emerge in the 1970s when the first of numerous heparin/HS-binding proteins were discovered. The groups of Klagsbrun and Folkman, in the early 1980s, championed the use of heparin affinity chromatography to purify EC mitogens.[14] This application was a major trigger of investigations into HS-mediated signaling by heparin-binding growth factors, which continue to the present.

3. The Structure, Synthesis, and Post-Synthetic Modification of HSPGs

3.1. *The HSPG core proteins*

There are multiple HSPG core proteins which contribute to the broad functional repertoire of these hybrid molecules (Fig.1A). Table 2 lists the major vertebrate HSPGs and indicates which are known to exhibit expression in ECs. Some core proteins carry two types of GAG chains — HS and chondroitin sulfate (CS); whereas others exclusively bear HS chains. The major carriers of non-surface bound HS in the extracellu-lar matrix appear to be agrin, perlecan, and collagen XVIII (reviewed in Refs. 11 and 15). Most cell surface HS is carried by glypicans, a family of at least six homologous glycophosphatidyl-inositol (GPI)-anchored proteins, syndecans and four related transmembrane pro-teins. However, a variety of additional integral membrane proteins, such as betaglycan and splice variants of CD44, are considered as "part-time" proteoglycans because they occasionally bear HS chains. ECs are presently known to express at least 11 distinct HSPG core proteins (Table 2). Specialized features of a given HSPG core pro-tein can: (1) provide direct interactions with cytoskeletal and/or signal-ing components, (2) define localization to intracellular or extracellular compartments, and (3) allow for cellular internalization, recycling, or transcellular transport (reviewed in Refs. 11, 16 and 17). Thus, the mul-tiplicity of core proteins does not merely reflect redundancy but rather amplifies functional diversity. The core protein together with its associ-ated cellular machinery constitute a platform of such profound utility

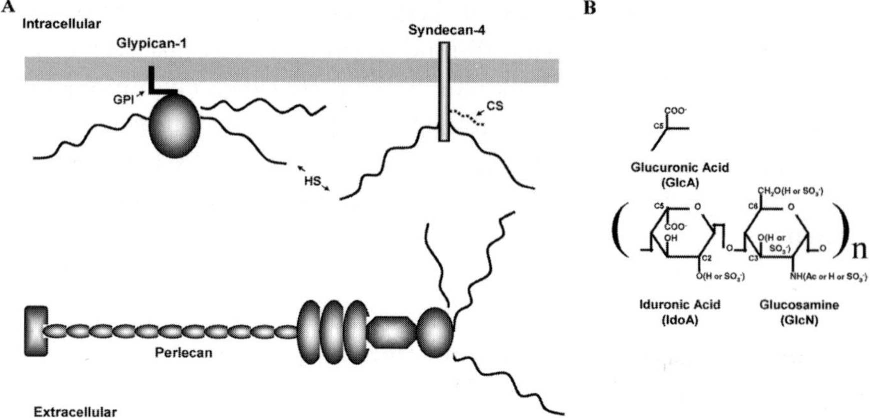

Fig. 1. Major HSPGs. **(A)** Major HSPGs of ECs. The EC cell surface and basement membrane are schematically depicted showing a major representative of an integral membrane (syndecan-4), a GPI-linked (glypican-1), and an extracellular (perlecan) HSPG. Syndecan-4 is presented as carrying two HS (unbroken line) and one CS (dotted line) chains, but can bear multiple permutations of these GAGs. **(B)** The structure of an HS disaccharide repeat. The same repeated disaccharide occurs in both HS and heparin. Remodeling of the glucosamine can involve addition of sulfates to the C_3 or C_6 positions. Furthermore, the N-acetyl group can be replaced with either a sulfate group or a proton (which generates a free amino group). The uronic acid can be remodeled by adding a sulfate at C_2, or by epimerization at C_5, which converts glucuronic acid (only C_5 region shown) to iduronic acid.

that almost all eukaryotic tissues and cells employ HSPGs in multiple roles.

3.2. *The structure of the HS chain*

Apart from the multiplicity of core proteins, the HS component of HSPGs further amplifies their complexity and functional diversity. HS is a heterogeneous linear polysaccharide consisting of a repeated disaccharide unit of glucuronic or iduronic acid alternating with glucosamine [hexuronic acid $\beta_{1\rightarrow4}$ N-acetylglucosamine (GlcNAc) $\alpha_{1\rightarrow4}$], that is partially decorated with N- and various O-sulfate groups (Fig.1B). The specific arrangement of the sulfate moieties, in large part, gives rise to distinct binding motifs that interact with an increasingly expansive list

Table 2. HSPG core proteins (derived from Refs. 11 and 15).

Core proteins	Endothelial expression known	GAG type (No. of chains)	Key features
I. Extracellular matrix			
Agrin	+	HS (3)	
Perlecan	+	HS/CS (3)	Large multidomain protein
Testican	+	HS/CS (2)	
Type XVIII collagen	+	HS/CS (3)	Cleavage product is endostatin
II. Cell surface			
Glypican family			GPI linked
Glypican-1	+	HS (3)	
Glypican-2		HS	
Glypican-3		HS	
Glypican-4		HS	
Glypican-5		HS	
Glypican-6		HS	
Syndecan family			Type I integral membrane proteins
Syndecan-1	+	HS/CS (3–5)	
Syndecan-2	+	HS/CS (3)	
Syndecan-3		HS/CS (3–5)	
Syndecan-4	+	HS/CS (3)	
Betaglycan	+	HS/CS	Part-time proteoglycan
CD44	+	HS/CS	Part-time proteoglycan
FGFR2	+	HS/CS (1)	Splice variants containing the "acid box"

of protein effectors (Table 1).[12,18,19] The HS chains are quite long (~40–80 nm) and range from 100 to 200 disaccharide units, respectively. Each chain is internally repetitive, containing short blocks of minimal sulfation alternating with blocks of highly sulfated motifs (reviewed in Refs. 11, 16 and 17). Multiple copies of different ligand binding motifs can occur on a single HS chain;[20] consequently, HS is exquisitely suited to function as a template for the assembly of multimolecular complexes, such as signaling complexes.

3.3. *The biosynthesis of HS*

HS synthesis occurs in the Golgi apparatus and involves an extensive series of post-translational modifications (Fig. 2). First, UDP-linked sugars are employed as substrates for polymerizing the HS chain. Then, this backbone is remodeled (largely by sulfotransferases) to create distinct HS motifs. HS sulfotransferases decorate the chain with critically

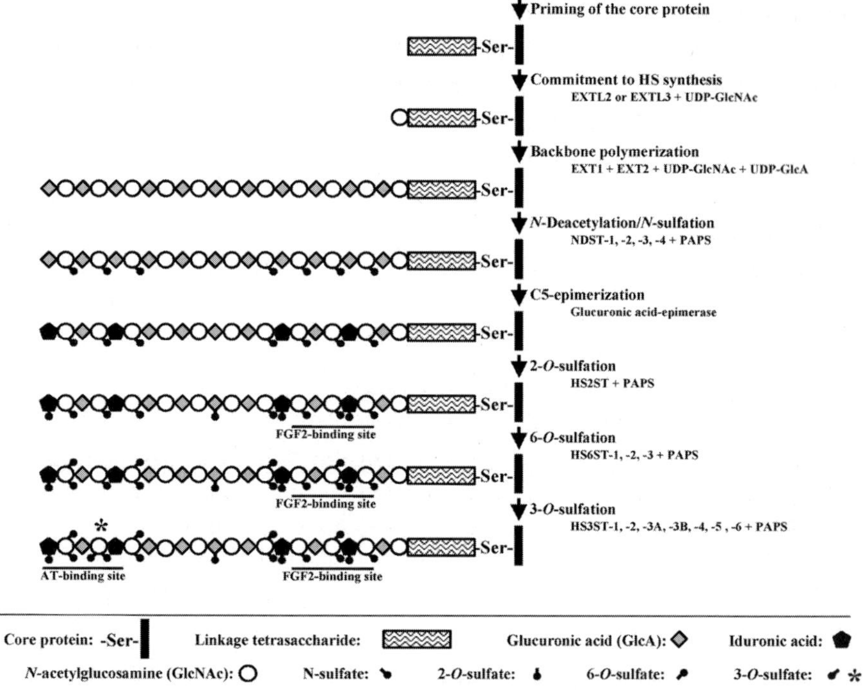

Fig. 2.   HS synthesis. Schematically depicted are the major steps of HS synthesis and their involved co-factors and enzymes. This process is relatively ordered and creates distinct sequence motifs. The early biosynthetic reactions create the minimal sequence required to bind an FGF2 monomer. Conversely, generation of the antithrombin (AT)-binding site requires additional later modification reactions. The very rare 3-O-sulfate moiety is indicated by ∗. Not shown are structures generated by gD-type HS3ST enzymes.

positioned sulfate groups by transferring a sulfuryl group from the universal "sulfate" donor 3′ phosphoadensosine-5′-phosophosulfate (PAPS) (reviewed in Ref. 15). Although many functions are conveyed by core proteins, they do not play a substantial role in defining the types of HS motifs that occur upon their HS chains.[21] Rather, the sequence-specific properties of the various modification enzymes is the critical factor that dictates the production of specific motifs.

Synthesis begins by the assembly of a short linkage tetrasaccharide at defined serine residues of the core proteins (Fig. 2). This step initiates both HS and CS synthesis, which explains why many HSPGs can also carry CS chains. CS exhibits different biological activities than HS due to its distinct structure; CS chains, compared to HS, contain galactosamine instead of GlcNAc, are sulfated in some distinct positions and are shorter. Commitment of the primed structure specifically down the HS pathway is determined by the addition of a GlcNAc residue by either EXTL2 or EXTL3, which are distinct isoforms of GlcNAc transferase I. Conversely, addition of a galactosamine residue triggers CS synthesis. The HS backbone is then polymerized by the simultaneous action of enzymes from the genes EXT1 and EXT2 (reviewed in Refs. 12 and 13).

The copolymer backbone is then remodeled by a semi-ordered series of reactions, which generate distinct HS motifs. First, the bifunctional enzyme HS N-deacetylase/N-sulfotransferase (NDST) deacetylates and N-sulfates subsets of N-acetylglucosamine residues.[22,23] Occasional residues escape sulfation, which results in a free amino group. Blocks containing a high density of N-sulfated glucosamine are then preferential substrates for the subsequent, less frequent modification reactions. The HS C_5 epimerase transforms occasional glucuronic acid residues into iduronic acid.[24,25] The HS 2-sulfotransferase (HS2ST) next produces 2-O-sulfated iduronic acid or to a much lesser degree 2-O-sulfated glucuronic acid.[26,27] Occasional glucosamine residues undergo 6-O-sulfation by HS 6-O-sulfotransferase (HS6ST).[28] "Late" modification also includes the generation of 3-O-sulfated glucosamine by HS 3-O-sulfotransferase (HS3ST, also known as 3OST).[21,29–31]

The NDST, HS6ST and HS3ST enzymes have multiple isozymes encoded by distinct genes. This multiplicity serves two purposes. First, multiple isoforms allow for cell type-specific production of distinct HS motifs (reviewed in Refs. 12 and 13). Second, individual isoforms can exhibit distinct sequence specificities, which further expands the structural diversity of HS.[31–33] These concepts are best illustrated by the largest multigene family of HS modification enzymes, the seven distinct isoforms of HS3ST. Despite the family's large size, 3-O-sulfates are the rarest HS modification (usually comprising < 0.5% of total sulfate moieties). Their rarity make 3-O-sulfates ideal for serving key regulatory roles.[21,34] The enzymes can be categorized as either AT-type or gD-type, based on their enzymatic specificities.[35] HS3ST1 preferentially modifies a specific pentasaccharide precursor structure to create HS with high affinity binding motifs for antithrombin (HS^{AT+}), and so is an AT-type enzyme. HS3ST1 is clearly regulatory in ECs, as its expression level governs the level of HS^{AT+} synthesis.[29] The interaction of antithrombin with this motif is thought to convey anti-inflammatory and anticoagulant properties to the endothelium. In contrast, HS3ST2, $HS3ST3_A$, $HS3ST3_B$, HS3ST4 and HS3ST6 are gD-type enzymes; they preferentially recognize a distinct precursor structure to create another 3-O-sulfated motif (HS^{gD+}), known to bind glycoprotein gD of herpes simplex virus-1. These isoforms are also regulatory, as cellular entry of this herpes simplex virus-1 can be dependent on cellular expression of gD-type enzymes.[36] Although the endogenous ligand for HS^{gD+} is not yet known, this 3-O-sulfated motif has recently been implicated in Notch signaling, described below. HS3ST5 is able to generate both motifs with equal efficiency, so falls into both classes.[37] Although this system appears to show redundancy, most isoforms exhibits unique cell type- and tissue-specific expression patterns. Sequence specificity and cell type-specific expression are also properties of the isoforms of NDST (NDST1 to NDST4) and HS6ST (HS6ST1 to HS6ST3) (reviewed in Refs. 12 and 13). Thus, individual cells produce distinct arrays of HS motifs through the expression of distinct combinations of HS sulfotransferases.

3.4. *The post-synthetic processing of HSPGs*

Four types of post-biosynthetic processing are known to further modify the properties of HSPGs. First, extracellular heparanases derived from leukocytes, tumor cells or the endothelium can degrade HS chains.[38] Second, nitric oxide can damage HS chains. Such HSPGs are repaired by undergoing cellular internalization, damaged chain regions are removed, fresh HS chains are re-synthesized on the core proteins, and the repaired HSPG is transported back to the cell surface.[39] Third, extracellular sulfatases, which remove specific sulfate moieties, can modify HS motifs. For example, 6-O-sulfate groups can be removed by the Sulf enzymes.[40] Fourth, proteoglycans bound to the cell membrane can be shed. Phospholipase activity liberates GPI-linked HSPGs whereas shedding of integral membrane HSPGs involves protease cleavage of the extracellular domain near the transmembrane domain.[11] Thus, the HSPG structure can be altered by multiple extracellular factors. The potential influence of such factors on HS-mediated cell signaling is discussed below.

4. Evolution of HSPGs

An evolutionary perspective reveals many fundamental features of HSPGs. *Bona fide* HS polysaccharide has not been found in plants, unicellular organisms, or prokaryotes. Rigorous structural studies have found that HS occurs in most metazoans, being absent only from the most primitive multicellular animal, the sponges. The two more complex phyla *Ctenophora* and *Cnidaria*, whose ancestors represent the earliest forms of Eumetazoa (true metazoans), clearly possess vertebrate-type HS structures (reviewed in Refs. 41 and 42). The long split between Eumetazoans and sponges, ~940 million years ago, testifies to the extreme antiquity of HS. On one hand, HS may have arisen by conferring a selection advantage through optimizing pre-existent processes. Potential candidates include processes which were firmly established in single cell organisms, such as cell signaling and adhesion. Additional candidates include totipotent stem cells, innate immunity, chemokine production, and apoptosis, which all occur in sponges. On the other hand, the emergence of the complex HS biosynthetic pathway may have involved selection for a unique process appearing at the divergence of Eumetazoans. In particular, the emergence of an integrated mechanism for whole organ homeostasis enabled this evolutionary split.[43] Eumetazoa does not include the sponges because they essentially function like a colony of animals. Sponges are comprised of a semi-autonomous groups of cells. Each group has an independent inlet to obtain seawater that contains food/oxygen and an independent outlet for removal of waste/carbon dioxide. The earliest Eumetazoans, in contrast, have a single inlet into and a single outlet draining a common gastrovascular cavity. This specialized sealed compartment is surrounded by an evolutionarily novel sheet of cells, the original "epithelium *sensu stricto*." This cell layer serves to regulate and separate the fluids in distinct extracellular compartments. Thus, Eumetazoans are characterized by a single circulatory system that enables whole body homeostasis.[43]

An increase in cellular diversity accompanied this change in body plan. Whereas sponges arise from a single germ layer, early Eumetazoans had two germ layers (endoderm and ectoderm) that enabled the formation of the new epithelium *sensu stricto*. Neurons were an additional novel early Eumetazoan cell type, which allowed for further coordination of whole body homeostasis. Thus, the emergence of Eumetazoans featured the simultaneous development of primitive circulatory and nervous systems. This common emergence might explain why both systems in higher organisms frequently employ similar signaling components such as FGFs, neuropilins, ephrins and Notch receptors, as elaborated below. From such primordial evolutionary roots, it is not surprising that HSPGs are involved in homeostatic mechanisms, serve to modulate cell type-specific phenotypes, occur in multiple organ systems and regulate embryonic development.

The advent of total genome sequencing has revealed the extent of the lower metazoan HS biosynthetic machinery as known for the roundworm (*C. elegans*) and the fruit fly (*Drosophila melanogaster*). These organisms possess genes encoding multiple core proteins and all of the various HS biosynthetic enzymes; thus all the components for HSPG production are present. Numerous studies show that the mammalian and invertebrate gene products are functionally equivalent (reviewed in Ref. 12). However, these invertebrates, in contrast to vertebrates, largely lack multigene families. They exhibit only single genes for NDST, HS6ST, syndecan, and glypican. A notable exception is that Drosophila shows two HS3ST genes which respectively encode gD-type and AT-type isoforms.[35] Thus, these two isoforms stem from the same predecessors of the two major functional groupings of the large mammalian 3-O-sulfortransferase multigene family.

5. HSPGs in Development

The roles of HS in development have been most extensively examined in *C. elegans* and *Drosophila*. These invertebrates are not useful for directly studying vascular development as they only have a rudimentary open circulatory system. Nevertheless, fundamental features of the roles of HS in development are revealed by the genetic investigations of these animals. Most importantly, these studies demonstrate a role for HSPGs in several signaling pathways that are known to be operable in vertebrate ECs. Thus, genetic analyses of such lower organisms exemplify potential ways in which HSPGs may function in mammalian ECs.

Studies of mutants lacking various core proteins show that each core is required for signaling events that are specific in time, place, cell type and tissue, and that HSPGs are required for the development of multiple organ systems (reviewed in Refs. 12 and 44). That each protein conveys discrete functions suggests the multitude of core proteins expressed in mammalian ECs should serve to expand endothelial functional diversity. Deletion of a single core protein in lower organisms is not usually lethal, due to the multiplicity of core proteins. However, early embryonic death occurs in *Drosophila* and *C. elegans* mutants that have core proteins but are completely lacking in HS (due to mutations in EXT enzymes, see Fig. 2, that prevent polymerization of the HS backbone). Such lethality shows that the HS component of HSPGs plays an essential role in development. The HS-deficient *Drosophila* mutants show defects in pathways mediated by three HS-binding signaling ligands (wingless, hedgehog, and decapentaplegic, which is a form of tumor growth factor-β (reviewed in Refs. 12 and 44). It is likely that mammalian ECs employ HS to regulate these pathways, as vertebrates have comparable ligands that are known to be involved in vasculogenesis and angiogenesis.

Genetic studies also show distinct roles for HS modification enzymes. Although global HS deficiency is lethal in *C. elegans*, mutants lacking either C_5-epimerase, HS2ST or HS6ST are viable. All such mutants have abnormal neuronal guidance; however, each enzyme deficiency affects a unique assortment of specific neurons. Thus, unique properties to HS are conveyed by different HS modifications in a cell type-specific fashion. These deficiencies interfere with developmental processes involving three receptors (Robo, Integrin and Ephrin) (reviewed in Refs. 12 and 44). Signaling through the vertebrates forms of all of these receptors is known to involve HSPGs, with integrins and ephrin receptors controlling key endothelial functions.

In *Drosophila*, deficiencies of specific HS modification enzymes leads to malformation of key organ systems and lethality. *Drosophila* mutants lacking a gD-type HS3ST isoform have disrupted signaling through the Notch receptor, which produces multiple developmental defects. The vertebrate Notch pathway conveys an arterial cell phenotype to non-committed ECs;[45] thus, a mammalian gD-type HS3ST isoform may participate in this form of EC differentiation. *Drosophila* mutants lacking HS6ST die from malformation of the tracheal airway system.[35,46] Development of the *Drosophila* trachea is analogous to vertebrate angiogenesis. Both processes control branching morphogenesis of tubular structures through comparable signaling components (including FGFs). The *Drosophila* HS6ST and FGF receptor (breathless) are co-expressed in tracheal cells; whereas, adjacent inducing cells express branchless (the *Drosophila* FGF). Mutants lacking branchless or breathless or HS6ST activity have stunted branching of the tracheal system that is phenotypically equivalent. Moreover, FGF-dependent activation of mitogen-activated protein kinase is impaired in HS6ST-deficient mutants.[46] Thus, 6-*O*-sulfates of HS are essential for FGF-induced branching morphogenesis in *Drosophila*. Mammalian FGF signaling also requires these moieties, as elaborated below. Clearly, such genetic investigations into the developmental roles of HS in lower organisms should facilitate the identification and delineation of HS-mediated signaling mechanisms that are operable in vertebrate ECs.

Roles of HS in vertebrate vessel development are now beginning to come to light. HS is critical for the signaling activity of VEGF164 and VEGF188, the splice variants of VEGF-A that contain HS-binding domains. Mice that exclusively express VEGF120 (non-HS-binding) exhibit abnormal angiogenesis with defective vessel branching.[47,48] That vessel branching requires HSPGs is indicated by zebrafish studies showing angiogenic sprouting requiring the syndecan-2 core protein.[49] Moreover, knock-in mice that express a perlecan core protein lacking HS chains exhibit defective FGF-stimulated angiogenesis.[50] Mice lacking various HS modification enzymes have recently been generated. Deficiencies in C_5-epimerase, NDST1, HS2ST and HS3ST1 alter HS structure and result in developmental defects that produce postnatal lethality.[51−56] Given the multiple roles of HS, evaluation of adult functions will require mice with cell type-specific deficiencies. Indeed, mice in which only ECs are deficient for NDST1 have recently been generated. They have a reduced infiltration of neutrophil into tissues, which involves reduced EC presentation of lumenal surface chemokines to leukocytes, EC transcytosis of chemokines, and L-selectin-mediated binding of leukocytes to ECs.[57] Thus, the involvement of specific HS structures in multiple aspects of vascular development and function should be revealed by future analyses of mice with global or cell type-specific deficiencies in HS modification enzymes.

6. HSPG Modulation of Ligand-Receptor Interactions

6.1. *HSPGs are co-receptors that augment ternary complex formation*

Multiple aspects of signaling involve HSPGs, which can serve as either positive or negative regulators (reviewed in Refs. 16 and 58). The various mechanisms have been most extensively established for the fibroblast growth factors (FGFs), which we review as a paradigm of how the HS chains of HSPGs operate in multiple fashions to regulate the activities of numerous heparin-binding growth factors.

As described in Chapter 3, FGF signaling involves a family of more than 20 ligands that activate five different tyrosine kinase receptors (FGFRs). The complexity of this system is further expanded by the existence of multiple splice variants for each FGFR. Signaling requires the formation of a ternary complex in which two FGFs bind and thereby dimerize two FGFRs. Receptor dimerization is essential for cytoplasmic domain cross-phosphorylation, which initiates intracellular signaling cascades (reviewed in Ref. 58). HSPGs modulate this system by acting as low affinity "co-receptors" that facilitate ligand-driven assembly of a functional ternary complex[59-61] (Fig. 3A).

Multiple endothelial HSPGs serve as co-receptors (including perlecan, glypican-1, and syndecans-1,-2, and-4), [62-64] which suggests the HS chain plays a critical role. One potential role of HS is to facilitate FGF dimerization into a configuration that is essential for receptor binding, as indicated by solution phase studies.[65] Mechanistic insights have been gained from X-ray crystallography of ligand:receptor complexes containing short HS (heparin) fragments. However, two distinct HS-FGF-FGFR ternary complexes have been crystallized so the precise means of ternary complex formation is disputed (reviewed in Ref. 58) (Fig.3A). In one case, the complex is symmetric, with each single FGF2-FGFR1 dimer being stabilized by the ends of two separate HS fragments.[66] In the other case, the complex is asymmetric, with the pair of FGF1-FGFR2 dimers being bridged by a single HS disaccharide.[67] Each complex was formed with different isoforms of ligand and receptor, which provides one possible explanation for these structural differences. Perhaps there are multiple ways to generate ternary complexes, given the large number of combinations of FGF ligands and receptors. Regardless, both models show that similar contacts are formed between the HS fragment(s) and HS-binding residues of the ligand and receptor. Thus, both models confirm that HS serves to crosslink FGF to the FGFR (reviewed in Ref. 58).

6.2. *HSPG co-receptors confer unique regulatory properties*

The deployment of an HSPG co-receptor influences growth factor signaling kinetics, strength and specificity and engenders several novel properties. Such distinct characteristics in part reflect fundamental differences between growth factors and HSPGs. For example, growth factors have a rapid turnover; whereas HSPGs are long-lived, exhibiting half-lives in the range of hours to days.[16] Consequently, HSPGs can drive sustained signaling events by stabilizing growth factors. For example, in the absence of HS, FGF2 binding to FGFRs produces

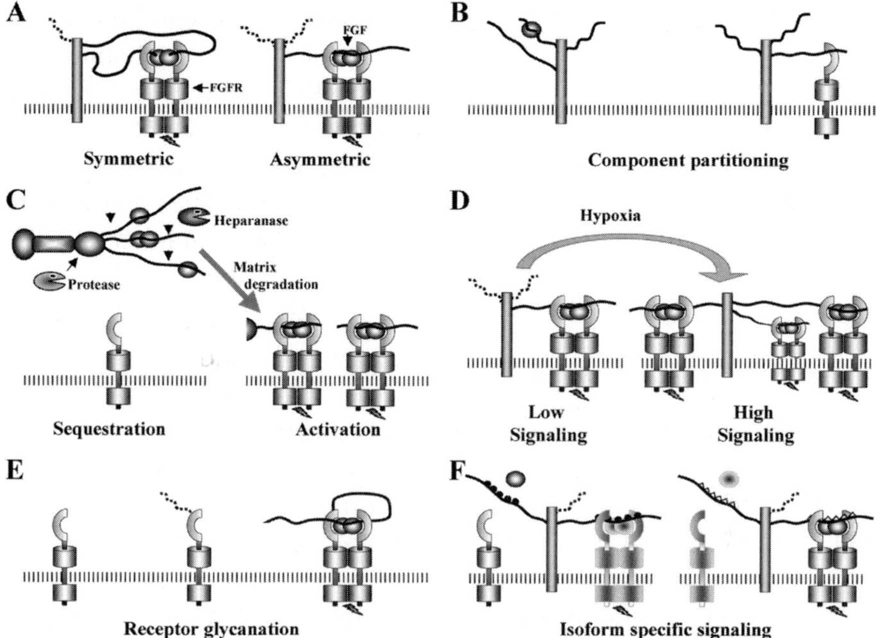

Fig. 3. HSPGs regulate signaling by multiple mechanisms. Schematically depicted are FGFs and the cell membrane with FGFRs (as indicated) and usually a syndecan-4 type HSPG, which can bear variable ratios of HS (solid lines) and CS (dotted lines) chains. Formation of an HS:FGF:FGFR ternary complex produces cell signaling (lightening bolt). For simplicity, asymmetric signaling complexes are predominantly depicted. **(A)** Potential forms of ternary complexes. The symmetric complex requires the free ends of two HS chains; whereas, the asymmetric complex requires only a single HS chain. **(B)** Extremely high HS levels potentially inhibit signaling by partitioning receptors and ligands onto separate HS chains, thereby preventing ternary complex formation. **(C)** Regulation of signaling by the extracellular matrix. Signaling can be inhibited by high levels of extracellular matrix HSPGs, such as perlecan (partially shown), which can sequester growth factors away from cell membrane receptors. Conditions such as tissue injury can liberate matrix-bound growth factors by cell-secreted proteases and heparanase, which initiate signaling by releasing HS:growth factor complexes that diffuse to the cell surface receptors. **(D)** Hypoxia enhances cellular responsiveness to FGF by increasing the HS content of heteroglycan-type HSPGs. **(E)** Certain FGFR splice variants can be non-glycanated or glycanated with either an HS or CS chain. The HS glycanated receptor can form a ternary complex in the absence of other HSPGs. **(F)** HS can regulate signaling of specific ligand:receptor isoforms. Shown are two cells exposed to two FGF isoforms and expressing two FGFR isoforms. HS sequence motifs (circles versus triangles) can determine which FGF:FGFR isoform combinations are functionally operable.

a transient phosphorylation of mitogen-activated protein kinase that does not lead to cell proliferation.[68] In contrast, sustained signaling that leads to mitogenesis requires the presence of HS.[69,70] Another difference is the nanomolar levels of growth factors can be greatly exceeded by the concentrations of pericellular HSPGs, which can achieve micromolar levels (reviewed in Ref. 16). HSPG concentration regulates signaling by dramatically influencing the stoichiometry of the ligand-receptor-HS ternary complex.

6.2.1. *Co-receptors engender stoichiometric control of signaling*

Since HS interacts with both FGFs and FGFRs, the levels of HSPG present on the cell surface tightly control the levels of functional HS:FGF:FGFR ternary complex. Quantitative modulation of signaling can be achieved; this occurs in developmental morphogen gradients, where the regional level of HSPG expression defines the local strength of growth factor signaling.[71] Much more extreme concentration effects can be demonstrated with HSPG-deficient cells, where the addition of exogenous HS/heparin restores FGF signaling. High concentrations of GAG inhibit FGF signaling but low concentrations are stimulatory.[62,72,73] It is thought that low HS/heparin concentrations promote ternary complex formation at the cell surface. In contrast, high GAG concentrations that exceed levels of receptors and ligands can prevent ternary complex formation by partitioning each component onto separate HS chains (Fig. 3B). Indeed, mitogenic signaling at high HS/heparin concentrations can be re-established by simply increasing the concentration of FGF.[74]

In vivo, high HS levels can clearly prevent ternary complex formation by a mechanism involving spatial segregation (Fig.3C). High levels of growth factors are bound to HSPGs of the endothelial basement membrane; however, the ligands are physically separated from their cell surface-bound receptors so signaling does not occur. Thus, growth factors are sequestered by the basement membrane. During vessel injury or remodeling, this reservoir of growth factors can be recruited. The actions of heparinase or proteases can release HS-growth factor complexes that diffuse to activate their cell surface receptors (reviewed in Refs. 12, 16 and 58).

6.2.2. *The effects of glycanation*

One means of regulating HS levels is to change glycanation, the type and number of GAG chains synthesized on a core protein. Glycanation control is germane to "heteroglycans" — where multiple GAG types occur on single core proteins. ECs express several heteroglycans (Table 2). For example, syndecan-4 exhibits three GAG attachment sites that can bear either HS or CS chains. Consequently, a variety of isoforms can be produced by a single cell type: pure CS-syndecan-4, pure HS-syndecan-4 and all possible permutations of heteroglycans.[75] The ratio of these forms is biosynthetically controlled in response to external stimuli. For example, exposure of ECs to hypoxia enhances cellular responsiveness to FGF2 by increasing the HS to CS ratio.[76] Core protein expression is not altered by hypoxia. Instead, hypoxia enhances the expression of EXTL2, which commits GAG attachment sites to synthesis of HS chains (Fig. 2). Major elevations in cell surface levels of HS occur because such biosynthetic control should increase the HS content on all heteroglycans expressed by a cell (Fig. 3D).

Glycanation of certain receptors results in additional signaling mechanisms (Fig. 3E). A single GAG chain occurs on a specific FGFR2 splice variant.[77] An alternate exon of the variant encodes a GAG attachment site that can bear either HS or CS chains. The HS-bearing FGFR2, compared to non-modified or CS modified receptors, exhibits high affinity for FGF1. In the absence of exogenous HS/heparin, only the HS modified receptor responds to FGF1 stimulation with sustained and enhanced signaling events that lead to mitogenesis.[77] Glycanation of FGFR3 may also be possible, as this receptor also has a similarly positioned potential GAG attachment site. Thus, the need for an HSPG co-receptor may be abrogated for select FGFRs that bear an HS chain. It follows that when cell surface levels of HSPGs are low, such receptors may play a dominant role in FGF signaling.

6.2.3. *HS sequence motifs regulate signaling*

The existence of distinct HS motifs provides multiple mechanisms for regulating signaling specificity. Specific HS motifs are critical for FGF signaling. For example, FGF2 binding requires an HS sequence with 2-*O*-sulfated iduronic acids.[78–80] A cell mutant deficient in HS2ST activity, which produces HS lacking 2-*O*-sulfated iduronic acid, confirms this necessity. Mutant cells can neither bind to nor respond to FGF2.[81] Successful signaling also requires 6-*O*-sulfates that form direct contacts with the FGFR.[66,67,82] As described above, this requirement is exemplified by a HS6ST-deficient *Drosophila* mutant, which consequently has defective FGF-driven tracheal morphogenesis.

HS motifs may also allow for temporal and cell type control of the specificity of ligand-receptor signaling. Across both the respective families of FGFs and FGFRs, the residues of the HS-binding regions are non-conserved (reviewed in Ref. 58). Consequently, distinct HS motifs are recognized by distinct isoforms of FGFs and FGFRs.[83] It follows that the structural composition of cellular HS will govern which ligand:receptor combinations are functionally operable and how efficiently they will signal.[84] Since HS structure varies between cell types, cell type-specific signaling can result. Two different cell types bearing the same FGFRs, but different HS motifs, can exhibit differential responsiveness to a given FGF isoform (Fig. 3F). Indeed, distinct FGF:FGFR combinations are activated by HS isolated from different EC phenotypes.[85] Physiological cues can also change the array of HS motifs on the cell surface, by altering the cellular expression of various HS sulfotransferases.[76] Thus, the growth factor responsiveness of a single cell type is modulated in a temporal fashion by regulation of HS synthesis.

The array of HS motifs expressed by a cell can also be modulated subsequent to HS synthesis. For example, endosulfatases (Sulf1 and Sulf2) secreted from cells can function in the extracellular environment to remove 6-*O*-sulfates from intact HS chains. These enzymes can influence FGF signaling, since 6-*O*-sulfates are required for HS interaction with certain FGFRs. Indeed, forced Sulf1 expression can inhibit formation of HS:FGF:FGFR ternary complexes and thereby prevents mesoderm induction and angiogenesis stimulated by FGF.[40] Thus, HS structure can be modulated to superimpose specificity control onto the myriad of FGFs and FGFRs.

7. HSPGs Enable Global Control of EC Phenotype

The ability of HS to interact with a vast number of growth factors serves to coordinate certain global phenotypes or processes. Such global trends are beginning to come to light. The endothelium of the adult is normally comprised of highly differentiated non-proliferating cells. As indicated above, the endothelial basement membrane contains high levels of HS, which sequesters a variety of heparin-binding growth factors and thereby limits their function (Fig. 3C). The importance of HS-bound growth factors stored in the basement membrane has recently been revealed by genetically engineered mice. On one hand, knock-in mice have been generated that express a HS-deficient form of perlecan, the major basement membrane HSPG. Mice expressing HS-deficient perlecan are viable but exhibit defective angiogenesis as evidenced by retarded FGF2-induced tumor growth and delayed wound healing.[50] These phenotypes are observed in the context of tissue damage, where the degradation of the basement membrane normally releases HS:growth factor complexes that stimulate EC proliferation. Thus, the low levels of basement membrane HS in these mice limits the reservoir of HS-bound growth factor that can be mobilized under injury. On the other hand, transgenic mice that overexpress heparanase, the heparan cleaving enzyme, should exhibit enhanced release of HS:growth factor complexes. Indeed, such mice show increased vascularization.[86] Combined, these results indicate that at least one function of HS of the endothelial basement membrane is to stimulate EC proliferation in response to tissue damage.

8. Future Therapeutic Directions

At present, HS-based therapeutics are only used for inhibition of coagulation. This application has evolved from the use of a very diverse mixture of molecules, unfractionated heparin, to the recent deployment of a pure synthetic pentasaccharide binding site for antithrombin, which is the active HS motif that conveys anticoagulant activity. The chemical synthesis of HS motifs is exceedingly complex and extremely difficult to scale up for industrial production. However, the recent cloning of HS degradation and biosynthetic enzymes should enable much more facile approaches that combine enzymatic and chemical methodologies. Thus, it is likely that the near future will see the therapeutic use of additional synthetic HS motifs. Potentially exploitable HS motifs should be identified through further studies of the sequence-specific nature of signaling by heparin-binding growth factors.

Given that angiogenesis is induced by the concerted action of a number of heparin-binding growth factors, this process is extremely amenable to targeting with HS-based therapeutics. Such therapeutics can serve to activate or inhibit this process. Inhibition of angiogenesis is desirable in the treatment of solid tumors, which stimulate tumor vascularization and tumor growth by secreting heparanase and heparin-binding growth factors. Optimal treatment will involve a mixture of HS motifs capable of binding specific growth factors but incapable of activating the corresponding receptors.[87,88] Treatment can also include an HS motif that inhibits tumor-secreted heparanase,[89] thereby preventing release of HS:growth factor complexes from the endothelial basement membrane.

Activation of angiogenesis is a potential means of treating ischemic disease by generating new blood vessels to bypass obstructed vessels. Endothelial cells downstream of an obstruction that experience hypoxia should exhibit enhanced HS levels and therefore enhanced sensitivity to growth factors. However, bypass vessel must originate upstream of the obstruction, where ECs are not hypoxic and so will likely show low responsiveness to growth factors. The need for endogenous cell surface HS can be circumvented by treating with a growth factor containing a covalently coupled HS fragment.[90] Moreover, specificity for a particular receptor isoform could be defined by the sequence of the attached HS motifs. Thus, HS:growth factor hybrids potentially afford far greater specificity than non-conjugated growth factors. Such specificity may even be capable of exploiting endothelial phenotypic diversity. For example, HS:growth factor hybrids can possibly allow for targeted activation of coronary versus peripheral vessels.

9. Conclusions

The deployment of HSPGs as co-receptors superimposes multiple levels of control over signaling by heparin-binding growth factors. On the global level, multiple signaling pathways can be regulated by altering the level of cell surface HS, which should influence the activity of multiple heparin-binding growth factors. Conversely, the signaling capacity of individual ligand:receptor combinations may be influenced by altered biosynthesis of specific HS sequence motifs. Multiple aspects of EC biology that are regulated by specific HS-motifs should be revealed by future studies of mice deficient in various HS-sulfotransferases. Such studies may additionally reveal novel therapeutic applications for synthetic HS motifs.

References

1. McLean, J (1916). The thromboplastic action of cephalin. *Am J Physiol* **41**: 250–257.
2. Röden, L (1989). Highlights in the history of heparin. In: Lane DA, Lindahl (eds.), *Heparin* (Edward Arnold, London), pp. 1–24.
3. Howell WH (1925). The purification of heparin and its presence in blood. *Am J Physiol* **71**: 553–562.
4. Contejean C (1895). Recherches sur les injections intraveineuses de peptone et leur influence sur la coagulabilite du sang chez le chien. *Arch Physiol Norm Pathol* **7**: 45–53.
5. Abildgaard U (1968). Highly purified antithrombin III with heparin cofactor activity prepared by disc gel electrophoresis. *Scand J Clin Lab Invest* **21**: 89–91.
6. Damus PS, Hicks M, Rosenberg RD (1973). Anticoagulant action of heparin. *Nature* **246**: 355–357.
7. Lam LH, Silbert JE, Rosenberg RD (1976). The separation of active and inactive forms of heparin. *Biochem Biophys Res Commun* **69**: 570–577.
8. Hook M, Bjork I, Hopwood J, Lindahl U (1976). Anticoagulant activity of heparin: separation of high-activity and low-activity heparin species by affinity chromatography on immobilized antithrombin. *FEBS Lett* **66**: 90–93.
9. Andersson LO, Barrowcliffe TW, Holmer E, *et al.* (1976). Anticoagulant properties of heparin fractionated by affinity chromatography on matrix-bound antithrombin III and by gel filtration. *Thromb Res* **9**: 575–583.
10. Jorpes JE, Holmgren H, Wilander O (1937). Über das Vorkommen von Heparin in den Gefässwänden und in den Augen. Ein beitrag zur Physiologie der Ehrlichschen Mastzellen. *Z Mikrosk Anat Forsch* **42**: 279–301.
11. Bernfield M, Gotte M, Park PW, *et al.* (1999). Functions of cell surface heparan sulfate proteoglycans. *Annu Rev Biochem* **68**: 729–777.
12. Esko JD, Selleck SB (2002). Order out of chaos: assembly of ligand binding sites in heparan sulfate. *Annu Rev Biochem* **71**: 435–471.
13. Rosenberg RD, Shworak NW, Liu J, *et al.* (1997). Heparan sulfate proteoglycans of the cardiovascular system. Specific structures emerge but how is synthesis regulated? *J Clin Invest* **99**: 2062–2070.

14. Shing Y, Folkman J, Sullivan R, *et al.* (1984). Heparin affinity: purification of a tumor-derived capillary endothelial cell growth factor. *Science* **223**: 1296–1299.

15. Silbert JE, Sugumaran G (2002). A starting place for the road to function. *Glycoconj J* **19**: 227–237.

16. Iozzo RV, San Antonio JD (2001). Heparan sulfate proteoglycans: heavy hitters in the angiogenesis arena. *J Clin Invest* **108**: 349–355.

17. Belting M (2003). Heparan sulfate proteoglycan as a plasma membrane carrier. *Trends Biochem Sci* **28**: 145–151.

18. Bottaro DP (2002). The role of extracellular matrix heparan sulfate glycosaminoglycan in the activation of growth factor signaling pathways. *Ann N Y Acad Sci* **961**: 158.

19. Sasisekharan R, Shriver Z, Venkataraman G Narayanasami U (2002). Roles of heparan-sulphate glycosaminoglycans in cancer. *Nat Rev Cancer* **2**: 521–528.

20. Zhang L, Yoshida K, Liu J, Rosenberg RD (1999). Anticoagulant heparan sulfate precursor structures in F9 embryonal carcinoma cells. *J Biol Chem* **274**: 5681–5691.

21. Shworak NW, Shirakawa M, Colliec-Jouault S, *et al.* (1994). Pathway-specific regulation of the synthesis of anticoagulantly active heparan sulfate. *J Biol Chem* **269**: 24941–24952.

22. Eriksson I, Sandbäck D, Ek B, *et al.* (1994). cDNA cloning and sequencing of mouse mastocytoma glucosaminyl *N*-deacetylase/*N*-sulfotransferase, an enzyme involved in the biosynthesis of heparin. *J Biol Chem* **269**: 10438–10443.

23. Hashimoto Y, Orellana A, Gil G, Hirschberg CB (1992). Molecular cloning and expression of rat liver *N*-heparan sulfate sulfotransferase. *J Biol Chem* **267**: 15744–15750.

24. Crawford BE, Olson SK, Esko JD, Pinhal MA (2001). Cloning, golgi localization, and enzyme activity of the full-length heparin/heparan sulfate-glucuronic acid C^5-epimerase. *J Biol Chem* **276**: 21538–21543.

25. Li J, Hagner-McWhirter A, Kjellen L, *et al.* (1997). Biosynthesis of heparin/heparan sulfate. cDNA cloning and expression of D-glucuronyl C^5-epimerase from bovine lung. *J Biol Chem* **272**: 28158–28163.

26. Rong J, Habuchi H, Kimata K, *et al.* (2001). Substrate specificity of the heparan sulfate hexuronic acid 2-*O*-sulfotransferase. *Biochemistry* **40**: 5548–5555.

27. Kobayashi M, Habuchi H, Yoneda M, *et al.* (1997). Molecular cloning and expression of Chinese hamster ovary cell heparan-sulfate 2-sulfotransferase. *J Biol Chem* **272**: 13980–13985.

28. Habuchi H, Kobayashi M, Kimata, K (1998). Molecular characterization and expression of heparan-sulfate 6-sulfotransferase. Complete cDNA cloning in human and partial cloning in Chinese hamster ovary cells. *J Biol Chem* **273**: 9208–9213.

29. Shworak NW, Fritze LM, Liu J, *et al.* (1996). Cell-free synthesis of anticoagulant heparan sulfate reveals a limiting activity which modifies a nonlimiting precursor pool. *J Biol Chem* **271**: 27063–27071.

30. Shworak NW, Liu J, Fritze LMS, *et al.* (1997). Molecular cloning and expression of mouse and human cDNAs encoding heparan sulfate D-glucosaminyl 3-*O*-sulfotransferase. *J Biol Chem* **272**: 28008–28019.

31. Shworak NW, Liu J, Petros LM, *et al.* (1999). Multiple isoforms of heparan sulfate D-glucosaminyl 3-*O*-sulfotransferase. Isolation, characterization, and expression of human cDNAs and identification of distinct genomic loci. *J Biol Chem* **274**: 5170–5184.
32. Aikawa J, Grobe K, Tsujimoto M, Esko JD (2001). Multiple isozymes of heparan sulfate/heparin GlcNAc *N*-deacetylase/GlcN *N*-sulfotransferase. Structure and activity of the fourth member, NDST4. *J Biol Chem* **276**: 5876–5882.
33. Habuchi H, Tanaka M, Habuchi O, *et al.* (2000). The occurrence of three isoforms of heparan sulfate 6-*O*-sulfotransferase having different specificities for hexuronic acid adjacent to the targeted *N*-sulfoglucosamine. *J Biol Chem* **275**: 2859–2868.
34. Colliec-Jouault S, Shworak NW, Liu J, *et al.* (1994). Characterization of a cell mutant specifically defective in the synthesis of anticoagulantly active heparan sulfate. *J Biol Chem* **269**: 24953–24958.
35. Kamimura K, Rhodes JM, Ueda R, *et al.* (2004). Regulation of Notch signaling by Drosophila heparan sulfate 3-*O*-sulfotransferase. *J Cell Biol* **166**: 1069–1079.
36. Shukla D, Liu J, Blaiklock P, *et al.* (1999). A novel role for 3-*O*-sulfated heparan sulfate in herpes simplex virus 1 entry. *Cell* **99**: 13–22.
37. Xia G, Chen J, Tiwari V, *et al.* (2002). Heparan sulfate 3-*O*-sulfotransferase isoform 5 generates both an antithrombin-binding site and an entry receptor for herpes simplex virus, type 1. *J Biol Chem* **277**: 37912–37919.
38. Vlodavsky I Friedmann Y (2001). Molecular properties and involvement of heparanase in cancer metastasis and angiogenesis. *J Clin Invest* **108**: 341–347.
39. Fransson LA, Belting M, Cheng F, *et al.* (2004). Novel aspects of glypican glycobiology. *Cell Mol Life Sci* **61**: 1016–1024.
40. Wang S, Ai X, Freeman SD, *et al.* (2004). QSulf1, a heparan sulfate 6-*O*-endosulfatase, inhibits fibroblast growth factor signaling in mesoderm induction and angiogenesis. *Proc Natl Acad Sci USA* **101**: 4833–4838.
41. DeAngelis PL (2002). Evolution of glycosaminoglycans and their glycosyltransferases: implications for the extracellular matrices of animals and the capsules of pathogenic bacteria. *Anat Rec* **268**: 317–326.
42. Medeiros GF, Mendes A, Castro RA, *et al.* (2000). Distribution of sulfated glycosaminoglycans in the animal kingdom: widespread occurrence of heparin-like compounds in invertebrates. *Biochim Biophys Acta* **1475**: 287–294.
43. Dewel RA (2000). Colonial origin for Emetazoa: major morphological transitions and the origin of bilaterian complexity. *J Morphol* **243**: 35–74.
44. Lee JS, Chien CB (2004). When sugars guide axons: insights from heparan sulphate proteoglycan mutants. *Nat Rev Genet* **5**: 923–935.
45. Lawson ND, Vogel AM, Weinstein BM (2002). *sonic hedgehog* and *vascular endothelial growth factor* act upstream of the Notch pathway during arterial endothelial differentiation. *Dev Cell* **3**: 127–136.
46. Kamimura K, Fujise M, Villa F, *et al.* (2001). Drosophila heparan sulfate 6-*O*-sulfotransferase (dHS6ST) gene. Structure, expression and function in the formation of the tracheal system. *J Biol Chem* **276**: 17014–17021.
47. Carmeliet P, Ng YS, Nuyens D, *et al.* (1999). Impaired myocardial angiogenesis and ischemic cardiomyopathy in mice lacking the vascular endothelial growth factor isoforms VEGF164 and VEGF188. *Nat Med* **5**: 495–502.

48. Ruhrberg C, Gerhardt H, Golding M, *et al.* (2002). Spatially restricted patterning cues provided by heparin-binding VEGF-A control blood vessel branching morphogenesis. *Genes Dev* **16**: 2684–2698.

49. Chen E, Hermanson S, Ekker SC (2004). Syndecan-2 is essential for angiogenic sprouting during zebrafish development. *Blood* **103**: 1710–1719.

50. Zhou Z, Wang J, Cao R, *et al.* (2004). Impaired angiogenesis, delayed wound healing and retarded tumor growth in perlecan heparan sulfate-deficient mice. *Cancer Res* **64**: 4699–4702.

51. Bullock SL, Fletcher JM, Beddington RS, Wilson VA (1998). Renal agenesis in mice homozygous for a gene trap mutation in the gene encoding heparan sulfate 2-sulfotransferase. *Genes Dev* **12**: 1894–1906.

52. Merry CL, Bullock SL, Swan DC, *et al.* (2001). The molecular phenotype of heparan sulfate in the $Hs2st^{-/-}$ mutant mouse. *J Biol Chem* **276**: 35429–35434.

53. Li JP, Gong F, Hagner-McWhirter A, *et al.* (2003). Targeted disruption of a murine glucuronyl C^5-epimerase gene results in heparan sulfate lacking *L*-iduronic acid and in neonatal lethality. *J Biol Chem* **278**: 28363–28366.

54. Ledin J, Staatz W, Li JP, *et al.* (2004). Heparan sulfate structure in mice with genetically modified heparan sulfate production. *J Biol Chem* **279**: 42732–42741.

55. Ringvall M, Ledin J, Holmborn K, *et al.* (2000). Defective heparan sulfate biosynthesis and neonatal lethality in mice lacking *N*-deacetylase/*N*-sulfotransferase-1. *J Biol Chem* **275**: 25926–25930.

56. HajMohammadi S, Enjyoji K, Princivalle M, *et al.* (2003). Normal levels of anticoagulant heparan sulfate are not essential for normal hemostasis. *J Clin Invest* **111**: 989–999.

57. Wang L, Fuster M, Sriramarao P, Esko JD (2005). Endothelial heparan sulfate deficiency impairs L-selectin- and chemokine-mediated neutrophil trafficking during inflammatory responses. *Nat Immunol* **6**: 902–910.

58. Pellegrini L (2001). Role of heparan sulfate in fibroblast growth factor signalling: a structural view. *Curr Opin Struct Biol* **11**: 629–634.

59. Olwin BB, Rapraeger A (1992). Repression of myogenic differentiation by aFGF, bFGF, and K-FGF is dependent on cellular heparan sulfate. *J Cell Biol* **118**: 631–639.

60. Ornitz DM, Leder P (1992). Ligand specificity and heparin dependence of fibroblast growth factor receptors 1 and 3. *J Biol Chem* **267**: 16305–16311.

61. Zhou FY, Kan M, Owens RT, *et al.* (1997). Heparin-dependent fibroblast growth factor activities: effects of defined heparin oligosaccharides. *Eur J Cell Biol* **73**: 71–80.

62. Bonneh-Barkay D, Shlissel M, Berman B, *et al.* (1997). Identification of glypican as a dual modulator of the biological activity of fibroblast growth factors. *J Biol Chem* **272**: 12415–12421.

63. Aviezer D, Hecht D, Safran M, *et al.* (1994). Perlecan, basal lamina proteoglycan, promotes basic fibroblast growth factor-receptor binding, mitogenesis, and angiogenesis. *Cell* **79**: 1005–1013.

64. Aviezer D, Iozzo RV, Noonan DM, Yayon, A (1997). Suppression of autocrine and paracrine functions of basic fibroblast growth factor by stable expression of perlecan antisense cDNA. *Mol Cell Biol* **17**: 1938–1946.

65. Moy FJ, Safran M, Seddon AP, *et al.* (1997). Properly oriented heparin-decasaccharide-induced dimers are the biologically active form of basic fibroblast growth factor. *Biochemistry* **36**: 4782–4791.

66. Schlessinger J, Plotnikov AN, Ibrahimi OA, *et al.* (2000). Crystal structure of a ternary FGF-FGFR-heparin complex reveals a dual role for heparin in FGFR binding and dimerization. *Mol Cell* **6**: 743–750.

67. Pellegrini L, Burke DF, von Delft F, *et al.* (2000). Crystal structure of fibroblast growth factor receptor ectodomain bound to ligand and heparin. *Nature* **407**: 1029–1034.

68. Fannon M, Nugent MA (1996). Basic fibroblast growth factor binds its receptors, is internalized, and stimulates DNA synthesis in Balb/c3T3 cells in the absence of heparan sulfate. *J Biol Chem* **271**: 17949–17956.

69. Delehedde M, Lyon M, Gallagher JT, *et al.* (2002). Fibroblast growth factor-2 binds to small heparin-derived oligosaccharides and stimulates a sustained phosphorylation of p42/44 mitogen-activated protein kinase and proliferation of rat mammary fibroblasts. *Biochem J* **366**: 235–244.

70. Delehedde M, Seve M, Sergeant N, *et al.* (2000). Fibroblast growth factor-2 stimulation of p42/44MAPK phosphorylation and IkappaB degradation is regulated by heparan sulfate/heparin in rat mammary fibroblasts. *J Biol Chem* **275**: 33905–33910.

71. Fujise M, Takeo S, Kamimura K, *et al.* (2003). Dally regulates Dpp morphogen gradient formation in the Drosophila wing. *Development* **130**: 1515–1522.

72. Jang JH, Wang F, Kan, M (1997). Heparan sulfate is required for interaction and activation of the epithelial cell fibroblast growth factor receptor-2IIIb with stromal-derived fibroblast growth factor-7. *In Vitro Cell Dev Biol Anim* **33**: 819–824.

73. Reich-Slotky R, Bonneh-Barkay D, Shaoul E, *et al.* (1994). Differential effect of cell-associated heparan sulfates on the binding of keratinocyte growth factor (KGF) and acidic fibroblast growth factor to the KGF receptor. *J Biol Chem* **269**: 32279–32285.

74. Padera R, Venkataraman G, Berry D, *et al.* (1999). FGF-2/fibroblast growth factor receptor/heparin-like glycosaminoglycan interactions: a compensation model for FGF-2 signaling. *FASEB J* **13**: 1677–1687.

75. Shworak NW, Shirakawa M, Mulligan RC Rosenberg RD (1994). Characterization of ryudocan glycosaminoglycan acceptor sites. *J Biol Chem* **269**: 21204–21214.

76. Li J, Shworak NW, Simons, M (2002). Increased responsiveness of hypoxic endothelial cells to FGF2 is mediated by HIF-1alpha-dependent regulation of enzymes involved in synthesis of heparan sulfate FGF2-binding sites. *J Cell Sci* **115**: 1951–1959.

77. Sakaguchi K, Lorenzi MV, Bottaro DP, Miki, T (1999). The acidic domain and first immunoglobulin-like loop of fibroblast growth factor receptor 2 modulate downstream signaling through glycosaminoglycan modification. *Mol Cell Biol* **19**: 6754–6764.

78. Habuchi H, Suzuki S, Saito T, *et al.* (1992). Structure of a heparan sulphate oligosaccharide that binds to basic fibroblast growth factor. *Biochem J* **285**: 805–813.

79. Maccarana M, Casu B, Lindahl, U (1993). Minimal sequence in heparin/heparan sulfate required for binding of basic fibroblast growth factor. *J Biol Chem* **268**: 23898–23905.

80. Turnbull JE, Fernig DG, Ke Y, *et al.* (1992). Identification of the basic fibroblast growth factor binding sequence in fibroblast heparan sulfate. *J Biol Chem* **267**: 10337–10341.

81. Bai X, Esko JD (1996). Mutant defective in heparan sulfate hexuronic acid 2-*O*-sulfation. *J Biol Chem* **271**: 17711–17717.

82. Guimond S, Maccarana M, Olwin BB, *et al.* (1993). Activating and inhibitory heparin sequences for FGF-2 (basic FGF). Distinct requirements for FGF-1, FGF-2 and FGF-4. *J Biol Chem* **268**: 23906–23914.

83. Guimond SE, Turnbull JE (1999). Fibroblast growth factor receptor signalling is dictated by specific heparan sulphate saccharides. *Curr. Biol* **9**: 1343–1346.

84. Ostrovsky O, Berman B, Gallagher J, *et al.* (2002). Differential effects of heparin saccharides on the formation of specific fibroblast growth factor (FGF) and FGF receptor complexes. *J Biol Chem* **277**: 2444–2453.

85. Knox S, Merry C, Stringer S, *et al.* (2002). Not all perlecans are created equal: interactions with fibroblast growth factor (FGF) 2 and FGF receptors. *J Biol Chem* **277**: 14657–14665.

86. Zcharia E, Metzger S, Chajek-Shaul T, *et al.* (2004). Transgenic expression of mammalian heparanase uncovers physiological functions of heparan sulfate in tissue morphogenesis, vascularization, and feeding behavior. *FASEB J* **18**: 252–263.

87. Presta M, Oreste P, Zoppetti G, *et al.* (2004). Antiangiogenic activity of semisynthetic biotechnological heparins. Low-molecular-weight-sulfated escherichia coli K5 polysaccharide derivatives as fibroblast growth factor antagonists. *Arterioscler Thromb Vasc Biol* **25**: 71–76.

88. Miao HQ, Ornitz DM, Aingorn E, *et al.* (1997). Modulation of fibroblast growth factor-2 receptor binding, dimerization, signaling, and angiogenic activity by a synthetic heparin-mimicking polyanionic compound. *J Clin Invest* **99**: 1565–1575.

89. Ferro V, Hammond E, Fairweather JK (2004). The development of inhibitors of heparanase, a key enzyme involved in tumour metastasis, angiogenesis and inflammation. *Mini Rev Med Chem* **4**: 693–702.

90. Pye DA, Gallagher JT (1999). Monomer complexes of basic fibroblast growth factor and heparan sulfate oligosaccharides are the minimal functional unit for cell activation. *J Biol Chem* **274**: 13456–13461.

6

Directional Cues in Angiogenesis

by Arie Horowitz

1. Introduction: Blood Vessels and Nerves Use Similar Guidance Cues

The investigation of guidance cues in the vascular system is a relatively new field — the first studies focusing on this topic came out only in the last six years. Guidance cues were no longer a new subject at that time, however, since they had been discovered earlier in the nervous system (reviewed in Ref. 1) and studied since intensively. It has become soon evident, however, that both systems largely use the same types of ligands and receptors for guidance. This commonality is not entirely surprising if one considers the morphological similarities between the two systems: they are both made up of a branching network which conducts bidirectionally either blood or electric signals, namely, arteries versus veins in the vascular system and sensory versus motor nerve fibers in the nervous system. Furthermore, the branches of both the vascular and the nervous systems are bi-layered and each is composed of two types of cells: mural and endothelial, or glia and neurons, respectively.

The structural similarity between the nervous and vascular systems extends to the growing neuron and the sprouting capillary. Axon path-finding is controlled by the growth cone, a lamellipodial structure present at the tip of growing axons which senses the gradients of chemoattractants and repellants.[2] Recent findings[3] revealed that a similar structure is present at the end of new capillaries, which are capped by specialized endothelial cells referred to as tip cells. Similar to growth cones, the tip cells probe their local environment by extending filopodia.

Though the morphological similarities between the vascular and nervous systems have been long noticed, their underlying reasons were only recently accounted for by invoking a mechanism of paracrine mutual guidance — each system secretes chemoattractants that act on its counterpart. For example, blood vessels secrete artemin[4] and neurotrophin[5] which recruit axons to grow alongside, while nerves secrete several VEGF isoforms[6] that guide blood vessels to follow along the same paths.

Numerous secreted factors such as the above are chemoattractants and therefore can be functionally considered as guidance cues. This term has evolved, however, to define a narrower group of molecules which includes both secreted and membrane proteins, some of which convey repellant rather than attractant signals. This review will focus on the members of this group, currently consisting of the semaphorins, neuropilins, plexins, ephrins and the Eph receptors, netrins and UNC5s, and slits and roundabouts (Robo).

The signaling of all the four classes of guidance cues described here is not strictly repulsive but bifunctional: it can be modulated by other factors and conditions so as to result in either repulsion or attraction. Cell repulsion is brought about by several changes in cell morphology, adhesion and migration: the cellular cytoskeleton is altered, lamellipodia are retracted, the number of stress fibers increases, and the cell contracts and reduces the area of its contact with the substrate. The cell retracts from sources of soluble repellants, or dissociates from cells presenting repellants on their plasma membrane. This review will describe, therefore, signaling events relevant to these processes (Fig. 1).

2. Semaphorin Signaling

One of the earliest indications for the presence of neuronal receptors in the vascular system was the discovery that neuropilin-1 (Npn-1),

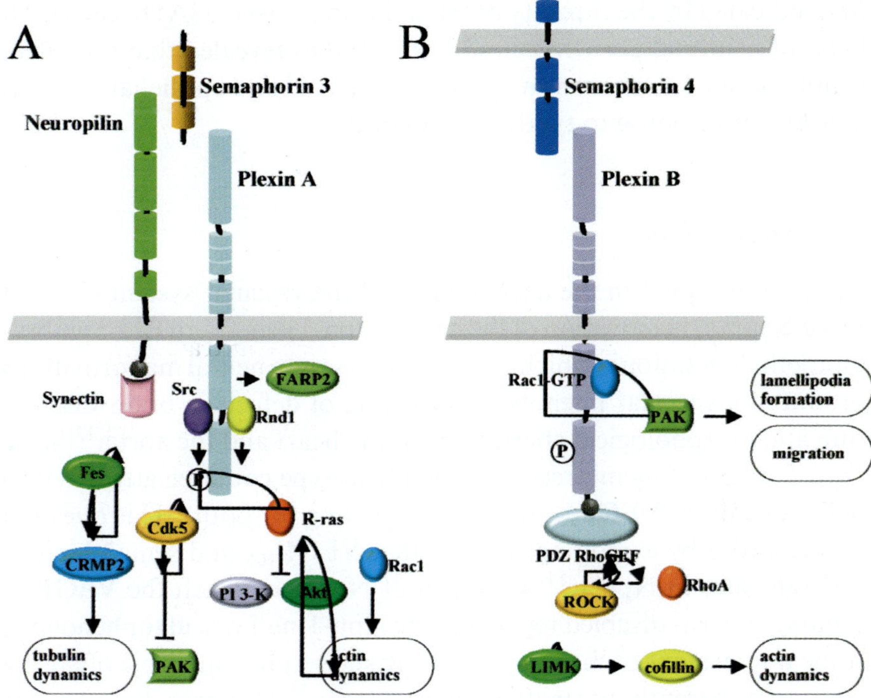

Fig. 1. **(A)** Signaling of secreted sema3 via class A plexin and neuropilin. **(B)** Signaling of membrane-bound semaphorins via class B plexin.

previously known as a semaphorin receptor in neurons, is expressed by endothelial cells and functions as a specific receptor of $VEGF_{165}$,[7] in tandem with the established tyrosine kinase VEGF receptor VEGFR-2. Neuropilins are expressed as two isoforms and participate with several members of the plexin family as well as other membrane receptors in binding secreted class 3 semaphorins.

Semaphorins consist of more than 20 members in vertebrates (see Chapter 1), that can be grouped in several classes based on sequence homology: class 3 semaphorins are secreted, while semaphorins of classes 4, 5 and 6 are transmembrane proteins and semaphorin 7A is glycosylphosphatidylinositol (GPI)-anchored in the plasma membrane (Fig. 1). The secreted class 3 semaphorins are probably the most extensively studied group. The repellant nature of semaphorin signaling was first indicated by the capacity of semaphorin 3A (sema3A) to cause collapse of neuronal growth cones.[8] Later studies revealed that this effect is not confined to neurons as both endothelial[9] and epithelial[10,11] cells contract in response to type 3 semaphorins.

2.1. *Neuropilins*

The role of Npn-1 in the development of the vascular system was first revealed upon inactivation of the mouse Npn-1 gene — an E12.5 embryonic lethal mutation[12] which in addition to severe neural malformations produced a vascular phenotype consisting of defects in brain vasculature and morphological aberrations of the heart and the aorta.[13] Since Npn-1 is a dual-ligand receptor, the phenotype could be attributed to defects in either $VEGF_{165}$ or sema3 signaling (or both). This question was resolved by exploiting the fact that $VEGF_{165}$ and sema3 bind to separate sites in Npn-1.[14] Knock-in of Npn-1 in which the $VEGF_{165}$ binding site was disabled replicated the Npn-1-null vascular phenotype, while impairment of the sema3 binding site resulted in viable offspring with an apparently normal vascular system.[15] However, the vascular effects of the latter mutation could have been abrogated by redundancy between Npn-1 and Npn-2 signaling. Indeed, Npn-2 deletion together with knock-in of Npn-1 that does not bind sema3 produced cardiac outflow tract defects. Strikingly, similar defects were observed in sema3C-null mice,[16] suggesting that sema3C modulates cardiac development via both Npn-1 and Npn-2.

Unlike Npn-1 deletion, Npn-2 null mice are viable and do not exhibit obvious vascular defects. While the relative significance of VEGF versus sema3 binding to Npn-2 for vascular development was not dissected as in the case of Npn-1, a double knock-out of both Npn-1 and Npn-2 caused embryonic lethality earlier than Npn-1 knock-out alone,[17] including a defect in the formation of the vascular plexus of embryonic yolk sac. Further analysis of the Npn-2-null mouse revealed an overall absence of lymphatic small vessels and capillaries.[18] The differences between the phenotypes of the Npn-1 and Npn-2-null mice reflect the mostly non-overlapping expression patterns of the two neuropilins. Npn-1 is expressed primarily in arteries and is absent from lymphatic vessels, where Npn-2 is the sole isoform.[18] Npn-2 is also the dominant isoform in veins.

Despite their high degree of homology, the two neuropilins differ in their affinities to class 3 semaphorins. While sema3A binds preferentially to Npn-1, sema3C and 3F have higher affinity to Npn-2.[19] With the exception of sema3E, sema3 signaling requires interaction with a heterodimer consisting of a plexin and a neuropilin molecule, wherein neuropilin provides the binding site.[20] The neuropilins play no part, however, in transducing the effects of sema3 binding into the cell.[21] The signaling downstream of the sema3 receptor complex is initiated by members of the plexin family, which unlike the neuropilins have a large cytoplasmic domain capable of binding several GTPases and possesses GTPase activating protein (GAP) catalytic activity towards R-Ras.[22]

2.2. Plexins

Several of the nine-member plexin family heterodimerize with either Npn-1, Npn-2, or both, forming together a receptor complex for secreted class 3 semaphorins.[23] Similar to other components of guidance cue signaling pathways, the plexins were initially characterized as morphogens of the nervous system. While most plexins are expressed in endothelial cells, no specific functions in the vascular system were attributed to individual members of the plexin family. The first and currently still the only plexin associated with distinct roles in the vascular system is plexin D1 (plexD1), which has widespread expression in vascular endothelial cells of both the embryo and adult mouse.[24,25] Similarly, plexD1 is expressed in the vasculature of the developing zebrafish, and its knock-down by anti-sense morpholinos resulted in defective branching and invasion of the interesegmental vessels into the somites, a region which these vessels normally avoid in the wild type (WT) zebrafish.[26] This phenotype was replicated by knock-down of the zebrafish sema3a1 and sema3a2, orthologs of mammalian sema3A, indicating that these semaphorins, which are expressed in the somites, are the likely plexD1 ligands, though the extent of defective patterning was much lower than the one produced by plexD1 knock-down. Thus the confinement of the developing vasculature to the interesegmental space is the combined effect of a repellant secreted by the surrounding tissue — the somites, and a receptor of the repellant expressed on the surface of the vascular endothelial cells. PlexD1 loss of function was determined (ibid) to underlie also the excessive sprouting of blood vessels previously observed in the chemically induced *out-of-bounds* zebrafish mutant.[27]

The phenotype of the PlexD1 deletion in the mouse resembled plexD1 knock-down in the zebrafish. It produced unregulated growth of interesegmental blood vessels,[25] and affected also the heart, which had a denser coronary vessel network and cardiac outflow tract defects. Interestingly, this phenotype resembled that of the sema3C-null mouse,[16] as well as those of endothelial-specific deletion of Npn-1 and knock-in of mutant Npn-1 that does not bind semaphorin (Npn-1$^{Sema-/-}$) in combination with Npn-2 knock-out.[15] The co-expression of plexD1, Npn-1 and Npn-2 in the endothelium of the aorta, combined with the expression of sema3C in the outflow tract myocardium suggested that a mechanism analogous to the one in the zebrafish underlies the morphogenesis of the mouse heart. A somewhat different plexD1-dependent mechanism appears to operate in the vasculature, as unlike in the heart, the plexD1 ligand affecting the patterning of the intersomitic blood vessels is sema3E.[11] The expression pattern of plexD1 and sema3E in the embryonic trunk are complementary to each other, as plexD1 is expressed in the intersomitic vasculature while sema3E is expressed in the somites. Surprisingly, and contrary to all other class 3 semaphorins, sema3E binding to plexD1 did not require a neuropilin co-receptor, since sema3E bound tissue sections of Npn-2-null and Npn-1$^{Sema-/-}$-expressing embryos.[11] The predominance of plexD1-sema3E signaling in the patterning of the vasculature is supported also by the above observation that the vascular patterning defects of sema3a1 and 3a2 knockdown were minor relative to plexD1 knock-down, and by the fact that such defects were not detected in all genetic backgrounds and were incompletely penetrant.[28]

The knowledge of the signaling downstream of the plexins is still fragmentary. Several plexin-binding proteins and their concomitant effectors are known, but the pathway leading from plexin to the activation of proteins that regulate actin dynamics is not fully mapped. In particular, there is still no data concerning the signaling downstream of plexD1, the plexin that appears to be the most relevant one to the morphogenesis of the cardiovascular system. *In vitro* experiments have shown that semaphorin treatment inhibits endothelial cell migration, adhesion, and *in vitro* tube formation.[9,28,29] Similar to their effect on neurons, secreted semaphorins induce the collapse of endothelial cells[30], a process that requires comprehensive re-arrangement of the actin cytoskeleton. Indeed, one of the major classes of effectors of semaphorin signaling via plexin are Rho-GTPases, proteins known to regulate actin dynamics. The signaling events downstream of the class A and class B plexins, representing receptors of secreted sema3 and of membrane-bound semaphorins, respectively, are probably the best studied ones. Based on these studies, it appears that the signaling pathway activated by the secreted semaphorins differs from the one activated by membrane-bound semaphorins. The differences are reflected primarily in the type of Rho-GTPases associated with each class of plexins.

A theme common to both plexin-A and plexin-B signaling is the antagonistic regulation of Rho-GTPases. Sema3 signaling requires the activities of Rac1 and Cdc42,[32,33] two Rho-GTPases that induce growth of lamellipodia or filopodia, respectively. Rac1 activation results from the dissociation of the Rac1 GEF FARP2 from the plexA1 cytoplasmic domain and its subsequent activation.[34] Class A plexins bind the Rho-like GTPase Rnd1, and the Rho-GTPase RhoD.[31,35] Rnd1, which binds also class B plexins,[36] is constitutively active and is known to induce stress fiber disassembly and cause cell detachment and rounding.[37] Similarly, Rnd1 binding triggers plexB1 GAP activity towards R-Ras, thus precipitating cell collapse.[22] Both the binding to and activation of plexA1 by Rnd1 are antagonized by GTP-bound active RhoD, a GTPase supporting the motility of early endosomes along actin filaments.[38] Though it is still unknown how RhoD blocks sema3A-induced cell collapse, its association with membrane traffic suggests that it may promote the removal of the sema3A receptors from the plasma membrane.

The downstream effectors of the GAP activity of class A and class B plexins, and the resulting deactivation of R-Ras are phosphatidylinositol 3-kinase (PI 3-K) and Akt, both of which are inactivated[39,40] upon sema4D binding to plexA1 and plexB1. PI 3-K inhibition inhibits β1 integrin, and, subsequently, cell migration.[40] These outcomes are similar to the effects of sema3A binding to plexA/Npn1 receptors.[28]

Class B plexins bind both Rac1 and RhoA. Rac1 binding is ligand- and activation state-dependent, i.e. binding occurs upon sema4D interaction with plexB1,[41] and is selective for GTP-bound Rac1.[42] Experiments in Drosophila suggest that binding to plexB inactivates Rac,[43] possibly by competing with Rac binding to its effector p21-activated kinase (PAK). Conversely, binding to plexB upregulates RhoA activity in Drosophila.[43] These coupled and opposite effects on Rac versus RhoA activity appear to be required for plexB-dependent morphogenesis of the Drosophila nervous system. While there is no evidence for RhoA binding to plexB in vertebrates, Rac1 does bind plexB, resulting in the inhibition of PAK.[44] Since PAK promotes the formation of lamellipodia and the disassembly of stress fibers,[45] its inhibition would result in the typical effects of semaphorin: retraction of lamellipodia and cell contraction.

Vertebrate class B plexins are linked in more than one way to Rho-GTPase signaling, as their carboxy-termini bind to the Postsynaptic density 95, Disk large, Zona occludens-1 (PDZ) domains of the RhoA guanine exchange factors (GEF) PDZ-RhoGEF (PRG) and leukemia-associated Rho GEF (LARG).[46–48] There is no consensus, however, over which plexB-dependent mechanisms regulate the activity of these RhoGEFs. Sema4D-induced neuronal growth cone collapse requires direct binding of PRG to plexB1,[46,48] is dependent on PRG activity,[48] and is enhanced by Rnd1 binding to plexB.[36] Ligand-induced dimerization of PlexB activated RhoA, but failed to do so when the PDZ-binding motif of plexB was deleted.[47] Dimerization of PRG and LARG has indeed been shown to stimulate their GTP exchange activity.[49] Targeting of both plexB and PRG to the plasma membrane appeared to require interaction between the two proteins, and facilitated the plexB-dependent activation of its associated RhoGEF.[48,50]

While semaphorin signaling via class B plexins clearly involves Rho-GTPases, there is no evidence to indicate that the plexB-associated RhoA is activated by RhoGEFs bound to the carboxy-terminus of plexB. Moreover, the signaling downstream of the GTPases activated by semaphorin binding to plexins are only partially known. Actin depolymerization by cofillin is probably one of the sema3A effects involved in the collapse of the actin cytoskeleton, as cofillin is activated by LIM-kinase dependent phosphorylation upon sema3A treatment.[51]

Similar to many plasma membrane receptors, plexins undergo tyrosine phosphorylation upon ligand binding. The non-receptor tyrosine kinase Fes is associated with plexA1, and is inhibited by Npn-1.[52] The manner in which Npn-1 inhibits Fes is not known. It could conceivably be allosteric in nature, e.g. blockage of the Fes catalytic site by the cytoplasmic domain of Npn-1. Sema3A binding releases the Npn-1 inhibition and activates Fes, which phosphorylates the cytoplasmic domain of plexA1 as well as collapsin response mediator 2 (CRMP2), a tubulin-binding cytoplasmic protein[53] and one of the first components of the semaphorin signaling pathway to be identified.[54] While Fes activity was required for sema3A-induced cell collapse, the Fes phosphorylation site in plexA1 was not identified, nor was this phosphorylation shown to be required for sema3A signaling. In contrast, a CRMP2 phosphorylation site was identified at serine 522, and its replacement by alanine impaired sema3A signaling.[55] Since CRMP2 promotes tubulin assembly,[56] it is likely that its phosphorylation on serine 522 antagonizes this activity and contributes to the collapse of the cellular cytoskeleton. While the collapse of the actin cytoskeleton is mediated by Rho-GTPases, it appears that CRMP2 regulates a separate branch of semaphorin signaling whose downstream effector are the microtubules. Class A plexins bind also Src and the closely related kinase Fyn.[52] Src and Fyn are activated upon sema3A binding and phosphorylate both plexA and Cdk5, a kinase expressed primarily in neurons but also in endothelial cells,[57] which phosphorylates and inhibits PAK.[58] Thus both plexA and plexB signaling appear to inactivate PAK, though via different pathways. Sema3A-dependent activation of Cdk5 resulted also in the phosphorylation of tau, which decreases its affinity to microtubules,[59] consequently reducing microtubule stability.

3. Ephrins and Eph Signaling

Ephrins are membrane-bound ligands of the tyrosine kinase Eph receptors. The tissue expression patterns of Eph receptors and ephrin ligands are complementary, similar to that of semaphorin and their plexin receptors, so that each of the two types of proteins is expressed along the boundaries of apposing domains.[60] Ephrins and their receptors participate in determining the body plan of the developing embryo by providing a repulsive signal during embryogenesis which prevents intermingling of structurally and functionally distinct domains. Ephrins are classified into two classes, depending on their interaction with the plasma membrane: class A ephrins (A1–A5) are GPI-anchored, while class B (B1–B3) are single-pass transmembrane proteins. The structure of the Eph receptors and ephrin ligands is described in detail in Chapter 2. Eph receptors generally, though not exclusively, bind only to one class of ephrins, and are classified accordingly (A1–A8, B1–B6). The interactions between Eph receptors and ephrin ligands within each class are promiscuous, with the exception of the EphB4 receptor in the vascular system, which binds with high affinity only to ephrin-B2. Since both receptors and ligands are attached to the plasma membrane, ephrin signaling requires cell-to-cell contact.

Similar to semaphorin and its receptors, Eph and ephrins were initially studied as neuronal guidance proteins (reviewed in Refs. 61 and 62). The role of the Eph-ephrin system in vascular morphogenesis became apparent when disruption of the mouse ephrin-B2 gene was found to result in defective arterial and venous angiogenesis and in embryonic lethality, though vasculogenesis of the major blood vessels was not affected.[63] The expression pattern of ephrin-B2 was complementary to that of one of its ligands, EphB4, such that they were expressed preferentially in arteries or in veins, respectively. It was subsequently observed that EphB4 is expressed exclusively in endothelial cells, and that an EphB4 loss-of-function mutation duplicates the ephrin-B2-null phenotype.[64] As discussed in more detail in Chapter 2, the reciprocal expression of ephrin-B2 and EphB4 functions as a mechanism for defining the identity of arteries and veins and for maintaining the boundary between them, as well as between blood vessels and their surrounding tissue.[63,65,66] This is similar to the function of the Eph receptor-ephrin ligand system in the nervous system, where it marks the boundary between the developing nerves and their surrounding tissue, thus constituting a guidance mechanism.[61]

The roles of receptor versus ligand are interchangeable in Eph-ephrin signaling, since ephrin also transduces intracellular signals upon engagement by Eph (reviewed in Ref. 67) (Fig. 2). Thus Eph receptors initiate forward signaling, while ephrins give rise to reverse signaling. This bidirectional signaling is similar to the interaction between plexA1 and the transmembrane semaphorin sema6D.[68] Despite the absence of a cytoplasmic domain, GPI-anchored class A ephrins also transduce intracellular signals upon interaction with EphA receptors.[69−72] The

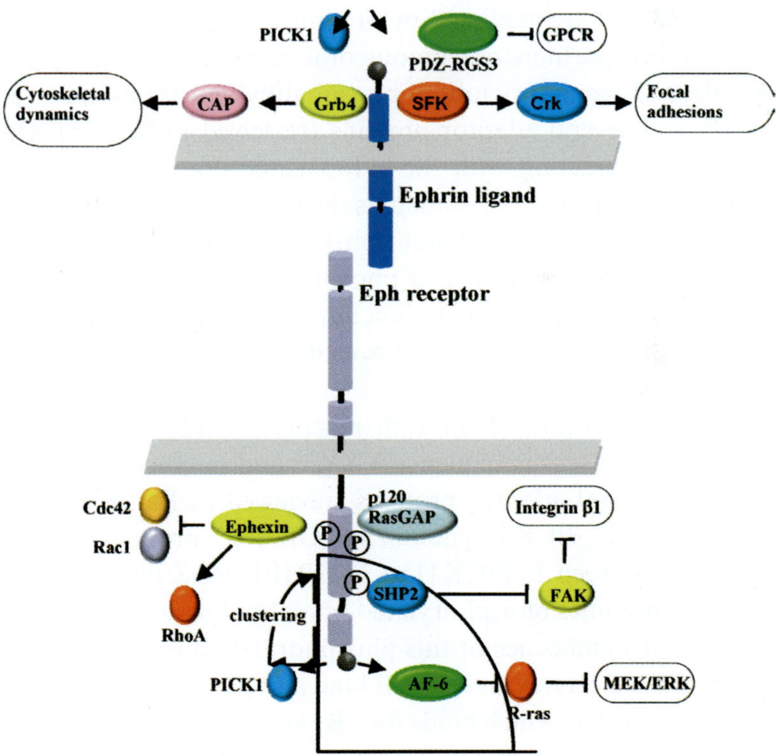

Fig. 2. Forward signaling via the Eph receptor and reverse signaling via the ephrin ligand.

mechanism by which this signaling occurs has not been fully elucidated (see below).

3.1. *Forward signaling*

Ephrin-stimulated forward signaling via Eph receptor requires clustering of the ephrins on the surface of the apposing cell.[73,74] Based on structural studies, it was proposed that the Eph signaling unit is a tetramer.[75] Eph tetramers undergo autophosphorylation on multiple tyrosines,[76] both in the juxtamembrane region, the kinase domain and the carboxy-terminus region. The phosphorylation of the juxtamembrane region releases its steric inhibition of the kinase domain,[77,78] and facilitates interaction with phosphotyrosines-binding proteins. Most of these interactions are mediated by Src homology 2 (SH2) domains of the Eph binding partners, and among others, include Src family kinases (SFK), the p85 regulatory unit of phosphatidylinositol 3-kinase, p120-RasGAP, and several adaptor proteins (reviewed in Ref. 76). One of the proteins interacting with the cytoplasmic domain of EphA and EphB receptors via an SH2 domain is the non-receptor tyrosine kinase Abl.[79] Recent results indicate that EphB4 stimulation by the ephrin-B2 ligand activates Abl, which then phosphorylates the adaptor protein Crk on Tyr221 and consequently uncouples it from the Crk-associated substrate p130(CAS).[80] In turn, dissociation of Crk from P130 (CAS) inhibits cell migration.[81]

The carboxy-termini of all Eph receptors conform to the PDZ-binding motif consensus sequence VXV,[82] and interact with the PDZ domains of several adaptor proteins (reviewed in Ref. 83). Some of these PDZ ligands, PICK1, syntenin and GRIP1,[82] are adaptor proteins that contain between 1 (PICK1) to 7 (GRIP1) PDZ domains. PICK1 clusters and becomes phosphorylated upon binding to EphB2, though the functional significance of this phosphorylation is not known. Eph receptors of both class A and class B bind and phosphorylate AF-6,[84] a large adaptor protein which binds Ras. By analogy to Bcr, another AF-6 tyrosine kinase,[85] AF-6 phosphorylation by Eph receptors may facilitate Ras binding to AF-6, thus segregating it away from its effector Raf-1 and consequently downregulating the Raf/MEK/ERK pathway.[82,84]

Similar to plexins, the main effectors of the Eph receptors are cytoskeleton and cell adhesion proteins. The functions of these proteins are regulated by the Eph receptors via several signaling pathways. The kinase domain of the EphA receptors binds ephexin, a RhoA, Rac1 and Cdc42-activating GEF,[86] and upon stimulation by the ephrin-A ligand activates ephexin via tyrosine phosphorylation by Src.[87,88] This phosphorylation increases the specificity of EphA-activated ephexin towards RhoA, resulting in relative inhibition of Cdc42 and Rac1. The coupling of RhoA activation and inhibition of Cdc42 and Rac1 facilitates cell collapse, similar to the effects of semaphorin. Additional RhoGEFs mediate Eph-induced rearrangement of the cytoskeleton.[89–92] Of these, the vascular smooth muscle-expressed ephexin homologue Vsm-RhoGEF, is specific to the vascular system and binds EphA receptors.[90] Similar to ephexin, Vsm-RhoGEF is activated by tyrosine phosphorylation upon EphA stimulation, subsequently activating RhoA and inducing stress fiber assembly. While most of the evidence suggests that Eph regulation of Rho-GTPases involves a tyrosine phosphorylation cascade, it appears that this is not always the case, as Cdc42 and Rac1 were inhibited even in cells expressing kinase-dead EphB3.[93] The manner in which the inhibition occurred was not determined, however, and a mechanism involving the activation of endogenous ephrin-B by the kinase-dead EphB3 could not be ruled out. Signaling via the ephrin ligand appears to play a role in the vascular system — overexpression of kinase-dead EphB4 was sufficient to alter the normal developmental program of the vascular system in the mouse,[94] suggesting that ephrin-B2 can function as a signaling receptor. Instead of the typical angiogenic patterning where a vascular network forms upon the emergence and interconnection of new vessels, the pre-existing blood vessels increased in diameter without sprouting new capillaries. This growth pattern was observed in developing and postnatal angiogenesis, as well as in tumors and normal growth. Given that the EphB4 receptor exerts a repulsive signal via ephrin-B2, it is possible that when the normal angiogenic program was suppressed by the expression of kinase-dead EphB4, endothelial cell proliferation, which is also activated by ephrin-B2,[95] was diverted into circumferential enlargement of the initial vascular network.

Eph receptors resemble plexins also in regulating integrin-dependent adhesion. This regulation is exerted via several pathways. When expressed in fibroblasts, activated EphB2 receptors phosphorylate a tyrosine in the effector domain of R-ras which blocks its binding to Raf-1 and, consequently, its ability to support integrin activation.[96,97] Alternatively, stimulation of endogenous epithelial cell EphA2 by ephrin-A1 induces binding of the tyrosine phosphatase SHP2 to EphA2, dephosphorylation and dissociation of focal adhesion kinase (FAK) from EphA2, followed by inhibition of integrin-dependent cell adhesion.[98] Conversely, Eph signaling can have an opposite effect on integrin-dependent attachment under certain conditions. Activation of EphB1 upon contact with a surface densely coated with ephrin-B1, augments integrin-mediated attachment of endothelial cells to the surface in a manner requiring the tyrosine kinase activity of EphB1.[99] Integrin activation by EphB1 appears to be mediated by the low molecular weight protein tyrosine phosphatase (LMW-PTP), by the adaptor protein Nck, and by a Nck-interacting kinase (NIK), though the pathway between NIK and integrin is not known.[100] A PI 3-K-dependent mechanism appears to mediate integrin activation in epithelial cells by EphA8,[101] where binding of the p110γ PI 3-K catalytic subunit to the juxtamembrane region of the cytoplasmic domain of EphA8 may facilitate the access of this catalytic subunit to its lipid substrate. It appears that all the known mechanisms of integrin regulation depend on the tyrosine kinase activity of the Eph receptor (e.g. Refs. 93, 98 and 100).

3.2. *Reverse signaling*

Reverse signaling via transmembrane ephrins B ligands mirrors to a large extent features of EphB receptor forward signaling. Similar to EphB receptors, the cytoplasmic tail of ephrin-B ligands undergoes phosphorylation at three conserved tyrosines.[102] As ephrins possess no catalytic activity, the tyrosines are not autophosphorylated but rather undergo phosphorylation by SFK following activation by EphB receptor ectodomains.[103] EphB binding to ephrin-B induces SFK recruitment to the cytoplasmic domain of the latter, possibly triggered by ephrin-B clustering. Phosphorylation of one of the tyrosines (Tyr^{317} in human ephrin-B1) generates an SH2-binding motif, to which the adaptor protein Grb4 binds and recruits several proteins via its three Src homology 3 (SH3) domains,[104] most of which are involved in regulating the actin cytoskeleton and focal adhesions. Among the Grb4-binding proteins, the Cbl-associated protein (CAP) is known to mediate binding of additional cytoskeletal proteins via its own SH3 domains, e.g. vinculin, FAK and PAK1. Grb4 has 68% homology to Nck, the adaptor protein that binds EphB1. The same study[104] has shown that ephrin-B1 reverse signaling results in disassembly of stress fibers, loss of focal adhesions, and cell rounding in a manner dependent on Grb4 binding to the cytoplasmic domain of ephrin-B1.

Several studies found that ephrin-B regulation of cell adhesion involves Crk,[105,106] an SH2/SH3 adaptor protein that couples integrin to cytoskeletal dynamics.[107] As a further instance of analogy between ephrin ligands and their Eph receptors, it has been recently shown that while EphB4 induces phosphorylation of the adaptor protein Crk in response to stimulation by the ephrin-B2 ligand,[80] activation of ephrin-B2 by EphB4 leads to tyrosine phosphorylation of the Crk family proteins CrkII and CrkL, possibly by a SFK.[108] In turn, this phosphorylation blocks the association of CrkII and CrkL with p130(CAS), resulting in reduction and alteration in the pattern of focal adhesions.

Though the expression of ephrin-B2 is highest in the arterial endothelium, it is also expressed on mural cells of both arteries and veins.[108] Inactivation of ephrin-B2 in mouse vascular smooth muscle cells caused the disruption of microvessel integrity, likely as a result of incomplete coverage of the nascent microvessels by vascular smooth muscle cells (VSMC). At the level of cell function, the absence of ephrin-B2 expression resulted in irregular cell shape and random protrusion of lamellipodia, indicating loss of directionality during migration. The altered cell morphology was accompanied by a reduced number of focal adhesions and a lower activation level of FAK.

Like EphB receptors, Ephrin-B ligands have carboxy-terminus PDZ-binding motifs of a similar consensus sequence (YKV), and bind several of the same adaptor proteins: PICK1, which clusters both Eph receptors and ephrin-B ligands, syntenin and GRIP1.[84] Ephrin-B1 is targeted to sphingolipid/cholesterol-enriched and caveolae-containing subdomains of the plasma membrane (commonly referred to as rafts) thought to be sites of signaling protein clusters,[109] and recruits to these regions two GRIP isoforms.[110] This recruitment could conceivably initiate the formation of a larger signaling complex containing ephrin effectors via the seven PDZ domains of GRIP. PDZ-dependent interactions of ephrin with cytoplasmic proteins appears to be required for reverse signaling. Mouse knock-in of ephrin-B1 lacking the PDZ-binding motif resulted in embryonic lethality caused by impaired migration of neural crest cells,[111] while a knock-in of a similarly truncated ephrin-B2 resulted in defective morphogenesis of the lymphatic system.[112] In the latter case, it was also shown that the phosphorylatable tyrosines in the ephrin-B2 cytoplasmic domain were not essential for lymphatic vasculogenesis. In addition to the above adaptor proteins, ephrin-B2 interacts via its PDZ-binding motif with PDZ-regulator of heterotrimeric G protein signaling 3 (PDZ-RGS3).[113] Expression of soluble EphB2, an ephrin-B2 receptor, inhibited cellular response to the chemoattractant SDF-1 during Xenopus development. Since SDF-1 signals through a G-coupled protein receptor, it is possible that ephrin-B2-activated PDZ-RGS3 contributed to the inhibition of SDF-1 signaling. It appears, therefore, that in addition to the other repulsive effects of ephrin, it counteracts chemoattractant signals.

GPI-anchored ephrin-A ligands are segregated into plasma membrane rafts.[70] Targeting of ephrin-A to such subdomains facilitates the clustering required for the activation of their cognate Eph receptors,[73,74] and brings the ephrin-A ligands in proximity to other signaling proteins that are compartmentalized into these subdomains.[109] Activated ephrin-A5 recruits the SFK Fyn to the raft subdomains and induces its activation via a still unknown mechanism that may hypothetically involve ephrin-A interaction in *cis* with a putative transmembrane receptor. This results in an overall increase in the tyrosine phosphorylation level of other raft proteins. Unlike Eph receptors, however, ephrin-A signaling augments integrin-dependent cell adhesion[69] and causes cell spreading rather than collapse.[72] Ephrin-A and EphA receptors are laterally segregated into separate domains on the plasma membrane, so that each population can interact with its receptors or ligands, respectively, in *trans*. Forced collocation of ephrin-A and EphA in the same subdomains by exchanging their ectodomains resulted in mutual interference in *cis* of the two types of proteins, and inhibited their signaling.[72] In fact, the same *cis* inhibition of EphA signaling by ephrin-A[114] can be invoked in order to attribute the ephrin-mediated increase in integrin-dependent cell adhesion to a relative increase in ephrin-A versus EphA signaling.[115]

Since Eph receptor-induced cell repulsion initially requires cell-to-cell contact, the *trans* interaction between Eph receptors and their ephrin ligands on the surfaces of opposing cells could hinder the repellant effects of Eph signaling. There are two known ways in which cells overcome this obstacle — either by trans-endocytosis, or by proteolytic cleavage of the ephrin ectodomain. EphB-receptor/ephrin-B complexes are engulfed by the EphB-presenting endothelial cells and fibroblasts in a process requiring Rac-dependent membrane ruffling which is not clathrin- or caveolin-dependent.[116] A concurrent study using fibroblasts and neurons detected bidirectional endocytosis of the full-length Eph receptor bound to full-length ephrin-B ligand into either host cell.[117] Removal of the EphA receptor/ephrin-A complex was first thought to occur via proteolytic cleavage by the ADAM10 metalloprotease which cleaves only EphA-bound ephrin-A ligand in *cis* . A subsequent study found, however, that ADAM10 associated with Eph receptors cleaves ephrin-A in *trans*.[118]

Unrelated to the endocytic removal of Eph-ephrin complexes EphB-ephrin-B signaling regulates clathrin-dependent endocytosis. EphB stimulated by ephrin-B inactivates the phosphatidylinositol 5'-phosphatase synaptojanin 1 and disrupts its association with endophilin by tyrosine phosphorylation. The dissociation from endophilin uncouples synaptojanin from the endocytic machinery, and together with the inhibition of its catalytic activity, downregulates clathrin-dependent endocytosis. It is not known, however, how this downregulation contributes to EphB receptor-induced cell repulsion.

4. Netrin and Slit Signaling

Netrins are a family of secreted proteins related to the extracellular matrix protein laminin. Similar to other guidance cues, it has been initially studied in the neural system (reviewed in Ref.119). Recent studies implicated netrin and one of its known receptors, uncoordinated 5b (Unc5b) in angiogenesis and in the development of the vascular system.[120–123] Unc5b, one of the four-member Unc5 family which is related to the Ig superfamily, is selectively expressed in the vascular system, including in endothelial tip cells.[120–123] Unlike other netrin receptors, Unc5b receptors transduce only repellant signals. Netrin binding to the second Ig domain of Unc5b relieves the autoinhibitory effect of this domain[124] and induces phosphorylation of Tyr^{482}.[125] The signaling pathway downstream of Unc5b has not been charted yet.

Deletion of Unc5b expression in the mouse and zebrafish caused aberrant growth of arterial branches, suggesting that netrin signaling via Unc5b is required for regulating the shape of the arterial tree.[126] At the cellular level, netrin suppressed endothelial cell migration and retracted the filopodia of tip cells of growing capillaries. These results contrasted with a previous study which reported that netrin-1 is a pro-angiogenic factor.[123] Further studies from the same group consistently showed that the effects of netrin-1 and -4 did not inhibit but rather promoted the formation of the vascular network in zebrafish and neo-vascularization in a mouse hindlimb ischemia model.[121] Surprisingly, none of the known netrin receptors, including Unc5b, were expressed at significant levels in the endothelial cell systems that responded to netrins, and netrin-4 did not bind to any of these receptors. To date, the discrepancy between the results of these two recent studies have not been resolved.

Mammalian slit is a family of secreted proteins consisting of four members that are composed of multiple EGF and leucine-rich domains (reviewed in Ref.127). They bind to roundabout (Robo), a four-member transmembrane receptor family composed of extracellular Ig and fibronectin type III motifs, and of conserved cytoplasmic (CC) domains. The significance of Robo to vascular morphogenesis was initially suggested by the discovery of an endothelial-specific member, Robo4.[128,129] The initial results suggested that Robo4 transmits repulsive signals, since endothelial cell migration was inhibited by slit.[129] Similar to other guidance cues, however, the nature of slit/Robo4 signaling was put in question by subsequent studies which found that slit stimulated endothelial cell migration and *in vitro* tube formation.[130] Furthermore, the actual identity of the Robo4 ligand has been questioned, since Robo4 is activated even when its amino-terminal Ig domain has been deleted,[131] and since unlike Robo4, slits are not expressed during embryonic development. *In vivo* studies in zebrafish indicated that Robo signaling is essential for the development of the vascular system since Robo4 knock-down caused defective sprouting and loss of intersomitic vessels.[132] An extension of these studies[133] showed that the migration of single angioblasts from zebrafish in which Robo4 expression was knocked down was more random than that of WT angioblasts.

Robo4 signaling in porcine aortic endothelial cells modulated the actin cytoskeleton by activating the Rho-GTPases Rac1 and Cdc42, and by inducing lamellipodia and filopodia.[133] These effects were fully or partially lost when cells were transfected with Robo4 mutants lacking the ectoplasmic domain, or both of the two CC domains, suggesting that ligand binding to the ectoplasmic domain and interactions of cellular proteins with the cytoplasmic domain are both required for Robo4 signaling. Transfection with Robo4 mutants that were unable to activate Rac1 and Cdc42 led to slower migration and reduced adhesion to fibronectin-coated substrate.

5. Open Questions

To a varying degree, the downstream signaling pathways of the guidance cues discussed here have been partially determined. There are still numerous knowledge gaps that need to be closed, and several contradictory findings that need to be reconciled.

The signaling downstream of plexD1 is particularly relevant to the morphogenesis of the vascular system, yet nothing is known to date concerning the immediate effectors. One intriguing aspect of plexD1 signaling is that it involves neuropilin when responding to sema3C, but not when responding to sema3E. There are several studies which have shown that neuropilin is endocytosed, but the mechanism of the endocytosis and its functional significance have not been determined. An attractive venue of investigation is the interaction of neuropilin with the PDZ domain protein synectin, which is an adaptor of the unconventional myosin VI.[134] Unlike the case of Eph-ephrin signaling, little is known about the reverse signaling via the membrane-bound semaphorins. Similarly, it is still not known how the interaction between plexin receptors in *trans* with membrane-bound semaphorin on apposing cells is terminated, allowing the cells to detach from each other.

A general question concerning Eph-ephrin signaling is the significance of *cis*, the interaction between membrane-bound receptor and the ligand e.g. Eph and ephrin. Resolution of this issue is of special interest for the signaling of ephrin-B2 when activated by kinase-dead EphB4, both of which are involved in vascular morphogenesis.

Two major issues still need to be resolved in netrin and Robo4 signaling. There are conflicting studies claiming that either netrin or Robo4 signaling is either repulsive or attractive, but no mechanism has been proposed for reversing the nature of the signaling, as in the case of semaphorin. Unc5b, the receptor thought to transduce netrin signaling in the vascular system is not expressed in several endothelial cell types that are responsive to netrin, suggesting there are more still unidentified netrin receptors. Similarly, Robo4 is likely to have ligands other than the currently known three slit isoforms, since its ligand binding site is different from the other Robo receptors.

Finally, the spatial and temporal patterns of Rho-GTPase activity downstream of the directional cue receptors reviewed here are still poorly known.

References

1. Huber AB, *et al.* (2003). Signaling at the growth cone: ligand-receptor complexes and the control of axon growth and guidance. *Annu Rev Neurosci* **26**: 509–563.
2. Song H, Poo M (2001). The cell biology of neuronal navigation. *Nat Cell Biol* **3**(3): E81–E88.
3. Gerhardt H, *et al.* (2003). VEGF guides angiogenic sprouting utilizing endothelial tip cell filopodia. *J Cell Biol* **161**(6): 1163–1177.
4. Honma Y, *et al.* (2002). Artemin is a vascular-derived neurotropic factor for developing sympathetic neurons. *Neuron* **35**(2): 267–282.
5. Kuruvilla R, *et al.* (2004). A neurotrophin signaling cascade coordinates sympathetic neuron development through differential control of TrkA trafficking and retrograde signaling. *Cell* **118**(2): 243–255.
6. Mukouyama YS, *et al.* (2002). Sensory nerves determine the pattern of arterial differentiation and blood vessel branching in the skin. *Cell* **109**(6): 693–6705.

7. Soker S, *et al.* (1998). Neuropilin-1 is expressed by endothelial and tumor cells as an isoform-specific receptor for vascular endothelial growth factor. *Cell* **92**(6): 735–745.
8. Luo Y, Raible D, Raper JA (1993). Collapsin: a protein in brain that induces the collapse and paralysis of neuronal growth cones. *Cell* **75**(2): 217–227.
9. Miao HQ, *et al.* (1999). Neuropilin-1 mediates collapsin-1/semaphorin III inhibition of endothelial cell motility: functional competition of collapsin-1 and vascular endothelial growth factor-165. *J Cell Biol* **146**(1): 233–242.
10. Castellani V, Falk J, Rougon G (2004). Semaphorin3A-induced receptor endocytosis during axon guidance responses is mediated by L1 CAM. *Mol Cell Neurosci* **26**(1): 89-100.
11. Gu C, *et al.* (2005). Semaphorin 3E and plexin-D1 control vascular pattern independently of neuropilins. *Science* **307**(5707): 265–268.
12. Kitsukawa T, *et al.* (1997). Neuropilin-semaphorin III/D-mediated chemorepulsive signals play a crucial role in peripheral nerve projection in mice. *Neuron* **19**(5): 995–1005.
13. Kawasaki T, *et al.* (1999). A requirement for neuropilin-1 in embryonic vessel formation. *Development* **126**(21): 4895–4902.
14. Gu C, *et al.* (2002). Characterization of neuropilin-1 structural features that confer binding to semaphorin 3A and vascular endothelial growth factor 165. *J Biol Chem* **277**(20): 18069–18076.
15. Gu C, *et al.* (2003). Neuropilin-1 conveys semaphorin and VEGF signaling during neural and cardiovascular development. *Dev Cell* **5**(1): 45–57.
16. Feiner L, *et al.* (2001). Targeted disruption of semaphorin 3C leads to persistent truncus arteriosus and aortic arch interruption. *Development* **128**(16): 3061–3070.
17. Takashima S, *et al.* (2002). Targeting of both mouse neuropilin-1 and neuropilin-2 genes severely impairs developmental yolk sac and embryonic angiogenesis. *Proc Natl Acad Sci USA* **99**(6): 3657–3662.
18. Yuan L, *et al.* (2002). Abnormal lymphatic vessel development in neuropilin 2 mutant mice. *Development* **129**(20): 4797–4806.
19. Chen H, *et al.* (1997). Neuropilin-2, a novel member of the neuropilin family, is a high affinity receptor for the semaphorins Sema E and Sema IV but not Sema III. *Neuron* **19**(3): 547–559.
20. Kolodkin AL, *et al.* (1997). Neuropilin is a semaphorin III receptor. *Cell* **90**(4): 753–762.
21. Takahashi T, *et al.* (1999). Plexin-neuropilin-1 complexes form functional semaphorin-3A receptors. *Cell* **99**(1): 59–69.
22. Oinuma I, *et al.* (2004). The Semaphorin 4D receptor Plexin-B1 is a GTPase activating protein for R-Ras. *Science* **305**(5685): 862–865.
23. Tamagnone L, *et al.* (1999). Plexins are a large family of receptors for transmembrane, secreted, and GPI-anchored semaphorins in vertebrates. *Cell* **99**(1): 71–80.
24. van der Zwaag B, *et al.* (2002). PLEXIN-D1, a novel plexin family member, is expressed in vascular endothelium and the central nervous system during mouse embryogenesis. *Dev Dyn* **225**(3): 336–343.

25. Gitler AD, Lu MM, Epstein JA (2004). PlexinD1 and semaphorin signaling are required in endothelial cells for cardiovascular development. *Dev Cell* **7**(1): 107–116.

26. Torres-Vazquez J, *et al.* (2004). Semaphorin-plexin signaling guides patterning of the developing vasculature. *Dev Cell* **7**(1): 117–123.

27. Childs S, *et al.* (2002). Patterning of angiogenesis in the zebrafish embryo. *Development* **129**(4): 973–982.

28. Serini G, *et al.* (2003). Class 3 semaphorins control vascular morphogenesis by inhibiting integrin function. *Nature* **424**(6947): 391–397.

29. Bielenberg DR, *et al.* (2004). Semaphorin 3F, a chemorepulsant for endothelial cells, induces a poorly vascularized, encapsulated, nonmetastatic tumor phenotype. *J Clin Invest* **114**(9): 1260–1271.

30. Miao HQ, *et al.* (2000). Neuropilin-1 expression by tumor cells promotes tumor angiogenesis and progression. *FASEB J* **14**(15): 2532–2539.

31. Rohm B, *et al.* (2000). Plexin/neuropilin complexes mediate repulsion by the axonal guidance signal semaphorin 3A. *Mech Dev* **93**(1–2): 95–104.

32. Jin Z, Strittmatter SM (1997). Rac1 mediates collapsin-1-induced growth cone collapse. *J Neurosci* **17**(16): 6256–6263.

33. Kuhn TB, *et al.* (1999). Myelin and collapsin-1 induce motor neuron growth cone collapse through different pathways: inhibition of collapse by opposing mutants of rac1. *J Neurosci* **19**(6): 1965–1975.

34. Toyofuku T, *et al.* (2005). FARP2 triggers signals for Sema3A-mediated axonal repulsion. *Nat Neurosci* **8**(12): 1712–1719.

35. Zanata SM, *et al.* (2002). Antagonistic effects of Rnd1 and RhoD GTPases regulate receptor activity in Semaphorin 3A-induced cytoskeletal collapse. *J Neurosci* **22**(2): 471–477.

36. Oinuma I, *et al.* (2003). Direct interaction of Rnd1 with Plexin-B1 regulates PDZ-RhoGEF-mediated Rho activation by Plexin-B1 and induces cell contraction in COS-7 cells. *J Biol Chem* **278**(28): 25671–25677.

37. Nobes CD, *et al.* (1998). A new member of the Rho family Rnd1, promotes disassembly of actin filament structures and loss of cell adhesion. *J Cell Biol* **141**(1): 187–197.

38. Murphy C, *et al.* (1996). Endosome dynamics regulated by a Rho protein. *Nature* **384**(6608): 427–432.

39. Ito Y, *et al.* (2006). Sema4D/plexin-B1 activates GSK-3beta through R-Ras GAP activity, inducing growth cone collapse. *EMBO Rep* **7**(7): 704–709.

40. Oinuma I, Katoh H, Negishi M (2006). Semaphorin 4D/Plexin-B1-mediated R-Ras GAP activity inhibits cell migration by regulating beta(1) integrin activity. *J Cell Biol* **173**(4): 601–613.

41. Vikis HG, *et al.* (2000). The semaphorin receptor plexin-B1 specifically interacts with active Rac in a ligand-dependent manner. *Proc Natl Acad Sci USA* **97**(23): 12457–12462.

42. Driessens MH, *et al.* (2001). Plexin-B semaphorin receptors interact directly with active Rac and regulate the actin cytoskeleton by activating Rho. *Curr Biol* **11**(5): 339–344.

43. Hu H, Marton TF, Goodman CS (2001). Plexin B mediates axon guidance in Drosophila by simultaneously inhibiting active Rac and enhancing RhoA signaling. *Neuron* **32**(1): 39-51.

44. Vikis HG, Li W, Guan KL (2002). The plexin-B1/Rac interaction inhibits PAK activation and enhances Sema4D ligand binding. *Genes Dev* **16**(7): 836–845.

45. Bokoch GM (2003). Biology of the p21-activated kinases. *Annu Rev Biochem* **72**: 743–781.

46. Aurandt J, *et al.* (2002). The semaphorin receptor plexin-B1 signals through a direct interaction with the Rho-specific nucleotide exchange factor LARG. *Proc Natl Acad Sci USA* **99**(19): 12085–12090.

47. Perrot V, Vazquez-Prado J, Gutkind JS (2002). Plexin B regulates Rho through the guanine nucleotide exchange factors leukemia-associated Rho GEF (LARG) and PDZ-RhoGEF. *J Biol Chem* **277**(45): 43115–43120.

48. Swiercz JM, *et al.* (2002). Plexin-B1 directly interacts with PDZ-RhoGEF/LARG to regulate RhoA and growth cone morphology. *Neuron* **35**(1): 51–63.

49. Chikumi H, *et al.* (2004). Homo- and hetero-oligomerization of PDZ-RhoGEF LARG and p115RhoGEF by their C-terminal region regulates their *in vivo* Rho GEF activity and transforming potential. *Oncogene* **23**(1): 233–240.

50. Hirotani M, *et al.* (2002). Interaction of plexin-B1 with PDZ domain-containing Rho guanine nucleotide exchange factors. *Biochem Biophys Res Commun* **297**(1): 32–37.

51. Aizawa H, *et al.* (2001). Phosphorylation of cofilin by LIM-kinase is necessary for semaphorin 3A-induced growth cone collapse. *Nat Neurosci* **4**(4): 367–373.

52. Mitsui N, *et al.* (2002). Involvement of Fes/Fps tyrosine kinase in semaphorin3A signaling. *Embo J* **21**(13): 3274–3285.

53. Deroanne C, *et al.* (2003). EphrinA1 inactivates integrin-mediated vascular smooth muscle cell spreading via the Rac/PAK pathway. *J Cell Sci* **116**(Pt 7): 1367–1376.

54. Goshima Y, *et al.* (1995). Collapsin-induced growth cone collapse mediated by an intracellular protein related to UNC-33. *Nature* **376**(6540): 509–514.

55. Brown M, *et al.* (2004). Alpha2-chimaerin, cyclin-dependent Kinase 5/p35, and its target collapsin response mediator protein-2 are essential components in semaphorin 3A-induced growth-cone collapse. *J Neurosci* **24**(41): 8994–9004.

56. Fukata Y, *et al.* (2002). CRMP-2 binds to tubulin heterodimers to promote microtubule assembly. *Nat Cell Biol* **4**(8): 583–591.

57. Sharma MR, Tuszynski GP, Sharma MC (2004). Angiostatin-induced inhibition of endothelial cell proliferation/apoptosis is associated with the down-regulation of cell cycle regulatory protein cdk5. *J Cell Biochem* **91**(2): 398–3409.

58. Nikolic M, *et al.* (1998). The p35/Cdk5 kinase is a neuron-specific Rac effector that inhibits Pak1 activity. *Nature* **395**(6698): 194–198.

59. Dehmelt L, Halpain S (2005). The MAP2/Tau family of microtubule-associated proteins. *Genome Biol* **6**(1): 204.

60. Gale NW, *et al.* (1996). Eph receptors and ligands comprise two major specificity subclasses and are reciprocally compartmentalized during embryogenesis. *Neuron* **17**(1): 9–19.

61. Kullander K, Klein R (2002). Mechanisms and functions of Eph and ephrin signalling. *Nat Rev Mol Cell Biol* **3**(7): 475–486.

62. Pasquale EB (2005). Eph receptor signalling casts a wide net on cell behaviour. *Nat Rev Mol Cell Biol* **6**(6): 462–475.

63. Wang HU, Chen ZF, Anderson DJ (1998). Molecular distinction and angiogenic interaction between embryonic arteries and veins revealed by ephrin-B2 and its receptor Eph-B4. *Cell* **93**(5): 741–753.

64. Gerety SS, *et al.* (1999). Symmetrical mutant phenotypes of the receptor EphB4 and its specific transmembrane ligand ephrin-B2 in cardiovascular development. *Mol Cell* **4**(3): 403–414.

65. Fuller T, *et al.* (2003). Forward EphB4 signaling in endothelial cells controls cellular repulsion and segregation from ephrinB2 positive cells. *J Cell Sci* **116**(Pt 12): 2461–2470.

66. Oike Y, *et al.* (2002). Regulation of vasculogenesis and angiogenesis by EphB/ephrin-B2 signaling between endothelial cells and surrounding mesenchymal cells. *Blood* **100**(4): 1326–1333.

67. Davy A, Soriano P (2005). Ephrin signaling *in vivo*: look both ways. *Dev Dyn* **232**(1): 1-10.

68. Toyofuku T, *et al.* (2004). Guidance of myocardial patterning in cardiac development by Sema6D reverse signalling. *Nat Cell Biol* **6**(12): 1204–1211.

69. Davy A, Robbins SM (2000). Ephrin-A5 modulates cell adhesion and morphology in an integrin-dependent manner. *EMBO J* **19**(20): 5396–5405.

70. Davy A, *et al.* (1999). Compartmentalized signaling by GPI-anchored ephrin-A5 requires the Fyn tyrosine kinase to regulate cellular adhesion. *Genes Dev* **13**(23): 3125–3135.

71. Huai J, Drescher U (2001). An ephrin-A-dependent signaling pathway controls integrin function and is linked to the tyrosine phosphorylation of a 120-kDa protein. *J Biol Chem* **276**(9): 6689–6694.

72. Marquardt T, *et al.* (2005). Coexpressed EphA receptors and ephrin-A ligands mediate opposing actions on growth cone navigation from distinct membrane domains. *Cell* **121**(1): 127–139.

73. Davis S, *et al.* (1994). Ligands for EPH-related receptor tyrosine kinases that require membrane attachment or clustering for activity. *Science* **266**(5186): 816–819.

74. Smith FM, *et al.* (2004). Dissecting the EphA3/Ephrin-A5 interactions using a novel functional mutagenesis screen. *J Biol Chem* **279**(10): 9522–9531.

75. Himanen JP, *et al.* (2001). Crystal structure of an Eph receptor-ephrin complex. *Nature* **414**(6866): 933–938.

76. Kalo MS, Pasquale EB (1999). Multiple *in vivo* tyrosine phosphorylation sites in EphB receptors. *Biochemistry* **38**(43): 14396–14408.

77. Binns KL, *et al.* (2000). Phosphorylation of tyrosine residues in the kinase domain and juxtamembrane region regulates the biological and catalytic activities of Eph receptors. *Mol Cell Biol* **20**(13): 4791–4805.

78. Wybenga-Groot LE, *et al.* (2001). Structural basis for autoinhibition of the Ephb2 receptor tyrosine kinase by the unphosphorylated juxtamembrane region. *Cell* **106**(6): 745–757.

79. Yu HH, *et al.* (2001). Multiple signaling interactions of Abl and Arg kinases with the EphB2 receptor. *Oncogene* **20**(30): 3995–34006.
80. Noren NK, *et al.* (2006). The EphB4 receptor suppresses breast cancer cell tumorigenicity through an Abl-Crk pathway. *Nat Cell Biol* **8**(8): 815–825.
81. Kain KH, Klemke RL (2001). Inhibition of cell migration by Abl family tyrosine kinases through uncoupling of Crk-CAS complexes. *J Biol Chem* **276**(19): 16185–16192.
82. Torres R, *et al.* (1998). PDZ proteins bind, cluster, and synaptically colocalize with Eph receptors and their ephrin ligands. *Neuron* **21**(6): 1453–1463.
83. Kalo MS, Pasquale EB (1999). Signal transfer by eph receptors. *Cell Tissue Res* **298**(1): 1–9.
84. Buchert M, *et al.* (1999). The junction-associated protein AF-6 interacts and clusters with specific Eph receptor tyrosine kinases at specialized sites of cell-cell contact in the brain. *J Cell Biol* **144**(2): 361–371.
85. Radziwill G, *et al.* (2003). The Bcr kinase downregulates Ras signaling by phosphorylating AF-6 and binding to its PDZ domain. *Mol Cell Biol* **23**(13): 4663–4672.
86. Shamah SM, *et al.* (2001). EphA receptors regulate growth cone dynamics through the novel guanine nucleotide exchange factor ephexin. *Cell* **105**(2): 233–244.
87. Knoll B, Drescher U (2004). Src family kinases are involved in EphA receptor-mediated retinal axon guidance. *J Neurosci* **24**(28): 6248–6257.
88. Sahin M, *et al.* (2005). Eph-dependent tyrosine phosphorylation of ephexin1 modulates growth cone collapse. *Neuron* **46**(2): 191–1204.
89. Cowan CW, *et al.* (2005). Vav family GEFs link activated Ephs to endocytosis and axon guidance. *Neuron* **46**(2): 205–217.
90. Ogita H, *et al.* (2003). EphA4-mediated Rho activation via Vsm-RhoGEF expressed specifically in vascular smooth muscle cells. *Circ Res* **93**(1): 23–31.
91. Penzes P, *et al.* (2003). Rapid induction of dendritic spine morphogenesis by trans-synaptic ephrinB-EphB receptor activation of the Rho-GEF kalirin. *Neuron* **37**(2): 263–274.
92. Tanaka M, Kamata R, Sakai R (2005).Phosphorylation of ephrin-B1 via the interaction with claudin following cell-cell contact formation. *EMBO J* **24**(21): 3700–3711.
93. Miao H, *et al.* (2005). Inhibition of integrin-mediated cell adhesion but not directional cell migration requires catalytic activity of EphB3 receptor tyrosine kinase. Role of Rho family small GTPases. *J Biol Chem* **280**(2): 923–932.
94. Erber R, *et al.* (2006). EphB4 controls blood vascular morphogenesis during postnatal angiogenesis. *EMBO J* **25**(3): 628–641.
95. Masood R, *et al.* (2005). Ephrin B2 expression in Kaposi sarcoma is induced by human herpesvirus type 8: phenotype switch from venous to arterial endothelium. *Blood* **105**(3): 1310–1318.
96. Zou JX, *et al.* (1999). An Eph receptor regulates integrin activity through R-Ras. *Proc Natl Acad Sci USA* **96**(24): 13813–13818.
97. Zhang Z, *et al.* (1996). Integrin activation by R-ras. *Cell* **85**(1): 61–69.

98. Miao H, *et al.* (2000). Activation of EphA2 kinase suppresses integrin function and causes focal-adhesion-kinase dephosphorylation. *Nat Cell Biol* **2**(2): 62–69.

99. Huynh-Do U, *et al.* (1999). Surface densities of ephrin-B1 determine EphB1-coupled activation of cell attachment through alphavbeta3 and alpha5beta1 integrins. *EMBO J* **18**(8): 2165–2173.

100. Becker E, *et al.* (2000). Nck-interacting Ste20 kinase couples Eph receptors to c-Jun N-terminal kinase and integrin activation. *Mol Cell Biol* **20**(5): 1537–1545.

101. Adams RH, *et al.* (2001). The cytoplasmic domain of the ligand ephrinB2 is required for vascular morphogenesis but not cranial neural crest migration. *Cell* **104**(1): 57-69.

102. Kalo MS, Yu HH, Pasquale EB (2001). *In vivo* tyrosine phosphorylation sites of activated ephrin-B1 and ephB2 from neural tissue. *J Biol Chem* **276**(42): 38940–38948.

103. Palmer A, *et al.* (2002). EphrinB phosphorylation and reverse signaling: regulation by Src kinases and PTP-BL phosphatase. *Mol Cell* **9**(4): 725–737.

104. Cowan CA, Henkemeyer M (2001). The SH2/SH3 adaptor Grb4 transduces B-ephrin reverse signals. *Nature* **413**(6852): 174–179.

105. Nagashima K, *et al.* (2002). Adaptor protein Crk is required for ephrin-B1-induced membrane ruffling and focal complex assembly of human aortic endothelial cells. *Mol Biol Cell* **13**(12): 4231–4242.

106. Sakamoto H, *et al.* (2004). Cell adhesion to ephrinb2 is induced by EphB4 independently of its kinase activity. *Biochem Biophys Res Commun* **321**(3): 681–687.

107. Chodniewicz D, Klemke RL (2004). Regulation of integrin-mediated cellular responses through assembly of a CAS/Crk scaffold. *Biochim Biophys Acta* **1692**(2–3): 63–76.

108. Foo SS, *et al.* (2006). Ephrin-B2 controls cell motility and adhesion during blood-vessel-wall assembly. *Cell* **124**(1): 161–173.

109. Simons K, Toomre D (2000). Lipid rafts and signal transduction. *Nat Rev Mol Cell Biol* **1**(1): 31–39.

110. Bruckner K, *et al.* (1999). EphrinB ligands recruit GRIP family PDZ adaptor proteins into raft membrane microdomains. *Neuron* **22**(3): 511–524.

111. Davy A, Aubin J, Soriano P (2004). Ephrin-B1 forward and reverse signaling are required during mouse development. *Genes Dev* **18**(5): 572–583.

112. Makinen T, *et al.* (2005). PDZ interaction site in ephrinB2 is required for the remodeling of lymphatic vasculature. *Genes Dev* **19**(3): 397–410.

113. Lu Q, *et al.* (2001). Ephrin-B reverse signaling is mediated by a novel PDZ-RGS protein and selectively inhibits G protein-coupled chemoattraction. *Cell* **105**(1): 69–79.

114. Carvalho RF, *et al.* (2006). Silencing of EphA3 through a cis interaction with ephrinA5. *Nat Neurosci* **9**(3): 322–330.

115. Halloran MC, Wolman MA (2006). Repulsion or adhesion: receptors make the call. *Curr Opin Cell Biol* **18**: 533–540.

116. Marston DJ, Dickinson S, Nobes CD (2003). Rac-dependent trans-endocytosis of ephrinBs regulates Eph-ephrin contact repulsion. *Nat Cell Biol* **5**(10): 879–888.

117. Zimmer M, *et al.* (2003). EphB-ephrinB bi-directional endocytosis terminates adhesion allowing contact mediated repulsion. *Nat Cell Biol* **5**(10): 869–878.
118. Janes PW, *et al.* (2005). Adam meets Eph: an ADAM substrate recognition module acts as a molecular switch for ephrin cleavage in trans. *Cell* **123**(2): 291–304.
119. Barallobre MJ, *et al.* (2005). The Netrin family of guidance factors: emphasis on Netrin-1 signalling. *Brain Res Brain Res Rev* **49**(1): 22–47.
120. Nguyen A, Cai H (2006). Netrin-1 induces angiogenesis via a DCC-dependent ERK1/2-eNOS feed-forward mechanism. *Proc Natl Acad Sci USA* **103**(17): 6530–6535.
121. Wilson BD, *et al.* (2006). Netrins promote developmental and therapeutic angiogenesis. *Science* **313**(5787): 640–644.
122. Artigiani S, *et al.* (2004). Plexin-B3 is a functional receptor for semaphorin 5A. *EMBO Rep* **5**(7): 710–714.
123. Park KW, *et al.* (2004). The axonal attractant Netrin-1 is an angiogenic factor. *Proc Natl Acad Sci USA* **101**(46): 16210–16215.
124. Kruger RP, *et al.* (2004). Mapping netrin receptor binding reveals domains of Unc5 regulating its tyrosine phosphorylation. *J Neurosci* **24**(48): 10826–10834.
125. Killeen M, *et al.* (2002). UNC-5 function requires phosphorylation of cytoplasmic tyrosine 482, but its UNC-40-independent functions also require a region between the ZU-5 and death domains. *Dev Biol* **251**(2): 348–366.
126. Lu X, *et al.* (2004). The netrin receptor UNC5B mediates guidance events controlling morphogenesis of the vascular system. *Nature* **432**(7014): 179–186.
127. Fujiwara M, Ghazizadeh M, Kawanami O (2006). Potential role of the Slit/Robo signal pathway in angiogenesis. *Vasc Med* **11**(2): 115–121.
128. Huminiecki L, *et al.* (2002). Magic roundabout is a new member of the roundabout receptor family that is endothelial specific and expressed at sites of active angiogenesis. *Genomics* **79**(4): 547–552.
129. Park KW, *et al.* (2003). Robo4 is a vascular-specific receptor that inhibits endothelial migration. *Dev Biol* **261**(1): 251–267.
130. Wang B, *et al.* (2003). Induction of tumor angiogenesis by Slit-Robo signaling and inhibition of cancer growth by blocking Robo activity. *Cancer Cell* **4**(1): 19–29.
131. Howitt JA, Clout NJ, Hohenester E (2004). Binding site for Robo receptors revealed by dissection of the leucine-rich repeat region of Slit. *EMBO J* **23**(22): 4406–4412.
132. Bedell VM, *et al.* (2005). roundabout4 is essential for angiogenesis *in vivo*. *Proc Natl Acad Sci USA* **102**(18): 6373–6378.
133. Kaur S, *et al.* (2006). Robo4 signaling in endothelial cells implies attraction guidance mechanisms. *J Biol Chem* **281**(16): 11347–11356.
134. Naccache SN, Hasson T, Horowitz A (2006). Binding of internalized receptors to the PDZ domain of GIPC/synectin recruits myosin VI to endocytic vesicles. *Proc Natl Acad Sci USA* **103**(34): 12735–12740.

7

Regulation of Angiogenesis and Arteriogenesis by Hypoxia-Inducible Factor-1

by Gregg L. Semenza

1. Oxygen Homeostasis: Phylogeny, Ontogeny, Physiology, and Pathobiology

The history of science and medicine can be viewed as a steady march towards reductionism, which has culminated in our increasingly more precise understanding of molecular mechanisms underlying the structure and function of organisms. With this reductionism, however, comes the attendant risk of failing to integrate these data into higher level systems, whether it is the role of particular molecules in different tissues or disease processes, the impact of disease on the individual as a whole, or the causes and consequences of disease within society at large. In respect of such pitfalls, this chapter begins with a broad perspective by introducing the concept of oxygen homeostasis as a fundamental principle of life.

The evolution of multicellular organisms on Earth was dependent upon two critical events. The first was the ability of primitive single-celled organisms to transduce solar energy into the chemical energy of carbon bonds. In this process of photosynthesis, CO_2 and H_2O are utilized to generate a compound containing multiple carbon bonds (glucose) and, as a side product, O_2. The resulting progressive rise in the atmospheric O_2 concentration led to the second major evolutionary event, which was the establishment of a symbiotic relationship between single-celled organisms and internalized primitive cells, which over several billion years became highly specialized (as mitochondria) to perform a series of chemical reactions in which glucose and O_2 were utilized to generate ATP through the process of oxidative phosphorylation. The dramatic increase in energy production attendant with the development of oxidative phosphorylation provided the resources required to undertake the daunting and costly task of coordinating individual cells to act in concert as a multicellular organism. The evolution of oxidative phosphorylation may also have been critical for the establishment of a chemical equilibrium on the planet, by regenerating the CO_2 and H_2O that are consumed during photosynthesis. Thus, the balanced equation for life on Earth as we know it can be written simply as: $CO_2 + H_2O + h\nu \rightarrow C_6H_{12}O_6 + O_2 \rightarrow ATP + CO_2 + H_2O$.

For simple metazoans, exemplified today by the roundworm *Caenorhabditis elegans*, which consists of approximately 10^3 cells, O_2 can diffuse to all cells at adequate concentrations for efficient mitochondrial respiration. However, as metazoans became increasingly larger and more complex, diffusion was no longer sufficient to insure all cells of adequate oxygenation. In organisms with multiple layers of cells, O_2 becomes limiting at a distance of 100–200 μm from the source, which, in the case of simple metazoans, was the exterior of the body. To solve this limitation of O_2 diffusion, organisms evolved with cell types that are specialized for O_2 delivery. In the fruit fly *Drosophila melanogaster*, a series of tracheal tubes conduct O_2 from the exterior to all cells, including those further removed from the source. The spacing of these tubules is such that the distance between any cell and the nearest tubule is less than that at which O_2 becomes diffusion limited.

In the case of mammals, the further dramatic increase in body size and cell number (more than ten orders of magnitude greater than *C. elegans*) necessitated even greater specialization through the development of the lung, an organ in which surface area for O_2 diffusion from the environment was greatly increased; the erythrocyte, a cell specialized for high affinity O_2 capture in the lung and release in peripheral tissues; the blood vessels, an organ for the delivery of O_2 and glucose to peripheral tissues and the removal of CO_2 and other metabolic wastes from them; and the heart, an organ for pumping erythrocytes through the blood vessels circuiting the lungs and peripheral tissues. During mammalian development, the circulatory system is the first functioning organ system. In mice, which have a gestational period of 20 days, establishment of the circulation must occur by embryonic day 9, when embryo size increases to the point where diffusion from the uterine cavity becomes insufficient to supply all cells with adequate amounts of O_2. Major defects in the development of the red blood cells, blood vessels, or heart result in embryonic lethality at this stage. In humans, the functioning circulatory system is established during the first trimester.

Although increased O_2 levels on Earth provided opportunities for increased energy production, the utilization of O_2 was not without its risks. The goal of oxidative phosphorylation is the orderly transfer of electrons through a series of acceptor cytochromes in order to generate a proton gradient within the inner mitochondrial membrane. The potential energy of this gradient is used to synthesize ATP. O_2 is utilized as the ultimate electron acceptor, resulting in the production of H_2O. However, this process is not 100% efficient, so that electron transfer to O_2 may occur at complex I or complex III, resulting in the generation of superoxide radicals. These reactive oxygen species can oxidize cellular proteins, lipids, and nucleic acids and, by doing so, interfere with their normal structure and function, resulting in cell death or carcinogenesis.

Due to the opposing perils of O_2 deprivation (hypoxia) and O_2 excess (hyperoxia), cellular O_2 concentrations are very tightly regulated. In the context of angiogenesis, this means that new blood vessels form rapidly under conditions of hypoxia but that any excessive vascularization resulting in hyperoxia stimulates vascular regression. Thus, O_2 delivery is matched to O_2 consumption. The molecular physiological mechanisms by which O_2 homeostasis is maintained are discussed in the following section. This ability to maintain O_2 homeostasis through angiogenesis is lost in many disease conditions, most notably ischemic cardiovascular disease. Hence, a more complete understanding of these mechanisms may lead to novel therapeutic approaches to the most common cause of mortality in Western societies.

2. Hypoxia-Inducible Factor 1: Master Regulator of O_2 Homeostasis

Blood O_2-carrying capacity is maintained by O_2-regulated production of erythropoietin (EPO), which stimulates the proliferation and survival of red blood cell progenitors. Analysis of *cis*-acting sequences required for increased transcription of the human *EPO* gene in response to hypoxia led to the identification,[1] biochemical purification,[2] and molecular cloning[3] of hypoxia-inducible factor 1 (HIF-1). EPO is produced primarily by a rare cell type in the kidney. However, HIF-1 is expressed in all cell types and functions as a master regulator of oxygen homeostasis by playing critical roles in embryonic development, postnatal physiology, and disease pathobiology. HIF-1 has been identified in all metazoan species that have been analyzed from *Caenorhabditis elegans* to *Homo sapiens*, suggesting that the appearance of HIF-1 represented an adaptation that was essential to metazoan evolution.

HIF-1 is a transcription factor that binds to specific *cis*-acting sequences, which are designated hypoxia response elements (HREs) and contain one or more copies of the core HIF-1 binding-site sequence 5′-(A/G)CGTG-3′.[4] HREs containing a single HIF-1 binding site also contain a second essential sequence which has the consensus 5′-CACAG-3′. In the *EPO* gene, mutation of either the 5′-ACGTG-3′ HIF-1 binding site or the 5′-CACAG-3′ sequence (Fig. 1) results in loss of HRE activity.[2] Thus, an HIF-1 binding site is necessary but not sufficient for HRE activity, which can only be demonstrated by transcriptional assays in which the insertion of a putative HRE into a reporter gene is shown to result in increased expression in hypoxic cells.[4,5]

The expression of over 40 genes is known to be activated at the transcriptional level by HIF-1 as determined by the most stringent criteria (Table 1), including the induction of gene expression in response to hypoxia, the presence of a functionally-essential HIF-1 binding site in the gene (Fig. 1), and an effect of HIF-1 gain-of-function or

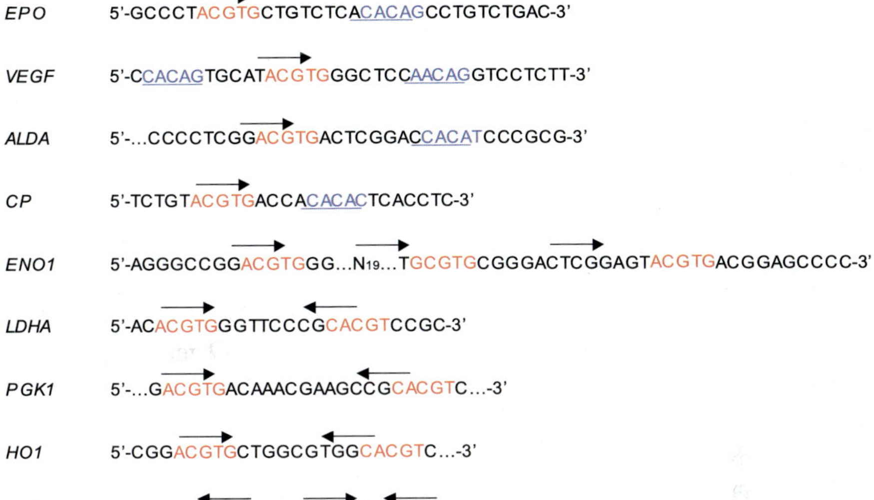

Fig. 1. Hypoxia response elements. Nucleotide sequences from the human genes encoding erythropoietin (EPO), vascular endothelial growth factor (VEGF), aldolase A (ALDA), ceruloplasmin (CP), enolase-1 (ENO1), and lactate dehydrogenase A (LDHA) are shown. The core HIF-1 binding site (5′-RCGTG-3′) is overlined with an arrow and the accessory sequence (consensus, 5′-CACAG-3′) is underscored. Ellipses indicate cases where the 5′ or 3′ end of the minimal hypoxia response element has not been determined and N_{19} denotes an additional 19 nucleotides that are not shown. See Table 1 for references.

loss-of-function on expression of the gene. This list of HIF-1 target genes probably underestimates the total number of genes regulated by HIF-1 by more than an order of magnitude.[7] The battery of genes regulated by HIF-1 is different in each cell type and, for some genes, expression can be induced or repressed by HIF-1 depending upon the cell type.[6] Among critical physiological processes regulated by HIF-1 target genes are angiogenesis, erythropoiesis and glycolysis, which represent examples of both cell-autonomous and non-cell-autonomous (systemic) adaptive responses to hypoxia. Additional candidate HIF-1 target genes have been identified by the analysis of HIF-1-deficient mouse embryonic stem cells[8] and fibroblasts,[9] tissue culture cells expressing constitutively-active forms of HIF-1α or HIF-1β,[10] and VHL-null renal carcinoma cells.[11–16]

Table 1. Genes that are directly regulated by HIF-1.

Gene product	References
ABCG2	86
α_{1B}-adrenergic receptor	87
Adrenomedullin	88
Aldolase A (ALDA)	64, 66
Carbonic anhydrase 9	15
CD18	89
Ceruloplasmin	90
Connective tissue growth factor	91
DEC1	15, 92
DEC2	92
Endocrine gland-derived VEGF	93
Endoglin	94
Endothelin 1	72
Enolase 1	4
Erythropoietin	1, 17
ETS-1	96
Furin	97
Glucose transporter 1 (GLUT1)	64, 66, 98
Glyceraldehyde-3-phosphate dehydrogenase	99, 100
Heme oxygenase 1	101
HGTD-P	102
ID2	103
Intestinal trefoil factor	104
Lactate dehydrogenase A (LDHA)	4, 64, 66
Lactase	105
Leptin	106
Multi-drug resistance 1 (ABCB1)	107
MT1-MMP	108
Nitric oxide synthase 2	109
Nitric oxide synthase 3	95
NIP3	110
NUR77	111
p35srj (CITED2)	112
Phosphoglycerate kinase 1	64, 66, 67
PFKFB3	113
Plasminogen activator inhibitor 1	114
Prolyl-4-hydroxylase a(I)	115

Table 1. (*Continued*).

Gene product	References
RORα	116, 117
RTP801	118
Telomerase (TERT)	119
Transferrin	120
Transferrin receptor	121, 122
Transforming growth factor β_3	71, 73
Vascular endothelial growth factor (VEGF)	5, 64, 66, 67
VEGF receptor 1 (Flt-1)	123

HIF-1 is a heterodimeric protein that is composed of HIF-1α and HIF-1β subunits.[2] The amino-terminal half of each subunit consists of basic helix-loop-helix and PAS domains that mediate heterodimerization and DNA binding.[3,17] The carboxyl-terminal half of HIF-1α contains two transactivation domains[18,19] that mediate interactions with co-activators such as CBP and p300.[20–26] Co-activators interact with both sequence-specific DNA binding proteins such as HIF-1 and with the general transcription factors associated with RNA polymerase II (reviewed in Ref. 27). Co-activators also have histone acetyltransferase activity that is required to make the DNA embedded in chromatin accessible to the polymerase complex for transcription into RNA.

The HIF-1β subunit is constitutively expressed, whereas the expression and activity of the HIF-1α subunit are precisely regulated by the cellular O_2 concentration.[3,28] HIF-1α accumulates instantaneously under hypoxic conditions and upon reoxygenation is rapidly degraded with a half-life of less than five minutes in post-hypoxic tissue culture cells.[3,29] This represents an overestimate of the half-life since it includes the time required for O_2 to diffuse out of the culture medium. In isolated perfused and ventilated lung preparations subjected to hypoxia and reoxygenation, the half-life of HIF-1α is less one minute.[30] No protein has been shown to have a shorter half-life.

In addition to HIF-1α, a structurally and functionally-related protein designated HIF-2αs which is the product of the *EPAS1* gene, can also heterodimerize with HIF-1β.[31] HIF-1α:HIF-1β and HIF-2α:HIF-1β heterodimers appear to have overlapping but distinct target gene specificities.[10,32] Unlike HIF-1α, HIF-2α is not expressed in all cell types and when expressed can be inactive as a result of cytoplasmic sequestration.[33] A third protein, designated HIF-3α, has also been identified.[34] Its role has not been well defined although a splice variant, designated IPAS, has been shown to bind to HIF-1α and inhibit its activity.[35,36]

The O_2-dependent degradation of HIF-1α involves ubiquitination and degradation by the 26S proteasome.[37-39] The von Hippel-Lindau tumor suppressor protein (VHL) is required for this process and renal carcinoma cells lacking functional VHL constitutively express HIF-1α and HIF-2α as well as mRNAs encoded by HIF-1 target genes, under non-hypoxic conditions.[40] VHL forms a complex with elongin B, elongin C, cullin 2, and RBX1 to form an E3 ubiquitin-protein ligase capable of functioning with E1 ubiquitin-activating and E2 ubiquitin-conjugating enzymes to mediate the ubiquitination of HIF-1α.[41] Proline (Pro) residue 564 is hydroxylated in an O_2-dependent manner and this modification is required for VHL binding.[42-44] Pro-402 represents a second site of hydroxylation and VHL binding.[45] Pro-402 and -564 are each contained within a similar amino acid sequence (LXXLAP; A, alanine; L, leucine; P, Pro; X, any amino acid). HIF-2α and HIF-3α expression are also regulated by prolyl hydroxylation and VHL binding.[40,46,47]

Three prolyl hydroxylases were identified in mammalian cells and shown to utilize O_2 as a substrate to generate 4-hydroxyproline at residue 402 and/or 564 of HIF-1α.[48-50] These proteins are homologues of EGL-9, which was identified as the HIF-1α prolyl hydroxylase in *C. elegans* by genetic studies.[49] Alternative designations for the three mammalian homologues include EGLN (EGL Nine homologue), PHD (Prolyl Hydroxylase Domain protein), and HPH (HIF-1α Prolyl Hydroxylase) 1–3. The hydroxylation reaction also requires α-ketoglutarate as a substrate and generates succinate as a side product.[49,51,52] Ascorbate is required as a co-factor. The prolyl hydroxylase catalytic site contains an Fe (II) ion that is coordinated by two histidine and one aspartate residue. Unlike heme-containing proteins, the Fe (II) in dioxygenases can be chelated or substituted by Co (II), rendering the enzyme inactive. Most importantly, these prolyl hydroxylases have a relatively high K_m for O_2 that is slightly above its atmospheric concentration, such that O_2 is rate limiting for enzymatic activity under physiological conditions.[46,49] As a result, changes in cellular O_2 concentration are directly transduced into changes in the rate at which HIF-1α is hydroxylated, ubiquitinated, and degraded.

Remarkably, HIF-1α transactivation domain function is regulated by O_2-dependent hydroxylation of asparagine residue 803, which blocks the binding of the co-activators CBP and p300.[53] FIH-1 (factor inhibiting HIF-1), which was identified as a protein that interacts with and inhibits the activity of the HIF-1α transactivation domain,[54] functions as the asparaginyl hydroxylase.[55,56] As in the case of the prolyl hydroxylases, FIH-1 appears to utilize O_2 and α-ketoglutarate and contain Fe (II) in its active site, although it has a K_m for O_2 that is three times lower than the prolyl hydroxylases.[57]

The more cells that are present in a tissue, the more O_2 is consumed. Thus, it is not surprising that major pathways that transduce proliferative and survival signals from growth factor receptors also induce HIF-1α expression in what can be viewed as a pre-emptive strategy for maintaining oxygen homeostasis. The increase in HIF-1α levels in response to growth factor stimulation differs in two important respects from the increase in HIF-1α levels in response to hypoxia. First, whereas hypoxia increases HIF-1α levels in all cell types, growth factor stimulation induces HIF-1α expression in a cell type-restricted manner. Second, whereas hypoxia is associated with decreased degradation of HIF-1α, growth factors, cytokines, and other signaling molecules stimulate HIF-1α synthesis via activation of the PI-3-kinase or MAP kinase pathways.[58-61]

Pulse-chase studies of MCF-7 breast cancer cells stimulated with heregulin demonstrated increased HIF-1α protein synthesis that was inhibited by treatment with rapamycin, a macrolide antibiotic which inhibits mTOR, a kinase that functions downstream of PI-3-kinase and AKT.[59] The effect of heregulin was mediated via the 5$'$-untranslated region of HIF-1α mRNA.[59] The known targets for phosphorylation by mTOR are regulators of protein synthesis. The translation of several dozen different mRNAs are known to be regulated by this pathway and specific sequences in the 5$'$-untranslated region may determine the degree to which the translation of any particular mRNA is modulated by mTOR signaling. HIF-1α protein expression is particularly sensitive to changes in the rate of synthesis because of its extremely short half-life under non-hypoxic conditions. In addition to effects on HIF-1α synthesis, activation of the RAF-MEK-ERK signaling pathway has also been shown to stimulate HIF-1α transactivation domain function.[62] This effect is due at least in part to ERK phosphorylation of the co-activator P300 with which the transactivation domains interact.[26]

As described earlier, O_2 delivery to cells of the developing embryo becomes limited by diffusion such that establishment of a functioning circulatory system is required for embryonic survival by embryonic day 9 (E9) in the mouse. In wild-type mouse embryos, HIF-1α expression increases dramatically between E8.5 and E9.5 whereas embryos that lack HIF-1α expression die between E9.5 and E10.5 with cardiac malformations, vascular regression, and massive cell death.[63–66] Thus, HIF-1α is essential for embryonic survival and vascularization.

3. Control of Angiogenic Growth Factor and Cytokine Production by HIF-1

Angiogenesis is mediated by a complex network of cell-cell communication, which begins when HIF-1 activity is induced within parenchymal cells in response to either hypoxia (e.g. ischemic tissue) or growth factor stimulation (e.g. tissue undergoing hyperplasia or hypertrophy). HIF-1 then mediates increased transcription of the *VEGF* gene,[5] leading to increased secretion of the VEGF protein. Binding of VEGF to its cognate receptor VEGF-R2 on endothelial cells induces endothelial cell activation, which is characterized by changes in gene expression that resulting in decreased interactions with pericytes/smooth muscle cells and increased cell proliferation, motility, invasiveness, and cell autonomous survival. The combination of VEGF and hypoxia induces the expression by vascular cells of a variety of angiogenic growth factors and cytokines in a temporal- and cell-type-specific manner. For example, endothelial cells produce endothelin 1 (EDN1), which binds to receptors on smooth muscle cells, inducing proliferation and hypertrophy.

As compared to growth/survival factors, cytokines primarily function to promote cell recruitment from distant sites and to regulate cell-cell interactions within the vasculature. Examples of cytokines regulated by HIF-1 are placental growth factor (PLGF) and stromal-derived growth factor 1 (SDF-1), which in concert with VEGF, promote the recruitment to ischemic tissue of bone marrow-derived cells expressing VEGF-R1 (for PLGF and VEGF) or CXCR4 (for SDF-1) on their cell surface.[6,67,68]

Another group of secreted molecules, including angiopoietin 1 (ANGPT1), angiopoietin 2 (ANGPT2), and platelet-derived growth factor B (PDGFB), regulate short-range interactions between endothelial cells and pericytes/smooth muscle cells, which are inhibited during vascular budding and then re-established once a new blood vessel branch has been formed. Several factors, including VEGF and transforming growth factor β, have effects on both the proliferation/survival and migration of endothelial cells. Both gain-of-function and loss-of-function studies have demonstrated that HIF-1 regulates the expression of the genes encoding ANGPT1, ANGPT2, EDN1, PDGFB, PLGF, SDF1, TGFB3, VEGF, and VEGFC both in primary cultures of vascular cells and *in vivo*.[6,7,66–74] Other secreted factors that are regulated by HIF-1 may promote angiogenesis in certain tissues or in response to specific physiological or pathological states include adrenomedullin, connective tissue growth factor, endocrine gland-derived vascular endothelial growth factor, erythropoietin, insulin-like growth factor 2, and leptin (see Table 1 for references).

4. Cell-Autonomous Effects of HIF-1 in Vascular Endothelial Cells

Gene expression profiles were compared in arterial endothelial cells cultured under non-hypoxic *versus* hypoxic conditions and in non-hypoxic cells infected with AdLacZ, an adenovirus encoding β-galactosidase, *versus* AdCA5, which encodes a constitutively-active form of HIF-1α.[7] Two hundred and forty-five gene probes showed at least 1.5-fold increase in expression in response to hypoxia and in response to AdCA5; 325 gene probes showed at least 1.5-fold decrease in expression in response to hypoxia and in response to AdCA5. The largest category of genes downregulated by both hypoxia and AdCA5-encoded proteins involved in cell growth/proliferation. Many genes upregulated by both hypoxia and AdCA5-encoded cytokines, growth factors, receptors, and other signaling proteins. These genes (Tables 2 and 3) include many that are known to play critical roles in angiogenesis and are discussed elsewhere in this volume. Transcription factors accounted for the largest group of HIF-1-regulated genes, indicating that HIF-1 controls a network of transcriptional responses to hypoxia in endothelial cells. Infection of endothelial cells with AdCA5 under non-hypoxic conditions was sufficient to induce increased basement membrane invasion and tube formation similar to the responses induced by hypoxia, indicating that HIF-1 mediates cell-autonomous activation of endothelial cells.[7]

These gain-of-function studies were complemented by loss-of-function studies in which HIF-1α expression in endothelial cells was selectively eliminated in mice that were homozygous for a floxed *Hif1a* allele and expressed cre recombinase from the *Tie2* gene promoter.[74] Loss of HIF-1α expression impaired endothelial cell proliferation, chemotaxis, and extracellular matrix invasion. Most strikingly, loss of HIF-1α expression in endothelial cells resulted in a profound impairment of tumor angiogenesis. All of the observed defects were associated with decreased VEGF expression and loss of autocrine signaling via VEGFR2 in HIF-1α-null endothelial cells. Thus, both gain-of-function and loss-of-function studies indicate that HIF-1 plays a cell-autonomous role in endothelial cells that is critical for tumor angiogenesis but less critical for angiogenesis during development which was apparently normal in the *Hif1a$^{flox/flox}$;Tie2-Cre* mice.[74] Although the studies involving these mice suggest that an important function of HIF-1 in endothelial cells is to mediate autocrine VEGF-VEGFR2 signaling, the microarray analysis[7] suggests that this is probably an oversimplification of the large network of gene expression that is controlled by HIF-1 and is likely to contribute to many aspects of endothelial cell biology.

5. Control of Angiogenesis and Arteriogenesis by HIF-1

To demonstrate that expression of a constitutively-active form of HIF-1α was sufficient to induce angiogenesis *in vivo*, mice were injected with AdCA5 and AdLacZ into the left and right eye, respectively. In contrast

Table 2. Selected genes induced by both hypoxia and AdCA5 in arterial endothelial cells.

Unigene	Gene name	Symbol	Hypoxia	AdCA5
Oxidoreductases				
Hs.302085	Prostaglandin I_2 (prostacyclin) synthase	PTGIS	11.5	16.4
Hs.302085	Prostaglandin I_2 (prostacyclin) synthase	PTGIS	3.1	5.0
Hs.443906	Prolyl hydroxylase domain-containing protein 3	EGLN3	4.6	5.0
Hs.318567	N-myc downstream-regulated gene 1	NDRG1	2.2	3.7
Hs.88474	Prostaglandin-endoperoxide synthase 1 (prostaglandin G/H synthase and cyclooxygenase)	PTGS1	2.3	2.8
Hs.302085	Prostaglandin I_2 (prostacyclin) synthase	PTGIS	2.5	2.3
Hs.88474	Prostaglandin-endoperoxide synthase 1 (prostaglandin G/H synthase and cyclooxygenase)	PTGS1	1.6	2.2
Hs.130946	Prolyl hydroxylase domain-containing protein 2	EGLN1	2.4	2.1
Hs.352733	Peptidylglycine alpha-amidating monooxygenase	PAM	1.7	1.9
Hs.10949	ERO1-like	ERO1L	1.7	1.8
Hs.179526	Thioredoxin interacting protein	TXNIP	2.6	1.7
Hs.352733	Peptidylglycine alpha-amidating monooxygenase	PAM	1.6	1.7
Hs.352733	Peptidylglycine alpha-amidating monooxygenase	PAM	1.8	1.7
Hs.83354	Lysyl oxidase-like 2	LOXL2	1.7	1.6
Collagens/Modifying Enzymes				
Hs.102267	Lysyl oxidase	LOX	2.7	7.1
Hs.102267	Lysyl oxidase	LOX	2.4	6.7
Hs.104772	Procollagen proline 4-hydroxylase, alpha 2	P4HA2	2.0	4.6

Table 2.   (*Continued*).

Unigene	Gene name	Symbol	Hypoxia	AdCA5
Hs.232115	Collagen, type 1, alpha 2	COL1A2	4.5	4.1
Hs.41270	Procollagen lysine hydroxylase 2	PLOD2	2.2	3.5
Hs.433695	Collagen, type V, alpha 1	COL5A1	2.3	3.0
Hs.76768	Procollagen proline 4-hydroxylase, alpha 1	PRHA1	3.4	2.6
Hs.433695	Collagen, type V, alpha 1	COL5A1	2.4	2.6
Hs.41270	Procollagen lysine hydroxylase 2	PLOD2	1.8	2.6
Hs.413175	Collagen, type XVIII, alpha 1	COL18A1	1.8	2.0
Hs.149809	Collagen, type IX, alpha 1	COL9A1	1.7	1.8
Hs.437173	Collagen, type IV, alpha 1	COL4A1	1.6	1.7
Hs.407912	Collagen, type IV, alpha 2	COL4A2	1.6	1.7
Hs.75093	Procollagen lysine hydroxylase	PLOD	2.0	1.5
Hs.413175	Collagen, type XVIII, alpha 1	COL18A1	1.6	1.5

Growth Factors/Cytokines

Unigene	Gene name	Symbol	Hypoxia	AdCA5
Hs.9613	Angiopoietin-like 4	ANGPTL4	17.5	14.9
Hs.25590	Stanniocalcin 1	STC1	3.1	12.2
Hs.9613	Inhibin, beta A (activin A, activin AB alpha)	INHBA	10.2	11.6
Hs.73793	Vascular endothelial growth factor	VEGF	6.7	8.8
Hs.279497	Inhibin, beta E	INHBE	2.2	4.8
Hs.155223	Stanniocalcin 2	STC2	2.3	4.3
Hs.511899	Endothelin 1	EDN1	1.9	4.0
Hs.2171	Growth differentiation factor 10	GDF10	1.5	3.3
Hs.450230	Insulin-like growth factor binding protein 3	IGFBP3	2.5	3.2
Hs.80420	Chemokine (C-X3-C motif) ligand 1	CX3CL1	2.1	2.9
Hs.79141	Vascular endothelial growth factor C	VEGFC	2.8	2.6
Hs.368996	Relaxin 1	RLN1	1.5	2.5
Hs.1976	Platelet-derived growth factor B polypeptide	PDGFB	1.9	2.5
Hs.73793	Vascular endothelial growth factor	VEGF	2.5	2.4
Hs.73793	Vascular endothelial growth factor	VEGF	1.9	2.2
Hs.252820	Placental growth factor	PGF	1.9	2.1
Hs.441047	Adrenomedullin	ADM	4.2	2.1
Hs.1976	Platelet-derived growth factor B polypeptide	PDGFB	2.3	1.7

Table 2.   (*Continued*).

Unigene	Gene name	Symbol	Hypoxia	AdCA5
Receptors				
Hs.439141	Gamma-aminobutyric acid (GABA) A receptor, pi	GABRP	3.4	8.1
Hs.231853	Chemokine orphan receptor 1	CMKOR1	1.7	6.4
Hs.370422	Very low density lipoprotein receptor	VLDLR	2.0	5.9
Hs.198288	Protein tyrosine phosphatase, receptor type, R	PTPRR	6.0	5.6
Hs.166206	Cubilin (intrinsic factor-cobalamin receptor)	CUBN	2.0	4.5
Hs.421986	Chemokine (C-X-C motif) receptor 4	CXCR4	3.1	2.9
Hs.197029	Adenosine A2a receptor	ADORA2A	2.3	2.8
Hs.438864	Natriuretic peptide receptor A/guanylate cyclase A	NPR1	1.8	2.7
Hs.438669	Insulin receptor	INSR	2.3	2.5
Hs.434375	Protein tyrosine phosphatase, receptor type, B	PTPRB	2.1	2.5
Hs.371974	Activin A receptor, type IB	ACVR1B	2.0	2.4
Hs.198288	Protein tyrosine phosphatase, receptor type, R	PTPRR	5.1	2.3
Hs.434375	Protein tyrosine phosphatase, receptor type, B	PTPRB	2.6	2.2
Hs.17109	Integral membrane protein 2A	ITM2A	1.8	1.8
Hs.127826	Erythropoietin receptor	EPOR	1.7	1.8
Hs.334131	Cadherin 2, type 1, N-cadherin (neuronal)	CDH2	1.6	1.8
Hs.51233	TNF receptor superfamily, member 10b	TNFRSF10B	1.6	1.7
Hs.279899	TNF receptor superfamily, member 14	TNFRSF14	1.5	1.7
Hs.170129	Opsin 3 (encephalopsin, panopsin)	OPN3	1.5	1.6
Hs.23581	Leptin receptor	LEPR	2.6	1.6
Hs.75216	Protein tyrosine phosphatase, receptor type, F	PTPRF	1.7	1.5
Hs.512235	Inositol-1,4,5-triphosphate receptor, type 2	ITPR2	1.8	1.5
Hs.83341	AXL receptor tyrosine kinase	AXL	1.5	1.5
Hs.23581	Leptin receptor	LEPR	1.6	1.5
Hs.211569	G protein-coupled receptor kinase 5	GRK5	1.5	1.5

Table 2. (*Continued*).

Unigene	Gene name	Symbol	Hypoxia	AdCA5
Other Signal Transduction				
Hs.106070	Cyclin-dependent kinase inhibitor 1C (p57, Kip2)	CDKN1C	3.6	17.2
Hs.512298	Phospholipase C, gamma 2	PLCG2	2.1	4.8
Hs.200598	AMP-activated protein kinase family member 5	ARK5	1.5	4.5
Hs.287460	PDZ domain-containing 3	PDZK3	4.4	4.2
Hs.106070	Cyclin-dependent kinase inhibitor 1C (p57, Kip2)	CDKN1C	1.6	3.9
Hs.213840	Ectonucleotide pyrophosphatase/phosphodiesterase 1	ENPP1	2.6	3.5
Hs.458389	SPRY domain-containing SOCS box protein SSB 1	SSB1	1.6	3.0
Hs.298654	Dual specificity phosphatase 6	DUSP6	2.3	2.9
Hs.44038	Pellino homolog 2	PELI2	1.9	2.8
Hs.301664	Ribosomal protein S6 kinase, 90 kDa, polypeptide 2	RPS6KA2	1.7	2.7
Hs.10862	Adenylate kinase 3	AK3	5.7	2.6
Hs.15099	Rho-related BTB domain-containing 1	RHOBTB1	1.6	2.6
Hs.288196	Calcium/calmodulin-dependent serine protein kinase	CASK	1.5	2.6
Hs.173380	CK2 interacting protein 1	CKIP-1	1.5	2.3
Hs.82294	Regulator of G-protein signaling 3	RGS3	1.9	2.3
Hs.112378	LIM and senescent cell antigen-like domain 1	LIMS1	1.6	2.2
Hs.10862	Adenylate kinase 3	AK3	3.8	2.1
Hs.80905	Ras association (RalGDS/AF-6) domain family 2	RASSF2	1.5	1.8
Hs.299883	FAD104	FAD104	1.6	1.8
Hs.446403	WAS protein family, member 2	WASF2	1.6	1.8
Hs.110571	Growth arrest and DNA damage-inducible, beta	GADD45B	2.3	1.7
Hs.13291	Cyclin G2	CCNG2	1.5	1.7
Hs.129836	Cyclin-dependent kinase (CDC2-like) 11	CDK11	1.5	1.7
Hs.367689	Triple functional domain (PTPRF interacting)	TRIO	1.5	1.7

Table 2.   (*Continued*).

Unigene	Gene name	Symbol	Hypoxia	AdCA5
Hs.301664	Ribosomal protein S6 kinase, 90 kDa, polypeptide 2	RPS6KA2	1.7	1.6
Hs.9651	Related RAS viral (r-ras) oncogene homolog	RRAS	2.1	1.6
Hs.301242	Myocytic induction/differentiation originator	MIDORI	1.9	1.5
Hs.497972	FK506 binding protein 9, 63 kDa	FKBP9	1.5	1.5
Transcription factors				
Hs.460889	V-maf oncogene homolog F	MAFF	2.1	5.0
Hs.387667	Peroxisome proliferative activated receptor, gamma	PPARG	16.8	3.9
Hs.232068	Transcription factor 8	TCF8	1.7	3.3
Hs.321707	Jumonji domain-containing 1	JMJD1	2.1	3.1
Hs.511938	Zinc finger and BTB domain-containing 1	ZBTB1	1.7	2.9
Hs.420830	Hypoxia inducible factor 3, alpha subunit	HIF3A	1.6	2.7
Hs.357901	SRY (sex determining region Y) box 4	SOX4	1.7	2.3
Hs.171825	Basic helix-loop-helix domain-containing, class B, 2	BHLHB2	3.4	2.3
Hs.324051	RelA-associated inhibitor	RAI	2.8	2.1
Hs.321164	Neuronal PAS domain protein 2	NPAS2	1.6	2.1
Hs.436100	Notch homolog 4	NOTCH4	2.6	2.1
Hs.118630	MAX interacting protein 1	MXI1	3.0	2.0
Hs.415033	Myocyte enhancer factor 2A	MEF2A	1.8	2.0
Hs.406491	Transducin-like enhancer of split 1 homolog	TLE1	1.5	1.9
Hs.75063	HIV-I enhancer binding protein 2	HIVEP2	1.8	1.9
Hs.318517	Transcription factor 7-like 1	TCF7L1	1.8	1.9
Hs.448341	Zinc finger protein 292	ZNF292	1.9	1.8
Hs.55967	Short stature homeobox 2	SHOX2	1.7	1.8
Hs.357901	SRY box 4	SOX4	1.5	1.8
Hs.179526	Thioredoxin interacting protein	TXNIP	2.6	1.7
Hs.250666	Hairy and enhancer of split 1	HES1	2.2	1.7
Hs.30209	Zinc fingers and homeoboxes 2	ZHX2	2.3	1.6

Table 2.   (*Continued*).

Unigene	Gene name	Symbol	Hypoxia	AdCA5
Hs.155024	B-cell CLL/lymphoma 6 (zinc finger protein 51)	BCL6	1.7	1.6
Hs.511950	Sirtuin (sir2 homolog) 3	SIRT3	1.7	1.6
Hs.250666	Hairy and enhancer of split 1	HES1	2.0	1.6
Hs.153863	MAD, mothers against decapentaplegic homolog 6	MADH6	1.9	1.6
Hs.171825	Basic helix-loop-helix domain containing, class B, 2	BHLHB2	2.2	1.5
Hs.77810	Nuclear factor of activated T-cells 4	NFATC4	1.9	1.5
Hs.59943	CAMP responsive element binding protein 3-like 2	CREB3L2	1.5	1.5

Date of the above table from Ref. 7.

to AdLacZ, which had no effect, AdCA5 induced neovascularization.[6] Lectin and anti-α-smooth muscle actin antibody staining demonstrated that these vessels contained both endothelial cells and smooth muscle cells or pericytes. The expression of mRNAs encoding ANGPT1, ANGPT2, PDGFB, PLGF, and VEGF was increased in AdCA5- as compared to AdLacZ-treated eyes.[6]

A link between HIF-1 expression and VEGF expression in ischemic retina has been well established,[75] as has a role for VEGF in ischemia-induced retinal neovascularization.[76] However, the ability of AdCA5 to induce sprouting of new vessels that extend into the vitreous cavity from the superficial capillary bed, as occurs in patients with diabetic retinopathy, contrasts with the inability of high levels of VEGF alone to do so.[77] These studies indicate that increased levels of a single VEGF isoform are not sufficient to cause new vessels to sprout from the superficial vessels. Neovascularization induced by AdCA5 was associated with a dramatic increased in PLGF mRNA and protein expression. PLGF has been shown to act synergistically with VEGF to stimulate neovascularization and both factors are required for neovascularization in the ischemic retina.[68] Thus, the combined effect of increased expression of PLGF and multiple VEGF isoforms may underlie neovascularization

Table 3. Selected genes repressed by both hypoxia and AdCA5 in arterial endothelial cells.

Unigene	Gene name	Symbol	Hypoxia	AdCA5
CDC/Cyclins				
Hs.408658	Cyclin E2	CCNE2	–1.7	–4.1
Hs.244723	Cyclin E1	CCNE1	–1.9	–3.9
Hs.1973	Cyclin F	CCNF	–2.1	–3.0
Hs.408658	Cyclin E2	CCNE2	–2.4	–2.8
Hs.334562	Cell division cycle 2	CDC2	–1.5	–2.2
Hs.405958	Cell division cycle 6	CDC6	–2.7	–2.1
Hs.334562	Cell division cycle 2	CDC2	–1.8	–2.1
Hs.28853	Cell division cycle 7	CDC7	–1.6	–2.1
Hs.334562	Cell division cycle 2	CDC2	–1.8	–1.9
MCM Proteins				
Hs.460184	Minichromosome maintenance deficient 4	MCM4	–2.4	–3.1
Hs.460184	Minichromosome maintenance deficient 4	MCM4	–2.2	–3.1
Hs.57101	Minichromosome maintenance deficient 2	MCM2	–2.2	–3.1
Hs.77171	Minichromosome maintenance deficient 5	MCM5	–2.4	–2.7
Hs.444118	Minichromosome maintenance deficient 6	MCM6	–1.9	–2.4
Hs.198363	Minichromosome maintenance deficient 10	MCM10	–2.2	–2.3
Hs.179565	Minichromosome maintenance deficient 3	MCM3	–1.5	–2.2
Hs.460184	Minichromosome maintenance deficient 4	MCM4	–1.8	–2.1
Hs.460184	Minichromosome maintenance deficient 4	MCM4	–2.0	–1.8
DNA/RNA Polymerases				
Hs.437186	Polymerase (RNA) III polypeptide K	POLR3K	–2.1	–4.3
Hs.441072	Polymerase (RNA) II polypeptide L	RNAPOL2	–2.0	–3.3
Hs.99185	Polymerase (DNA directed), epsilon 2	POLE2	–2.0	–2.6
Hs.282387	Polymerase (RNA) III (32 kD)	RPC32	–1.9	–1.9
Hs.306791	Polymerase (DNA directed), delta 2	POLD2	–1.5	–1.7
Hs.290921	Polymerase (DNA directed), gamma	POLG	–1.7	–1.6

Table 3. (*Continued*).

Unigene	Gene name	Symbol	Hypoxia	AdCA5
Replication Factors				
Hs.79018	Chromatin assembly factor 1, subunit A	CHAF1A	–6.0	–10.0
Hs.433180	DNA replication complex GINS protein PSF2	PFS2	–2.2	–3.4
Hs.194665	DNA replication helicase 2-like	DNA2L	–1.7	–3.3
Hs.443227	Replication factor C5	RFC5	–1.9	–3.2
Hs.115474	Replication factor C (activator 1) 3	RFC3	–1.9	–2.9
Hs.443227	Replication factor C5	RFC5	–1.6	–2.3
Hs.49760	Origin recognition complex, subunit 6-like	ORC6L	–1.9	–1.9
Hs.139226	Replication factor C (activator 1) 2	RFC2	–1.6	–1.6
Other Cell Cycle				
Hs.23348	S-phase kinase-associated protein 2	SKP2	–2.4	–3.1
Hs.79078	MAD2 mitotic arrest deficient-like 1	MAD2L1	–1.7	–2.9
Hs.511945	Block of proliferation 1	BOP1	–1.7	–2.8
Hs.42650	ZW10 interactor	ZWINT	–2.0	–2.5
Hs.445084	M-phase phosphoprotein 9	MPHOSPH9	–1.7	–2.5
Hs.75337	Nucleolar and coiled-body phosphoprotein 1	NOLC1	–1.6	–2.4
Hs.374491	Proliferation-associated 2G4	PA2G4	–1.6	–2.1
Hs.152759	Activator of S phase kinase	ASK	–2.7	–2.0
Hs.344037	Protein regulator of cytokinesis 1	PRC1	–2.0	–1.6
Hs.433008	NIMA (never in mitosis a)-related kinase 4	NEK4	–1.5	–1.5
Ribonucleotide Metabolism				
Hs.226390	Ribonucleotide reductase M2 polypeptide	RRM2	–2.4	–7.2
Hs.464813	Dihydrofolate reductase	DHFR	–2.0	–4.5
Hs.251871	CTP synthase	CTPS	–1.8	–3.2
Hs.226390	Ribonucleotide reductase M2 polypeptide	RRM2	–1.8	–2.4
Hs.56	Phosphoribosyl pyrophosphate synthetase 1	PRPS1	–1.6	–2.3
Hs.2057	Uridine monophosphate synthetase	UMPS	–1.6	–2.3
Hs.464813	Dihydrofolate reductase	DHFR	–1.5	–2.3

Table 3. (*Continued*).

Unigene	Gene name	Symbol	Hypoxia	AdCA5
Hs.90280	5-aminoimidazole-4-carboxamide ribonucleotide transformylase	ATIC	−1.5	−2.2
Hs.177766	Poly(ADP-ribose) polymerase	ADPRT	−1.7	−2.0
Hs.215766	GTP-binding protein	GTPBP4	−1.5	−2.0
Hs.409412	ADP-ribosyltransferase (NAD+; Poly(ADP-ribose) polymerase)-like 2	ADPRTL2	−1.5	−1.8
Hs.2057	Uridine monophosphate synthetase	UMPS	−1.5	−1.7
Hs.383396	Ribonucleotide reductase M1 polypeptide	RRM1	−1.6	−1.5
Ribosomal/Mitochondrial Biogenesis				
Hs.411125	Mitochondrial ribosomal protein S12	MRPS12	−1.7	−9.9
Hs.435643	Ts translation elongation factor, mitochondrial	TSFM	−3.7	−6.5
Hs.511949	Ribosomal RNA processing 4	RRP4	−2.0	−4.4
Hs.71827	Ribosome biogenesis regulator homologue	RRS1	−1.9	−3.1
Hs.132748	Ribosomal protein L39	RPL39	−1.8	−2.5
Hs.256583	Interleukin enhancer binding protein 3	ILF3	−2.0	−2.3
Hs.511949	Ribosomal RNA processing 4	RRP4	−1.7	−2.3
Hs.101414	Zinc finger protein 500	ZNF500	−1.6	−2.3
Hs.431307	Mitochondrial ribosomal protein L40	MRPL40	−1.6	−2.1
Hs.376064	Nucleolar protein (KKED repeat)	NOL5A	−1.9	−2.0
Hs.376681	Mitofusin 2	MFN2	−1.7	−1.7
Hs.424264	Mitochondrial ribosomal protein L46	MRPL46	−1.5	−1.5
RNA Binding/Metabolism				
Hs.443960	DEAD/H box polypeptide 11	DDX11	−4.1	−3.9
Hs.434901	Small nuclear ribonucleoprotein polypeptide A′	SNRPA1	−1.6	−3.0
Hs.166463	Heterogeneous nuclear ribonucleoprotein U	HNRPU	−1.7	−2.9
Hs.153768	RNA, U3 small nucleolar interacting protein 2	RNU3IP2	−1.7	−2.4
Hs.363492	DEAD box polypeptide 18	DDX18	−1.5	−2.4

Table 3.   (*Continued*).

Unigene	Gene name	Symbol	Hypoxia	AdCA5
Hs.434901	Small nuclear ribonucleoprotein polypeptide A′	SNRPA1	−1.9	−2.1
Hs.416994	POP7 (processing of precursor) homolog	RPP20	−1.8	−2.1
Hs.30174	Small nuclear RNA activating complex 5	SNAPC5	−1.7	−2.1
Hs.372673	Heterogeneous nuclear ribonucleoprotein D-like	HNRPDL	−1.6	−2.1
Hs.130098	DEAD box polypeptide 23	DDX23	−1.5	−2.1
Hs.7174	DEAH (Asp-Glu-Ala-His) box polypeptide 35	DHX35	−2.1	−2.0
Hs.73965	Splicing factor, arginine/serine-rich 2	SFRS2	−1.6	−2.0
Hs.438726	Heterogeneous nuclear ribonucleoprotein D	HNRPD	−2.2	−1.9
Hs.434901	Small nuclear ribonucleoprotein polypeptide A′	SNRPA1	−1.9	−1.9
Hs.438726	DEAD box polypeptide 21	DDX21	−1.7	−1.9
Hs.372673	Heterogeneous nuclear ribonucleoprotein D-like	HNRPDL	−1.6	−1.7
Hs.86948	Small nuclear ribonucleoprotein D1 polypeptide	SNRPD1	−1.6	−1.7
Hs.356549	Small nuclear ribonucleoprotein D3 polypeptide	SNRPD3	−1.5	−1.7
Hs.511756	Ribonuclease P1	RNASEP1	−1.6	−1.6
Hs.139120	Ribonuclease P (30 kD)	RPP30	−1.6	−1.6
Ubiquitin/Proteasome				
Hs.152978	Proteasome activator subunit 3	PSME3	−1.6	−2.9
Hs.396393	Ubiquitin-conjugating enzyme E2S	UBE2S	−1.6	−2.9
Hs.25223	Ubiquitin carboxyl-terminal hydrolase L5	UCHL5	−1.6	−2.1
Hs.443379	Proteasome 26S subunit, non-ATPase, 11	PSMD11	−1.5	−2.1
Hs.152978	Proteasome activator subunit 3	PSME3	−1.6	−1.8
Hs.119563	Proteasome activator subunit 4	PSME4	−1.6	−1.8
Hs.12820	Ubiquitin specific protease 39	USP39	−1.6	−1.6
Hs.35086	Ubiquitin specific protease 1	USP1	−1.5	−1.5

Table 3. (*Continued*).

Unigene	Gene name	Symbol	Hypoxia	AdCA5
Transcription Factors				
Hs.446451	Myeloid/lymphoid leukemia; translocated to, 10	MLLT10	–2.2	–5.1
Hs.54089	BRCA1 associated RING domain 1	BARD1	–1.6	–3.6
Hs.194143	Breast cancer 1, early onset	BRCA1	–2.7	–3.2
Hs.181128	ELK1, member of ETS oncogene family	ELK1	–2.4	–3.0
Hs.75133	Transcription factor A, mitochondrial	TFAM	–1.6	–3.0
Hs.448398	MYC-associated zinc finger protein	MAZ	–1.6	–2.8
Hs.194143	Breast cancer 1, early onset	BRCA1	–2.0	–2.4
Hs.106415	Peroxisome proliferative activated receptor, delta	PPARD	–2.1	–2.3
Hs.408528	Retinoblastoma 1	RB1	–1.8	–2.3
Hs.436187	Thyroid hormone receptor interactor 13	TRIP13	–1.9	–2.1
Hs.75133	Transcription factor 6-like 1	TF6M	–1.6	–1.7
Hs.75133	Transcription factor A, mitochondrial	TFAM	–1.6	–1.7
Hs.410900	Inhibitor of DNA binding 1	ID1	–1.6	–1.5

in the superficial capillary bed (angiogenesis) induced by intravitreous injection of AdCA5.

To investigate the role of HIF-1 in mediating recovery of blood flow following arterial occlusion, a novel non-operative rabbit model of arterial occlusion was utilized.[78] A catheter was threaded from the carotid artery into the left femoral artery and coils were introduced which occlude the lumen. As a result, the potentially confounding effects of an operative wound at the site of occlusion were eliminated. Angiograms were performed immediately before and after coil placement to verify complete arterial occlusion (Fig. 1A–1C). A total of 6×10^8 plaque forming units (pfu) of AdCA5 or AdLacZ was injected into the adductor muscle (1×10^8 pfu at each of six sites) in the distribution of the occluded artery.

Calf blood pressure (BP) was measured preoperatively and on postoperative day 14, at which time angiography was repeated (Fig. 1D), the animal was sacrificed, and adductor muscle tissue was excised and fixed. Compared to AdLacZ-treated rabbits, AdCA5-treated animals showed a significant improvement in the calf BP ratio (coiled:non-coiled limb) and a significant decrease in the time required for complete perfusion of the femoral circulation after contrast injection (Fig. 2).

Anti-CD31 immunohistochemistry revealed a significant increase in capillary:myocyte ratio in adductor muscle for AdCA5-treated animals (Fig. 3). More importantly, whereas the total number of arteries was not changed, AdCA5 increased the total luminal area of arteries $> 100\,\mu$m in diameter, as determined by image analysis of sections of adductor muscle stained with anti-α-smooth muscle actin antibodies (Fig. 4). Thus, AdCA5 administration significantly stimulated the remodeling of pre-existing collateral vessels (arteriogenesis) (Fig. 5), which is the major determinant of blood flow following arterial occlusion.[78] AdCA5 treatment induced increased expression of PDGF-B, PLGF, MCP-1, SDF-1, and VEGF mRNA in adductor muscle.[78] Both MCP-1 and PLGF have been shown to promote arteriogenesis in previous studies.[79,80]

Taken together these two studies demonstrate that AdCA5 can induce both angiogenesis and arteriogenesis. In both cases, AdCA5 was shown to induce the expression of multiple angiogenic growth factors and cytokines. As a result, HIF-1α gene therapy may have advantages over gene therapy approaches that only increase the expression of a single angiogenic growth factor or cytokine — a strategy that has been unsuccessful in clinical trials. Further studies are required to determine whether AdCA5 administration may represent a novel treatment option for patients with extensive peripheral vascular disease who are not candidates for conventional therapies.

6. Control of Tumor Angiogenesis by HIF-1

Human colon cancer cells transfected with an expression vector encoding HIF-1α manifest a dramatic increase in tumor xenograft growth and angiogenesis, with significant increases in tumor vascular volume and vascular permeability demonstrated *in vivo* by magnetic resonance

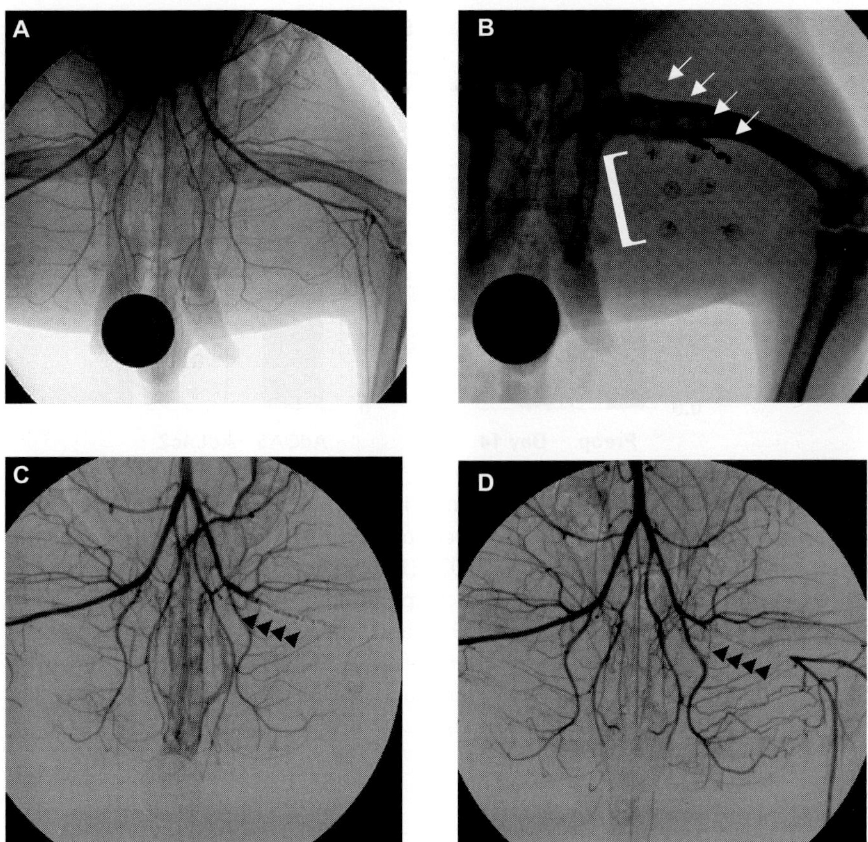

Fig. 2.   Representative pelvic and hindlimb angiograms of an AdCA5-treated rabbit. (**A**) Digital angiogram, day 0, before coiling. (**B**) Anterior-posterior digital spot image showing coils (arrows) in the left femoral artery and six needles (bracketed area) placed on the sites for adenoviral injection. (**C**) Digital subtraction angiogram, day 0, immediately after coiling. Arrowheads indicate location of coils. (**D**) Digital subtraction angiogram of an AdCA5-treated rabbit (day 14) showing collateral vessel development in the medial thigh. Arrowheads indicate location of the coils. (Reprinted from Ref. 78, copyright 2005 with permission from the European Society of Cardiology.)

imaging.[81] Conversely, human gastric cancer cells transfected with an expression vector encoding a dominant negative form of HIF-1α (HIF-1αDN) manifest a striking reduction in tumor growth, either as subcutaneous xenografts or after orthotopic transplantation into the gastric

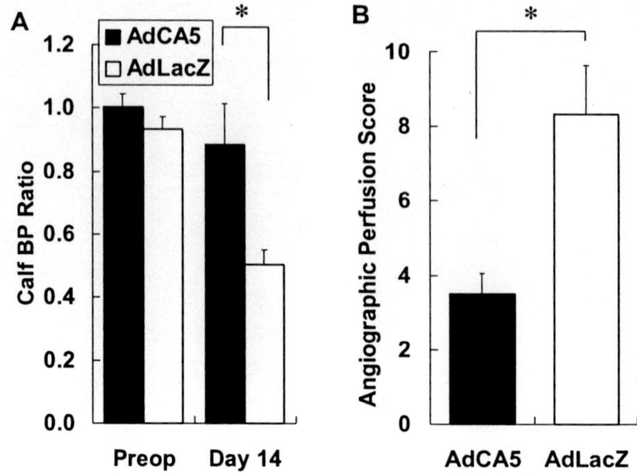

Fig. 3. Quantitative effect of AdCA5 injection on blood flow and collateral vessel development after unilateral femoral artery occlusion. **(A)** Calf blood pressure ratio performed on day 0 and day 14 ($^*p = 0.02$). **(B)** Angiographic perfusion score on day 14($^*p < 0.01$). The frame number on which opacification of the left femoral artery just distal to the occlusion occurred was and the frame number on which contrast opacified the bifurcation of the right femoral artery were determined. The difference between the two values reflects the difference in angiographic perfusion between the left and right calf. (Reprinted from Ref. 78, copyright 2005 with permission from the European Society of Cardiology.)

wall of immunodeficient mice.[82] Histological analysis of tumor sections revealed a dramatic reduction in vessel luminal area within tumors derived from cells expressing HIF-1αDN. Not only were the vessels markedly smaller, but pericyte coverage of the endothelium was also dramatically decreased. The vascularization of xenografts derived from HIF-1α-null mouse embryonic stem cells is also markedly impaired.[66,67] These data from experimental models are consistent with immunohistochemical analyses of biopsy sections which demonstrate that HIF-1α overexpression is associated with increased tumor microvessel density, increased tumor VEGF levels, and increased patient mortality in many different human cancers (Table 4).

Many of the novel molecularly targeted anti-cancer agents have anti-angiogenic effects. These effects appear to be due in part to their inhibition of HIF-1.[83] As discussed earlier, HIF-1 activity is induced both by

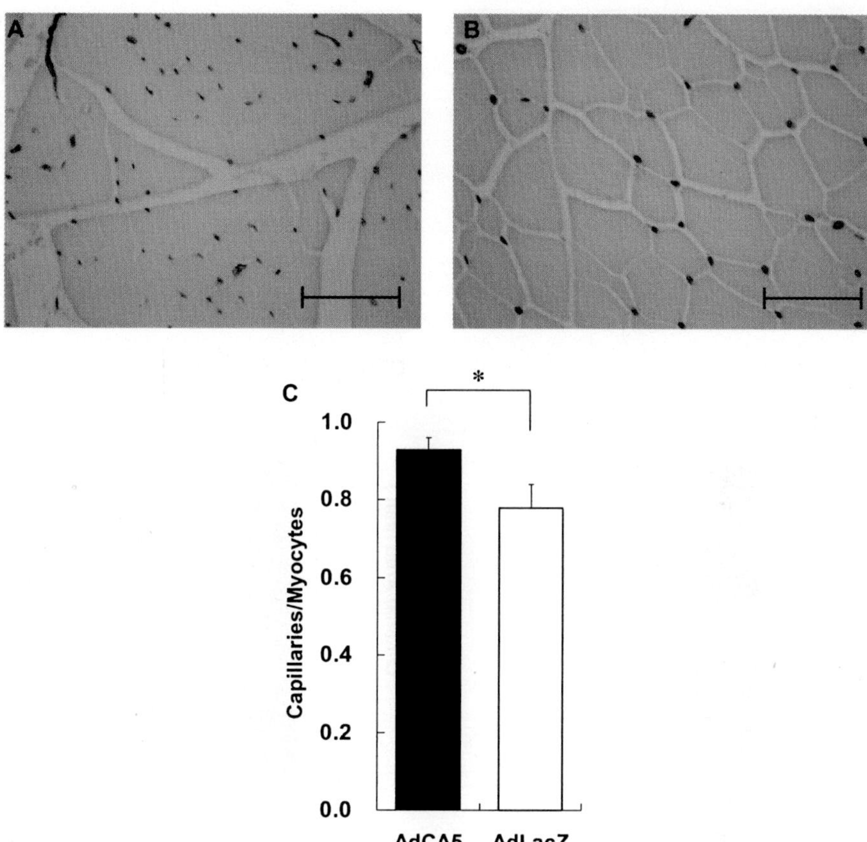

Fig. 4. Capillary morphometry of sections from left adductor muscles harvested on day 14. CD-31 staining of the adductor muscle injected with AdCA5 (**A**) and AdLacZ (**B**). Magnification: × 400. Scale bars: 100 μm. (**C**) Capillary/myocyte ratio on day 14 (* $p <$ 0.05). (Reprinted from Ref. 78, copyright 2005 with permission from the European Society of Cardiology.)

hypoxia and by growth factor signal transduction. A critical hallmark of cancer cells is the acquisition of independence from external sources of growth factor due to the establishment of autocrine signaling pathways in which the tumor expresses both the growth factor and its receptor. HIF-1 appears to play an important role in these pathways because HIF-1 is both upstream and downstream of the receptors. A consequence of receptor (e.g. EGFR, IGF-R1) activation by ligand is an

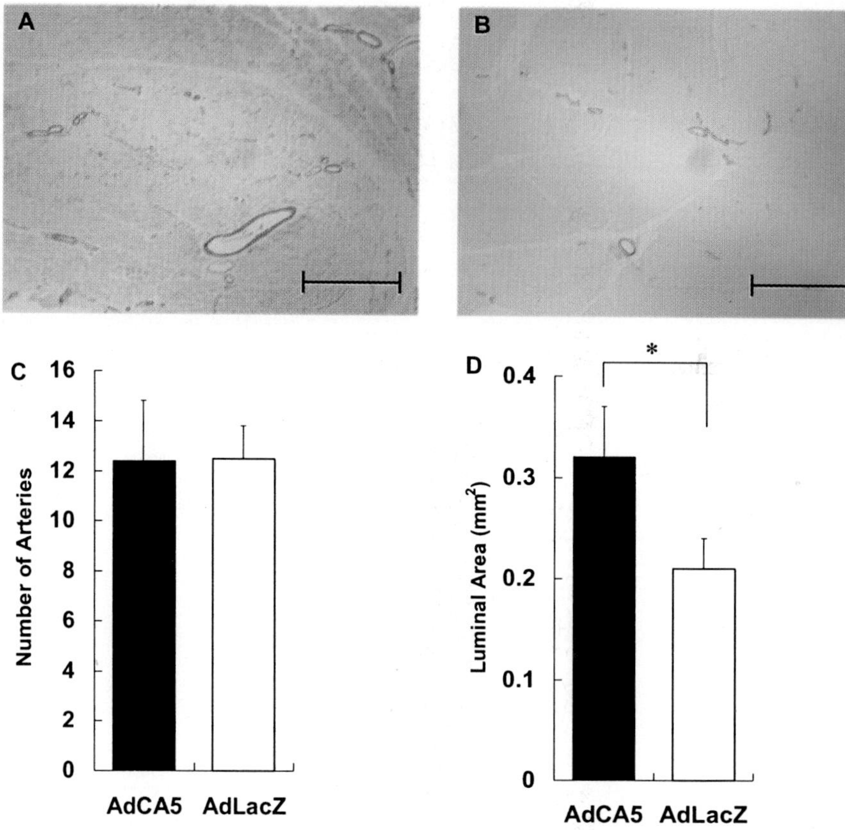

Fig. 5. Arterial morphometry of sections from left adductor muscles harvested on day 14. α-smooth muscle actin immunohistochemistry was performed on sections of adductor muscles injected with AdCA5 (**A**) and AdLacZ (**B**). Magnification: ×40. Scale bars: 1 mm. (**C**) Mean number of arteries with a diameter greater than 100 μm in the adductor muscles. (**D**) Total luminal area of the same arteries (*$p < 0.05$). (Reprinted from Ref. 78, copyright 2005 with permission from the European Society of Cardiology.)

increase in HIF-1α synthesis, which is mediated via the PI-3-kinase and MAP kinase signal-transduction pathways, as described above. A consequence of increased HIF-1 activity is increased expression of genes encoding growth factors (including IGF-2 and TGF-α), which completes the autocrine loop. An additional consequence of increased HIF-1 activity is the production of VEGF and other angiogenic growth factors and cytokines.

Table 4. Effect of HIF-1a overexpression in human cancers.

Tumor type	Association	Reference
Astrocytoma, diffuse	Mortality, MVD	124
Bladder, superficial urothelial	Mortality (w/mutant p53), grade, MVD	125
Bladder, transitional cell	Mortality, grade, MVD, VEGF	126
Brain, glioma	Grade, MVD	127
Breast	Grade, VEGF, MVD (in DCIS)	128
Breast, c-erbB-2-positive	Mortality	129
Breast, LN-positive	Mortality	130
Breast, LN-negative	Mortality	131
Cervix, early-stage	Mortality	132
Cervix, RTX	Mortality	133
Cervix, IB-IIIB, RTX	Radiation resistance, mortality	134
Colon	Invasion, metastasis, MVD, VEGF	135
Esophagus, SCC	MVD, VEGF, venous invasion	136
Esophagus, early stage	PDT response (w/BCL2 overexpression)	137
Esophagus	VEGF	138
Endometrial	Mortality, VEGF, MVD	139
GIST, stomach	Mortality, metastasis, MVD, VEGF	140
Head and neck-SCC	MVD, mortality (HIF-2a)	141
Lung, NSCLC	VEGF, PD-ECGF, FGF2	142
Malignant melanoma	VEGF, mortality (HIF-2a)	143
Oligodendroglioma	Mortality, MVD	144
Oropharynx-SCC	Mortality, radiation resistance	145
Ovarian	MVD, mortality (w/mutant p53)	146
Pancreas	TNM stage, MVD	147
Pancreas	MVD, VEGF, metastasis	148
Wilms	VEGF	149

Small molecule inhibitors of growth factor receptors or downstream signaling molecules block a major stimulus for HIF-1α synthesis.[83] The resulting decrease in angiogenesis may lead to tumor hypoxia, which will induce HIF-1 activity via the reduced activity of the prolyl and asparaginyl hydroxylases that negatively regulate HIF-1, thus providing a mechanism by which hypoxia may promote treatment failure. These findings suggest that an inhibitor of HIF-1 activity may be useful as a therapeutic agent in combination with signal transduction inhibitors that target either the tumor cells or the endothelium. Angiogenesis inhibitors, such as antibodies against VEGF or small-molecule inhibitors of VEGF receptors, may be particularly effective against tumor vessels that lack pericyte coverage .[84] Based upon the results of expressing HIF-1αDN in gastric cancer cells, inhibition of HIF-1 activity may inhibit pericyte coverage and thus promote the effects of angiogenesis inhibitors. A recent microarray analysis of the transcriptional programs affected by treatment of endothelial cells with the anti-angiogenic agent endostatin revealed a significant reduction in the expression of HIF-1 target genes,[85] a result that is consistent with the finding that HIF-1 controls both the production of angiogenic growth factors/cytokines and cell-autonomous responses of endothelial cells to angiogenic signals.[7,74]

References

1. Semenza GL, Wang GL (1992). A nuclear factor induced by hypoxia via de novo protein synthesis binds to the human erythropoietin gene enhancer at a site required for transcriptional activation. *Mol Cell Biol* **12**: 5447–5454.
2. Wang GL, Semenza GL (1995). Purification and characterization of hypoxia-inducible factor 1. *J Biol Chem* **270**: 1230–1237.
3. Wang GL, Jiang BH, Rue EA, Semenza GL (1995). Hypoxia-inducible factor 1 is a basic-helix-loop-helix-PAS heterodimer regulated by cellular O_2 tension. *Proc Natl Acad Sci USA* **92**: 5510–5514.
4. Semenza GL, Jiang BH, Leung SW, Passantino R, Concordet JP, Maire P, Giallongo A (1996). Hypoxia response elements in the aldolase A, enolase 1, and lactate dehydrogenase A gene promoters contain essential binding sites for hypoxia-inducible factor 1. *J Biol Chem* **271**: 32529–32537.
5. Forsythe JA, Jiang BH, Iyer NV, Agani F, Leung SW, Koos RD, Semenza GL (1996). Activation of vascular endothelial growth factor gene transcription by hypoxia-inducible factor 1. *Mol Cell Biol* **16**: 4604–4613.
6. Kelly BD, Hackett SF, Hirota K, Oshima Y, Cai Z, Berg-Dixon S, Rowan A, Yan Z, Campochiaro PA, Semenza GL (2003). Cell type-specific regulation of angiogenic growth factor gene expression and induction of angiogenesis in non-ischemic tissue by a constitutively active form of hypoxia-inducible factor 1. *Circ Res* **93**: 1074–1081.
7. Manalo DJ, Rowan A, Lavoie T, Natarajan L, Kelly BD, Ye SQ, Garcia JG, Semenza GL (2005). Transcriptional regulation of vascular endothelial cell responses to hypoxia by HIF-1. *Blood* **105**: 659–669.

8. Greijer A, van der Groep P, Kemming D, Shvarts A, Semenza G, Meijer G, van de Wiel M, Belien J, van Diest P, van der Wall E (2005). Up-regulation of gene expression by hypoxia is mediated predominantly by hypoxia-inducible factor 1 (HIF-1). *J Pathol* **206**: 291–304.

9. Vengellur A, Woods BG, Ryan HE, Johnson RS, LaPres JJ (2003). Gene expression profiling of the hypoxia signaling pathway in hypoxia-inducible factor 1α null mouse embryonic fibroblasts. *Gene Expr* **11**: 181–197.

10. Hu CJ, Wang LY, Chodosh LA, Keith B, Simon MC (2003). Differential roles of hypoxia-inducible factor 1α (HIF-1α) and HIF-2α in hypoxic gene regulation. *Mol Cell Biol* **23**: 9361–9374.

11. Jiang Y, Zhang W, Kondo K, Klco JM, St Martin TB, Dufault MR, Madden SL, Kaelin WG Jr, Nacht M (2003). Gene expression profiling in a renal cell carcinoma cell line: dissecting VHL and hypoxia-dependent pathways. *Mol Cancer Res* **1**: 453–462.

12. Staller P, Sulitkova J, Lisztwan J, Moch H, Oakeley EJ, Krek W (2003). Chemokine receptor CXCR4 downregulated by von Hippel-Lindau tumour suppressor pVHL. *Nature* **425**: 307–311.

13. Stassar MJ, Devitt G, Brosius M, Rinnab L, Prang J, Schradin T, Simon J, Petersen S, Kopp-Schneider A, Zoller M (2001). Identification of human renal cell carcinoma associated genes by suppression subtractive hybridization. *Br J Cancer* **85**: 1372–1382.

14. Takahashi M, Rhodes DR, Furge KA, Kanayama H, Kagawa S, Haab BB, Teh BT (2001). Gene expression profiling of clear cell renal cell carcinoma: gene identification and prognostic classification. *Proc Natl Acad Sci USA* **98**: 9754–9759.

15. Wykoff CC, Pugh CW, Maxwell PH, Harris AL, Ratcliffe PJ (2000). Identification of novel hypoxia dependent and independent target genes of the von Hippel-Lindau (VHL) tumour suppressor by mRNA differential expression profiling. *Oncogene* **19**: 6297–6305.

16. Wykoff CC, Sotiriou C, Cockman ME, Ratcliffe PJ, Maxwell P, Liu E, Harris AL (2004). Gene array of VHL mutation and hypoxia shows novel hypoxia-induced genes and that cyclin D1 is a VHL target gene. *Br J Cancer* **90**: 1235–1243.

17. Jiang BH, Rue E, Wang GL, Roe R, Semenza GL (1996). Dimerization, DNA binding and transactivation properties of hypoxia-inducible factor 1. *J Biol Chem* **271**: 17771–17778.

18. Jiang BH, Zheng JZ, Leung SW, Roe R, Semenza GL (1997). Transactivation and inhibitory domains of hypoxia-inducible factor 1α. Modulation of transcriptional activity by oxygen tension. *J Biol Chem* **272**: 19253–19260.

19. Pugh CW, O'Rourke JF, Nagao M, Gleadle JM, Ratcliffe PJ (1997). Activation of hypoxia-inducible factor-1; definition of regulatory domains within the α subunit. *J Biol Chem* **272**: 11205–11214.

20. Arany Z, Huang LE, Eckner R, Bhattacharya S, Jiang C, Goldberg MA, Bunn HF, Livingston DM. An essential role for p300/CBP in the cellular response to hypoxia. *Proc Natl Acad Sci USA* **93**: 12969–12973.

21. Dames SA, Martinez-Yamout M, De Guzman RN, Dyson HJ, Wright PE (2002). Structural basis for HIF-1α/CBP recognition in the cellular hypoxic response. *Proc Natl Acad Sci USA* **99**: 5271–5276.

22. Freedman SJ, Sun ZY, Poy F, Kung AL, Livingston DM, Wagner G, Eck MJ (2002). Structural basis for recruitment of CBP/p300 by hypoxia-inducible factor-1α. *Proc Natl Acad Sci USA* **99**: 5367–5372.

23. Kallio PJ, Okamoto K, O'Brien S, Carrero P, Makino Y, Tanaka H, Poellinger L (1998). Signal transduction in hypoxic cells: inducible nuclear translocation and recruitment of the CBP/p300 co-activator by the hypoxia-inducible factor-1α. *EMBO J* **17**: 6573–6586.

24. Ruas JL, Poellinger L, Pereira T (2002). Functional analysis of hypoxia-inducible factor-1α-mediated transactivation. Identification of amino acid residues critical for transcriptional activation and/or interaction with CREB-binding protein. *J Biol Chem* **277**: 38723–38730.

25. Sang N, Fang J, Srinivas V, Leshchinsky I, Caro J (2002). Carboxyl-terminal transactivation activity of hypoxia-inducible factor 1gs governed by a von Hippel-Lindau protein-independent, hydroxylation-regulated association with p300/CBP. *Mol Cell Biol* **22**: 2984–2992.

26. Sang N, Stiehl DP, Bohensky J, Leshchinsky I, Srinivas V, Caro J (2003). MAPK signaling up-regulates the activity of hypoxia-inducible factors by its effects on p300. *J Biol Chem* **278**: 14013–14019.

27. Semenza GL (2000). *Transcription Factors and Human Disease* (Oxford University Press, New York).

28. Jiang BH, Semenza GL, Bauer C, Marti HH (1996). Hypoxia-inducible factor 1 levels vary exponentially over a physiologically relevant range of O_2 tension. *Am J Physiol* **271**: C1172–C1180.

29. Jewell UR, Kvietikova I, Scheid A, Bauer C, Wenger RH, Gassmann M (2001). Induction of HIF-1α in response to hypoxia is instantaneous. *FASEB J* **15**: 1312–1314.

30. Yu AY, Frid MG, Shimoda LA, Wiener CM, Stenmark K, Semenza GL (1998). Temporal, spatial and oxygen-regulated expression of hypoxia-inducible factor-1 in the lung. *Am J Physiol* **275**: L818–L826.

31. Tian H, McKnight SL, Russell DW (1997). Endothelial PAS domain protein 1 (EPAS1), a transcription factor selectively expressed in endothelial cells. *Genes Dev* **11**: 72–82.

32. Sowter HM, Raval RR, Moore JW, Ratcliffe PJ, Harris AL (2003). Predominant role of hypoxia-inducible transcription factor (HIF)-1α versus HIF-2α in regulation of the transcriptional response to hypoxia. *Cancer Res* **63**: 6130–6134.

33. Park SK, Dadak AM, Haase VH, Fontana L, Giaccia AJ, Johnson RS (2003). Hypoxia-induced gene expression occurs solely through the action of hypoxia-inducible factor 1α (HIF-1α): role of cytoplasmic trapping of HIF-2α. *Mol Cell Biol* **23**: 4959–4971.

34. Gu YZ, Moran SM, Hogenesch JB, Wartman L, Bradfield CA (1998). Molecular characterization and chromosomal localization of a third α-class hypoxia inducible factor subunit, HIF-3α. *Gene Expr* **7**: 205–213.

35. Makino Y, Kanopka A, Wilson WJ, Tanaka H, Poellinger L (2002). Inhibitory PAS domain protein (IPAS) is a hypoxia-inducible splicing variant of the hypoxia-inducible factor-3α locus. *J Biol Chem* **277**: 32405–32408.

36. Makino Y, Cao R, Svensson K, Bertilsson G, Asman M, Tanaka H, Cao Y, Berkenstam A (2001). Poellinger L. Inhibitory PAS domain protein is a negative regulator of hypoxia-inducible gene expression. *Nature* **414**: 550–554.

37. Huang LE, Gu J, Schau M, Bunn HF (1998). Regulation of hypoxia-inducible factor 1α is mediated by an O_2-dependent degradation domain via the ubiquitin-proteasome pathway. *Proc Natl Acad Sci USA* **95**: 7987–7992.

38. Kallio PJ, Wilson WJ, O'Brien S, Makino Y, Poellinger L (1999). Regulation of the hypoxia-inducible transcription factor 1α by the ubiquitin-proteasome pathway. *J Biol Chem* **274**: 6519–6525.

39. Salceda S, Caro J (1997). Hypoxia-inducible factor 1α (HIF-1α) protein is rapidly degraded by the ubiquitin-proteasome system under normoxic conditions. Its stabilization by hypoxia depends on redox-induced changes. *J Biol Chem* **272**: 22642–22647.

40. Maxwell PH, Wiesener MS, Chang GW, Clifford SC, Vaux EC, Cockman ME, Wykoff CC, Pugh CW, Maher ER, Ratcliffe PJ (1999). The tumour suppressor protein VHL targets hypoxia-inducible factors for oxygen-dependent proteolysis. *Nature* **399**: 271–275.

41. Kamura T, Sato S, Iwai K, Czyzyk-Krzeska M, Conaway RC, Conaway JW (2000). Activation of HIF-1α ubiquitination by a reconstituted von Hippel-Lindau (VHL) tumor suppressor complex. *Proc Natl Acad Sci USA* **97**: 10430–10435.

42. Ivan M, Kondo K, Yang H, Kim W, Valiando J, Ohh M, Salic A, Asara JM, Lane WS, Kaelin WG Jr (2001). HIF-α targeted for VHL-mediated destruction by proline hydroxylation: implications for O2 sensing. *Science* **292**: 464–468.

43. Jaakkola P, Mole DR, Tian YM, Wilson MI, Gielbert J, Gaskell SJ, Kriegsheim Av, Hebestreit HF, Mukherji M, Schofield CJ, Maxwell PH, Pugh CW, Ratcliffe PJ (2001). Targeting of HIF-α to the von Hippel-Lindau ubiquitylation complex by O2-regulated prolyl hydroxylation. *Science* **292**: 468–472.

44. Yu F, White SB, Zhao Q, Lee FS (2001). HIF-1α binding to VHL is regulated by stimulus-sensitive proline hydroxylation. *Proc Natl Acad Sci USA* **98**: 9630–9635.

45. Masson N, Willam C, Maxwell PH, Pugh CW, Ratcliffe PJ (2001). Independent function of two destruction domains in hypoxia-inducible factor-α chains activated by prolyl hydroxylation. *EMBO J* **20**: 5197–5206.

46. Hirsila M, Koivunen P, Gunzler V, Kivirikko KI, Myllyharju J (2003). Characterization of the human prolyl 4-hydroxylases that modify the hypoxia-inducible factor. *J Biol Chem* **278**: 30772–30780.

47. Maynard MA, Qi H, Chung J, Lee EH, Kondo Y, Hara S, Conaway RC, Conaway JW, Ohh M (2003). Multiple splice variants of the human HIF-3α locus are targets of the von Hippel-Lindau E3 ubiquitin ligase complex. *Biol Chem* **278**: 11032–11040.

48. Bruick RK, McKnight SL (2001). A conserved family of prolyl-4-hydroxylases that modify HIF. *Science* **294**: 1337–1340.

49. Epstein AC, Gleadle JM, McNeill LA, Hewitson KS, O'Rourke J, Mole DR, Mukherji M, Metzen E, Wilson MI, Dhanda A, Tian YM, Masson N, Hamilton DL, Jaakkola P, Barstead R, Hodgkin J, Maxwell PH, Pugh CW, Schofield CJ, Ratcliffe PJ (2001). C. elegans EGL-9 and mammalian homologs define a family of dioxygenases that regulate HIF by prolyl hydroxylation. *Cell* **107**: 43–54.

50. Ivan M, Haberberger T, Gervasi DC, Michelson KS, Gunzler V, Kondo K, Yang H, Sorokina I, Conaway RC, Conaway JW, Kaelin WG Jr (2002). Biochemical purification and pharmacological inhibition of a mammalian prolyl hydroxylase acting on hypoxia-inducible factor. *Proc Natl Acad Sci U S A* **99**: 13459–13464.

51. McNeill LA, Hewitson KS, Gleadle JM, Horsfall LE, Oldham NJ, Maxwell PH, Pugh CW, Ratcliffe PJ, Schofield CJ (2002). The use of dioxygen by HIF prolyl hydroxylase (PHD1). *Bioorg Med Chem Lett* **12**: 1547–1550.

52. Mole DR, Schlemminger I, McNeill LA, Hewitson KS, Pugh CW, Ratcliffe PJ, Schofield CJ (2003). 2-oxoglutarate analogue inhibitors of HIF prolyl hydroxylase. *Bioorg Med Chem Lett* **13**: 2677–2680.

53. Lando D, Peet DJ, Whelan DA, Gorman JJ, Whitelaw ML (2002). Asparagine hydroxylation of the HIF transactivation domain a hypoxic switch. *Science* **295**: 858–861.

54. Mahon PC, Hirota K, Semenza GL (2001). FIH-1: a novel protein that interacts with HIF-1alpha and VHL to mediate repression of HIF-1 transcriptional activity. *Genes Dev* **15**: 2675–2686.

55. Hewitson KS, McNeill LA, Riordan MV, Tian YM, Bullock AN, Welford RW, Elkins JM, Oldham NJ, Bhattacharya S, Gleadle JM, Ratcliffe PJ, Pugh CW, Schofield CJ (2002). Hypoxia-inducible factor (HIF) asparagine hydroxylase is identical to factor inhibiting HIF (FIH) and is related to the cupin structural family. *J Biol Chem* **277**: 26351–26355.

56. Lando D, Peet DJ, Gorman JJ, Whelan DA, Whitelaw ML, Bruick RK (2002). FIH-1 is an asparaginyl hydroxylase enzyme that regulates the transcriptional activity of hypoxia-inducible factor. *Genes Dev* **16**: 1466–1471.

57. Koivunen P, Hirsila M, Gunzler V, Kivirikko KI, Myllyharju J (2004). Catalytic properties of the asparaginyl hydroxylase (FIH) in the oxygen sensing pathway are distinct from those of its prolyl 4-hydroxylases. *J Biol Chem* **279**: 9899–9904.

58. Fukuda R, Hirota K, Fan F, Jung YD, Ellis LM, Semenza GL (2002). Insulin-like growth factor 1 induces hypoxia-inducible factor 1-mediated vascular endothelial growth factor expression, which is dependent on MAP kinase and phosphatidylinositol 3-kinase signaling in colon cancer cells. *J Biol Chem* **277**: 38205–38211.

59. Laughner E, Taghavi P, Chiles K, Mahon PC, Semenza GL (2001). HER2 (neu) signaling increases the rate of hypoxia-inducible factor 1α (HIF-1α) synthesis: novel mechanism for HIF-1-mediated vascular endothelial growth factor expression. *Mol Cell Biol* **21**: 3995–4004.

60. Treins C, Giorgetti-Peraldi S, Murdaca J, Semenza GL, Van Obberghen E (2002). Insulin stimulates hypoxia-inducible factor 1 through a phosphatidylinositol 3-kinase/target of rapamycin-dependent signaling pathway. *J Biol Chem* **277**: 27975–27981.

61. Zhong H, Chiles K, Feldser D, Laughner E, Hanrahan C, Georgescu MM, Simons JW, Semenza GL (2000). Modulation of hypoxia-inducible factor 1α expression by the epidermal growth factor/phosphatidylinositol 3-kinase/PTEN/AKT/FRAP pathway in human prostate cancer cells: implications for tumor angiogenesis and therapeutics. *Cancer Res* **60**: 1541–1545.

62. Richard DE, Berra E, Gothie E, Roux D, Pouyssegur J (1999). p42/p44 mitogen-activated protein kinases phosphorylate hypoxia-inducible factor 1α (HIF-1α) and enhance the transcriptional activity of HIF-1. *J Biol Chem* **274**: 32631–32637.

63. Compernolle V, Brusselmans K; Franco D, Moorman A, Dewerchin M, Collen D, Carmeliet P (2003). Cardia bifida, defective heart development and abnormal neural crest migration in embryos lacking hypoxia-inducible factor-1α. *Cardiovasc Res* **60**: 569–579.

64. Iyer NV, Kotch LE, Agani F, Leung SW, Laughner E, Wenger RH, Gassmann M, Gearhart JD, Lawler AM, Yu AY, Semenza G (1998). Cellular and developmental control of O_2 homeostasis by hypoxia-inducible factor 1α. *Genes Dev* **12**: 149–162.

65. Kotch LE, Iyer NV, Laughner E, Semenza GL (1999). Defective vascularization of HIF-1α -null embryos is not associated with VEGF deficiency but with mesenchymal cell death. *Dev Biol* **209**: 254–267.

66. Ryan HE, Lo J, Johnson RS (1998). HIF-1α is required for solid tumor formation and embryonic vascularization. *EMBO J* **17**: 3005–3015.

67. Carmeliet P, Dor Y, Herbert JM, Fukumura D, Brusselmans K, Dewerchin M, Neeman M, Bono F, Abramovitch R, Maxwell P, Koch CJ, Ratcliffe P, Moons L, Jain RK, Collen D, Keshet E (1998). Role of HIF-1α in hypoxia-mediated apoptosis, cell proliferation and tumour angiogenesis. *Nature* **394**: 485–490.

68. Carmeliet P, Moons L, Luttun A, Vincenti V, Compernolle V, De Mol M, Wu Y, Bono F, Devy L, Beck H, Scholz D, Acker T, DiPalma T, Dewerchin M, Noel A, Stalmans I, Barra A, Blacher S, Vandendriessche T, Ponten A, Eriksson U, Plate KH, Foidart JM, Schaper W, Charnock-Jones DS, Hicklin DJ, Herbert JM, Collen D, Persico MG (2001). Synergism between vascular endothelial growth factor and placental growth factor contributes to angiogenesis and plasma extravasation in pathological conditions. *Nat Med* **7**: 575–583.

69. Ceradini DJ, Kulkarni AR, Callaghan MJ, Tepper OM, Bastidas N, Kleinman ME, Capla JM, Galiano RD, Levine JP, Gurtner GC (2004). Progenitor cell trafficking is regulated by hypoxic gradients through HIF-1 induction of SDF-1. *Nat Med* **10**: 858–864.

70. Camenisch G, Stroka DM, Gassmann M, Wenger RH (2001). Attenuation of HIF-1 DNA-binding activity limits hypoxia-inducible endothelin-1 expression. *Pflugers Arch* **443**: 240–249.

71. Caniggia I, Mostachfi H, Winter J, Gassmann M, Lye SJ, Kuliszewski M, Post M (2000). Hypoxia-inducible factor-1 mediates the biological effects of oxygen on human trophoblast differentiation through TGFbeta (3) *J Clin Invest* **105**: 577–587.

72. Hu J, Discher DJ, Bishopric NH, Webster KA (1998). Hypoxia regulates expression of the endothelin-1 gene through a proximal hypoxia-inducible factor-1 binding site on the antisense strand. *Biochem Biophys Res Commun* **245**: 894–899.

73. Nishi H, Nakada T, Hokamura M, Osakabe Y, Itokazu O, Huang LE, Isaka K (2004). Hypoxia-inducible factor-1 transactivates transforming growth factor β_3 in trophoblast. *Endocrinology* **145**: 4113–4118.

74. Tang N, Wang L, Esko J, Giordano FJ, Huang Y, Gerber HP, Ferrara N, Johnson RS (2004). Loss of HIF-1α in endothelial cells disrupts a hypoxia-driven VEGF autocrine loop necessary for tumorigenesis. *Cancer Cell* **6**: 485–495.

75. Ozaki H, Yu AY, Della N, Ozaki K, Luna JD, Yamada H, Hackett SF, Okamoto N, Zack DJ, Semenza GL, Campochiaro PA (1999). Hypoxia inducible factor-1α is increased in ischemic retina: temporal and spatial correlation with VEGF expression. *Invest Ophthalmol Vis Sci* **40**: 182–189.

76. Miller JW, Adamis AP, Shima DT, D'Amore PA, Moulton RS, O'Reilly MS, Folkman J, Dvorak HF, Brown LF, Berse B, *et al.* (1994). Vascular endothelial growth factor/vascular permeability factor is temporally and spatially correlated with ocular angiogenesis in a primate model. *Am J Pathol* **145**: 574–584.

77. Ohno-Matsui K, Hirose A, Yamamoto S, Saikia J, Okamoto N, Gehlbach P, Duh EJ, Hackett S, Chang M, Bok D, Zack DJ, Campochiaro PA (2002). Inducible expression of vascular endothelial growth factor in adult mice causes severe proliferative retinopathy and retinal detachment. *Am J Pathol* **160**: 711–719.

78. Patel TH, Kimura H, Weiss CR, Semenza GL, Hofmann LV (2005). Constitutively active HIF-1α improves perfusion and arterial remodeling in an endovascular model of limb ischemia. *Cardiovasc Res* **24**; [Epub ahead of print].

79. Ito WD, Arras M, Winkler B, Scholz D, Schaper J, Schaper W (1997). Monocyte chemotactic protein-1 increases collateral and peripheral conductance after femoral artery occlusion. *Circ Res* **80**: 829–837.

80. Pipp F, Heil M, Issbrucker K, Ziegelhoeffer T, Martin S, van den Heuvel J, Weich H, Fernandez B, Golomb G, Carmeliet P, Schaper W, Clauss M (2003). VEGFR-1-selective VEGF homologue PlGF is arteriogenic: evidence for a monocyte-mediated mechanism. *Circ Res* **92**: 378–385.

81. Ravi R, Mookerjee B, Bhujwalla ZM, Sutter CH, Artemov D, Zeng Q, Dillehay LE, Madan A, Semenza GL, Bedi A (2000). Regulation of tumor angiogenesis by p53-induced degradation of hypoxia-inducible factor 1α. *Genes Dev* **14**: 34–44.

82. Stoeltzing O, McCarty MF, Wey JS, Fan F, Liu W, Belcheva A, Bucana CD, Semenza GL, Ellis LM (2004). Role of hypoxia-inducible factor 1α in gastric cancer cell growth, angiogenesis, and vessel maturation. *J Natl Cancer Inst* **96**: 946–956.

83. Semenza GL (2003). Targeting HIF-1 for cancer therapy. *Nat Rev Cancer* **3**: 721–732.

84. McCarty MF, Wey J, Stoeltzing O, Liu W, Fan F, Bucana C, Mansfield PF, Ryan AJ, Ellis LM (2004). ZD6474, a vascular endothelial growth factor receptor tyrosine kinase inhibitor with additional activity against epidermal growth factor receptor tyrosine kinase, inhibits orthotopic growth and angiogenesis of gastric cancer. *Mol Cancer Ther* **3**: 1041–1048.

85. Abdollahi A, Hahnfeldt P, Maercker C, Grone HJ, Debus J, Ansorge W, Folkman J, Hlatky L, Huber PE (2004). Endostatin's antiangiogenic signaling network. *Mol Cell* **13**: 649–663.

86. Krishnamurthy P, Ross DD, Nakanishi T, Bailey-Dell K, Zhou S, Mercer KE, Sarkadi B, Sorrentino BP, Schuetz JD (2004). The stem cell marker Bcrp/ABCG2 enhances hypoxic cell survival through interactions with heme. *J Biol Chem* **279**: 24218–24225.

87. Eckhart AD, Yang N, Xin X, Faber JE (1997). Characterization of the α_{1B}-adrenergic receptor gene promoter region and hypoxia regulatory elements in vascular smooth muscle. *Proc Natl Acad Sci USA* **94**: 9487–9492.

88. Cormier-Regard S, Nguyen SV, Claycomb WC (1998). Adrenomedullin gene expression is developmentally regulated and induced by hypoxia in rat ventricular cardiac myocytes. *J Biol Chem* **273**: 17787–17792.

89. Kong T, Eltzschig HK, Karhausen J, Colgan SP, Shelley CS (2004). Leukocyte adhesion during hypoxia is mediated by HIF-1-dependent induction of β_2 integrin gene expression. *Proc Natl Acad Sci USA* **101**: 10440–10445.

90. Mukhopadhyay CK, Mazumder B, Fox PL (2000). Role of hypoxia-inducible factor-1 in transcriptional activation of ceruloplasmin by iron deficiency. *J Biol Chem* **275**: 21048–21054.

91. Higgins DF, Biju MP, Akai Y, Wutz A, Johnson RS, Haase VH (2004). Hypoxic induction of Ctgf is directly mediated by HIF-1. *Am J Physiol Renal Physiol* **287**: F1223– F1232.

92. Miyazaki K, Kawamoto T, Tanimoto K, Nishiyama M, Honda H, Kato Y (2002). Identification of functional hypoxia response elements in the promoter region of the DEC1 and DEC2 genes. *J Biol Chem* **277**: 47014–47021.

93. LeCouter J, Kowalski J, Foster J, Hass P, Zhang Z, Dillard-Telm L, Frantz G, Rangell L, DeGuzman L, Keller GA, Peale F, Gurney A, Hillan KJ, Ferrara N (2001). Identification of an angiogenic mitogen selective for endocrine gland endothelium. *Nature* **412**: 877–884.

94. Sanchez-Elsner T, Botella LM, Velasco B, Langa C, Bernabeu C (2002). Endoglin expression is regulated by transcriptional cooperation between the hypoxia and transforming growth factor-beta pathways. *J Biol Chem* **277**: 43799–43808.

95. Coulet F, Nadaud S, Agrapart M, Soubrier F (2003). Identification of hypoxia-response element in the human endothelial nitric-oxide synthase gene promoter. *J Biol Chem* **278**: 46230–46240.

96. Oikawa M, Abe M, Kurosawa H, Hida W, Shirato K, Sato Y (2001). Hypoxia induces transcription factor ETS-1 via the activity of hypoxia-inducible factor-1. *Biochem Biophys Res Commun* **289**: 39–43.

97. McMahon S, Grondin F, McDonald PP, Richard DE, Dubois CM (2005). Hypoxia-enhanced expression of the proprotein convertase furin is mediated by hypoxia-inducible factor-1: impact on the bioactivation of proproteins. *J Biol Chem* **280**: 6561–6569.

98. Wood SM, Wiesener MS, Yeates KM, Okada N, Pugh CW, Maxwell PH, Ratcliffe PJ (1998). Selection and analysis of a mutant cell line defective in the hypoxia-inducible factor-1α-subunit (HIF-1α). Characterization of HIF-1α-dependent and -independent hypoxia-inducible gene expression. *J Biol Chem* **273**: 8360–8368.

99. Graven KK, Yu Q, Pan D, Roncarati JS, Farber HW (1999). Identification of an oxygen responsive enhancer element in the glyceraldehyde-3-phosphate dehydrogenase gene. *Biochim Biophys Acta* **1447**: 208–218.

100. Lu S, Gu X, Hoestje S, Epner DE (2002). Identification of an additional hypoxia responsive element in the glyceraldehyde-3-phosphate dehydrogenase gene promoter. *Biochim Biophys Acta* **1574**: 152–156.

101. Lee PJ, Jiang BH, Chin BY, Iyer NV, Alam J, Semenza GL, Choi AM (1997). Hypoxia-inducible factor-1 mediates transcriptional activation of the heme oxygenase-1 gene in response to hypoxia. *J Biol Chem* **272**: 5375–5381.

102. Lee MJ, Kim JY, Suk K, Park JH (2004). Identification of the hypoxia-inducible factor 1α-responsive HGTD-P gene as a mediator in the mitochondrial apoptotic pathway. *Mol Cell Biol* **24**: 3918–3927.

103. Lofstedt T, Jogi A, Sigvardsson M, Gradin K, Poellinger L, Pahlman S, Axelson H (2004). Induction of ID2 expression by hypoxia-inducible factor-1: a role in dedifferentiation of hypoxic neuroblastoma cells. *J Biol Chem* **279**: 39223–39231.

104. Furuta GT, Turner JR, Taylor CT, Hershberg RM, Comerford K, Narravula S, Podolsky DK, Colgan SP (2001). Hypoxia-inducible factor 1-dependent induction of intestinal trefoil factor protects barrier function during hypoxia. *J Exp Med* **193**: 1027–1034.

105. Lee SY, Madan A, Furuta GT, Colgan SP, Sibley E (2002). Lactase gene transcription is activated in response to hypoxia in intestinal epithelial cells. *Mol Genet Metab* **75**: 65–69.

106. Grosfeld A, Hauguel-De Mouzon S, Berra E, Pouyssegur J, Guerre-Millo M (2002). Hypoxia-inducible factor 1 transactivates the human leptin gene promoter. *J Biol Chem* **277**: 42953–42957.

107. Comerford KM, Wallace TJ, Karhausen J, Louis NA, Montalto MC, Colgan SP (2002). Hypoxia-inducible factor-1-dependent regulation of the multidrug resistance (MDR1) gene. *Cancer Res* **62**: 3387–3394.

108. Petrella BL, Lohi J, Brinckerhoff CE (2005). Identification of membrane type-1 matrix metalloproteinase as a target of hypoxia-inducible factor-2α in von Hippel-Lindau renal cell carcinoma. *Oncogene* **24**: 1043–1052.

109. Melillo G, Musso T, Sica A, Taylor LS, Cox GW, Varesio L (1995). A hypoxia-responsive element mediates a novel pathway of activation of the inducible nitric oxide synthase promoter. *J Exp Med* **182**: 1683–1693.

110. Bruick RK (2000). Expression of the gene encoding the proapoptotic Nip3 protein is induced by hypoxia. *Proc Natl Acad Sci USA* **97**: 9082–9087.

111. Choi JW, Park SC, Kang GH, Liu JO, Youn HD (2004). Nur77 activated by hypoxia-inducible factor-1α overproduces proopiomelanocortin in von Hippel-Lindau-mutated renal cell carcinoma. *Cancer Res* **64**: 35–39.

112. Bhattacharya S, Michels CL, Leung MK, Arany ZP, Kung AL, Livingston DM (1999). Functional role of p35srj, a novel p300/CBP binding protein, during transactivation by HIF-1. *Genes Dev* **13**: 64–75.

113. Minchenko A, Leshchinsky I, Opentanova I, Sang N, Srinivas V, Armstead V, Caro J (2002). Hypoxia-inducible factor-1-mediated expression of the 6-phosphofructo-2-kinase/fructose-2,6-bisphosphatase-3 (PFKFB3) gene. Its possible role in the Warburg effect. *J Biol Chem* **277**: 6183–6187.

114. Kietzmann T, Roth U, Jungermann K (1999). Induction of the plasminogen activator inhibitor-1 gene expression by mild hypoxia via a hypoxia response element binding the hypoxia-inducible factor-1 in rat hepatocytes. *Blood* **94**: 4177–4185.

115. Takahashi Y, Takahashi S, Shiga Y, Yoshimi T, Miura T (2000). Hypoxic induction of prolyl 4-hydroxylase α(I) in cultured cells. *J Biol Chem* **275**: 14139–14146.
116. Chauvet C, Bois-Joyeux B, Berra E, Pouyssegur J, Danan JL (2004). The gene encoding human retinoic acid-receptor-related orphan receptor α is a target for hypoxia-inducible factor 1. *Biochem J* **384**: 79–85.
117. Miki N, Ikuta M, Matsui T (2004). Hypoxia-induced activation of the retinoic acid receptor-related orphan receptor α4 gene by an interaction between hypoxia-inducible factor-1 and Sp1. *J Biol Chem* **279**: 15025–15031.
118. Shoshani T, Faerman A, Mett I, Zelin E, Tenne T, Gorodin S, Moshel Y, Elbaz S, Budanov A, Chajut A, Kalinski H, Kamer I, Rozen A, Mor O, Keshet E, Leshkowitz D, Einat P, Skaliter R, Feinstein E (2002). Identification of a novel hypoxia-inducible factor 1-responsive gene, RTP801, involved in apoptosis. *Mol Cell Biol* **22**: 2283–2293.
119. Nishi H, Nakada T, Kyo S, Inoue M, Shay JW, Isaka K (2004). Hypoxia-inducible factor 1 mediates upregulation of telomerase (hTERT). *Mol Cell Biol* **24**: 6076–6083.
120. Rolfs A, Kvietikova I, Gassmann M, Wenger RH (1997). Oxygen-regulated transferrin expression is mediated by hypoxia-inducible factor-1. *J Biol Chem* **272**: 20055–20062.
121. Lok CN, Ponka P (1999). Identification of a hypoxia response element in the transferrin receptor gene. *J Biol Chem* **274**: 24147–24152.
122. Tacchini L, Bianchi L, Bernelli-Zazzera A, Cairo G (1999). Transferrin receptor induction by hypoxia. HIF-1-mediated transcriptional activation and cell-specific post-transcriptional regulation. *J Biol Chem* **274**: 24142–24146.
123. Gerber HP, Condorelli F, Park J, Ferrara N (1997). Differential transcriptional regulation of the two vascular endothelial growth factor receptor genes. Flt-1, but not Flk-1/KDR, is up-regulated by hypoxia. *J Biol Chem* **272**: 23659–23667.
124. Korkolopoulou P, Patsouris E, Konstantinidou AE, Pavlopoulos PM, Kavantzas N, Boviatsis E, Thymara I, Perdiki M, Thomas-Tsagli E, Angelidakis D, Rologis D, Sakkas D (2004). Hypoxia-inducible factor 1α/vascular endothelial growth factor axis in astrocytomas. Associations with microvessel morphometry, proliferation and prognosis. *Neuropathol Appl Neurobiol* **30**: 267–278.
125. Theodoropoulos VE, Lazaris AC, Kastriotis I, Spiliadi C, Theodoropoulos GE, Tsoukala V, Patsouris E, Sofras F (2005). Evaluation of hypoxia-inducible factor 1α overexpression as a predictor of tumour recurrence and progression in superficial urothelial bladder carcinoma. *BJU Int* **95**: 425–431.
126. Theodoropoulos VE, Lazaris ACh, Sofras F, Gerzelis I, Tsoukala V, Ghikonti I, Manikas K, Kastriotis I (2005). Hypoxia-inducible factor 1α expression correlates with angiogenesis and unfavorable prognosis in bladder cancer. *BJU Int* **95**: 425–431.
127. Zagzag D, Zhong H, Scalzitti JM, Laughner E, Simons JW, Semenza GL (2000). Expression of hypoxia-inducible factor 1α in brain tumors: association with angiogenesis, invasion, progression. *Cancer* **88**: 2606–2618.
128. Bos R, Zhong H, Hanrahan CF, Mommers EC, Semenza GL, Pinedo HM, Abeloff MD, Simons JW, van Diest PJ, van der Wall E (2001). Levels of hypoxia-inducible factor-1α during breast carcinogenesis. *J Natl Cancer Inst* **93**: 309–314.

129. Giatromanolaki A, Koukourakis MI, Simopoulos C, Polychronidis A, Gatter KC, Harris AL, (2004) Sivridis E. c-erbB-2 related aggressiveness in breast cancer is hypoxia inducible factor-1α dependent. *Clin Cancer Res* **10**: 7972–7977.

130. Schindl M, Schoppmann SF, Samonigg H, Hausmaninger H, Kwasny W, Gnant M, Jakesz R, Kubista E, Birner P, Oberhuber G (2002). Overexpression of hypoxia-inducible factor 1α is associated with an unfavorable prognosis in lymph node-positive breast cancer. *Clin Cancer Res* **8**: 1831–1837.

131. Bos R, van der Groep P, Greijer AE, Shvarts A, Meijer S, Pinedo HM, Semenza GL, van Diest PJ, van der Wall E (2003). Levels of hypoxia-inducible factor-1α independently predict prognosis in patients with lymph node negative breast carcinoma. *Cancer* **97**: 1573–1581.

132. Birner P, Schindl M, Obermair A, Plank C, Breitenecker G, Oberhuber G (2000). Overexpression of hypoxia-inducible factor 1α is a marker for an unfavorable prognosis in early-stage invasive cervical cancer. *Cancer Res* **60**: 4693–4696.

133. Burri P, Djonov V, Aebersold DM, Lindel K, Studer U, Altermatt HJ, Mazzucchelli L, Greiner RH, Gruber GG (2003). Significant correlation of hypoxia-inducible factor-1α with treatment outcome in cervical cancer treated with radical radiotherapy. *Int J Radiat Oncol Biol Phys* **56**: 494–501.

134. Bachtiary B, Schindl M, Potter R, Dreier B, Knocke TH, Hainfellner JA, Horvat R, Birner P (2003). Overexpression of hypoxia-inducible factor 1α indicates diminished response to radiotherapy and unfavorable prognosis in patients receiving radical radiotherapy for cervical cancer. *Clin Cancer Res* **9**: 2234–2240.

135. Kuwai T, Kitadai Y, Tanaka S, Onogawa S, Matsutani N, Kaio E, Ito M, Chayama K (2003). Expression of hypoxia-inducible factor-1α is associated with tumor vascularization in human colorectal carcinoma. *Int J Cancer* **105**: 176–181.

136. Kimura S, Kitadai Y, Tanaka S, Kuwai T, Hihara J, Yoshida K, Toge T, Chayama K (2004). Expression of hypoxia-inducible factor (HIF)-1α is associated with vascular endothelial growth factor expression and tumour angiogenesis in human esophageal squamous cell carcinoma. *Eur J Cancer* **40**: 1904–1912.

137. Koukourakis MI, Giatromanolaki A, Skarlatos J, Corti L, Blandamura S, Piazza M, Gatter KC, Harris AL (2001). Hypoxia inducible factor (HIF-1α and HIF-2α) expression in early esophageal cancer and response to photodynamic therapy and radiotherapy. *Cancer Res* **61**: 1830–1832.

138. Matsuyama T, Nakanishi K, Hayashi T, Yoshizumi Y, Aiko S, Sugiura Y, Tanimoto T, Uenoyama M, Ozeki Y, Maehara T (2005). Expression of hypoxia-inducible factor-1α in esophageal squamous cell carcinoma. *Cancer Sci* **96**: 176–182.

139. Sivridis E, Giatromanolaki A, Gatter KC, Harris AL, Koukourakis MI (2002). Tumor and Angiogenesis Research Group. Association of hypoxia-inducible factors 1α and 2α with activated angiogenic pathways and prognosis in patients with endometrial carcinoma. *Cancer* **95**: 1055–1063.

140. Takahashi R, Tanaka S, Hiyama T, Ito M, Kitadai Y, Sumii M, Haruma K, Chayama K (2003). Hypoxia-inducible factor-1α expression and angiogenesis in gastrointestinal stromal tumor of the stomach. *Oncol Rep* **10**: 797–802.

141. Koukourakis MI, Giatromanolaki A, Sivridis E, Simopoulos C, Turley H, Talks K, Gatter KC, Harris AL (2002). Hypoxia-inducible factor (HIF1A and HIF2A),

angiogenesis, chemoradiotherapy outcome of squamous cell head-and-neck cancer. *Int J Radiat Oncol Biol Phys* **53**: 1192–1202.

142. Giatromanolaki A, Koukourakis MI, Sivridis E, Turley H, Talks K, Pezzella F, Gatter KC, Harris AL (2001). Relation of hypoxia inducible factor 1α, 2α in operable non-small cell lung cancer to angiogenic/molecular profile of tumours and survival. *Br J Cancer* **85**: 881–890.

143. Giatromanolaki A, Sivridis E, Kouskoukis C, Gatter KC, Harris AL, Koukourakis MI (2003). Hypoxia-inducible factors 1α and 2α are related to vascular endothelial growth factor expression and a poorer prognosis in nodular malignant melanomas of the skin. *Melanoma Res* **13**: 493–501.

144. Birner P, Gatterbauer B, Oberhuber G, Schindl M, Rossler K, Prodinger A, Budka H, Hainfellner JA (2001). Expression of hypoxia-inducible factor-1α in oligodendrogliomas: its impact on prognosis and on neoangiogenesis. *Cancer* **92**: 165–171.

145. Aebersold DM, Burri P, Beer KT, Laissue J, Djonov V, Greiner RH, Semenza GL (2001). Expression of hypoxia-inducible factor-1α: a novel predictive and prognostic parameter in the radiotherapy of oropharyngeal cancer. *Cancer Res* **61**: 2911–2916.

146. Birner P, Schindl M, Obermair A, Breitenecker G, Oberhuber G (2001). Expression of hypoxia-inducible factor 1α in epithelial ovarian tumors: its impact on prognosis and on response to chemotherapy. *Clin Cancer Res* **7**: 1661–1668.

147. Kitada T, Seki S, Sakaguchi H, Sawada T, Hirakawa K, Wakasa K (2003). Clinicopathological significance of hypoxia-inducible factor-1α expression in human pancreatic carcinoma. *Histopathology* **43**: 550–555.

148. Shibaji T, Nagao M, Ikeda N, Kanehiro H, Hisanaga M, Ko S, Fukumoto A, Nakajima Y (2003). Prognostic significance of HIF-1α overexpression in human pancreatic cancer. *Anticancer Res* **23**: 4721–4727.

149. Karth J, Ferrer FA, Perlman E, Hanrahan C, Simons JW, Gearhart JP, Rodriguez R (2000). Coexpression of hypoxia-inducible factor 1α and vascular endothelial growth factor in Wilms' tumor. *J Pediatr Surg* **35**: 1749–1753.

8

Redox State and Regulation of Angiogenic Responses

by Masuko Ushio-Fukai
and R. Wayne Alexander

1. Introduction

Angiogenesis, the formation of new blood vessels from the pre-existing vasculature, is involved in physiological processes including embryonic development and wound repair as well as in pathological conditions such as ischemic heart and limb disease, cancer, diabetic retinopathy, or chronic inflammation including atherosclerosis.[1] This tightly regulated process occurs through degradation of extracellular matrix, migration and proliferation, and tube formation of endothelial cells (ECs). Vascular endothelial growth factor (VEGF) is a potent growth factor and stimulates proliferation, migration and tube formation of ECs and angiogenesis *in vivo*.[2]

Reactive oxygen species (ROS) have been described as bacterial killing in host defenses.[3] Accumulating evidence suggests that non-phagocytic cells including cardiovascular cells can produce ROS such as superoxide ($O_2^{\bullet-}$) and hydrogen peroxide (H_2O_2) which play a role as signaling molecules in physiological and pathophysiological responses.[4] High concentrations of ROS cause cell death and apoptosis[5–8] and oxidative stress is associated with the pathogenesis of various cardiovascular diseases including hypertension, heart failure and cardiac hypertrophy, atherosclerosis, and diabetes[9] Low (micromolar or submicromolar) levels of ROS, which can be produced during tissue ischemia/hypoxia or ischemic preconditioning,[10–12] stimulate EC proliferation and migration[13,14] as well as angiogenesis *in vivo*.[10–12,14–18] The precise underlying mechanisms, however, are not completely understood.

Although multiple ROS-generating systems have been described, NAD(P)H oxidase (NOX) enzymes are one of the major source of ROS in endothelial and other vascular cells.[9] Vascular cells may also be exposed to exogenous ROS produced extrinsically by phagocytes or circulating enzymes such as xanthine oxidase.[19] Endogenous ROS derived from NAD(P)H oxidase serve as signaling molecules to activate multiple intracellular signaling pathways leading to cell growth, migration, and modification of the extracellular matrix, which are fundamental responses contributing to angiogenesis. Of importance are angiogenic growth factors such as VEGF and angiopoietin-1, which stimulate, through their receptor tyrosine kinases (RTK), NAD(P)H oxidase-derived ROS production, which are involved in EC migration and proliferation.[20,21] We and others showed that NAD(P)H oxidase plays an essential role *in vivo* in VEGF-mediated postnatal neovascularization,[20] hindlimb ischemia,[12] as well as ischemic retinopathy.[18] Moreover, the peptide hormone angiotensin II (Ang II), a major stimulus for vascular NAD(P)H oxidase,[22–25] plays an important role in angiogenesis. In this chapter we shall summarize recent progress that has been made in the rapidly emerging area of redox signaling and the role of ROS in angiogenesis.

2. Reactive Oxygen Species (ROS) in the Vasculature

Macrophages and monocytes have been assumed to be the source of most of the ROS in the vessel wall. However, virtually all vascular cells such as endothelial cells, vascular smooth muscle cells (VSMCs) and adventitial fibroblasts produce ROS, in varying amounts and in response to diverse stimuli.[9] Cells also use different enzymes to produce and scavenge ROS in different circumstances. Ultimately, it is the balance of pro-oxidant and antioxidant enzyme activity that dictates both intracellular and extracellular ROS levels. ROS can then act in an autocrine or paracrine fashion to modulate cellular function. In the vasculature, multiple enzymatic systems produce ROS, including the NAD(P)H oxidases, xanthine oxidase, myeloperoxidase, uncoupled endothelial nitric oxide synthase (eNOS), the cytochrome p450s, cyclooxygenases, and mitochondrial electron transport chain.[9]

Cellular redox state is an important determinant regulating cell growth, survival, apoptosis and gene expression. One of the most important product of ROS in the vasculature is superoxide ($O_2^{\bullet-}$), which is formed by the univalent reduction of oxygen.[4] Although $O_2^{\bullet-}$ can itself exert effects on vascular function, at high concentrations its reaction with nitric oxide ($NO^{\bullet}$) to generate peroxynitrite ($ONOO^-$), a highly reactive ROS, can be particularly deleterious. Superoxide is rapidly dismutated by superoxide dismutase (SOD) which consists of three isoforms, including extracellular SOD (ecSOD), cytosolic copper/zinc SOD (Cu/ZnSOD), and the mitochondrial-restricted manganese SOD (MnSOD), thereby resulting in producing the more stable and diffusible ROS, hydrogen peroxide (H_2O_2). Of importance, H_2O_2 functions as a signaling molecule to regulate various biological responses in the vasculature.[26] H_2O_2 is then converted enzymatically into H_2O by catalase and glutathione peroxidase (GPx), or undergoes Fenton reaction to form hydroxyl radical ($^{\bullet}OH$) in the presence of heavy metals, or metabolized by myeloperoxidase (MPO) to form hypochlorous acid (HOCl) (Fig. 1).

3. ROS and Angiogenesis

Accumulating evidence reveals that ROS are emerging as important regulators of angiogenesis. Exogenous ROS stimulates induction of VEGF by VSMCs,[27] skeletal myotubes,[28] fibroblasts,[29] keratinocytes,[29,30] retinal pigment epithelial cells,[31] human U937 macrophage, rat peritoneal macrophages and RAW264.7 cell lines[32] as well as ECs.[33] High concentrations of ROS inhibit while low concentrations of ROS stimulate EC proliferation.[13,14,34] In ECs, H_2O_2 also induces angiogenic-related responses including cell migration,[34] cytoskeletal reorganization[35] and

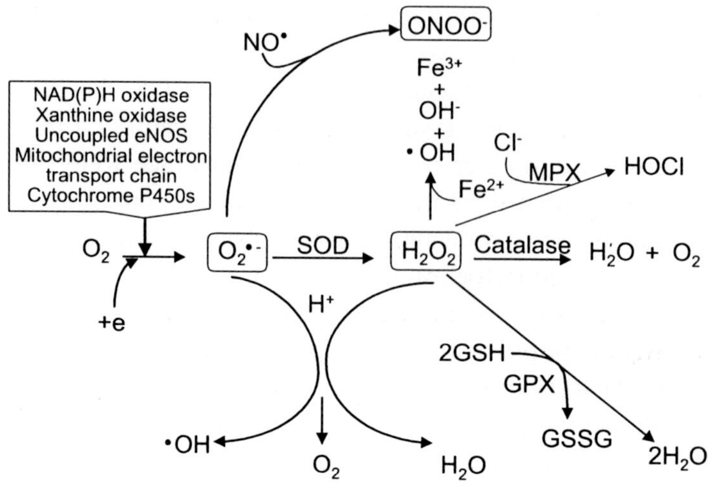

Fig. 1. Schema of superoxide generation. See text for details.

tubular morphogenesis.[36] H_2O_2 is also involved in the T-lymphocyte activation and induction of angiogenesis during tumor growth.[37]

Hypoxia/reoxygenation, which produces ROS, elicits capillary tube formation in human microvascular ECs grown on a 3D reconstituted extracellular matrix (Matrigel).[38] Scratch wounding of confluent monolayers of ECs stimulates H_2O_2 accumulation in actively migrating cells at the wound edge area, which is required for EC migration towards the site of injury.[39,40] Adhesion of activated polymorphonuclear leukocytes (PMNs) to ECs promotes angiogenesis through an increase in H_2O_2.[41] Advanced glycation end-products (AGEs) stimulate VEGF expression in RAW264.7 macrophages through an increase in ROS, thereby contributing to the development of angiopathy in diabetes mellitus.[42] Leptin, a circulating hormone secreted mainly from adipose tissue, functions as an angiogenic factor[43] and increases VEGF mRNA expression and EC proliferation through an increase in ROS.[44]

Antioxidants, green tea catechins and vitamin E, inhibit angiogenesis-related responses of human microvascular ECs via suppression of IL-8 production.[45] Recently, Oak *et al.*[46] reported that natural polyphenols, which have antioxidant properties, inhibit key angiogenic processes such as proliferation and migration of ECs and VSMCs as well the expression of two major pro-angiogenic factors, VEGF and matrix metalloproteinase (MMP)-2.[46] Pigment epithelium-derived factor (PEDF), a potent natural inhibitor of angiogenesis [47] with antioxidant properties, blocks angiogenic effects of leptin through inhibiting ROS production in ECs [44] as well as the H_2O_2-induced increase in VEGF mRNA in retinal pericytes.[48] Thus, PEDF might be a novel therapeutic antioxidant factor for treatment of angiogenesis-dependent pathophysiologies such as diabetic retinopathy and cancer.

In vivo, there is strong correlation between ROS production with neovascularization and VEGF expression in the eyes of diabetics [49–51] and in balloon injured arteries.[27] ROS are increased during the reperfusion of the ischemic retina *in vivo*, which contributes to VEGF mRNA expression through increasing its mRNA stability.[31] Short exposure to hypoxia/reoxygenation in hearts produces ROS, which contributes to myocardial angiogenesis.[10,11] Of note, arteriogenesis is an important aspect of the development of collateral circulation. Gu *et al.* reported that brief coronary artery occulusion and reperfusion of dogs cause ROS production which contribute to coronary collateral development.[52] Moreover, it has been proposed that ROS play an important role in wound healing and repair processes *in vivo*.[53,54]

The antioxidant, pyrrolidine dithiocarbamate[55] and the major green tea extract, epigallocatechin-3-gallate (EGCG) which has antioxidant properties,[56] inhibit retinal neovascularization in the mouse. Similarly, EGCG prevents the growth of new blood vessels in a murine Matrigel model.[57] The natural compounds in red wine and grapes block angiogenesis in the chick embryo chorioallantoic membrane (CAM) or cornea models of mice.[46,58] The SOD, its membrane permeable mimetic tempol, catalase, and the NAD(P)H oxidase inhibitors, 4-(2-aminoetyyl)-benzenesulfonyl fluoride (AEBSF) and apocynin, but not the xanthine/xanthine oxidase inhibitor allopurinol, decrease angiogenesis and the expression and activity of iNOS in the CAM model.[59] The thiol antioxidant, N-acetylcysteine (NAC) attenuates EC invasion and angiogenesis in a tumor model *in vivo*.[60] Moreover, overexpression of PEDF inhibits angiogenesis and melanoma growth *in vivo*.[61] Wheeler *et al.*[62] demonstrated that overexpression of ecSOD using adenovirus inhibits tumor vascularization and growth of B16 melanomas in mice. By contrast, FGF-induced angiogenesis as well as C6 glioma tumor development are enhanced in Cu/ZnSOD transgenic mice,[63] which may be due to the increase in intracellular H_2O_2 levels through enhanced Cu/ZnSOD expression. Consistent with this, Grzenkowicz-Wydra *et al.*[64] have shown that gene transfer of Cu/ZnSOD in NIH3T3 fibroblasts enhances VEGF synthesis by activation of hypoxia-inducible factor response element (HRE) as well as SP1 recognition site of VEGF promoter through an increase in H_2O_2.

4. NAD(P)H Oxidase: A Major Source of ROS in the Vasculature

As noted, ROS are generated in mammalian cells from a number of sources including the mitochondrial electron transport system, xanthine oxidase, the cytochrome p450, the NAD(P)H oxidase and nitric oxide synthase.[9] The NAD(P)H oxidase (Nox) family of enzymes have now been accepted as one of the major sources for ROS in the vasculature.[9,20] Vascular NAD(P)H oxidase is a multi-subunit enzyme complex that differs structurally and biochemically from the phagocytic NAD(P)H oxidase. The phagocytic oxidase releases large amounts of $O_2^{\bullet-}$ in bursts, whereas the vascular NAD(P)H oxidase(s) continuously produce low levels of $O_2^{\bullet-}$ in unstimulated cells, and which can be stimulated acutely by various agonists and growth factors.[9] ECs express NAD(P)H oxidase subunits that are identical to those found in phagocytes, including the membrane bound gp91phox (also known as Nox2) and p22phox, the cytosolic components p40phox, p47phox and p67phox, and Rac1.[65,66] Upon stimulation, cytosolic components translocate to the membrane to form a multimeric protein complex, leading to production of ROS.[65] Recently, novel gp91phox (Nox2) homologues, termed Nox1, Nox3, Nox4, Nox5.[67] have been identified in non-phagocytic cells including vascular cells,[68,69] suggesting the presence of multiple NAD(P)H oxidase forms in these cells.

Abid *et al.* demonstrated that ROS derived from NAD(P)H oxidase are required for EC proliferation and migration.[70] Studies using knockout mice, inhibitory peptides or antisense oligonucleotide have established that Nox2 is a critical component of ROS-generating NAD(P)H

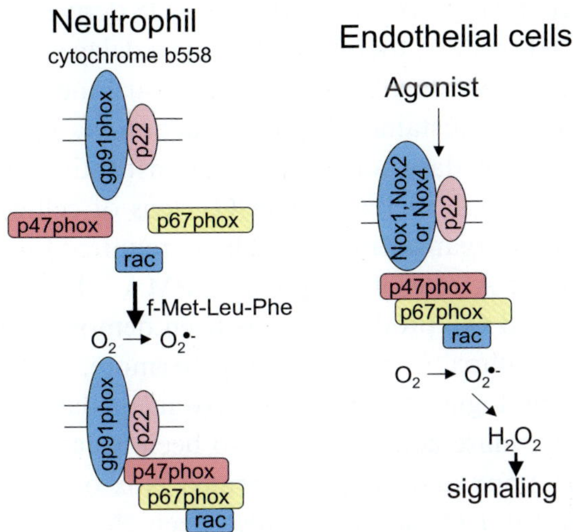

Fig. 2. Reactive oxygen species generation in neutrophils and endothelial cells. See text for details.

oxidase in ECs [20,71–74] (Fig. 2). VEGF, angiopoietin-1, and atrial natriuretic peptide (ANP) stimulate ROS production in ECs through a Rac1- or Nox2-dependent mechanisms.[20,21,74–76] In human umbilical vein endothelial cells (HUVECs), both Nox2 expression and O_2^- formation are increased by oxidized LDL[77] and endothelin-1,[78] but are decreased by 17β-estradiol.[79] Nox1 is upregulated by oscillatory shear stress, mediating ROS-dependent leukocyte adhesion to endothelium.[80] Furthermore, Nox1 regulates apoptosis and stimulates branching morphogenesis in sinusoidal ECs.[81] In addition to Nox2 and Nox1,[80] Nox4 are abundantly expressed in ECs and seem to be important for basal O_2^- production[82] (Fig. 2). Most recently, Yamagishi *et al.*[83] demonstrated that Ang II stimulation of HUVECs increases Nox2, Nox4 and p22phox mRNAs as well as NAD(P)H oxidase activity and that Ang II-induced ROS generation is inhibited by antisense DNAs targeted to each of the NAD(P)H oxidase components. Similar to Nox2, Nox1 and Nox4 form complexes with p22phox,[84] and p22phox is required for Nox1-dependent O_2^- formation.[85] A recent study showed that p22phox expression correlates well with expression of Nox4 in human arteries and that of Nox2 in veins.[86] Djordgevic *et al.*[87] reported that expression of p22phox is regulated by ROS derived from p22phox-based NAD(P)H oxidase, thereby increasing a delayed ROS generation in ECs, which represents a positive feedforward mechanism whereby thrombin stimulates sustained ROS production via upregulation of a critical NAD(P)H oxidase component. A constitutively active form of Rac1 induces, through increase of H_2O_2, loss of cell-cell adhesion[88] and cytoskeletal reorganization[89] which are required for the migratory responses of ECs. A role of p47phox in PMA-, TNFα- and oscillatory shear-induced O_2^- production has been demonstrated using ECs isolated from p47phox$^{-/-}$ mice.[90,91] Furthermore, evolutionary more distinct Nox homologues, termed DUOX (<u>du</u>al <u>ox</u>idase) 1 and DUOX2, which have peroxidase activity have also been isolated.[92] In addition, isoforms of p47phox and p67phox termed Nox organizer 1 (Noxo1) and Nox activator 1 (Noxa1) have also been characterized and have been shown to regulate Nox1 activity.[93–96] Of importance, Ago *et al.*[97] have recently shown that cerebral artery ECs express Nox1, Nox2 and cytosolic components p67phox and, to a lesser extent, p47phox, Noxo1, and Noxa1. These suggest the presence of multiple NAD(P)H oxidase forms in vascular cells.[3,98,99]

It is possible that the function of each NAD(P)H oxidase component is dependent on its distinctive subcellular localization, and is subject to specific regulations by selective agonists. In unstimulated ECs, NAD(P)H oxidase components exist as pre-assembled complexes in a predominantly perinuclear location associated with the intracellular cytoskeleton.[72] In VSMCs, Nox1 localizes to caveolae while Nox4 is found in focal adhesions.[100] In ECs; however, Nox 4 localizes at endoplasmic reticulum[101] while Nox2 is found at perinuclear cytoskeletal structure.[72] Gu *et al.*[102] reported that p47phox plays a role in TNFα-induced c-terminal Jun kinase activation and that p47phox localizes to the cytoskeletal elements in HUVEC cell line ECV304. After agonist stimulation including VEGF, p47phox translocates to the membrane ruffles through association with WAVE1 in ECs, thereby activating NAD(P)H oxidase.[101,103] Qian *et al.*[104] showed that arsenic-induced NAD(P)H oxidase activation and EC migration are dependent on the actin cytoskeleton. Using a monolayer scratch assay with confluent ECs, we have demonstrated that ROS production is increased at the margin of the scratch area and Nox2 translocates to the leading edge, where it co-localizes and associates with both actin and IQGAP1, an actin- and Rac1-binding scaffold protein, in migrating ECs.[105] Thus, endothelial NAD(P)H oxidases seem to be associated with actin cytoskeleton, thereby regulating EC spreading, motility and cell-cell adhesions, which may contribute to angiogenesis.

5. Role of NAD(P)H Oxidase in Angiogenesis

Although other sources of ROS such as cytochrome p450[106] and mitochondria[107] are involved, it has become evident that NAD(P)H oxidase plays an important role in angiogenesis. VEGF stimulation of ECs activates Rac1-dependent NAD(P)H oxidase to produce ROS[20,76,88,108] and Nox2-derived ROS are involved in VEGF-stimulated angiogenic-related responses such as EC migration and proliferation[20,40,109] (Fig. 3). Furthermore, ethanol stimulation induces H_2O_2 production

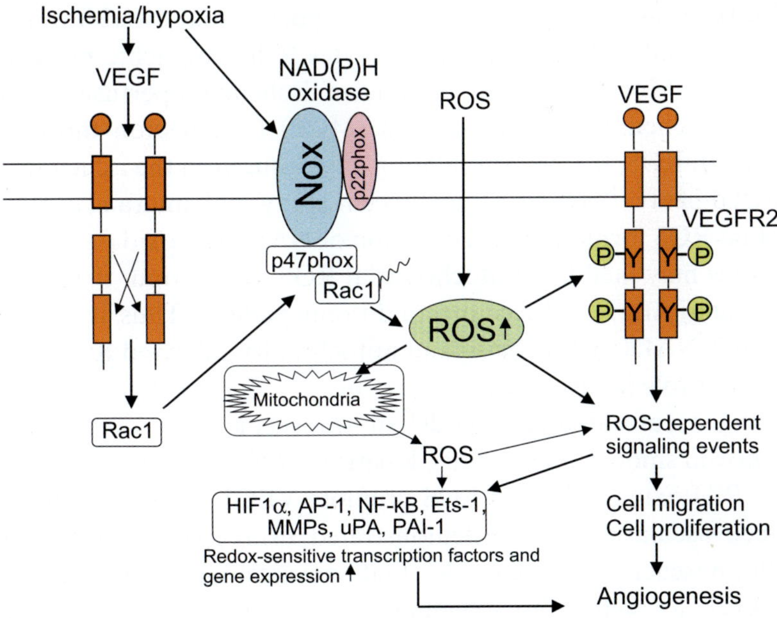

Fig. 3. Role of NADPH in reactive oxygen species generation. See text for details.

through activation of Cdc42, which promotes actin cytoskeletal reorganization, cell motility and tube formation in ECs, an *in vitro* model of angiogenesis.[110] Of importance, Cdc42 is involved in arsenic-induced NAD(P)H oxidase activation and cell migration through regulating actin reorganization in ECs.[104]

These *in vitro* data strongly suggest that NAD(P)H oxidase may play an important role in postnatal angiogenesis *in vivo*. Expression of VEGF and Nox2 and production of ROS are increased during angiogenesis in ischemic retinopathy, and inhibition of NAD(P)H oxidase blocks VEGF overexpression and neovascularization.[18] Similarily, VEGF and Nox2 expression as well as ROS production are increased in ischemic hindlimbs, and post-ischemic neovascularization is impaired in Nox2 knockout mice.[12] Nox4 is upregulated and prominently expressed in newly formed capillaries in brain ischemia-induced angiogenesis in mice.[111]

In cardiomyocytes NAD(P)H oxidase is an important source of ROS.[112–115] Interestingly, the expression of Nox2 and p22phox is increased in parallel with the elevation of lipid peroxidation in myocardial infarct sites.[116] Krijnen *et al.*[117] reported that Nox2 expression is upregulated in human cardiomyocytes following acute myocardial infarction. Moreover, short periods of ischemia/reperfusion induce monocyte chemoattractant protein (MCP)-1 expression through an increase in ROS, thereby stimulating angiogenesis in the ischemic non-infarcted heart.[118] Of note, most recent study by Kimura *et al.*[119] suggest that ROS formation via activation of NAD(P)H oxidase in cardiac myocytes may facilitate mitochondrial ROS production, thereby contributing to Ang II-induced preconditioning effects. Thus, it is possible that both NAD(P)H oxidase- and mitochondrial-derived ROS play an important role in myocardial angiogenesis.

Using a carotid injury model of transgenic mice overexpressing p22phox in smooth muscle cells, Khatri *et al.*[120] reported that vascular NAD(P)H oxidase-derived ROS promote VEGF expression and intimal neovascularization. We demonstrated, using a sponge implant and hindlimb ischemic mouse model, that neovessel formation in response to VEGF as well as to ischemia is significantly inhibited both in wild-type mice treated with antioxidants and in Nox2$^{-/-}$ mice.[12,20]

Moreover, inhibition of NAD(P)H oxidase activity with apocynin and gp91ds-tat blocks ischemia-induced increase in ROS production, VEGF expression and retinal neovascularization in a mouse model of ischemic retinopathy.[18] These results suggest that ROS derived from Nox2-containing NAD(P)H oxidase are important in postnatal angiogenesis *in vivo*. Consistent with our findings, HMG CoA reductase inhibitors, statins, which reduce vascular NAD(P)H oxidase activity through inhibiting Rac1[115,121−125] have been shown to inhibit angiogenesis dose-dependently *in vivo*.[126]

The renin-angiotensin system has been implicated in angiogenesis. Ang II is a potent stimulator for NAD(P)H oxidase in various cardiovascular cells including VSMCs,[22−25,127] ECs,[128−132] adventitia,[133] cardiac myocytes[113,134−136] and isolated hearts subjected to ischemia/reperfusion.[137] *In vitro* AT$_1$ receptor stimulation induces migration of VSMC and monocytes, and promotes EC proliferation.[138,139] Ang II potentiates the VEGF-induced tube formation of bovine retinal ECs.[140] In primary culture of myofibroblasts isolated from adult rat infarcted heart, Ang II stimulation increases expression of VEGF and the VEGF receptor, which may contribute to angiogenesis at this site.[141] *In vivo*, Ang II has been shown to be involved in ischemia- and VEGF-induced angiogenesis[142−145] through upregulation of VEGF or VEGF receptor.[140,146−148] Chymase, an alternative Ang II-generating enzyme, is also involved in angiogenesis in a hamster sponge implant model.[149] Moreover, AT$_1$ receptor and angiotensin I-converting enzyme play an important role in tumor-associated angiogenesis in a murine model[150−152]. Ang II has been shown to be involved in coronary capillary angiogenesis at the insulin-resistant stage of a non-insulin diabetes mellitus (NIDDM) rat model.[153] Given that Ang II is a potent stimulator for vascular NAD(P)H oxidase, one may speculate that ROS derived from oxidase may play a role in Ang II-induced angiogenesis.

While non-transformed cells respond to growth factors/cytokines with the regulated production of ROS, tumor cells frequently overproduce H_2O_2. Arbiser *et al.*[154] demonstrated that overexpression of Nox1 into a prostate cancer cell line increases VEGF, VEGF receptor expression and MMP activity through increase of H_2O_2, which contributes to the vascularization of tumors. Most recently, Lim *et al.*[155] have shown that both Nox1 and H_2O_2 are increased in human prostate cancer tissue and in an animal model.

Endothelial progenitor cells (EPCs) also contribute to postnatal neovascularization.[156,157] Recent evidence suggests that ROS derived from NAD(P)H oxidase may also regulate vasculogenesis. Dernbach *et al.*[158] reported that EPCs express higher level of the antioxidant enzymes catalase, glutathione peroxidase and MnSOD than EC to protect against oxidative stress. Indeed, the survival and migratory capacity of EPCs is reduced by knockdown of antioxidant enzymes using siRNA.[158] Furthermore, redox state modulates self-renewal and differentiation of EPCs.[159,160] Short-term exposure of Ang II, a potent stimulator of vascular NAD(P)H oxidase, potentiates VEGF-induced proliferation and network formation of EPCs,[161] while its long-term exposure accelerates senescence of EPC through induction of Nox2 and oxidative stress.[162] Thus, the amount of ROS is likely to determine the fate and function of EPCs. Most recently, Sauer *et al.*[163] demonstrated that NAD(P)H oxidase-derived ROS are involved in EC differentiation and angiogenesis of mouse embryonic stem cells after direct current electrical field stimulation, further suggesting an important role of ROS in the function of EPCs.

Hyperglycemia is a primary cause of macro- and micro-vascular complications in diabetes. Furthermore, impaired reparative angiogenesis impedes proper post-ischemic healing and wound closure in diabetic patients. This defect was attributed to the shortage of, or insensitivity to, angiogenic growth factors including VEGF.[164,165] EPCs, which play a critical role in forming new vessels, are also dysfunctional in hyperglycemia.[166] As discussed, low concentrations of ROS acts as signaling molecules and are necessary for reparative angiogenesis and wound healing,[53,167] while excess amount of ROS (oxidative stress) contributes to the pathologensis of atherosclerosis and diabetes, in part by inactivating nitric oxide and causing EC dysfunction.[168−171] Of importance, antioxidants accelerate diabetic wound healing.[172] Evidence suggests that NAD(P)H oxidase is involved in increased production of ROS in diabetic patients and mice as well as ECs cultured under high glucose conditions.[9,169,173,174] Hyperglycemia also contributes to an impairment of EPC count and function, at least in part, through inhibition of

Akt and eNOS phosphorylation by unknown mechanisms.[166] Thus, diabetes-associated over-activation of NAD(P)H oxidase may contribute to disturbing proper angiogenic growth factor signaling and function in ECs and EPCs, which may result in impaired angiogenesis/vasculogenesis in response to ischemic injury.

6. ROS as Signaling Molecules in Angiogenesis

Signal transduction by ROS, "redox signaling" is a rapidly growing area of investigation. NAD(P)H oxidase is activated by numerous stimuli including growth factors such as VEGF, transforming growth factor-β1, cytokines, shear stress, hypoxia and G-protein coupled receptor agonists including Ang II in ECs.[9,175] ROS produced via activation of NAD(P)H oxidase stimulate diverse redox signaling pathways leading to angiogenesis (Fig. 3). VEGF binds to two tyrosine kinase receptors, VEGF receptor-1 (VEGFR1, also termed Flt-1) and VEGFR2 (also termed KDR/Flk1) in ECs. The mitogenic and chemotactic effects of VEGF in ECs are mediated mainly through VEGFR2.[2,176,177] VEGFR2 is activated through ligand-stimulated receptor dimerization and transphosphorylation (autophosphorylation) of tyrosine residues in the cytoplasmic kinase domain. At present, tyrosine residues 951 and 996 in the kinase insert domain, and 1054 and 1059 in the kinase catalytic domain have been identified as autophosphorylation sites for VEGFR2 in a bacterial expression system.[178] This event is followed by activation of diverse downstream signaling pathways such as mitogen-activated protein kinases (MAPKs), Akt/protein kinase B and endothelial NOS (eNOS), which are essential for VEGF-induced EC migration and proliferation.[179–182] ROS are important mediators for VEGF-mediated angiogenic signaling including VEGF receptor autophosphorylation, cSrc activation and VE-cadherin phosphorylation in ECs.[20,76,88,108] Antioxidant green tea catechins suppress VEGF-induced tube formation in EC through inhibiting VE-cadherin tyrosine phosphorylation and Akt activation.[183] We have demonstrated that VEGF-induced VEGFR2 autophosphorylation is inhibited by the thiol antioxidant NAC, various NAD(P)H oxidase inhibitors, dominant negative Rac1 and Nox2 antisense oligonucleotides.[20] These results suggest that VEGFR2 is one of the proximal molecular targets of ROS derived from the Nox2-containing NAD(P)H oxidase in cultured EC (Fig. 3). Moreover, Tie-2 receptors (Tie-2 R) are RTKs activated by angiopoietin-1 (Ang-1) and selectively expressed in ECs. They play an important role in embryonic development and promote differentiation, tube formation, migration, adherence and survival of ECs. Harfouche et al.[21] demonstrated that ROS derived from Rac1-dependent, Nox2-based NAD(P)H oxidase mediate Ang-1/Tie-2 R signaling linked to EC chemotaxis. However, ROS are not involved in Ang-1-induced Tie-2 R autophosphorylation, suggesting that mechanisms of autorphosphorylation of VEGFR2 and Tie-2 R are different.

A role of ROS in tyrosine phosphorylation of RTKs has been previously reported. We showed that Ang II induces ROS-dependent tyrosine phosphorylation of the epidermal growth factor (EGF) receptor (EGF-R) in VSMCs.[24] Rao[184] reported that H_2O_2 induces tyrosine phosphorylation of the EGF-R, which in turn recruits the Shc-Grb2-Sos adaptor protein complex to the receptor to activate subsequent signaling cascades. Furthermore, EGF-, PDGF- and insulin-induced autophosphorylation of their receptors are inhibited by catalase, which scavenges intracellular H_2O_2,[185–187] suggesting that endogenously-produced H_2O_2 plays an important role in RTK phosphorylation. We have shown that ROS derived from NAD(P)H oxidase are involved in Ang II-induced transmodulation of EGF-R,[23] which serves as a scaffold for various signaling molecules and plays an important role in cross-talk between G-protein coupled receptor and RTK signaling in VSMCs.[188] Of note, Ang II-induced angiogenesis has been shown to be mediated through induction of VEGF and angiopoietin expression via heparin-binding EGF-like growth factor (HB-EGF)-mediated EGF-R transmodulation in ECs.[189] Thus, it is likely that ROS derived from NAD(P)H oxidase may play a role in Ang II-induced angiogenesis via regulating EGF-R phosphorylation.

Identifying the molecular targets of ROS in agonist-stimulated signal transduction is critically important to the understanding of the mechanisms of redox signaling in the vasculature. Accumulating evidence suggests that protein tyrosine phosphatases (PTPs) are direct targets of ROS, and PTPs negatively regulate RTK activity and downstream signaling.[190–192] Most PTPs have low-pK_a cysteine residues within their active site, that exist as thiolate anions at neutral pH.[193] The reversible oxidative inhibition of active cysteine residues in PTPs by ROS is an important mechanism through which ROS shift the balance of protein tyrosine phosphorylation/dephosphorylation towards enhanced tyrosine phosphorylation in growth factor signaling [190,192,194]. Thus, ROS produced after RTK engagement form a feedback loop that, through the oxidative inhibition of PTPs, promotes the tyrosine phosphorylation/activation and the downstream signaling of RTKs. Furthermore, the lipid phosphatase tumor suppressor PTEN,[195] membrane phosphatase CD45,[196] and transmembrane receptor RPTPα[197] are also susceptible to H_2O_2-dependent oxidative inactivation *in vivo*. Indeed, Connor *et al.*[198] demonstrated that mitochondrial H_2O_2 production by MnSOD overexpression induces oxidative inactivation of PTEN, thereby promoting EC sprouting in a 3D *in vitro* angiogenesis assay as well as *in vivo* blood vessel formation in CAM assay. It should be noted that MnSOD can serve as a source of the potent signaling molecule, H_2O_2, from the mitochondria. Of importance, VEGF-induced ROS have been shown to be involved in induction of MnSOD through activation of Rac1-dependent NAD(P)H oxidase in ECs,[75] which can represent a feedforward mechanism by which ROS-triggered ROS formation play an important role in angiogenesis.

A reversible oxidation of PTPs such as PTP1B, low molecular weight (LMW)-PTP and SHP-2 during RTK stimulation with EGF, insulin and platelet-derived growth factor[187,195,199–201] have been reported. In case of VEGFR2, several PTPs including SHP-1, SHP-2 and LMW-PTP (HCPTPA) have been shown to inducibly associate with VEGFR2 after VEGF stimulation.[202–204] Although a role of SHP-1 in VEGF signaling remains unclear, Guo *et al.*[203] reported that TNF-α inhibits VEGF signaling and cell proliferation by facilitating recruitment of SHP-2 to the VEGFR2 in ECs. Huang *et al.*[204] demonstrated that overexpression of HCPTPA inhibits VEGF-induced VEGFR2 autophosphorylation, cell migration and proliferation. High cell density-enhanced PTP1 (DEP-1)/CD148 has been reported to participate in inhibition of VEGFR2 phosphorylation in confluent, contact-inhibited ECs.[205] A small molecule inhibitor of PTP1B has been shown to enhance VEGF-induced VEGFR2 activation, migration and proliferation of EC as well as neovascularization in matrigel plugs of mice.[206] As with VEGFR2, SHP-2 binds to the activated, autophosphorylated Tie-2 R following Ang-1 stimulation in ECs,[207] which in turn inhibits phosphatidylinositol 3′ kinase-dependent signaling pathways linked to EC migration.[208,209] Thus, it will be important to determine which PTPs are reversibly oxidized by VEGF- or Ang-1-induced ROS, thereby promoting RTK-mediated redox signaling in angiogenesis.

In addition to ROS, NO also plays an important role in VEGF signaling and postnatal angiogenesis.[210] NO enhances VEGF synthesis in several cell types and is required for execution of VEGF angiogenic effect in ECs. Cross-talk between NAD(P)H oxidase, EC growth as well as eNOS enzyme activity and expression has been demonstrated.[211] It is thus important to understand how both ROS and NO regulate VEGF signaling and angiogenesis. We have demonstrated that ROS, but not NO, are involved in VEGF-induced VEGFR2 autophosphorylation in HUVECs.[20] Furthermore, exogenous H_2O_2 or Ang II-stimulated increase of ROS potently activates eNOS, which in turn promotes NO production in ECs.[130,175] H_2O_2 stimulates EC proliferation, migration and cGMP production, which are reversed by a guanylate cyclase inhibitor, while antioxidants block cell proliferation and migration through downregulation of eNOS activity.[212] Thus, NAD(P)H oxidase-derived H_2O_2 seems to play a significant role in promoting angiogenic responses such as EC proliferation and migration at least in part via regulating eNOS/NO pathways.[213] This notion is further supported by the recent report using an *in vivo* mice model that exercise training increases both oxidative stress and eNOS expression, and both are inhibited in transgenic mice overexpressing human catalase in the vascular endothelium.[214] This result suggests that endogenously-produced H_2O_2 is involved in the endothelial adaptation to exercise by upregulation of eNOS *in vivo*. Whether similar mechanisms apply to VEGF signaling and postnatal angiogenesis *in vivo* require further investigation.

7. Angiogenesis-Dependent Transcription Factors and Genes Regulated by ROS

In addition to PTPs, transcription factors with low-pK_a cysteine residues such as the nuclear factor-κB (NF-κB),[215] AP-1,[216] hypoxia-inducible factor1α (HIF-1α),[217] p53[218] and p21Ras[219] can be oxidized by H_2O_2. Low concentrations of H_2O_2 stimulate induction of the transcription factor Ets-1, which is required for EC proliferation and tube formation.[34] Inflammation plays an important role in angiogenesis. H_2O_2 induces an increase in MCP-1 mRNA levels in an AP-1- and NF-κB-dependent manner in ECs.[118] In addition, H_2O_2-induced increase in NF-κB binding to DNA is involved in IL-8 production, which is required for tubular morphogenesis in human microvascular ECs.[36] Ang II-induced ROS derived from NAD(P)H oxidase are involved in upregulation of MCP-1 and NF-κB,[220] vascular cell-adhesion molecule-1 (VCAM-1)[221] and STAT1.[222] It should be noted that all of these redox-sensitive genes and transcription factors are activated by VEGF, raising the possibility that they are regulated by ROS generated after VEGF stimulation (Fig. 3). Consistent with this notion, VEGF-induced ROS have been shown to be involved in induction of NFκB in VSMCs.[223]

HIF-1 is a heterodimeric basic helix-loop-helix transcription factor composed of HIF-1α and HIF-1β aryl hydrocarbon nuclear translocator subunits (see Chapter 7). HIF-1 expression is induced by hypoxia, growth factors, and activation of oncogenes. In response to hypoxia, HIF-1 activates the expression of many angiogenesis-related genes including VEGF and erythropoietin. Of importance, accumulating evidence suggest that ROS derived from NAD(P)H oxidase are involved in induction of HIF-1α under normoxia and hypoxia in vascular cells.[224–230] Under normoxia, agonist-induced ROS may serve as signaling molecules to upregulate HIF-1α possibly by modulating upstream signaling pathways such as hydroxylases or kinases and phosphatases. In contrast, the role of ROS in hypoxia-induced HIF-1α regulation is less well understood. Turcotte *et al.*[231] demonstrated that hypoxia stimulates ROS production, thereby increasing Rho GTPase expression which is required for HIF-1α accumulation in renal cell carcinoma. It has been shown that NAD(P)H oxidase or a cytochrome b-type NAD(P)H oxidoreductase or mitochondria may produce ROS under hypoxia. The difference may be due to variabilities in the experimental conditions, cell type, the amount of oxygen available or the measuring techniques.[107] Gorlach *et al.*[229] showed that overexpression of Rac1 increases expression of HIF-1 through ROS-dependent mechanisms. These data suggest that Rac1/NAD(P)H oxidase/ROS pathways are important for upregulation of HIF-1 and VEGF expression in response to VEGF and hypoxia (Fig. 3) Of note, the redox protein thioredoxin-1 increases HIF-1α protein expression in cancer cells, thereby promoting VEGF production and tumor angiogenesis.[232] It has been shown that anti-angiogenic therapy reduces both plaque growth and intimal neovascularization in apolipoprotein-E deficient (ApoE$^{-/-}$) mice. Dietary supplementation with antioxidants vitamins C and E reduces vascular VEGF and VEGFR-2 expression in ApoE$^{-/-}$mice[233].

Other important ROS-dependent genes and proteins associated with angiogenesis are urokinase plasminogen activator (uPA) and MMPs (Fig. 3). ROS activate and increase the expression of MMPs.[234] Ets-1, which is induced by H_2O_2,[34] regulates the expression of genes involved in extracellular matrix degradation, including uPA and MMP-1. Grote *et al.*.[235] have shown that the mechanical stress-induced increase in MMP-2 activation and mRNA expression are inhibited in VSMCs derived from p47pho$x^{-/-}$ mice. Furthermore, lysophosphatidylcholine (lysoPC) increases the secretion of MMP-2 through activation of NAD(P)H oxidase in cultured ECs.[236] Given that VEGF stimulates induction of MMP-1 and -2 expression in ECs,[237] this mechanism may also be mediated through ROS derived from NAD(P)H oxidase. Moreover, HIF-1 mediates upregulation of plasminogen activator inhibitor-1 (PAI-1) expression under hypoxia and ROS have also been implicated in PAI-1 gene expression. Thus, overexpression of Rac1 upregulates PAI-1 expression through an increase in ROS.[229] Of note, lysoPC also induces uPA and its cell surface receptor in human macrophages, in a ROS-dependent manner, which may contribute to the intimal neovascularization in atherosclerotic plaque.[238]

8. Conclusion

Although the loss of VEGFR2 is an embryonic lethal because of vascular abnormalities, Nox2$^{-/-}$ mice do not have such a phenotype, suggesting that VEGFR2 activated by Nox2-derived ROS is not involved in vascular development. Rather, it is likely that oxidase-derived ROS play an important role in postnatal angiogenesis during pathological conditions such as ischemia, chronic inflammation such as atherosclerosis, diabetes, and cancer, in which VEGF expression and ROS production are increased. Moreover, although our *in vitro* data strongly support the role of endothelial NAD(P)H oxidase in VEGF signaling and angiogenic-related responses, an impairment of inflammatory response in phagocytic cells as well as of EPC function in Nox2$^{-/-}$ mice cannot be excluded. Studies focused on the relative importance of NAD(P)H oxidase in ECs, inflammatory cells and EPCs in neovascularization *in vivo* and defining the role of other components of NAD(P)H oxidase in redox signaling linked to angiogenesis are subjects of future investigation. Moreover, significant additional work on the mechanisms of the activation of NAD(P)H oxidases by various angiogenesis factors and identification of molecular targets of oxidase-derived ROS in angiogenesis will be required. The development of specific inhibitors of NAD(P)H oxidases can provide useful tools to elucidate the roles of these enzymes experimentally in angiogenic-related responses *in vitro* and angiogenesis *in vivo*. These studies may provide novel insight into the components of NAD(P)H oxidase as potential therapeutic targets for treatment of angiogenesis-dependent diseases and for promoting neovascularization in ischemic heart and limb diseases.

References

1. Folkman J (1995). Angiogenesis in cancer, vascular, rheumatoid and other disease. *Nat Med* **1**: 27–31.
2. Zachary I, Gliki G (2001). Signaling transduction mechanisms mediating biological actions of the vascular endothelial growth factor family. *Cardiovasc Res* **49**: 568–581.
3. Geiszt M, Leto TL (2004). The Nox family of NAD(P)H oxidases: host defense and beyond. *J Biol Chem* **279**: 51715–51718.
4. Droge W (2002). Free radicals in the physiological control of cell function. *Physiol Rev* **82**: 47–95.

5. Taylor BM, Fleming WE, Benjamin CW, Wu Y, Mathews WR, Sun FF (1996). The mechanism of cytoprotective action of lazaroids I: inhibition of reactive oxygen species formation and lethal cell injury during periods of energy depletion. *J Pharmacol Exp Ther* **276**: 1224–1231.

6. Hogg N, Browning J, Howard T, Winterford C, Fitzpatrick D, Gobe G (1999). Apoptosis in vascular endothelial cells caused by serum deprivation, oxidative stress and transforming growth factor-beta. *Endothelium* **7**: 35–49.

7. Bresgen N, Karlhuber G, Krizbai I, Bauer H, Bauer HC, Eckl PM (2003). Oxidative stress in cultured cerebral endothelial cells induces chromosomal aberrations, micronuclei, and apoptosis. *J Neurosci Res* **72**: 327–333.

8. Warren MC, Bump EA, Medeiros D, Braunhut SJ (2000). Oxidative stress-induced apoptosis of endothelial cells. *Free Radic Biol Med* **29**: 537–547.

9. Griendling KK, Sorescu D, Ushio-Fukai M (2000). NAD(P)H oxidase: role in cardiovascular biology and disease. *Circ Res* **86**: 494–501.

10. Maulik N (2002). Redox signaling of angiogenesis. *Antioxid Redox Signal* **4**: 805–815.

11. Maulik N (2002). Redox regulation of vascular angiogenesis. *Antioxid Redox Signal* **4**: 783–784.

12. Tojo T, Ushio-Fukai M, Yamaoka-Tojo M, Ikeda S, Patrushev NA, Alexander RW (2005). Role of gp91phox (Nox2)-containing NAD(P)H oxidase in angiogenesis in response to hindlimb ischemia. *Circulation* **111**: 2347–2355.

13. Stone JR, Collins T (2002). The role of hydrogen peroxide in endothelial proliferative responses. *Endothelium* **9**: 231–238.

14. Luczak K, Balcerczyk A, Soszynski M, Bartosz G (2004). Low concentration of oxidant and nitric oxide donors stimulate proliferation of human endothelial cells *in vitro* . *Cell Biol Int* **28**: 483–486.

15. Chen W, Gabel S, Steenbergen C, Murphy E (1995). A redox-based mechanism for cardioprotection induced by ischemic preconditioning in perfused rat heart. *Circ Res* **77**: 424–429.

16. Skyschally A, Schulz R, Gres P, Korth HG, Heusch G (2003). Attenuation of ischemic preconditioning in pigs by scavenging of free oxyradicals with ascorbic acid. *Am J Physiol Heart Circ Physiol* **284**: H698–H703.

17. Tanaka K, Weihrauch D, Kehl F, Ludwig LM, LaDisa JF Jr, Kersten JR, Pagel PS, Warltier DC (2002). Mechanism of preconditioning by isoflurane in rabbits: a direct role for reactive oxygen species. *Anesthesiology* **97**: 1485–1490.

18. Al-Shabrawey M, Bartoli M, El-Remessy AB, Platt DH, Matragoon S, Behzadian MA, Caldwell RW, Caldwell RB (2005). Inhibition of NAD(P)H oxidase activity blocks vascular endothelial growth factor overexpression and neovascularization during ischemic retinopathy. *Am J Pathol* **167**: 599–607.

19. Landmesser U, Harrison DG, Drexler H (2005). Oxidant stress-a major cause of reduced endothelial nitric oxide availability in cardiovascular disease. *Eur J Clin Pharmacol* 1–7.

20. Ushio-Fukai M, Tang Y, Fukai T, Dikalov S, Ma Y, Fujimoto M, Quinn MT, Pagano PJ, Johnson C, Alexander RW (2002). Novel role of gp91phox-containing NAD(P)H oxidase in vascular endothelial growth factor-induced signaling and angiogenesis. *Circ Res* **91**: 1160–1167.

21. Harfouche R, Malak NA, Brandes RP, Karsan A, Irani K, Hussain SN (2005). Roles of reactive oxygen species in angiopoietin-1/tie-2 receptor signaling. *FASEB J* **19**: 1728–1730.

22. Griendling KK, Sorescu D, Lassègue B, Ushio-Fukai M (2000). Modulation of protein kinase activity and gene expression by reactive oxygen species and their role in vascular physiology and pathophysiology. *Arterioscler Thromb Vasc Biol* **20**: 2175–2183.

23. Ushio-Fukai M, Griendling KK, Becker PL, Hilenski L, Halleran S, Alexander RW (2001). Epidermal growth factor receptor transactivation by angiotensin II requires reactive oxygen species in vascular smooth muscle cells. *Arterioscler Thromb Vasc Biol* **21**: 489–495.

24. Ushio-Fukai M, Alexander RW, Akers M, Yin Q, Fujio Y, Walsh K, Griendling KK (1999). Reactive oxygen species mediate the activation of Akt/Protein kinase B by angiotensin II in vascular smooth muscle cells. *J Biol Chem* **274**: 22699–22704.

25. Ushio-Fukai M, Alexander RW, Akers M, Griendling KK (1998). p38MAP kinase is a critical component of the redox-sensitive signaling pathways by angiotensin II: role in vascular smooth muscle cell hypertrophy. *J Biol Chem* **273**: 15022–15029.

26. Cai H (2005). Hydrogen peroxide regulation of endothelial function: origins, mechanisms, and consequences. *Cardiovasc Res* **68**: 26–36.

27. Ruef J, Hu ZY, Yin LY, Wu Y, Hanson SR, Kelly AB, Harker LA, Rao GN, Runge MS, Patterson C (1997). Induction of vascular endothelial growth factor in balloon-injured baboon arteries. *Circ Res* **81**: 24–33.

28. Kosmidou I, Xagorari A, Roussos C, Papapetropoulos A (2001). Reactive oxygen species stimulate VEGF production from C(2)C(12) skeletal myotubes through a PI3K/Akt pathway. *Am J Physiol Lung Cell Mol Physiol* **280**: L585–L592.

29. Cisowski J, Loboda A, Jozkowicz A, Chen S, Agarwal A, Dulak J (2005). Role of heme oxygenase-1 in hydrogen peroxide-induced VEGF synthesis: effect of HO-1 knockout. *Biochem Biophys Res Commun* **326**: 670–676.

30. Brauchle M, Funk JO, Kind P, Werner S (1996). Ultraviolet B and H_2O_2 are potent inducers of vascular endothelial growth factor expression in cultured keratinocytes. *J Biol Chem* **271**: 21793–21797.

31. Kuroki M, Voest EE, Amano S, Beerepoot LV, Takashima S, Tolentino M, Kim RY, Rohan RM, Colby KA, Yeo KT, *et al.* (1996). Reactive oxygen intermediates increase vascular endothelial growth factor expression *in vitro* and *in vivo* . *J Clin Invest* **98**: 1667–1675.

32. Cho M, Hunt TK, Hussain MZ (2001). Hydrogen peroxide stimulates macrophage vascular endothelial growth factor release. *Am J Physiol Heart Circ Physiol* **280**: H2357–H2363.

33. Chua CC, Hamdy RC, Chua BH (1998). Upregulation of vascular endothelial growth factor by H_2O_2 in rat heart endothelial cells. *Free Radic Biol Med* **25**: 891–897.

34. Yasuda M, Ohzeki Y, Shimizu S, Naito S, Ohtsuru A, Yamamoto T, Kuroiwa Y (1999). Stimulation of *in vitro* angiogenesis by hydrogen peroxide and the relation with ETS-1 in endothelial cells. *Life Sci* **64**: 249–258.

35. Vepa S, Scribner WM, Parinandi NL, English D, Garcia JG, Natarajan V (1999). Hydrogen peroxide stimulates tyrosine phosphorylation of focal adhesion kinase in vascular endothelial cells. *Am J Physiol* **277**: L150–L158.

36. Shono T, Ono M, Izumi H, Jimi SI, Matsushima K, Okamoto T, Kohno K, Kuwano M (1996). Involvement of the transcription factor NF-kappaB in tubular morphogenesis of human microvascular endothelial cells by oxidative stress. *Mol Cell Biol* **16**: 4231–4239.

37. Monte M, Davel LE, Sacerdote de Lustig E (1997). Hydrogen peroxide is involved in lymphocyte activation mechanisms to induce angiogenesis. *Eur J Cancer* **33**: 676–682.

38. Lelkes PI, Hahn KL, Sukovich DA, Karmiol S, Schmidt DH (1998). On the possible role of reactive oxygen species in angiogenesis. *Adv Exp Med Biol* **454**: 295–310.

39. Moldovan L, Moldovan NI, Sohn RH, Parikh SA, Goldschmidt-Clermont PJ (2000). Redox changes of cultured endothelial cells and actin dynamics. *Circ Res* **86**: 549–557.

40. Ikeda S, Ushio-Fukai M, Zuo L, Tojo T, Alexander RW (2005). Novel role of ARF6 in vascular endothelial growth factor signaling and angiogenesis. *Circ Res* **96**: 467–475.

41. Yasuda M, Shimizu S, Tokuyama S, Watana T, Kiuchi Y, Yamamoto T (2000). A novel effect of polymorphonuclear leukocytes in the facilitation of angiogenesis [In Process Citation]. *Life Sci* **66**: 2113–2121.

42. Urata Y, Yamaguchi M, Higashiyama Y, Ihara Y, Goto S, Kuwano M, Horiuchi S, Sumikawa K, Kondo T (2002). Reactive oxygen species accelerate production of vascular endothelial growth factor by advanced glycation end products in RAW264.7 mouse macrophages. *Free Radic Biol Med* **32**: 688–701.

43. Sierra-Honigmann MR, Nath AK, Murakami C, Garcia-Cardena G, Papapetropoulos A, Sessa WC, Madge LA, Schechner JS, Schwabb MB, Polverini PJ, *et al.* (1998). Biological action of leptin as an angiogenic factor [see comments]. *Science* **281**: 1683–1686.

44. Yamagishi S, Amano S, Inagaki Y, Okamoto T, Takeuchi M, Inoue H (2003). Pigment epithelium-derived factor inhibits leptin-induced angiogenesis by suppressing vascular endothelial growth factor gene expression through anti-oxidative properties. *Microvasc Res* **65**: 186–190.

45. Tang FY, Meydani M (2001). Green tea catechins and vitamin E inhibit angiogenesis of human microvascular endothelial cells through suppression of IL-8 production. *Nutr Cancer* **41**: 119–125.

46. Oak MH, El Bedoui J, Schini-Kerth VB (2005). Antiangiogenic properties of natural polyphenols from red wine and green tea. *J Nutr Biochem* **16**: 1–8.

47. Dawson DW, Volpert OV, Gillis P, Crawford SE, Xu H, Benedict W, Bouck NP (1999). Pigment epithelium-derived factor: a potent inhibitor of angiogenesis. *Science* **285**: 245–248.

48. Amano S, Yamagishi S, Inagaki Y, Nakamura K, Takeuchi M, Inoue H, Imaizumi T (2005). Pigment epithelium-derived factor inhibits oxidative stress-induced apoptosis and dysfunction of cultured retinal pericytes. *Microvasc Res* **69**: 45–55.

49. Ellis EA, Guberski DL, Somogyi-Mann M, Grant MB (2000). Increased H2O2, vascular endothelial growth factor and receptors in the retina of the BBZ/Wor diabetic rat. *Free Radic Biol Med* **28**: 91–101.

50. Ellis EA, Grant MB, Murray FT, Wachowski MB, Guberski DL, Kubilis PS, Lutty GA (1998). Increased NADH oxidase activity in the retina of the BBZ/Wor diabetic rat. *Free Radic Biol Med* **24**: 111–120.

51. Caldwell RB, Bartoli M, Behzadian MA, El-Remessy AE, Al-Shabrawey M, Platt DH, Liou GI, Caldwell RW (2005). Vascular endothelial growth factor and diabetic retinopathy: role of oxidative stress. *Curr Drug Targets* **6**: 511–524.

52. Gu W, Weihrauch D, Tanaka K, Tessmer JP, Pagel PS, Kersten JR, Chilian WM, Warltier DC (2003). Reactive oxygen species are critical mediators of coronary collateral development in a canine model. *Am J Physiol Heart Circ Physiol* **285**: H1582–H1589.

53. Sen CK (2003). The general case for redox control of wound repair. *Wound Repair Regen* **11**: 431–438.

54. Roy S, Khanna S, Nallu K, Hunt TK, Sen CK (2006). Dermal wound healing is subject to redox control. *Mol Ther* **13**: 211–220.

55. Yoshida A, Yoshida S, Ishibashi T, Kuwano M, Inomata H (1999). Suppression of retinal neovascularization by the NF-kappaB inhibitor pyrrolidine dithiocarbamate in mice. *Invest Ophthalmol Vis Sci* **40**: 1624–1629.

56. Cao Y, Cao R (1999). Angiogenesis inhibited by drinking tea [letter]. *Nature* **398**: 381.

57. Dona M, Dell'Aica I, Calabrese F, Benelli R, Morini M, Albini A, Garbisa S (2003). Neutrophil restraint by green tea: inhibition of inflammation, associated angiogenesis, and pulmonary fibrosis. *J Immunol* **170**: 4335–4341.

58. Brakenhielm E, Cao R, Cao Y (2001). Suppression of angiogenesis, tumor growth, and wound healing by resveratrol, a natural compound in red wine and grapes. *FASEB J* **15**: 1798–1800.

59. Polytarchou C, Papadimitriou E (2004). Antioxidants inhibit angiogenesis *in vivo* through down-regulation of nitric oxide synthase expression and activity. *Free Radic Res* **38**: 501–508.

60. Cai T, Fassina G, Morini M, Aluigi MG, Masiello L, Fontanini G, D'Agostini F, De Flora S, Noonan DM, Albini A (1999). N-acetylcysteine inhibits endothelial cell invasion and angiogenesis. *Lab Invest* **79**: 1151–1159.

61. Abe R, Shimizu T, Yamagishi S, Shibaki A, Amano S, Inagaki Y, Watanabe H, Sugawara H, Nakamura H, Takeuchi M, *et al.* (2004). Overexpression of pigment epithelium-derived factor decreases angiogenesis and inhibits the growth of human malignant melanoma cells *in vivo* . *Am J Pathol* **164**: 1225–1232.

62. Wheeler MD, Smutney OM, Samulski RJ (2003). Secretion of extracellular superoxide dismutase from muscle transduced with recombinant adenovirus inhibits the growth of B16 melanomas in mice. *Mol Cancer Res* **1**: 871–881.

63. Marikovsky M, Nevo N, Vadai E, Harris-Cerruti C (2002). Cu/Zn superoxide dismutase plays a role in angiogenesis. *Int J Cancer* **97**: 34–41.

64. Grzenkowicz-Wydra J, Cisowski J, Nakonieczna J, Zarebski A, Udilova N, Nohl H, Jozkowicz A, Podhajska A, Dulak J (2004). Gene transfer of CuZn superoxide

dismutase enhances the synthesis of vascular endothelial growth factor. *Mol Cell Biochem* **264**: 169–181.

65. Babior BM (2000). The NADPH oxidase of endothelial cells. *IUBMB Life* **50**: 267–269.

66. Li JM, Shah AM (2004). Endothelial cell superoxide generation: regulation and relevance for cardiovascular pathophysiology. *Am J Physiol Regul Integr Comp Physiol* **287**: R1014–R1030.

67. Lambeth JD, Cheng G, Arnold R, Edens WA (2000). Novel homologs of gp91phox. *Trends in Biochem Sci* **25**: 459–461.

68. Cheng G, Cao Z, Xu X, van Meir EG, Lambeth JD (2001). Homologs of gp91phox: cloning and tissue expression of Nox3, Nox4, and Nox5. *Gene* **269**: 131–140.

69. Maturana A, Arnaudeau S, Ryser S, Banfi B, Hossle JP, Schlegel W, Krause KH, Demaurex N (2001). Heme histidine ligands within gp91(phox) modulate proton conduction by the phagocyte NADPH oxidase. *J Biol Chem* **276**: 30277–30284.

70. Abid MR, Kachra Z, Spokes KC, Aird WC (2000). NADPH oxidase activity is required for endothelial cell proliferation and migration. *FEBS Lett* **486**: 252–256.

71. Gorlach A, Brandes RP, Nguyen K, Amidi M, Dehghani F, Busse R (2000). A gp91phox containing NADPH oxidase selectively expressed in endothelial cells is a major source of oxygen radical generation in the arterial wall. *Circ Res* **87**: 26–32.

72. Li J-M, Shah AM (2002). Intracellular localization and preassembly of the NADPH oxidase complex in cultured endothelial cells. *J Biol Chem* **277**: 19952–19960.

73. Frey RS, Rahman A, Kefer JC, Minshall RD, Malik AB (2002). PKCzeta regulates TNF-alpha-induced activation of NADPH oxidase in endothelial cells. *Circ Res* **90**: 1012–1019.

74. Furst R, Brueckl C, Kuebler WM, Zahler S, Krotz F, Gorlach A, Vollmar AM, Kiemer AK (2005). Atrial natriuretic peptide induces mitogen-activated protein kinase phosphatase-1 in human endothelial cells via Rac1 and NAD(P)H oxidase/Nox2-activation. *Circ Res* **96**: 43–53.

75. Abid MR, Tsai JC, Spokes KC, Deshpande SS, Irani K, Aird WC (2001). Vascular endothelial growth factor induces manganese-superoxide dismutase expression in endothelial cells by a Rac1-regulated NADPH oxidase-dependent mechanism. *FASEB J* **15**: 2548–2550.

76. Colavitti R, Pani G, Bedogni B, Anzevino R, Borrello S, Waltenberger J, Galeotti T (2002). Reactive oxygen species as downstream mediators of angiogenic signaling by vascular endothelial growth factor receptor-2/KDR. *J Biol Chem* **277**: 3101–3108.

77. Rueckschloss U, Galle J, Holtz J, Zerkowski HR, Morawietz H (2001). Induction of NAD(P)H oxidase by oxidized low-density lipoprotein in human endothelial cells: antioxidative potential of hydroxymethylglutaryl coenzyme A reductase inhibitor therapy. *Circulation* **104**: 1767–1772.

78. Duerrschmidt N, Wippich N, Goettsch W, Broemme HJ, Morawietz H (2000). Endothelin-1 induces NAD(P)H oxidase in human endothelial cells. *Biochem Biophys Res Commun* **269**: 713–717.

79. Wagner AH, Schroeter MR, Hecker M (2001). 17beta-estradiol inhibition of NADPH oxidase expression in human endothelial cells. *FASEB J* **15**: 2121–2130.

80. Sorescu GP, Song H, Tressel SL, Hwang J, Dikalov S, Smith DA, Boyd NL, Platt MO, Lassegue B, Griendling KK, *et al.* (2004). Bone morphogenic protein 4 produced in endothelial cells by oscillatory shear stress induces monocyte adhesion by stimulating reactive oxygen species production from a nox1-based NADPH oxidase. *Circ Res* **95**: 773–779.

81. Kobayashi S, Nojima Y, Shibuya M, Maru Y (2004). Nox1 regulates apoptosis and potentially stimulates branching morphogenesis in sinusoidal endothelial cells. *Exp Cell Res* **300**: 455–462.

82. Ago T, Kitazono T, Ooboshi H, Iyama T, Han YH, Takada J, Wakisaka M, Ibayashi S, Utsumi H, Iida M (2004). Nox4 as the major catalytic component of an endothelial NAD(P)H oxidase. *Circulation* **109**: 227–233.

83. Yamagishi S, Takeuchi M, Matsui T, Nakamura K, Imaizumi T, Inoue H (2005). Angiotensin II augments advanced glycation end product-induced pericyte apoptosis through RAGE overexpression. *FEBS Lett* **579**: 4265–4270.

84. Ambasta RK, Kumar P, Griendling KK, Schmidt HH, Busse R, Brandes RP (2004). Direct interaction of the novel Nox proteins with p22phox is required for the formation of a functionally active NADPH oxidase. *J Biol Chem* **279**: 45935–45941.

85. Takeya R, Sumimoto H (2003). Molecular mechanism for activation of superoxide-producing NADPH oxidases. *Mol Cells* **16**: 271–277.

86. Guzik TJ, Sadowski J, Kapelak B, Jopek A, Rudzinski P, Pillai R, Korbut R, Channon KM (2004). Systemic regulation of vascular NAD(P)H oxidase activity and nox isoform expression in human arteries and veins. *Arterioscler Thromb Vasc Biol* **24**: 1614–1620.

87. Djordjevic T, Pogrebniak A, BelAiba RS, Bonello S, Wotzlaw C, Acker H, Hess J, Gorlach A (2005). The expression of the NADPH oxidase subunit p22phox is regulated by a redox-sensitive pathway in endothelial cells. *Free Radic Biol Med* **38**: 616–630.

88. van Wetering S, van Buul JD, Quik S, Mul FP, Anthony EC, ten Klooster JP, Collard JG, Hordijk PL (2002). Reactive oxygen species mediate Rac-induced loss of cell-cell adhesion in primary human endothelial cells. *J Cell Sci* **115**: 1837–1846.

89. Moldovan L, Irani K, Moldovan NI, Finkel T, Goldschmidt-Clermont PJ (1999). The actin cytoskeleton reorganization induced by Rac1 requires the production of superoxide. *Antioxid Redox Signal* **1**: 29–43.

90. Li JM, Mullen AM, Yun S, Wientjes F, Brouns GY, Thrasher AJ, Shah AM (2002). Essential role of the NADPH oxidase subunit p47(phox) in endothelial cell superoxide production in response to phorbol ester and tumor necrosis factor-alpha. *Circ Res* **90**: 143–150.

91. Hwang J, Saha A, Boo YC, Sorescu GP, McNally JS, Holland SM, Dikalov S, Giddens DP, Griendling KK, Harrison DG, *et al.* (2003). Oscillatory shear stress stimulates endothelial production of O2- from p47phox-dependent NAD(P)H oxidases, leading to monocyte adhesion. *J Biol Chem* **278**: 47291–47298.

92. Lambeth JD (2002). Nox/Duox family of nicotinamide adenine dinucleotide (phosphate) oxidases. *Curr Opin Hematol* **9**: 11–17.

93. Banfi B, Clark RA, Steger K, Krause KH (2003). Two novel proteins activate superoxide generation by the NADPH oxidase NOX1. *J Biol Chem* **278**: 3510–3513.

94. Geiszt M, Lekstrom K, Witta J, Leto TL (2003). Proteins homologous to p47phox and p67phox support superoxide production by NAD(P)H oxidase 1 in colon epithelial cells. *J Biol Chem* **278**: 20006–20012.

95. Takeya R, Ueno N, Kami K, Taura M, Kohjima M, Izaki T, Nunoi H, Sumimoto H (2003). Novel human homologues of p47phox and p67phox participate in activation of superoxide-producing NADPH oxidases. *J Biol Chem* **278**: 25234–25246.

96. Cheng G, Lambeth JD (2004). NOXO1, regulation of lipid binding, localization, and activation of Nox1 by the Phox homology (PX) domain. *J Biol Chem* **279**: 4737–4742.

97. Ago T, Kitazono T, Kuroda J, Kumai Y, Kamouchi M, Ooboshi H, Wakisaka M, Kawahara T, Rokutan K, Ibayashi S, *et al.* (2005). NAD(P)H oxidases in rat basilar arterial endothelial cells. *Stroke* **36**: 1040–1046.

98. Griendling KK (2004). Novel NAD(P)H oxidases in the cardiovascular system. *Heart* **90**: 491–493.

99. Lambeth JD (2004). NOX enzymes and the biology of reactive oxygen. *Nat Rev Immunol* **4**: 181–189.

100. Hilenski LL, Clempus RE, Quinn MT, Lambeth JD, Griendling KK (2004). Distinct subcellular localizations of Nox1 and Nox4 in vascular smooth muscle cells. *Arterioscler Thromb Vasc Biol* **24**: 677–683.

101. Van Buul JD, Fernandez-Borja M, Anthony EC, Hordijk PL (2005). Expression and localization of NOX2 and NOX4 in primary human endothelial cells. *Antioxid Redox Signal* **7**: 308–317.

102. Gu Y, Xu YC, Wu RF, Souza RF, Nwariaku FE, Terada LS (2002). TNFalpha activates c-Jun amino terminal kinase through p47(phox). *Exp Cell Res* **272**: 62–74.

103. Wu RF, Gu Y, Xu YC, Nwariaku FE, Terada LS (2003). Vascular endothelial growth factor causes translocation of p47phox to membrane ruffles through WAVE1. *J Biol Chem* **278**: 36830–36840.

104. Qian Y, Liu KJ, Chen Y, Flynn DC, Castranova V, Shi X (2005). Cdc42 regulates arsenic-induced NADPH oxidase activation and cell migration through actin filament reorganization. *J Biol Chem* **280**: 3875–3884.

105. Ikeda S, Yamaoka-Tojo M, Hilenski L, Patrushev NA, Anwar GM, Quinn MT, Ushio-Fukai M (2005). IQGAP1 regulates reactive oxygen species-dependent endothelial cell migration through interacting with Nox2. *Arterioscler Thromb Vasc Biol* **25**: 2295–2300.

106. Fleming I, Michaelis UR, Bredenkotter D, Fisslthaler B, Dehghani F, Brandes RP, Busse R (2001). Endothelium-derived hyperpolarizing factor synthase (Cytochrome P450 2C9) is a functionally significant source of reactive oxygen species in coronary arteries. *Circ Res* **88**: 44–51.

107. Kietzmann T, Gorlach A (2005). Reactive oxygen species in the control of hypoxia-inducible factor-mediated gene expression. *Semin Cell Dev Biol* **16**: 474–486.
108. Lin MT, Yen ML, Lin CY, Kuo ML (2003). Inhibition of vascular endothelial growth factor-induced angiogenesis by resveratrol through interruption of Src-dependent vascular endothelial cadherin tyrosine phosphorylation. *Mol Pharmacol* **64**: 1029–1036.
109. Yamaoka-Tojo M, Ushio-Fukai M, Hilenski L, Dikalov SI, Chen YE, Tojo T, Fukai T, Fujimoto M, Patrushev NA, Wang N, *et al.* (2004). IQGAP1, a novel vascular endothelial growth factor receptor binding protein, is involved in reactive oxygen species-dependent endothelial migration and proliferation. *Circ Res* **95**: 276–283.
110. Qian Y, Luo J, Leonard SS, Harris GK, Millecchia L, Flynn DC, Shi X (2003). Hydrogen peroxide formation and actin filament reorganization by Cdc42 are essential for ethanol-induced *in vitro* angiogenesis. *J Biol Chem* **278**: 16189–16197.
111. Vallet P, Charnay Y, Steger K, Ogier-Denis E, Kovari E, Herrmann F, Michel JP, Szanto I (2005). Neuronal expression of the NADPH oxidase NOX4, and its regulation in mouse experimental brain ischemia. *Neuroscience* **132**: 233–238.
112. MacCarthy PA, Grieve DJ, Li JM, Dunster C, Kelly FJ, Shah AM (2001). Impaired endothelial regulation of ventricular relaxation in cardiac hypertrophy: role of reactive oxygen species and NADPH oxidase. *Circulation* **104**: 2967–2974.
113. Bendall JK, Cave AC, Heymes C, Gall N, Shah AM (2002). Pivotal role of a gp91(phox)-containing NADPH oxidase in angiotensin II-induced cardiac hypertrophy in mice. *Circulation* **105**: 293–296.
114. Li JM, Gall NP, Grieve DJ, Chen M, Shah AM (2002). Activation of NADPH oxidase during progression of cardiac hypertrophy to failure. *Hypertension* **40**: 477–484.
115. Maack C, Kartes T, Kilter H, Schafers HJ, Nickenig G, Bohm M, Laufs U (2003). Oxygen free radical release in human failing myocardium is associated with increased activity of rac1-GTPase and represents a target for statin treatment. *Circulation* **108**: 1567–1574.
116. Fukui T, Yoshiyama M, Hanatani A, Omura T, Yoshikawa J, Abe Y (2001). Expression of p22-phox and gp91-phox, essential components of NADPH oxidase, increases after myocardial infarction. *Biochem Biophys Res Commun* **281**: 1200–1206.
117. Krijnen PA, Meischl C, Hack CE, Meijer CJ, Visser CA, Roos D, Niessen HW (2003). Increased Nox2 expression in human cardiomyocytes after acute myocardial infarction. *J Clin Pathol* **56**: 194–199.
118. Lakshminarayanan V, Lewallen M, Frangogiannis NG, Evans AJ, Wedin KE, Michael LH, Entman ML (2001). Reactive oxygen intermediates induce monocyte chemotactic protein-1 in vascular endothelium after brief ischemia. *Am J Pathol* **159**: 1301–1311.
119. Kimura S, Zhang GX, Nishiyama A, Shokoji T, Yao L, Fan YY, Rahman M, Suzuki T, Maeta H, Abe Y (2005). Role of NAD(P)H oxidase- and

mitochondria-derived reactive oxygen species in cardioprotection of ischemic reperfusion injury by angiotensin II. *Hypertension* **45**: 860–866.

120. Khatri JJ, Johnson C, Magid R, Lessner SM, Laude KM, Dikalov SI, Harrison DG, Sung HJ, Rong Y, Galis ZS (2004). Vascular oxidant stress enhances progression and angiogenesis of experimental atheroma. *Circulation* **109**: 520–525.

121. Wassmann S, Laufs U, Baumer AT, Muller K, Ahlbory K, Linz W, Itter G, Rosen R, Bohm M, Nickenig G (2001). HMG-CoA reductase inhibitors improve endothelial dysfunction in normocholesterolemic hypertension via reduced production of reactive oxygen species. *Hypertension* **37**: 1450–1457.

122. Wassmann S, Laufs U, Baumer AT, Muller K, Konkol C, Sauer H, Bohm M, Nickenig G (2001). Inhibition of geranylgeranylation reduces angiotensin II-mediated free radical production in vascular smooth muscle cells: involvement of angiotensin AT1 receptor expression and Rac1 GTPase. *Mol Pharmacol* **59**: 646–654.

123. Wassmann S, Laufs U, Muller K, Konkol C, Ahlbory K, Baumer AT, Linz W, Bohm M, Nickenig G (2002). Cellular antioxidant effects of atorvastatin *in vitro* and *in vivo*. *Arterioscler Thromb Vasc Biol* **22**: 300–305.

124. Vecchione C, Brandes RP (2002). Withdrawal of 3-hydroxy-3-methylglutaryl coenzyme A reductase inhibitors elicits oxidative stress and induces endothelial dysfunction in mice. *Circ Res* **91**: 173–179.

125. Laufs U, Kilter H, Konkol C, Wassmann S, Bohm M, Nickenig G (2002). Impact of HMG CoA reductase inhibition on small GTPases in the heart. *Cardiovasc Res* **53**: 911–920.

126. Weis M, Heeschen C, Glassford AJ, Cooke JP (2002). Statins have biphasic effects on angiogenesis. *Circulation* **105**: 739–745.

127. Touyz RM, Yao G, Schiffrin EL (2003). c-Src induces phosphorylation and translocation of p47phox: role in superoxide generation by angiotensin II in human vascular smooth muscle cells. *Arterioscler Thromb Vasc Biol* **23**: 981–987.

128. Lang D, Mosfer SI, Shakesby A, Donaldson F, Lewis MJ (2000). Coronary microvascular endothelial cell redox state in left ventricular hypertrophy: the role of angiotensin II. *Circ Res* **86**: 463–469.

129. Landmesser U, Cai H, Dikalov S, McCann L, Hwang J, Jo H, Holland SM, Harrison DG (2002). Role of p47(phox) in vascular oxidative stress and hypertension caused by angiotensin II. *Hypertension* **40**: 511–515.

130. Cai H, Li Z, Dikalov S, Holland SM, Hwang J, Jo H, Dudley SC Jr, Harrison DG (2002). NAD(P)H oxidase-derived hydrogen peroxide mediates endothelial nitric oxide production in response to angiotensin II. *J Biol Chem* **277**: 48311–48317.

131. Li JM, Shah AM (2003). Mechanism of endothelial cell NADPH oxidase activation by angiotensin II. Role of the p47phox subunit. *J Biol Chem* **278**: 12094–12100.

132. Rueckschloss U, Duerrschmidt N, Morawietz H (2003). NADPH oxidase in endothelial cells: impact on atherosclerosis. *Antioxid Redox Signal* **5**: 171–180.

133. Cifuentes ME, Rey FE, Carretero OA, Pagano PJ (2000). Upregulation of p67(phox) and gp91(phox) in aortas from angiotensin II-infused mice. *Am J Physiol Heart Circ Physiol* **279**: H2234–H2240.

134. Takemoto M, Node K, Nakagami H, Liao Y, Grimm M, Takemoto Y, Kitakaze M, Liao JK (2001). Statins as antioxidant therapy for preventing cardiac myocyte hypertrophy. *J Clin Invest* **108**: 1429–1437.

135. Grishko V, Pastukh V, Solodushko V, Gillespie M, Azuma J, Schaffer S (2003). Apoptotic cascade initiated by angiotensin II in neonatal cardiomyocytes: role of DNA damage. *Am J Physiol Heart Circ Physiol* **285**: H2364–H2372.

136. Nakagami H, Takemoto M, Liao JK (2003). NADPH oxidase-derived superoxide anion mediates angiotensin II-induced cardiac hypertrophy. *J Mol Cell Cardiol* **35**: 851–859.

137. Oudot A, Vergely C, Ecarnot-Laubriet A, Rochette L (2003). Angiotensin II activates NADPH oxidase in isolated rat hearts subjected to ischaemia-reperfusion. *Eur J Pharmacol* **462**: 145–154.

138. Bell L, Madri JA (1990). Influence of the angiotensin system on endothelial and smooth muscle cell migration. *Am J Pathol* **137**: 7–12.

139. Kintscher U, Wakino S, Kim S, Fleck E, Hsueh WA, Law RE (2001). Angiotensin II induces migration and Pyk2/paxillin phosphorylation of human monocytes. *Hypertension* **37**: 587–593.

140. Otani A, Takagi H, Suzuma K, Honda Y (1998). Angiotensin II potentiates vascular endothelial growth factor-induced angiogenic activity in retinal microcapillary endothelial cells. *Circ Res* **82**: 619–628.

141. Chintalgattu V, Nair DM, Katwa LC (2003). Cardiac myofibroblasts: a novel source of vascular endothelial growth factor (VEGF) and its receptors Flt-1 and KDR. *J Mol Cell Cardiol* **35**: 277–286.

142. Munzenmaier DH, Greene AS (1996). Opposing actions of angiotensin II on microvascular growth and arterial blood pressure. *Hypertension* **27**: 760–765.

143. Sasaki K, Murohara T, Ikeda H, Sugaya T, Shimada T, Shintani S, Imaizumi T (2002). Evidence for the importance of angiotensin II type 1 receptor in ischemia-induced angiogenesis. *J Clin Invest* **109**: 603–611.

144. Emanueli C, Salis MB, Stacca T, Pinna A, Gaspa L, Madeddu P (2002). Angiotensin AT(1) receptor signalling modulates reparative angiogenesis induced by limb ischaemia. *Br J Pharmacol* **135**: 87–92.

145. Walsh DA, Hu DE, Wharton J, Catravas JD, Blake DR, Fan TP (1997). Sequential development of angiotensin receptors and angiotensin I converting enzyme during angiogenesis in the rat subcutaneous sponge granuloma. *Br J Pharmacol* **120**: 1302–1311.

146. Amaral SL, Linderman JR, Morse MM, Greene AS (2001). Angiogenesis induced by electrical stimulation is mediated by angiotensin II and VEGF. *Microcirculation* **8**: 57–67.

147. Tamarat R, Silvestre JS, Kubis N, Benessiano J, Duriez M, deGasparo M, Henrion D, Levy BI (2002). Endothelial nitric oxide synthase lies downstream from angiotensin II-induced angiogenesis in ischemic hindlimb. *Hypertension* **39**: 830–835.

148. Shimizu T, Okamoto H, Chiba S, Matsui Y, Sugawara T, Akino M, Nan J, Kumamoto H, Onozuka H, Mikami T, *et al.* (2003). VEGF-mediated angiogenesis is impaired by angiotensin type 1 receptor blockade in cardiomyopathic hamster hearts. *Cardiovasc Res* **58**: 203–212.

149. Muramatsu M, Katada J, Hayashi I, Majima M (2000). Chymase as a proangiogenic factor. A possible involvement of chymase-angiotensin-dependent pathway in the hamster sponge angiogenesis model. *J Biol Chem* **275**: 5545–5552.

150. Yoshiji H, Kuriyama S, Kawata M, Yoshii J, Ikenaka Y, Noguchi R, Nakatani T, Tsujinoue H, Fukui H (2001). The angiotensin-I-converting enzyme inhibitor perindopril suppresses tumor growth and angiogenesis: possible role of the vascular endothelial growth factor. *Clin Cancer Res* **7**: 1073–1078.

151. Fujita M, Hayashi I, Yamashina S, Itoman M, Majima M (2002). Blockade of angiotensin AT1a receptor signaling reduces tumor growth, angiogenesis, and metastasis. *Biochem Biophys Res Commun* **294**: 441–447.

152. Fujita M, Hayashi I, Yamashina S, Fukamizu A, Itoman M, Majima M (2005). Angiotensin type 1a receptor signaling-dependent induction of vascular endothelial growth factor in stroma is relevant to tumor-associated angiogenesis and tumor growth. *Carcinogenesis* **26**: 271–279.

153. Jesmin S, Hattori Y, Sakuma I, Mowa CN, Kitabatake A (2002). Role of ANG II in coronary capillary angiogenesis at the insulin-resistant stage of a NIDDM rat model. *Am J Physiol Heart Circ Physiol* **283**: H1387–H1397.

154. Arbiser JL, Petros J, Klafter R, Govindajaran B, McLaughlin ER, Brown LF, Cohen C, Moses M, Kilroy S, Arnold RS, *et al.* (2002). Reactive oxygen generated by Nox1 triggers the angiogenic switch. *Proc Natl Acad Sci USA* **99**: 715–720.

155. Lim SD, Sun C, Lambeth JD, Marshall F, Amin M, Chung L, Petros JA, Arnold RS (2005). Increased Nox1 and hydrogen peroxide in prostate cancer. *Prostate* **62**: 200–207.

156. Hibbert B, Olsen S, O'Brien E (2003). Involvement of progenitor cells in vascular repair. *Trends Cardiovasc Med* **13**: 322–326.

157. Asahara T, Kawamoto A (2004). Endothelial progenitor cells for postnatal vasculogenesis. *Am J Physiol Cell Physiol* **287**: C572–C579.

158. Dernbach E, Urbich C, Brandes RP, Hofmann WK, Zeiher AM, Dimmeler S (2004). Anti-oxidative stress-associated genes in circulating progenitor cells: evidence for enhanced resistance against oxidative stress. *Blood* **104**: 3591–3597.

159. Smith J, Ladi E, Mayer-Proschel M, Noble M (2000). Redox state is a central modulator of the balance between self-renewal and differentiation in a dividing glial precursor cell. *Proc Natl Acad Sci USA* **97**: 10032–10037.

160. Noble M, Smith J, Power J, Mayer-Proschel M (2003). Redox state as a central modulator of precursor cell function. *Ann NY Acad Sci* **991**: 251–271.

161. Imanishi T, Hano T, Nishio I (2004). Angiotensin II potentiates vascular endothelial growth factor-induced proliferation and network formation of endothelial progenitor cells. *Hypertens Res* **27**: 101–108.

162. Imanishi T, Hano T, Nishio I (2005). Angiotensin II accelerates endothelial progenitor cell senescence through induction of oxidative stress. *J Hypertens* **23**: 97–104.

163. Sauer H, Bekhite MM, Hescheler J, Wartenberg M (2005). Redox control of angiogenic factors and CD31-positive vessel-like structures in mouse embryonic stem cells after direct current electrical field stimulation. *Exp Cell Res* **304**: 380–390.

164. Jeffcoate WJ, Price P, Harding KG (2004). Wound healing and treatments for people with diabetic foot ulcers. *Diabetes Metab Res Rev* **20**(Suppl 1): S78–S89.

165. Jeffcoate WJ, Harding KG (2003). Diabetic foot ulcers. *Lancet* **361**: 1545–1551.

166. Krankel N, Adams V, Linke A, Gielen S, Erbs S, Lenk K, Schuler G, Hambrecht R (2005). Hyperglycemia reduces survival and impairs function of circulating blood-derived progenitor cells. *Arterioscler Thromb Vasc Biol* **25**: 698–703.

167. Ushio-Fukai M, Alexander RW (2004). Reactive oxygen species as mediators of angiogenesis signaling: role of NAD(P)H oxidase. *Mol Cell Biochem* **264**: 85–97.

168. Kim YK, Lee MS, Son SM, Kim IJ, Lee WS, Rhim BY, Hong KW, Kim CD (2002). Vascular NADH oxidase is involved in impaired endothelium-dependent vasodilation in OLETF rats, a model of type 2 diabetes. *Diabetes* **51**: 522–527.

169. Guzik TJ, Mussa S, Gastaldi D, Sadowski J, Ratnatunga C, Pillai R, Channon KM (2002). Mechanisms of increased vascular superoxide production in human diabetes mellitus: role of NAD(P)H oxidase and endothelial nitric oxide synthase. *Circulation* **105**: 1656–1662.

170. Cosentino F, Hishikawa K, Katusic ZS, Luscher TF (1997). High glucose increases nitric oxide synthase expression and superoxide anion generation in human aortic endothelial cells. *Circulation* **96**: 25–28.

171. Cosentino F, Eto M, De Paolis P, van der Loo B, Bachschmid M, Ullrich V, Kouroedov A, Delli Gatti C, Joch H, Volpe M, *et al.* (2003). High glucose causes upregulation of cyclooxygenase-2 and alters prostanoid profile in human endothelial cells: role of protein kinase C and reactive oxygen species. *Circulation* **107**: 1017–1023.

172. Galeano M, Torre V, Deodato B, Campo GM, Colonna M, Sturiale A, Squadrito F, Cavallari V, Cucinotta D, Buemi M, *et al.* (2001). Raxofelast, a hydrophilic vitamin E-like antioxidant, stimulates wound healing in genetically diabetic mice. *Surgery* **129**: 467–477.

173. Inoguchi T, Li P, Umeda F, Yu HY, Kakimoto M, Imamura M, Aoki T, Etoh T, Hashimoto T, Naruse M, *et al.* (2000). High glucose level and free fatty acid stimulate reactive oxygen species production through protein kinase C-dependent activation of NAD(P)H oxidase in cultured vascular cells. *Diabetes* **49**: 1939–1945.

174. Luo JD, Wang YY, Fu WL, Wu J, Chen AF (2004). Gene therapy of endothelial nitric oxide synthase and manganese superoxide dismutase restores delayed wound healing in type 1 diabetic mice. *Circulation* **110**: 2484–2493.

175. Cai H, Griendling KK, Harrison DG (2003). The vascular NAD(P)H oxidases as therapeutic targets in cardiovascular diseases. *Trends Pharmacol Sci* **24**: 471–478.

176. Cross MJ, Claesson-Welsh L (2001). FGF and VEGF function in angiogenesis: signalling pathways, biological responses and therapeutic inhibition. *Trends Pharmacol Sci* **22**: 201–207.

177. Matsumoto T, Claesson-Welsh L (2001). VEGF receptor signal transduction. *Sci STKE* **2001**: RE21.

178. Dougher-Vermazen M, Hulmes JD, Bohlen P, Terman BI (1994). Biological activity and phosphorylation sites of the bacterially expressed cytosolic domain of the KDR VEGF-receptor. *Biochem Biophys Res Commun* **205**: 728–738.

179. Morales-Ruiz M, Fulton D, Sowa G, Languino LR, Fujio Y, Walsh K, Sessa WC (2000). Vascular endothelial growth factor-stimulated actin reorganization and migration of endothelial cells is regulated via the Serine/Threonine kinase Akt. *Circ Res* **86**: 892–896.

180. Rousseau S, Houle F, Kotanides H, Witte L, Waltenberger J, Landry J, Huot J (2000). Vascular endothelial growth factor (VEGF)-driven actin-based motility is mediated by VEGFR2 and requires concerted activation of stress-activated protein kinase 2 (SAPK2/p38) and geldanamycin-sensitive phosphorylation of focal adhesion kinase. *J Biol Chem* **275**: 10661–10672.

181. Rousseau S, Houle F, Huot J (2000). Integrating the VEGF signals leading to actin-based motility in vascular endothelial cells. *Trends Cardiovasc Med* **10**: 321–327.

182. Dimmeler S, Dernbach E, Zeiher AM (2000). Phosphorylation of the endothelial nitric oxide synthase at ser-1177 is required for VEGF-induced endothelial cell migration. *FEBS Lett* **477**: 258–262.

183. Tang FY, Nguyen N, Meydani M (2003). Green tea catechins inhibit VEGF-induced angiogenesis *in vitro* through suppression of VE-cadherin phosphorylation and inactivation of Akt molecule. *Int J Cancer* **106**: 871–878.

184. Rao GN (1996). Hydrogen peroxide induces complex formation of SHC-Grb2-SOS with receptor tyrosine kinase and activates ras and extracellular signal-regulated protein kinases group of mitogen activated protein kinases. *Oncogene* **13**: 713–719.

185. Bae YS, Kang SW, Seo MS, Baines IC, Takle E, Chock PB, Rhee SG (1997). Epidermal growth factor (EGF)-induced generation of hydrogen peroxide. Role in EGF receptor-mediated tyrosine phosphorylation. *J Biol Chem* **272**: 217–221.

186. Sundaresan M, Zu-Xi Y, Ferrans VJ, Irani K, Finkel T (1995). Requirement for generation of H_2O_2 for platelet-derived growth factor signal transduction. *Science* **270**: 296–299.

187. Mahadev K, Wu X, Zilbering A, Zhu L, Lawrence JT, Goldstein BJ (2001). Hydrogen peroxide generated during cellular insulin stimulation is integral to activation of the distal insulin signaling cascade in 3T3-L1 adipocytes. *J Biol Chem* **276**: 48662–48669.

188. Eguchi S, Numaguchi K, Iwasaki H, Matsumoto T, Yamakawa T, Utsunomiya H, Motley ED, Kawakatsu H, Owada KM, Hirata Y, *et al.* (1998). Calcium-dependent epidermal growth factor receptor transactivation mediates the angiotensin II-induced mitogen-activated protein kinase activation in vascular smooth muscle cells. *J Biol Chem* **273**: 8890–8896.

189. Fujiyama S, Matsubara H, Nozawa Y, Maruyama K, Mori Y, Tsutsumi Y, Masaki H, Uchiyama Y, Koyama Y, Nose A, *et al.* (2001). Angiotensin AT(1) and AT(2) receptors differentially regulate angiopoietin-2 and vascular endothelial growth factor expression and angiogenesis by modulating heparin binding-epidermal growth factor (EGF)-mediated EGF receptor transactivation. *Circ Res* **88**: 22–29.

190. Rhee SG, Bae YS, Lee SR, Kwon J (2000). Hydrogen peroxide: a key messenger that modulates protein phosphorylation through cysteine oxidation. *Sci STKE* **2000**: PE1.

191. Ostman A, Bohmer FD (2001). Regulation of receptor tyrosine kinase signaling by protein tyrosine phosphatases. *Trends Cell Biol* **11**: 258–266.

192. Chiarugi P, Cirri P (2003). Redox regulation of protein tyrosine phosphatases during receptor tyrosine kinase signal transduction. *Trends Biochem Sci* **28**: 509–514.

193. Denu JM, Dixon JE (1998). Protein tyrosine phosphatases: mechanisms of catalysis and regulation. *Curr Opin Chem Biol* **2**: 633–641.

194. Finkel T (1999). Signal transduction by reactive oxygen species in non-phagocytic cells. *J Leukoc Biol* **65**: 337–340.

195. Lee SR, Kwon KS, Kim SR, Rhee SG (1998). Reversible inactivation of protein-tyrosine phosphatase 1B in A431 cells stimulated with epidermal growth factor. *J Biol Chem* **273**: 15366–15372.

196. Rider DA, Sinclair AJ, Young SP (2003). Oxidative inactivation of CD45 protein tyrosine phosphatase may contribute to T lymphocyte dysfunction in the elderly. *Mech Ageing Dev* **124**: 191–198.

197. Blanchetot C, Tertoolen LG, den Hertog J (2002). Regulation of receptor protein-tyrosine phosphatase alpha by oxidative stress. *EMBO J* **21**: 493–503.

198. Connor KM, Subbaram S, Regan KJ, Nelson KK, Mazurkiewicz JE, Bartholomew PJ, Aplin AE, Tai YT, Aguirre-Ghiso J, Flores SC, *et al.* (2005). Mitochondrial H2O2 regulates the angiogenic phenotype via PTEN oxidation. *J Biol Chem* **280**: 16916–16924.

199. Mahadev K, Zilbering A, Zhu L, Goldstein BJ (2001). Insulin-stimulated hydrogen peroxide reversibly inhibits protein-tyrosine phosphatase 1b *in vivo* and enhances the early insulin action cascade. *J Biol Chem* **276**: 21938–21942.

200. Chiarugi P, Fiaschi T, Taddei ML, Talini D, Giannoni E, Raugei G, Ramponi G (2001). Two vicinal cysteines confer a peculiar redox regulation to low molecular weight protein tyrosine phosphatase in response to platelet-derived growth factor receptor stimulation. *J Biol Chem* **276**: 33478–33487.

201. Meng TC, Fukada T, Tonks NK (2002). Reversible oxidation and inactivation of protein tyrosine phosphatases *in vivo* . *Mol Cell* **9**: 387–399.

202. Kroll J, Waltenberger J (1997). The vascular endothelial growth factor receptor KDR activates multiple signal transduction pathways in porcine aortic endothelial cells. *J Biol Chem* **272**: 32521–32527.

203. Guo DQ, Wu LW, Dunbar JD, Ozes ON, Mayo LD, Kessler KM, Gustin JA, Baerwald MR, Jaffe EA, Warren RS, *et al.* (2000). Tumor necrosis factor employs a protein-tyrosine phosphatase to inhibit activation of KDR and vascular endothelial cell growth factor-induced endothelial cell proliferation. *J Biol Chem* **275**: 11216–11221.

204. Huang L, Sankar S, Lin C, Kontos CD, Schroff AD, Cha EH, Feng SM, Li SF, Yu Z, Van Etten RL, *et al.* (1999). HCPTPA, a protein tyrosine phosphatase that regulates vascular endothelial growth factor receptor-mediated signal transduction and biological activity. *J Biol Chem* **274**: 38183–38188.

205. Grazia Lampugnani M, Zanetti A, Corada M, Takahashi T, Balconi G, Breviario F, Orsenigo F, Cattelino A, Kemler R, Daniel TO, *et al.* (2003). Contact inhibition of VEGF-induced proliferation requires vascular endothelial cadherin, beta-catenin, and the phosphatase DEP-1/CD148. *J Cell Biol* **161**: 793–804.

206. Soeda S, Shimada T, Koyanagi S, Yokomatsu T, Murano T, Shibuya S, Shimeno H (2002). An attempt to promote neo-vascularization by employing a newly synthesized inhibitor of protein tyrosine phosphatase. *FEBS Lett* **524**: 54–58.

207. Huang L, Turck CW, Rao P, Peters KG (1995). GRB2 and SH-PTP2: potentially important endothelial signaling molecules downstream of the TEK/TIE2 receptor tyrosine kinase. *Oncogene* **11**: 2097–2103.

208. Kontos CD, Stauffer TP, Yang WP, York JD, Huang L, Blanar MA, Meyer T, Peters KG (1998). Tyrosine 1101 of Tie2 is the major site of association of p85 and is required for activation of phosphatidylinositol 3-kinase and Akt. *Mol Cell Biol* **18**: 4131–4140.

209. Jones N, Chen SH, Sturk C, Master Z, Tran J, Kerbel RS, Dumont DJ (2003). A unique autophosphorylation site on Tie2/Tek mediates Dok-R phosphotyrosine binding domain binding and function. *Mol Cell Biol* **23**: 2658–2668.

210. Murohara T, Asahara T (2002). Nitric oxide and angiogenesis in cardiovascular disease. *Antioxid Redox Signal* **4**: 825–831.

211. Bayraktutan U (2004). Nitric oxide synthase and NAD(P)H oxidase modulate coronary endothelial cell growth. *J Mol Cell Cardiol* **36**: 277–286.

212. Polytarchou C, Papadimitriou E (2005). Antioxidants inhibit human endothelial cell functions through down-regulation of endothelial nitric oxide synthase activity. *Eur J Pharmacol* **510**: 31–38.

213. Marumo T, Noll T, Schini-Kerth VB, Harley EA, Duhault J, Piper HM, Busse R (1999). Significance of nitric oxide and peroxynitrite in permeability changes of the retinal microvascular endothelial cell monolayer induced by vascular endothelial growth factor. *J Vasc Res* **36**: 510–515.

214. Lauer N, Suvorava T, Ruther U, Jacob R, Meyer W, Harrison DG, Kojda G (2005). Critical involvement of hydrogen peroxide in exercise-induced up-regulation of endothelial NO synthase. *Cardiovasc Res* **65**: 254–262.

215. Schreck R, Rieber P, Baeuerle PA (1991). Reactive oxygen intermediates as apparently widely used messengers in the activation of the NF-kappa B transcription factor and HIV-1. *EMBO J* **10**: 2247–2258.

216. Okuno H, Akahori A, Sato H, Xanthoudakis S, Curran T, Iba H (1993). Escape from redox regulation enhances the transforming activity of Fos. *Oncogene* **8**: 695–701.

217. Wang DL, Wung BS, Shyy Y-J, Lin C-F, Chao Y-J, Usami S, Chien S (1995). Mechanical strain induces monocyte chemotactic protein-1 gene expression in endothelial cells. *Circ Res* **77**: 294–302.

218. Rainwater R, Parks D, erson ME, Tegtmeyer P, Mann K (1995). Role of cysteine residues in regulation of p53 function. *Mol Cell Biol* **15**: 3892–3903.

219. Lander HM, Ogiste JS, Teng KK, Novogrodsky A (1995). p21ras as a common signaling target of reactive free radicals and cellular redox stress. *J Biol Chem* **270**: 21195–21198.

220. Tummala PE, Chen XL, Sundell CL, Laursen JB, Hammes CP, Alexander RW, Harrison DG, Medford RM (1999). Angiotensin II induces vascular cell adhesion molecule-1 expression in rat vasculature: a potential link between the renin-angiotensin system and atherosclerosis. *Circulation* **100**: 1223–1229.

221. Chen XL, Tummala PE, Olbrych MT, Alexander RW, Medford RM (1998). Angiotensin II induces monocyte chemoattractant protein-1 gene expression in rat vascular smooth muscle cells. *Circ Res* **83**: 952–959.

222. Schieffer B, Luchtefeld M, Braun S, Hilfiker A, Hilfiker-Kleiner D, Drexler H (2000). Role of NAD(P)H oxidase in angiotensin II-induced JAK/STAT signaling and cytokine induction. *Circ Res* **87**: 1195–1201.

223. Wang Z, Castresana MR, Newman WH (2001). Reactive oxygen and NF-kappaB in VEGF-induced migration of human vascular smooth muscle cells. *Biochem Biophys Res Commun* **285**: 669–674.

224. Richard DE, Berra E, Pouyssegur J (2000). Nonhypoxic pathway mediates the induction of hypoxia-inducible factor 1alpha in vascular smooth muscle cells. *J Biol Chem* **275**: 26765–26771.

225. Gorlach A, Diebold I, Schini-Kerth VB, Berchner-Pfannschmidt U, Roth U, Brandes RP, Kietzmann T, Busse R (2001). Thrombin activates the hypoxia-inducible factor-1 signaling pathway in vascular smooth muscle cells: Role of the p22(phox)-containing NADPH oxidase. *Circ Res* **89**: 47–54.

226. Haddad JJ, Land SC (2001). A non-hypoxic, ROS-sensitive pathway mediates TNF-alpha-dependent regulation of HIF-1alpha. *FEBS Lett* **505**: 269–274.

227. Yang ZZ, Zhang AY, Yi FX, Li PL, Zou AP (2003). Redox regulation of HIF-1alpha levels and HO-1 expression in renal medullary interstitial cells. *Am J Physiol Renal Physiol* **284**: F1207–F1215.

228. Haddad JJ (2003). Science review: redox and oxygen-sensitive transcription factors in the regulation of oxidant-mediated lung injury: role for hypoxia-inducible factor-1alpha. *Crit Care* **7**: 47–54.

229. Gorlach A, Berchner-Pfannschmidt U, Wotzlaw C, Cool RH, Fandrey J, Acker H, Jungermann K, Kietzmann T (2003). Reactive oxygen species modulate HIF-1 mediated PAI-1 expression: involvement of the GTPase Rac1. *Thromb Haemost* **89**: 926–935.

230. BelAiba RS, Djordjevic T, Bonello S, Flugel D, Hess J, Kietzmann T, Gorlach A (2004). Redox-sensitive regulation of the HIF pathway under non-hypoxic conditions in pulmonary artery smooth muscle cells. *Biol Chem* **385**: 249–257.

231. Turcotte S, Desrosiers RR, Beliveau R (2003). HIF-1alpha mRNA and protein upregulation involves Rho GTPase expression during hypoxia in renal cell carcinoma. *J Cell Sci* **116**: 2247–2260.

232. Welsh SJ, Bellamy WT, Briehl MM, Powis G (2002). The redox protein thioredoxin-1 (Trx-1) increases hypoxia-inducible factor 1alpha protein expression: Trx-1 overexpression results in increased vascular endothelial growth factor production and enhanced tumor angiogenesis. *Cancer Res* **62**: 5089–5095.

233. Nespereira B, Perez-Ilzarbe M, Fernandez P, Fuentes AM, Paramo JA, Rodriguez JA (2003). Vitamins C and E downregulate vascular VEGF and VEGFR-2 expression in apolipoprotein-E-deficient mice. *Atherosclerosis* **171**: 67–73.

234. Ivan E, Khatri JJ, Johnson C, Magid R, Godin D, Nandi S, Lessner S, Galis ZS (2002). Expansive arterial remodeling is associated with increased neointimal macrophage foam cell content: the murine model of macrophage-rich carotid artery lesions. *Circulation* **105**: 2686–2691.

235. Grote K, Flach I, Luchtefeld M, Akin E, Holland SM, Drexler H, Schieffer B (2003). Mechanical stretch enhances mRNA expression and proenzyme release of matrix metalloproteinase-2 (MMP-2) via NAD(P)H oxidase-derived reactive oxygen species. *Circ Res* **92**: e80–e86.

236. Inoue N, Takeshita S, Gao D, Ishida T, Kawashima S, Akita H, Tawa R, Sakurai H, Yokoyama M (2001). Lysophosphatidylcholine increases the secretion of matrix metalloproteinase 2 through the activation of NADH/NADPH oxidase in cultured aortic endothelial cells. *Atherosclerosis* **155**: 45–52.

237. Wary KK, Thakker GD, Humtsoe JO, Yang J (2003). Analysis of VEGF-responsive genes involved in the activation of endothelial cells. *Mol Cancer* **2**: 25.

238. Oka H, Kugiyama K, Doi H, Matsumura T, Shibata H, Miles LA, Sugiyama S, Yasue H (2000). Lysophosphatidylcholine induces urokinase-type plasminogen activator and its receptor in human macrophages partly through redox-sensitive pathway. *Arterioscler Thromb Vasc Biol* **20**: 244–250.

9

Angiogenesis and Arteriogenesis in Cardiac Hypertrophy

*by Robert J. Tomanek
and Eduard I. Dedkov*

1. Introduction

A continuous O_2 supply is necessary for the myocardium since its anaerobic capacity is limited. Coronary flow and O_2 utilization are linearly coupled and blood flow may increase four-to-five fold when myocardial work is extreme.[1] Accordingly, the myocardium has a rich supply of microvessels, i.e. capillaries and arterioles. When myocytes enlarge in response to increased work, vascular density will decrease unless an appropriate stimulus for angiogenesis is triggered. A limitation or lack of capillary growth will increase diffusion distance for oxygen, while inadequate arteriolar growth will limit maximal myocardial perfusion. Since maximal flow will be decreased due to the larger heart mass, coronary reserve (the difference between maximal and resting flows) will be compromised. Thus, for cardiac hypertrophy to be effective as a compensator for increased work, adequate angiogenesis and arteriogenesis must also occur.

In some models of cardiac hypertrophy (e.g. thyroxine-induced) angiogenesis and/or arteriogenesis are well documented. In other models, vascular growth may be limited or virtually non-existent. Thus, some stimuli that evoke cardiac hypertrophy are associated with factors that promote vascular growth. Accordingly, this review examines various models of cardiac hypertrophy and possible mechanisms that underlie angiogenesis and/or arteriogenesis.

2. Assessing Coronary Angiogenesis and Arteriogenesis

Since most of vascular resistance resides in arterioles, growth of this component (arteriogenesis) will facilitate a better maximal myocardial perfusion. Growth of these vessels may occur via formation of new arterioles or by remodeling of existing arterioles to increase their diameters (Fig. 1). For optimal O_2 diffusion, angiogenesis, i.e. sprouting or splitting (intussusception) of capillaries, needs to occur in order to attenuate or prevent increases in capillary domains, i.e. the tissue served by one capillary. Thus, when ventricular mass increases, regardless of the stimulus, normalization of (1) maximal myocardial perfusion and (2) capillary domains are necessary for adequate O_2 delivery, especially during periods of high metabolic demand.

Maximal myocardial perfusion evaluated during pharmacologically-induced maximal vasodilation provides an estimate of the extent of the vascular bed. Radioactive or fluorescent microspheres are injected to estimate perfusion and values are adjusted for perfusion pressure and expressed as "conductance." Another way of appraising the growth is to express the perfusion data as "minimal coronary vascular resistance" (pressure/flow). Morphometric approaches enable the extent of specific components of the coronary vasculature to be quantified. Numerical density (number of vessels/mm^2 tissue) is a commonly used estimate. Length density (aggregate length of vessels in a volume of tissue) is a more accurate gauge of vascularity and is not affected by orientation of plane of sectioning. The same is true of volume density.

3. Pressure Overload-Induced Hypertrophy

Hypertension and aortic or pulmonary artery coarctation are the most common causes of cardiac hypertrophy encountered in the clinical

Compensatory Angiogenesis and Arteriogenesis

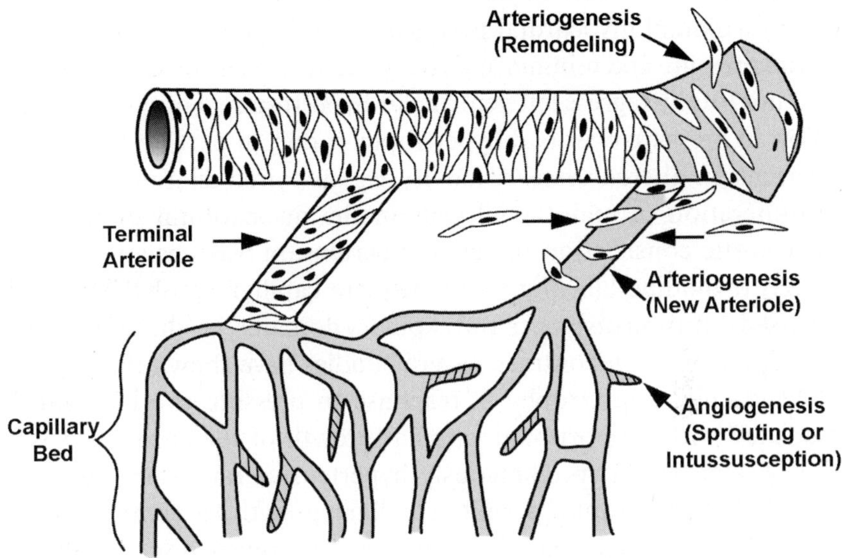

Fig. 1. Vascular growth in response to cardiac hypertrophy may include: (1) angiogenesis, i.e. growth of capillaries by sprouting or intussusception (partitioning to bifurcate a capillary), and (2) arteriogenesis, i.e. creation of a new arteriole via recruitment of smooth muscle cells to an endothelial tube, or remodeling of an existing arteriole or artery to increase its diameter.

setting. In humans, long-term pressure overload-induced hypertrophy is associated with a decrease in coronary reserve, which has generally been attributed to an absence or inadequate growth of the coronary vasculature. Many experimental studies on animal models with various forms of pressure overload have also concluded that angiogenesis, if it occurs, does not compensate for the increase in ventricular mass (reviewed in Refs. 2 and 3). A decline in coronary reserve and/or maximal myocardial perfusion has also been documented in a variety of animal species, e.g. rat, cat, pig and dog.[4–10] However, hypertension, as well as cardiac hypertrophy, contribute to the decline in coronary reserve.[11] Work from our lab showed that cardiac hypertrophy and arterial pressure could be dissociated.[12,13] Normalization of blood pressure with hydralazine in spontaneously hypertensive rats markedly reduced minimal coronary vascular resistance and normalized coronary reserve, despite the persistence of cardiac hypertrophy.[13] Subsequently, we explored the effect of reversing established cardiac hypertrophy and elevated blood pressure on maximal myocardial perfusion.[14] ACE inhibition normalized arterial pressure and minimal coronary vascular resistance, even though it did not totally regress left ventricular hypertrophy. Thus, impairment of flow in pressure overload hypertrophy is due, in part, to the chronic hypertension-altered vessel reactivity, e.g. impaired dilation.[11]

Proliferation of endothelial cells in the myocardium of rats with either aortic constriction or renal hypertension was absent as shown by [^{3}H] thymidine labeling.[15]. Inadequate capillary growth results in increased diffusion distances and capillary domains, i.e. the tissue served by a capillary.[2,16] In contrast, several studies have shown that right or left ventricular hypertrophy in response to pressure overload can be associated with proportional vascular growth of the coronary vasculature (Table 1). In spontaneously hypertensive rats, stabilization of myocardial hypertrophy permits capillary growth to compensate for the additional cardiac mass.[17,18] Moreover, coronary reserve and minimal coronary vascular resistance normalize over time.[5,19] The angiogenesis associated with stabilized hypertrophy correlates with an elevation in VEGF mRNA which has been shown to occur at 28 and 32 weeks of age in SHR.[20] Dogs with renal hypertension (one kidney, one clip) of six weeks duration were found to have moderate (27%) left ventricular hypertrophy and a 67% increase in minimal coronary vascular resistance.[10] However, when the renal hypertension was prolonged to seven months, dogs exhibited normal LV-MCVR, arteriolar numeral densities, despite the fact that LV weight was higher than controls.[21]

Right ventricular hypertrophy evoked by pulmonary artery banding in dogs,[22] and swine[23,24] was associated with normal or decreased MCVR. Arteriogenesis, as indicated by arteriolar densities that were similar to the controls, occurred in a model of progressive pulmonary artery constriction that caused a 91% increase in RV weight.[24] Capillary density, however, was lower in the pressure overload group, a finding that supports the concept that growth of capillaries and arterioles is not necessarily parallel. In sum, there are many reports in the literature that

Table 1. Angiogenesis and arteriogenesis in pressure overload cardiac hypertrophy.

Ref. no.	Type of overload and magnitude of hypertrophy	Species	Findings
18	SHR LV weight ↑ 24%–27%	Rat	Capillary proliferation between 21 and 45 days: normal N_A, capillary/myocyte ratio increased
22	P.A. band RVW ↑ 2.5×	Young dogs	Lower MCVR
23	P.A. band RVW/BW ↑ 112%	Young swine	Normal MCVR and arteriolar density
24	Progressive P.A. constriction	Adult mini pigs	Normal MCVR and arteriolar density
21	Renal hypertension (1K, 1C) LVW/BW ↑ 46%	Dogs	Normal MCVR Normal arteriolar density
19	SHR LVW/BW ↑ 15%, 25%, 29%	Rat	Peak flow velocity, repayment/delete ratio ↓ at 3 and 7 months, normal at 12 months
17	Spontaneous hypertension	Rat	Capillary density ↓ at peak hypertrophy (7 months) then normalized at 12 months
5	Spontaneous hypertension	Rat	Coronary reserve normalized when hypertrophy stabilized

LV = Left ventricle; RV = right ventricle; P.A. = pulmonary artery; MCVR = minimal coronary vascular resistance; 1K = one kidney; 1C = one clip; SHR = spontaneously hypertensive rat.

have documented a lack of compensatory vascular growth in cardiac hypertrophy associated with pressure overload. Nevertheless, there is some documentation that coronary angiogenesis and arteriogenesis can occur in some of these models as shown by data from several species. Although the reasons for the discrepancy between these studies are not evident, the stabilization of hypertrophy, its duration, and the specificity of the type of pressure overload are likely factors that facilitate angiogenesis and arteriogenesis.

4. Volume Overload-Induced Cardiac Hypertrophy

Volume overload-induced cardiac hypertrophy is characterized by increased diastolic stress and cardiomyocyte hypertrophy that is largely due to longitudinal growth.[25,26] Compensated hypertrophy in this model is usually devoid of ventricular dysfunction, especially when the volume overload is due to an aortocaval fistula.[27] Most experimental studies have shown that vascular growth in this model is proportional to the myocardial hypertrophy (Table 2). When a decrease in capillary length or numerical density was reported, it was limited to the subepicardium or was not marked. Anemia, which also induces a volume overload, was found to evoke a marked increase in capillary volume density.[28] Most importantly, we documented normal arteriolar length density, despite substantial hypertrophy in rats with five months of volume overload, indicating that vascular growth is proportional to the increase in myocardial mass.[26]

Vascular growth, especially that of resistance vessels, appears to be the anatomical basis for the normal maximal myocardial perfusion in volume overloaded hearts noted in several studies. Myocardial blood flow per gram and endocardial/epicardial flow ratio during exercise was found to be normal in dogs with A-V shunt.[29] Other studies reported that minimal coronary vascular resistance (during adenosine infusion) was also similar in dogs with A-V shunts and their controls.[30,31] However, redistribution of myocardial blood flow, that is a decrease in endocardial/epicardial flow ratio, has been noted in dogs during maximal vasodilation and during exercise.[31,32] The lower flow to the endocardial portion of the myocardium during maximal perfusion may be,

Table 2. Angiogenesis and arteriogenesis in volume overload cardiac hypertrophy.

Reference no.	Model and magnitude of hypertrophy	Species	Vascular growth
26	A-V shunt 61% $\uparrow$ RV weight 55% $\uparrow$ LV weight	Rat	Arteriolar L_V normal Capillary L_V $\downarrow$ only in subendo Capillary diameter $\uparrow$ only in subendo
16	A-V shunt 70% $\uparrow$ L_V weight	Rat	Capillary L_A normal
60	A-V shunt 39% $\uparrow$ in heart weight/daily weight	Rat	Capillary N_A $\downarrow$ only in subendocardium
31	A-V shunt LVW/BW $\uparrow$ 27%	Dog	Capillary N_A $\downarrow$ only in subendocardium
127	A-V shunt	Dog	Capillary N_A normal
28	Anemia RVW $\uparrow$ 65% LVW $\uparrow$ 47%	Rat	Capillary V_V $\uparrow$ 65% L_V, 34% RV
	Interatrial septal defect 28% $\uparrow$ RVW/BW	Cat	Capillary N_A normal Capillary diameter $\uparrow$

L_V = Length density; N_A = numerical density; RVW, LVW = right and left ventricular weights; BW = body weight; A-V = atrioventricular.

in part, due to the decrease in capillarity in this region previously noted. However, Badke *et al.* reported that the lower endo/epi flow ratio in dogs with aortocaval fistulas was normalized when the fistula was closed.[32] This finding suggested that the perfusion abnormality was due to hemodynamic alterations and not to hypertrophy *per se*. In dogs with experimentally-induced chronic tricuspid regurgitation, myocardial blood flow during adenosine infusion was found to be slightly lower than in controls.[33]

Data from humans are less definitive. Coronary reactive hyperemia was used to assess coronary reserve in patients with right ventricular hypertrophy secondary to atrial septal defects who suffered a 50% decrease in peak-to-resting blood flow ratio.[34] In children with ventral septal defect, coronary flow velocity reserve was reduced because resting flow velocity was increased.[35] However, average peak flow velocity was not reduced, indicating that vascular growth paralleled the magnitude of hypertrophy. Moreover, Strauer concluded that patients with volume overload due to aortic insufficiency do not necessarily have a reduced coronary reserve.[36]

Taken together the experimental data on volume overload-induced cardiac hypertrophy supports the conclusion that this model of hypertrophy provides a stimulus for angiogenesis and arteriogenesis. A major difference between volume and pressure overload is that the former experiences a greater stretch during diastole which may serve as the mechanical trigger for vascular growth (see later discussion).

5. Thyroxine-Induced Hypertrophy

Administration of thyroid hormone leads to rapid cardiac hypertrophy and enhanced ventricular systolic function[37–46] and substantial angiogenesis.[38,39,41–46] Studies in our lab and those by others have documented both capillary and arteriolar growth in this model of cardiac enlargement.[38,39,41–44,47,48] The finding that arteriolar length density in pigs with thyroid hormone-induced hypertrophy (LV weight/body weight increased 47%) is normal indicates a substantial arteriogenesis.[43] Minimal coronary resistance during adenosine administration was found to be lower in the thyroxine-treated group. Thus, maximal flow was higher than in controls, a finding in concert with data on rats treated with thyroxine that have a higher maximal myocardial perfusion than controls.[41] We subsequently showed that age does not preclude the growth of resistance vessels by providing data that maximal flow in thyroxine-induced hypertrophy is similar in senescent rats and their non-hypertrophic age group controls.[45] Taken together, these studies indicate that thyroxine-stimulated cardiac hypertrophy is accompanied by a similar or greater magnitude of angiogenesis and arteriogenesis that facilitate a normal or higher maximal coronary perfusion.

The mechanism underlying vascular growth in this model is not a reaction to cardiac hypertrophy since capillary angiogenesis precedes ventricular enlargement.[39] That capillary growth parallels or exceeds the magnitude of the hypertrophy is supported by many studies with a wide range of cardiac enlargement (17%–82%).[38,41,42,44–47,49] Several lines of evidence suggest that both cardiac hypertrophy and coronary vascular growth are secondary to hemodynamic and metabolic effects. Heterotopically transplanted, non-loaded hearts do not undergo cardiac hypertrophy despite chronically elevated thyroxine levels in the host.[50] Similarly, neovascularization is similar in unloaded but beating rat hearts grafted *in oculo* in hyperthyroid, euthyroid and hypothyroid adult rats.[51] The stimulus for angiogenesis in this model is most likely mechanical since the enhanced metabolic demands in hyperthyroidism evoke greater myocardial perfusion, which is known to cause DNA synthesis in endothelial cells.[52] Increased coronary flow is associated with myocardial capillary growth as demonstrated by chronic administration of adenosine or HWA-285, a xanthine derivative, to rabbits, or dipyridamole to rats or rabbits.[44,53–55] Another mechanical stimulus candidate is the increased left ventricular end diastolic volume, which subjects the ventricle to an enhanced stretch.[56] Thus, the hyperthyroid state creates a volume overload, similar to that which occurs in aorto-caval shunt or valve regurgitation. The angiogenic response in thyroid hormone-induced hypertrophy is consistent with that of surgically-induced volume overload, as detailed in the previous section.

6. Hypoxia-Induced Hypertrophy

Many studies in rats have documented capillary neogenesis in the right ventricle (RV) enlarged by exposure to hypobaria from the time of birth. Turek *et al.* found that capillary numerical density remained stable as RV muscle fiber diameters increased by 46%, a finding that indicates a growth of capillaries proportional to the increase in myocardial mass.[57] Even when RV mass increases dramatically, e.g. more than twofold, capillary density does not decline.[58] The hypertrophic response in adult rats to hypobaric hypoxia was less, i.e. 96%, but capillary growth was also proportional. One study documented an increase in capillary density despite a 2.6-fold increase in RV weight.[59] Therefore, capillary growth may exceed even a massive increase in RV mass stimulated by chronic hypoxia.

A comparison of guinea pigs either born at high altitude or subjected to high altitude after birth also showed normal capillary densities and diffusion distances.[60] The hypothesis that hypoxia can stimulate capillary growth in the heart hypertrophied by pressure overload was tested in rats.[61] When spontaneously hypertensive or aortic constricted rats were exposed to six weeks of hypobaric hypoxia, both groups demonstrated an increase in capillary density without a further increase in ventricular mass. Thus, the angiogenic response was not dependent on the development of myocardial hypertrophy.

In sum, chronic hypoxia appears to be a stimulus for myocardial angiogenesis which, in turn, reduces diffusion distances for O_2. This occurs when hypoxia is the stimulus for cardiac hypertrophy (as seen in the right ventricle) or when hypoxia is evoked after the development of cardiac hypertrophy by pressure overload (as seen in the left ventricle.) The mechanism underlying right ventricular capillary growth associated with hypoxia may be the persistence of elevated VEGF.[62] Although VEGF mRNA increases five-fold in the left ventricle during exposure to hypoxia, it returns to baseline levels by the end of the first day. In contrast, VEGF mRNA is elevated 2.5–4.5-fold in the right ventricle throughout the 30-day period of hypoxic exposure. It is well established that HIF-1 mediates the transcriptional response to O_2 levels, and that HIF-α is important in VEGF release.[63] VEGF mRNA is also stimulated by adenosine, while adenosine A2 receptor antagonists reduce the VEGF mRNA stimulated by hypoxia.[64] Thus, the increase in blood flow that occurs during hypoxia is a factor in the angiogenic response. The angiopoietins may also contribute to vascular growth during hypoxia since Tie-2 expression is enhanced in microvascular endothelial cells exposed to low O_2.[65]

7. Exercise-Induced Hypertrophy

Endurance training (aerobic exercise) increases cardiac work and O_2 demand and has been shown by many studies to stimulate coronary angiogenesis (see reviews in Refs. 66 to 68). In the absence of cardiac hypertrophy, angiogenesis increases vascular density and may increase maximal myocardial perfusion/unit heart weight above that of controls. Most studies have shown that when exercise training evokes cardiac hypertrophy that sufficient coronary angiogenesis occurs and is fully compensatory.[8,44,69–78] Capillary growth, the parameter most often addressed by these studies, paralleled the increase in cardiac mass in the left or right ventricle was documented in rats,[69,77,79,80] pigeons[81] and pigs.[82]

That capillary growth, as evidenced by an increase in numerical capillary density, may exceed the magnitude of ventricular hypertrophy induced by exercise training has also been reported in several studies.[70,76,80,83] It appears that the magnitude of hypertrophy is not the key determinant of capillary growth since a 65% exercise-induced increase in heart mass had a minimal effect on intercapillary distance; a 27% increase in capillary diameter maintained volume density despite a 22% reduction in numerical density.[84] However, two studies have shown that capillary density in the right ventricle and interventricular septum declines after strenuous exercise training.[18,77] In contrast, training that was more moderate in intensity and of a shorter duration (seven weeks) enhanced capillary density in the right ventricle.[70]

The supposition that exercise-induced cardiac hypertrophy provides a stimulus for growth of pre-capillary vessels sufficient to preserve coronary reserve is supported by data from both myocardial perfusion and morphometric studies. Coronary flow capacity was increased in rats[78] and pigs with exercise-induced cardiac hypertrophy.[43,85] That enhancement of coronary vascular reserve associated with exercise training is independent of cardiac hypertrophy was demonstrated by Buttrick *et al.*[73] An eight week swimming program enhanced vascular reserve in both male and female rats even though only the females developed cardiac hypertrophy. Arteriolar growth was also documented in pigs with cardiac enlargement ranging from <15% to >30%. Taken together, these studies indicate that exercise training of the appropriate

intensity provides stimuli(us) for coronary angiogenesis and arteriogenesis. The most likely stimulus for angiogenesis/arteriogenesis in this
model appears to be increased blood flow and shear stress.

8. Myocardial Infarction-Induced Hypertrophy

Acute myocardial infarction (MI) results in the sudden death of
a great number of cardiac myocytes. To preserve cardiac function,
non-infarcted myocytes of the surviving myocardium, which experienced a chronic functional overload, undergo a reactive compensatory
hypertrophy.[57,86–98] Experimental studies have shown that surviving
cardiac myocytes grow in length, as well as in diameter; the magnitude of this reactive hypertrophy is mainly determined by the size
of infarct.[57,88,90,92–101] The progressive increase in myocyte transverse
areas affects a reduction in capillary density, and consequently, increases
O_2 diffusion distance.[90–92,95,97,98,102–104] However, the decreased capillary density is not caused by loss of capillaries *per se* since capillary to myocyte ratio in the surviving, hypertrophied myocardium is
unchanged. Moreover the fact that some capillary growth occurs is
supported by work documenting absolute increases in the capillary
bed, i.e. increased aggregate capillary length and increases in capillary/myocyte ratio (Table 3). Taken together, these findings demonstrate that the reduction in capillary parameters often detected in
the surviving myocardium of post-infarcted hearts is not a consequence of the absence of capillary growth, but it is rather a failure of sufficient capillary bed expansion to match cardiomyocyte
enlargement.

Although one study reported that maximal myocardial perfusion is normalized three weeks after myocardial infarction,[105] data
from our laboratory and others document marked reductions in
maximal myocardial perfusion as well as in coronary reserve, in
the surviving myocardium of rats between three and eight weeks
after infarction.[93,101,105–109] Since a strong correlation between minimal coronary resistance and myocyte cross-sectional area was
documented,[93] it was assumed that the depression in myocardial perfusion was primarily related to the ongoing hypertrophic process. This

Table 3.   Angiogenesis and arteriogenesis in myocardial infarction-induced hypertrophy.

Reference no(s).	Magnitude of hypertrophy	Species and age	Vascular growth
101	VW and VW/BW unchanged LV myocyte CSA in ↑ subendo and ↑subepi	Rat 12 months	LV arteriolar L_V ↑
96	HW, LVW and LVW/BW unchanged LV myocyte CSA in ↑subendo and ↑subepi	Rat 9 weeks	Capillary N_A in ↑ subendo and ↑ subepi C/M ratio in ↑ subendo and ↑subepi
105	HW and HW/BW unchanged in 3-week MI	Rat 270–320 g	Normal MCVP in 3-week MI
90 and 92	LVW unchanged LV myocyte CSA ↑19%–40% LV myocyte volume/nucleus ↑61%–126% LV myocyte length/nucleus ↑37%–64%	Rat 80 days	*LV aggregate capillary length ↑7%–44%
91	LVW unchanged, LVW/BW ↑8% LV myocyte diameter ↑6% LV myocyte CSA ↑12% LV myocyte volume/nucleus ↑28% LV myocyte length/nucleus ↑14%	Rat 300 g	*LV aggregate capillary length ↑11%

HW = Heart weight; RVW, LVW = right and left ventricular weight; VW = total ventricular weight (RVW + LVW); BW = body weight; MCVP = maximal coronary vascular perfusion; CSA = cross-sectional area; N_A = numerical density; L_V = length density; *LV aggregate capillary length was compared between surviving myocardium of post-MI rats and myocardium of sham-operated rats, which destined to survive after MI.

idea was also confirmed experimentally, since pharmacological prevention (with captopril, enalapril, losartan) of myocardial hypertrophy was able to completely restore a maximal myocardial perfusion.[105,110] However, a recent finding showing that angiotensin II type 1 receptor blockade (valsartan), which reduced cardiac hypertrophy but did not limit cardiac interstitial fibrosis, is not able to improve coronary vasodilator reserve and minimal coronary vascular resistance.[108] suggests that interstitial fibrosis can be an additional determinant regulating myocardial perfusion in post-infarcted hearts. The contribution of fibrosis is also suggested by recent findings from our laboratory which demonstrated that although arteriolar growth in the remaining myocardium of post-MI rats exceeded that detected in sham-operated animals, it was not associated with a proportional improvement in maximal myocardial perfusion and vasodilator reserve.[101] Therefore, fibrosis may also contribute to a decline in maximal myocardial perfusion.

One apparent contributor to the limited post-infarction angiogenesis is the failure of growth factor increases to persist in the surviving myocardium. We found that increases in bFGF, VEGF and Tie-2 were transient, i.e. elevated only during the first few days after infarction.[100] Similarly, we reported that increased levels of VEGF, flt-1, flk-1 in the surviving myocardium returned to control levels by seven days.[111] Others have found increases in VEGF protein or mRNA to be limited to the ischemic border zone.[112–114] Thus, the surviving myocardium distal to the border undergoes compensatory hypertrophy, but limited vascular growth, likely due to the failure of elevated growth factors to persist during the growth period. When rats with infarctions underwent heart rate reduction, we found they had higher VEGF, Flt-1 and bFGF levels than rats with infarction and non-reduced heart rates.[109] These experiments support the idea that stretch resulting from increased diastolic dimensions could stimulate expression of angiogenic ligands and receptors.

In sum, the limited angiogenesis after infarction does not compensate for the reactive cardiac hypertrophy characteristic of the post-infarction period. Failure of sufficient angiogenesis to occur is largely due to limited and transient increases in growth factors and their receptors.

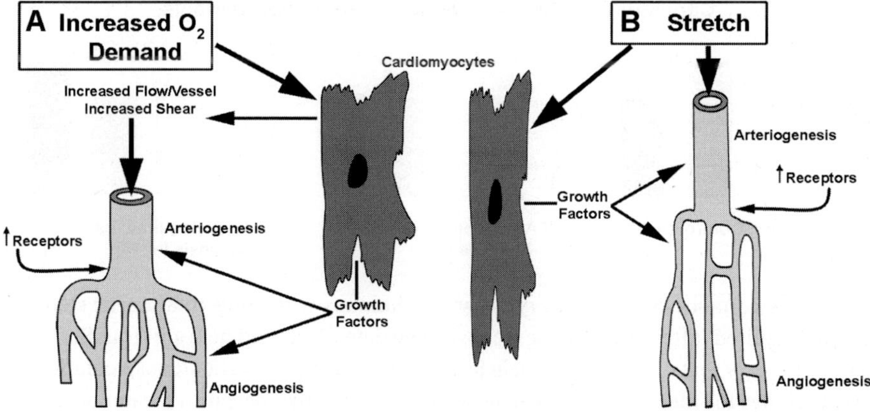

Fig. 2. Primary stimuli that may trigger the cascades that lead to angiogenesis and arteriogenesis in some models of cardiac hypertrophy. The hypertrophic response to work increases oxygen demand in cardiomyocytes (**A**). This relative hypoxia or increased metabolism triggers growth factors that facilitate angiogenesis. An increase in blood flow in response to increased O_2 demand may also trigger growth factors that are important for arteriogenesis and angiogenesis. Stretch (**B**), which is increased in some models of cardiac hypertrophy (volume overload, exercise) is known to upregulate growth factors in the cardiomyocyte and receptors on endothelial cells since both cell types are subjected to stretch. This mechanical influence may also facilitate angiogenesis and arteriogenesis.

9. Modulators of Angiogenesis During Hypertrophy

As indicated in the previous sections, vascular growth during cardiac enlargement is, at least in part, model dependent. There is convincing evidence that some factors that stimulate compensatory vascular growth are operable in certain models. Figure 2 illustrates that both metabolic and mechanical factors are likely primary stimuli that trigger growth factor expression. Another important determinant of the extent of angiogenesis in cardiac hypertrophy is age, as documented in the next few paragraphs.

Embryonic myocardial vascularization is initiated by vasculogenesis, i.e. formation of vascular tubes from progenitor cells. Endothelial cells as well as vascular smooth muscle cells arise from the epicardium (see Ref. 115 for review). The rate and extent of vascularization appear

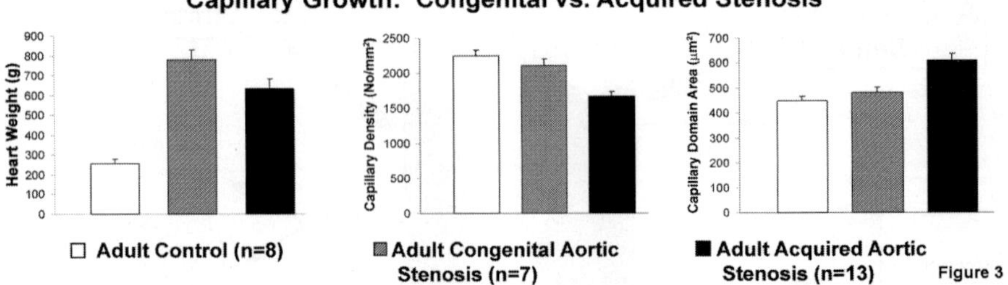

Fig. 3. A comparison of capillary growth in humans with congenital aortic stenosis versus acquired (after birth) aortic stenosis. The data illustrated here are from Ref. 117. The data show that capillary growth in hearts with congenital aortic stenosis completely compensates for the marked cardiac hypertrophy. Note that despite a greater degree of cardiac hypertrophy in the hearts of the congenital aortic stenosis group, capillary density and domain area are normal. In contrast, hypertrophy that develops after birth is not characterized by a capillary growth that matches the magnitude of hypertrophy.

to be dictated, at least in part, by the increase in ventricular mass.[116] The supposition that the developing heart in early life has a greater potential for angiogenesis is supported by data from both humans and animals. Rakusan and colleagues presented evidence that congenital aortic stenosis and coarctation are characterized by capillary growth that is proportional to the increase in heart mass (Fig. 3).[117] This is in contrast to adults with acquired aortic stenosis who experience a marked decrease in capillary density. Data from this study show that capillary density in hearts hypertrophied by congenital pressure over-load have capillary densities virtually identical to controls. In order to maintain normal capillary densities, angiogenesis must be 2.5- to 3.0-fold higher than the controls since heart weight exceeds that of controls by this magnitude during infancy, childhood and adulthood.

Animal studies support the data from humans. The degree of ventricular hypertrophy does not appear to be a determinant of compensatory capillary growth in neonatal rats subjected to a gradual pressure overload.[118] When aortic constriction was imposed on days 2 and 6 of postnatal life, heart weight increased 24% and 55%, respectively, yet capillary densities in both groups were similar to the controls. Similar data were obtained from young lambs subjected to aortic banding. These data indicate that the extent of myocardial angiogenesis during postnatal development is proportional to, rather than limited by, the magnitude of myocardial growth. Several studies have shown that the angiogenic response is stronger in young than in older animals during the development of hypertrophy evoked by various stimuli.[41,42,45,80,119,120] However, the angiogenic response may be species dependent. For example, vascular growth was limited or non-existent in young dogs with pressure overload hypertrophy as indicated by decreased maximal coronary perfusion and increased minimal coronary vascular resistance.[121,122]

The embryonic/fetal heart is also characterized by a robust angiogenic response to the induction of cardiac hypertrophy. When we constricted the outflow track in chicks (*in ovo*) prior to coronary vascularization, ventricular mass increased by 64%.[116] Since vascular volume and numerical densities were normal, the findings of these experiments revealed that the rate and magnitude of coronary vascular growth adapts to the rate and magnitude of myocardial growth.

Attempts to utilize exercise training to stimulate angiogenesis in the hearts of rats hypertrophied by hypertension have not been successful.[15,69,74,123] Part of the failure may have been due to the strenuous nature of the training and a further increase in arterial pressure. However, when exercise training was begun in six-week-old spontaneously hypertensive rats, to correspond to a period of developing hypertension and cardiac hypertrophy, capillary growth paralleled the increase in cardiac mass.[119] Thus, either age or the timing of the stimulus appear to be a determinant of the angiogenic response.

10. Stimuli of Angiogenesis During Hypertrophy

This review has noted that compensatory angiogenesis and arteriogenesis are characteristic of some models of cardiac hypertrophy, i.e. thyroxine, exercise and hypoxia. In volume overload, vascular growth occurs, though not necessarily uniformly, in the ventricular wall. Hypertrophy in response to pressure overload is problematic. In most circumstances vascular growth is non-existent or limited and maximal perfusion is compromised. However, there are exceptions, as noted earlier, when angiogenesis and arteriogenesis are stimulated in this model. As previously noted, both metabolic and mechanical stimuli affect growth factors and their receptors and thereby play key roles in the angiogenic and arteriogenic responses in the heart (Fig. 2).

Hypoxia is known to upregulate VEGF via activation of hypoxia inducible factor (HIF-1α)[124] and also upregulates other growth factor genes, namely VEGF receptors, neuropilins, angiopoetin 2, PDGF-BB and IL-8 (reviewed in Ref. 125). As seen in Fig. 2A, the increased O_2 demand in the heart undergoing hypertrophy necessitates an increase in blood flow. This is likely to be a consequence of adenosine which has also been found to stimulate VEGF mRNA.[64] From these findings, we can conclude that hypoxia may serve as a direct metabolic factor for angiogenesis or an indirect factor by increasing blood flow which provides mechanical stimuli that trigger the angiogenic cascade. Increased flow, including elevated shear stress and wall tension, is a well-recognized stimulator of vascular growth.[126] Shear stress activates a variety of molecules, e.g. transcription factors, proteins, in both vascular smooth muscle and endothelial cells.[125] Laminar shear stress provides stimuli for endothelial cell proliferation by enhancement of NO and Ca^{2+} entry into cells, activation of phosphatidylinositol, and cell shape changes (reviewed in Ref. 126). There is a well-established link between increased blood flow and vessel growth, whether via growth factor enhancement or by direct mechanical effects on vascular cells. Thus, hypoxia may influence vascular growth directly via growth factor upregulation or indirectly by increasing blood flow. Chronic interventions that increase blood flow and cause coronary angiogenesis include dipyridamole/adenosine/xanthine derivatives, alcohol, thyroxine, exercise and hypoxia (reviewed in Ref. 126). Thus, the mechanical effects of enhanced blood flow may play a role in three of the models of hypertrophy discussed in this review, namely, hyperthyroidism, hypoxia, and exercise training.

Both shear stress and cyclic stretch have been implicated in vascular growth.[126] An appropriate example is volume overload-induced cardiac hypertrophy.[16,26,28,31,60,127] That stretch plays a role in vascular growth has support from *in vitro* studies. Work from our lab has documented that cyclic stretch of cardiomyocytes causes upregulation of VEGF and possibly other angiogenic factors; conditioned media from these stretched cells stimulates endothelial cells to proliferate, migrate and form tubes.[128] Furthermore, stretch of coronary endothelial cells enhances tie-1, tie-2 and Flk-1.[129] Volume overload is characterized by increased diastolic dimensions, which necessarily cause stretch of the ventricular cardiomyocytes and vasculature. As documented in this review and illustrated in Fig. 2B, this model of hypertrophy is associated with angiogenesis and arteriogenesis.

Hemodynamic forces not only play a role in the formation of microvessels, but are known to affect remodeling of larger vessels (reviewed in Ref. 130). High flow is known to drive endothelial cell proliferation in arterial remodeling.[131] For example, Kassab and colleagues documented an increased luminal diameter in the right coronary artery in pigs with right ventricular hypertrophy induced by pulmonary hypertension.[132]

11. Summary

The development and persistence of cardiac hypertrophy may or may not include compensatory growth of the coronary vasculature. Ideally, growth occurs at two levels, i.e. there is a numerical increase in both arterioles and capillaries. The former maintains adequate maximal coronary perfusion, while the latter prevents expansion of capillary domains and diffusion distance for O_2. It is obvious that angiogenesis and arteriogenesis are not linked to cardiac hypertrophy, but are generally model specific. Therefore, one needs to consider stimuli that are activated along with the factors evoking myocardial growth. These include metabolic stimuli, such as hypoxia and mechanical stimuli, i.e. stretch, shear stress. Other factors that may modify angiogenesis and arteriogenesis include age, altered hemodynamic states that negatively impact on coronary vessels and the intensity of the stimulus. Learning more about the angiogenic and arteriogenic mechanism that can be activated in various types of cardiac hypertrophy can provide the basis for effective non-invasive therapies.

References

1. Olsson R, Bugni W (1986). *Coronary Circulation* (Raven Press, New York), Ch. 48, pp. 987–1037.
2. Tomanek RJ (1990). Response of the coronary vasculature to myocardial hypertrophy. *J Am Coll Cardiol* **15**: 528–533.
3. Rakusan K (1999). Vascularization of the heart during normal and pathological growth. *Coron Angiogenesis, Adv Oregan Biol* **7**: 129–153.
4. Alfaro A, Schaible TF, Malhotra A, Yipintsoi T, Scheuer J (1983). Impaired coronary flow and ventricular function in hearts of hypertensive rats. *Cardiovasc Res* **17**: 553–561.
5. Wangler RD, Peters KG, Marcus ML, Tomanek RJ (1982). Effects of duration and severity of arterial hypertension and cardiac hypertrophy on coronary vasodilator reserve. *Circ Res* **51**: 10–18.
6. Yamamoto J, Tsuchiya M, Saito M, Ikeda M (1985). Cardiac contractile and coronary flow reserves in deoxycorticosterone acetate-salt hypertensive rats. *Hypertension* **7**: 569–577.
7. Breisch EA, White FC, Bloor CM (1984). Myocardial characteristics of pressure overload hypertrophy. A structural and functional study. *Lab Invest* **51**: 333–342.
8. Breisch EA, White FC, Nimmo LE, Bloor CM (1986). Cardiac vasculature and flow during pressure-overload hypertrophy. *Am J Physiol* **251**: H1031–H1037.
9. Mueller TM, Marcus ML, Kerber RE, Young JA, Barnes RW, Abboud FM (1978). Effect of renal hypertension and left ventricular hypertrophy on the coronary circulation in dogs. *Circ Res* **42**: 543–549.
10. Tomanek RJ, Palmer PJ, Peiffer GL, Schreiber KL, Eastham CL, Marcus ML (1986). Morphometry of canine coronary arteries, arterioles, and capillaries during hypertension and left ventricular hypertrophy. *Circ Res* **58**: 38–46.
11. Harrison DG, Marcus ML, Dellsperger KC, Lamping KG, Tomanek RJ (1991). Pathophysiology of myocardial perfusion in hypertension. *Circulation* **83**: III14–III18.
12. Tomanek RJ, Davis JW, Anderson SC (1979). The effects of alpha-methyldopa on cardiac hypertrophy in spontaneously hypertensive rats: ultrastructural, stereological, and morphometric analysis. *Cardiovasc Res* **13**: 173–182.
13. Tomanek RJ, Wangler RD, Bauer CA (1985).Prevention of coronary vasodilator reserve decrement in spontaneously hypertensive rats. *Hypertension* **7**: 533–540.
14. Canby CA, Tomanek RJ (1989). Role of lowering arterial pressure on maximal coronary flow with and without regression of cardiac hypertrophy. *Am J Physiol* **257***: H1110–H1118.
15. Ljungqvist A, Unge G (1973). The proliferative activity of the myocardial tissue in various forms of experimental cardiac hypertrophy. *Acta Pathol Microbiol Scand [A]* **81**: 233–240.
16. Batra S, Rakusan K (1991). Geometry of capillary networks in volume overloaded rat heart. *Microvasc Res* **42**: 39–50.

17. Tomanek RJ, Searls JC, Lachenbruch PA (1982). Quantitative changes in the capillary bed during developing, peak, and stabilized cardiac hypertrophy in the spontaneously hypertensive rat. *Circ Res* **51**: 295–304.

18. Anversa P, Melissari M, Beghi C, Olivetti G (1984). Structural compensatory mechanisms in rat heart in early spontaneous hypertension. *Am J Physiol* **246**: H739–H746.

19. Peters KG, Wangler RD, Tomanek RJ, Marcus ML (1984). Effects of long-term cardiac hypertrophy on coronary vasodilator reserve in SHR rats. *Am J Cardiol* **54**: 1342–1348.

20. McAinsh AM, Geyer M, Fandrey J, Ruegg JC, Wiesner RJ (1998). Expression of vascular endothelial growth factor during the development of cardiac hypertrophy in spontaneously hypertensive rats. *Mol Cell Biochem* **187**: 141–146.

21. Tomanek RJ, Schalk KA, Marcus ML, Harrison DG (1989). Coronary angiogenesis during long-term hypertension and left ventricular hypertrophy in dogs. *Circ Res* **65**: 352–359.

22. Botham MJ, Lemmer JH, Gerren RA, Long RW, Behrendt DM, Gallagher KP (1984). Coronary vasodilator reserve in young dogs with moderate right ventricular hypertrophy. *Ann Thorac Surg* **38**: 101–107.

23. Manohar M (1985). Transmural coronary vasodilator reserve, and flow distribution during tachycardia in conscious young swine with right ventricular hypertrophy. *Cardiovasc Res* **19**: 104–112.

24. White FC, Nakatani Y, Nimmo L, Bloor CM (1992). Compensatory angiogenesis during progressive right ventricular hypertrophy. *Am J Cardiovasc Pathol* **4**: 51–68.

25. Liu Z, Hilbelink DR, Gerdes AM (1991). Regional changes in hemodynamics and cardiac myocyte size in rats with aortocaval fistulas. 2. Long-term effects. *Circ Res* **69**: 59–65.

26. Chen Y, Torry RJ, Baumbach GL, Tomanek RJ (1994). Proportional arteriolar growth accompanies cardiac hypertrophy induced by volume overload. *Am J Physiol* **267**: H2132–H2137.

27. Carabello BA, Nakano K, Corin W, Biederman R, Spann JF Jr (1989). Left ventricular function in experimental volume overload hypertrophy. *Am J Physiol* **256**: H974–H981.

28. Olivetti G, Lagrasta C, Quaini F, Ricci R, Moccia G, Capasso JM, Anversa P (1989). Capillary growth in anemia-induced ventricular wall remodeling in the rat heart. *Circ Res* **65**: 1182–1192.

29. Hultgren PB, Bove AA (1981). Myocardial blood flow and mechanics in volume overload-induced left ventricular hypertrophy in dogs. *Cardiovasc Res* **15**: 522–528.

30. Gascho JA, Mueller TM, Eastham C, Marcus ML (1982). Effect of volume overload hypertrophy on the coronary circulation awake dogs. *Cardiovasc Res* **16**: 288–292.

31. Thomas DP, Phillips SJ, Bove AA (1984) Myocardial morphology and blood flow distribution in chronic volume overload hypertrophy in dogs. *Basic Res Cardiol* **79**: 379–388.

32. Badke FR, White FC, Le Winter M, Covell J, Andres J, Bloor C (1981). Effects of experimental volume overload hypertrophy on myocardial blood flow and cardiac function. *Am J Physiol* **241**: H564–H570.

33. Bauman RP, Rembert JC Greenfield JC Jr (1998). Myocardial blood flow in awake dogs with chronic tricuspid regurgitation. *Basic Res Cardiol* **93**: 63–69.

34. Doty DB, Wright CB, Hiratzka LF, Eastham CL, Marcus ML (1984). Coronary reserve in volume-induced right ventricular hypertrophy from atrial septal defect. *Am J Cardiol* **54**: 1059–1063.

35. Harada K, Aoki M, Toyono M, Tamura M (2004). Coronary flow velocity and coronary flow velocity reserve in children with ventricular septal defect. *Tohoku J Exp Med* **202**: 77–85.

36. Strauer BE (1992). The concept of coronary flow reserve. *J Cardiovasc Pharmacol* **19**(Suppl 5): S67–S80.

37. Friberg P (1988). Diastolic characteristics and cardiac energetics of isolated hearts exposed to volume and pressure overload. *Cardiovasc Rese* **22**: 329–339.

38. Tomanek RJ, Doty MK, Sandra A (1998). Early coronary angiogenesis in response to thyroxine: growth characteristics and upregulation of basic fibroblast growth factor. *Circ Res* **82**: 587–593.

39. Tomanek RJ, Busch TL (1998) Coordinated capillary and myocardial growth in response to thyroxine treatment. *Anat Rec* **251**: 44–49.

40. Gay RG, Raya TE, Lancaster LD, Lee RW, Morkin E, Goldman S (1988). Effects of thyroid state on venous compliance and left ventricular performance in rats. *Am J Physiol* **254**: H81–H88.

41. Chilian WM, Wangler RD, Peters KG, Tomanek RJ, Marcus ML (1985). Thyroxine-induced left ventricular hypertrophy in the rat. Anatomical and physiological evidence for angiogenesis. *Circ Res* **57**: 591–598.

42. Wachtlova M, Ostadal B, Mares V (1985). Thyroxine-induced cardiomegaly in rats of different age. *Physiol Bohemoslov* **34**: 385–394.

43. Breisch EA, White FC, Hammond HK, Flynn S, Bloor CM (1989). Myocardial characteristics of thyroxine stimulated hypertrophy. A structural and functional study. *Basic Res Cardiol* **84**: 345–358.

44. Mall G, Zimmer G, Baden S, Mattfeldt T (1990). Capillary neoformation in the rat heart — stereological studies on papillary muscles in hypertrophy and physiologic growth. *Basic Res Cardiol* **85**: 531–540.

45. Tomanek RJ, Connell PM, Butters CA, Torry RJ (1995). Compensated coronary microvascular growth in senescent rats with thyroxine-induced cardiac hypertrophy. *Am J Physiol* **268**: H419–H425.

46. Heron MI, Kolar F, Papousek F, Rakusan K (1997). Early and late effect of neonatal hypo- and hyperthyroidism on coronary capillary geometry and long-term heart function in rat. *Cardiovasc Res* **33**: 230–240.

47. Heron MI, Rakusan K (1994). Geometry of coronary capillaries in hyperthyroid and hypothyroid rat heart. *Am J Physiol* **267**: H1024–H1031.

48. Heron MI, Rakusan K (1996). Short- and long-term effects of neonatal hypo- and hyperthyroidism on coronary arterioles in rat. *Am J Physiol* **271**: H1746–H1754.

49. Craft-Cormney C, Hansen JT (1980). Early ultrastructural changes in the myocardium following thyroxine-induced hypertrophy. *Virchows Arch B Cell Pathol Incl Mol Pathol* **33**: 267–273.

50. Klein I, Hong C (1986). Effects of thyroid hormone on cardiac size and myosin content of the heterotopically transplanted rat heart. *J Clin Invest* **77**: 1694–1698.

51. Rongish BJ, Torry RJ, Tomanek RJ (1995). Coronary neovascularization of embryonic rat hearts cultured in oculo is independent of thyroid hormones. *Am J Physiol* **268**: H811–H816.

52. Davies PF, Remuzzi A, Gordon EJ, Dewey CF Jr, Gimbrone MA Jr (1986). Turbulent fluid shear stress induces vascular endothelial cell turnover *in vitro*. *Proc Natl Acad Sci* USA **83**: 2114–2117.

53. Ziada AM, Hudlicka O, Tyler KR, Wright AJ (1984). The effect of long-term vasodilatation on capillary growth and performance in rabbit heart and skeletal muscle. *Cardiovasc Res* **18**: 724–732.

54. Mall G, Schikora I, Mattfeldt T, Bodle R (1987). Dipyridamole-induced neoformation of capillaries in the rat heart. Quantitative stereological study on papillary muscles. *Lab Invest* **57**: 86–93.

55. Torry RJ, O'Brien DM, Connell PM, Tomanek RJ (1992). Dipyridamole-induced capillary growth in normal and hypertrophic hearts. *Am J Physiol* **262**: H980–H986.

56. Fadel BM, Ellahham S, Ringel MD, Lindsay J Jr, Wartofsky L, Burman KD (2000). Hyperthyroid heart disease. *Clin Cardiol* **23**: 402–408.

57. Turek Z, Grandtner M, Kreuzer F (1972). Cardiac hypertrophy, capillary and muscle fiber density, muscle fiber diameter, capillary radius and diffusion distance in the myocardium of growing rats adapted to a simulated altitude of 3500 m. *Pflugers Arch* **335**: 19–28.

58. Pietschmann M, Bartels H (1985). Cellular hyperplasia and hypertrophy, capillary proliferation and myoglobin concentration in the heart of newborn and adult rats at high altitude. *Respir Physiol* **59**: 347–360.

59. Moravec M, Turek Z, Moravec J (2002). Persistence of neoangiogenesis and cardiomyocyte divisions in right ventricular myocardium of rats born and raised in hypoxic conditions. *Basic Res Cardiol* **97**: 153–160.

60. Rakusan K, Turek Z, Kreuzer F (1981). Myocardial capillaries in guinea pigs native to high altitude (Junin, Peru, 4,105 m). *Pflugers Arch* **391**: 22–24.

61. Lund DD, Tomanek RJ (1980). The effects of chronic hypoxia on the myocardial cell of normotensive and hypertensive rats. *Anat Rec* **196**: 421–430.

62. Partovian C, Adnot S, Eddahibi S, Teiger E, Levame M, Dreyfus P, Raffestin B, Frelin C (1998). Heart and lung VEGF mRNA expression in rats with monocrotaline- or hypoxia-induced pulmonary hypertension. *Am J Physiol* **275**: H1948–H1956.

63. Ryan HE, Lo J, Johnson RS (1998). HIF-1 alpha is required for solid tumor formation and embryonic vascularization. *EMBO J* **17**: 3005–3015.

64. Takagi H, King GL, Ferrara N, Aiello LP (1996). Hypoxia regulates vascular endothelial growth factor receptor KDR/Flk gene expression through adenosine A2 receptors in retinal capillary endothelial cells. *Invest Ophthalmol Vis Sci* **37**: 1311–1321.

65. William C, Koehne P, Jurgensen J, Grafe M, Wagner K, Bachmann S, Frei U, Eckardt K (2000). Tie2 receptor is stimulated by hypoxia and proinflammatory cytokines in human endothelial cells. *Circ Res* **87**: 370.

66. Laughlin MH, McAllister RM (1992). Exercise training-induced coronary vascular adaptation. *J Appl Physiol* **73**: 2209–2225.

67. Tomanek RJ (1994). Exercise-induced coronary angiogenesis: a review. *Med Sci Sports Exerc* **26**: 1245–1251.

68. Brown M (2003). Exercise and coronary vascular remodeling in the healthy heart. *Exp Physiol* **88**(5): 645–658.

69. Marcus KD, Tipton CM (1985). Exercise training and its effects with renal hypertensive rats. *J Appl Physiol* **59**: 1410–1415.

70. Anversa P, Levicky V, Beghi C, McDonald SL, Kikkawa Y (1983). Morphometry of exercise-induced right ventricular hypertrophy in the rat. *Circ Res* **52**: 57–64.

71. White FC, Witzel G, Breisch EA, Bloor CM, Nimmo LE (1988). Regional capillary and myocyte distribution in normal and exercise trained male and female rat hearts. *Am J Cardiovasc Pathol* **2**: 247–253.

72. Mattfeldt T, Kramer KL, Zeitz R, Mall G (1986). Stereology of myocardial hypertrophy induced by physical exercise. *Virchows Arch A Pathol Anat Histopathol* **409**: 473–484.

73. Buttrick PM, Levite HA, Schaible TF, Ciambrone G, Scheuer J (1985). Early increases in coronary vascular reserve in exercised rats are independent of cardiac hypertrophy. *J Appl Physiol* **59**: 1861–1865.

74. Rakusan K, Wicker P (1990). Morphometry of the small arteries and arterioles in the rat heart: effects of chronic hypertension and exercise. *Cardiovasc Res* **24**: 278–284.

75. Laughlin MH, Hale CC, Novela L, Gute D, Hamilton N, Ianuzzo CD (1991). Biochemical characterization of exercise-trained porcine myocardium. *J Appl Physiol* **71**: 229–235.

76. Tittle K, Knacke K, Brauer B, Otto H (1966). Der Einfluss Korperlicker Belastengen Uterschiedlicker dauer und Intersitof auf die kapillarisierung der herzund skelettmuskulatur bei albino ratten. *XVI Int World Congress Sports Medicine,* pp. 181–188.

77. Loud AV, Beghi C, Olivetti G, Anversa P (1984). Morphometry of right and left ventricular myocardium after strenuous exercise in preconditioned rats. *Lab Invest* **51**: 104–111.

78. Spear KL, Koerner JE, Terjung RL (1978). Coronary blood flow in physically trained rats. *Cardiovasc Res* **12**: 135–143.

79. Mall G, Mattfeldt T, Hasslacher C, Mann J (1986). Morphological reaction patterns in experimental cardiac hypertrophy — a quantitative stereological study. *Basic Res Cardiol* **81**(Suppl 1): 193–201.

80. Bloor CM, Leon AS (1970). Interaction of age and exercise on the heart and its blood supply. *Lab Invest* **22**: 160–165.

81. Rakusan K, Ost'adal B, Wachtlova M (1971). The influence of muscular work on the capillary density in the heart and skeletal muscle of pigeon (Columbia livia dom.). *Can J Physiol Pharmacol* **49**: 167–170.

82. White FC, McKirnan MD, Breisch EA, Guth BD, Liu YM, Bloor CM (1987). Adaptation of the left ventricle to exercise-induced hypertrophy. *J Appl Physiol* **62**: 1097–1100.

83. Leon AS, Bloor CM (1968). Effects of exercise and its cessation on the heart and its blood supply. *J Appl Physiol* **24**: 485–490.

84. Frenzel H, Schwartzkopff B, Holtermann W, Schnurch HG, Novi A, Hort W (1988). Regression of cardiac hypertrophy: morphometric and biochemical studies in rat heart after swimming training. *J Mol Cell Cardiol* **20**: 737–751.

85. Laughlin MH, Overholser KA, Bhatte MJ (1989). Exercise training increases coronary transport reserve in miniature swine. *J Appl Physiol* **67**: 1140–1149.

86. Norman TD, Coers CR (1960). Cardiac hypertrophy after coronary artery ligation in rats. *Arch Pathol* **69**: 181–184.

87. Reiner L, Freudenthal RR, Suarez FH (1968). Cardiac weights of rats after experimental myocardial infarction. *Arch Pathol* **86**: 465–474.

88. Rubin SA, Fishbein MC, Swan HJ (1983). Compensatory hypertrophy in the heart after myocardial infarction in the rat. *J Am Coll Cardiol* **1**: 1435–1441.

89. Anversa P, Beghi C, Levicky V, McDonald SL, Kikkawa Y, Olivetti G (1985). Effects of strenuous exercise on the quantitative morphology of left ventricular myocardium in the rat. *J Mol Cell Cardiol* **17**: 587-595.

90. Anversa P, Beghi C, Kikkawa Y, Olivetti G (1985). Myocardial response to infarction in the rat. Morphometric measurement of infarct size and myocyte cellular hypertrophy. *Am J Pathol* **118**: 484–492.

91. Anversa P, Beghi C, Kikkawa Y, Olivetti G (1986). Myocardial infarction in rats. Infarct size, myocyte hypertrophy, and capillary growth. *Circ Res* **58**: 26–37.

92. Anversa P, Capasso JM, Sonnenblick EH, Olivetti G (1990). Mechanisms of myocyte and capillary growth in the infarcted heart. *Eur Heart J* **11**(Suppl B): 123–132.

93. Karam R, Healy BP, Wicker P (1990). Coronary reserve is depressed in postmyocardial infarction reactive cardiac hypertrophy. *Circulation* **81**: 238–246.

94. Kozlovskis PL, Gerdes AM, Smets M, Moore JA, Bassett AL, Myerburg RJ (1991). Regional increase in isolated myocyte volume in chronic myocardial infarction in cats. *J Mol Cell Cardiol* **23**: 1459–1466.

95. Olivetti G, Capasso JM, Meggs LG, Sonnenblick EH, Anversa P (1991). Cellular basis of chronic ventricular remodeling after myocardial infarction in rats. *Circ Res* **68**: 856–869.

96. Xie Z, Gao M, Batra S, Koyama T (1997). The capillarity of left ventricular tissue of rats subjected to coronary artery occlusion. *Cardiovasc Res* **33**: 671–676.

97. Van Kerckhoven R, Saxena PR, Schoemaker RG (2002). Restored capillary density in spared myocardium of infarcted rats improves ischemic tolerance. *J Cardiovasc Pharmacol* **40**: 370–380.

98. Van Kerckhoven R, van Veghel R, Saxena PR, Schoemaker RG (2004). Pharmacological therapy can increase capillary density in post-infarction remodeled rat hearts. *Cardiovasc Res* **61**: 620–629.

99. Kramer CM, Rogers WJ, Park CS, Seibel PS, Shaffer A, Theobald TM, Reichek N, Onodera T, Gerdes AM (1998). Regional myocyte hypertrophy parallels regional myocardial dysfunction during post-infarct remodeling. *J Mol Cell Cardiol* **30**: 1773–1778.

100. Zheng W, Weiss RM, Wang X, Zhou R, Arlen AM, Lei L, Lazartigues E, Tomanek RJ (2004). DITPA stimulates arteriolar growth and modifies myocardial postinfarction remodeling. *Am J Physiol Heart Circ Physiol* **286**: H1994–H2000.
101. Dedkov EI, Christensen LP, Weiss RM, Tomanek RJ (2005). Reduction of heart rate by chronic beta1-adrenoceptor blockade promotes growth of arterioles and preserves coronary perfusion reserve in postinfarcted heart. *Am J Physiol Heart Circ Physiol* **288**: H2684–H2693.
102. Turek Z, Grandtner M, Kubat K, Ringnalda BE, Kreuzer F (1978). Arterial blood gases, muscle fiber diameter and intercapillary distance in cardiac hypertrophy of rats with an old myocardial infarction. *Pflugers Arch* **376**: 209–215.
103. Przyklenk K, Groom AC (1983) Microvascular evidence for a transition zone around a chronic myocardial infarct in the rat. *Can J Physiol Pharmacol* **61**: 1516–1522.
104. Olivetti G, Ricci R, Beghi C, Guideri G, Anversa P (1986). Response of the border zone to myocardial infarction in rats. *Am J Pathol* **125**: 476–483.
105. Kalkman EA, Bilgin YM, van Haren P, van Suylen RJ, Saxena PR, Schoemaker RG (1996). Determinants of coronary reserve in rats subjected to coronary artery ligation or aortic banding. *Cardiovasc Res* **32**: 1088–1095.
106. Nelissen-Vrancken HJ, Debets JJ, Snoeckx LH, Daemen MJ, Smits JF (1996). Time-related normalization of maximal coronary flow in isolated perfused hearts of rats with myocardial infarction. *Circulation* **93**: 349–355.
107. Kalkman EA, van Haren P, Saxena PR, Schoemaker RG (1997). Regionally different vascular response to vasoactive substances in the remodelled infarcted rat heart; aberrant vasculature in the infarct scar. *J Mol Cell Cardiol* **29**: 1487–1497.
108. Gervais M, Richer C, Fornes P, De Gasparo M, Giudicelli JF (1999). Valsartan and coronary haemodynamics in early post-myocardial infarction in rats. *Fundam Clin Pharmacol* **13**: 635–645.
109. Lei L, Zhou R, Zheng W, Christensen LP, Weiss RM, Tomanek RJ (2004). Bradycardia induces angiogenesis, increases coronary reserve, and preserves function of the postinfarcted heart. *Circulation* **110**: 796–802.
110. Schieffer B, Wirger A, Meybrunn M, Seitz S, Holtz J, Riede UN, Drexler H (1994). Comparative effects of chronic angiotensin-converting enzyme inhibition and angiotensin II type 1 receptor blockade on cardiac remodeling after myocardial infarction in the rat. *Circulation* **89**: 2273–2282.
111. Li J, Brown LF, Hibberd MG, Grossman JD, Morgan JP, Simons M (1996). VEGF, flk-1, and flt-1 expression in a rat myocardial infarction model of angiogenesis. *Am J Physiol* **270**: H1803–H1811.
112. Jingjing L, Srinivasan B, Bian X, Downey HF, Roque RS (2000). Vascular endothelial growth factor is increased following coronary artery occlusion in the dog heart. *Mol Cell Biochem* **214**: 23–30.
113. Kawata H, Yoshida K, Kawamoto A, Kurioka H, Takase E, Sasaki Y, Hatanaka K, Kobayashi M, Ueyama T, Hashimoto T, Dohi K (2001). Ischemic

preconditioning upregulates vascular endothelial growth factor mRNA expression and neovascularization via nuclear translocation of protein kinase C epsilon in the rat ischemic myocardium. *Circ Res* **88**: 696–704.

114. Heba G, Krzeminski T, Porc M, Grzyb J, Ratajska A, Dembinska-Kiec A (2001). The time course of tumor necrosis factor-alpha, inducible nitric oxide synthase and vascular endothelial growth factor expression in an experimental model of chronic myocardial infarction in rats. *J Vasc Res* **38**: 288–300.

115. Tomanek RJ (2005). Formation of the coronary vasculature during development. *Angiogenesis* **8**: 273–284.

116. Tomanek RJ, Hu N, Phan B, Clark EB (1999). Rate of coronary vascularization during embryonic chicken development is influenced by the rate of myocardial growth. *Cardiovasc Res* **41**: 663–671.

117. Rakusan K, Flanagan MF, Geva T, Southern J, Van Praagh R (1992). Morphometry of human coronary capillaries during normal growth and the effect of age in left ventricular pressure overload hypertrophy. *Circulation* **86**: 38–46.

118. Kolar F, Papousek F, Pelouch V, Ostadal B, Rakusan K (1998). Pressure overload induced in newborn rats: effects on left ventricular growth, morphology, and function. *Pediatr Res* **43**: 521–526.

119. Crissman RP, Rittman B, Tomanek RJ (1985). Exercise-induced myocardial capillary growth in the spontaneously hypertensive rat. *Microvasc Res* **30**: 185–194.

120. Unge G, Carlsson S, Ljungqvist A, Tornling G, Adolfsson J (1979). The proliferative activity of myocardial capillary wall cells in variously aged swimming-exercised rats. *Acta Pathol Microbiol Scand [A]* **87**: 15–17.

121. Bache RJ, Alyono D, Sublett E, Dai XZ (1986). Myocardial blood flow in left ventricular hypertrophy developing in young and adult dogs. *Am J Physiol* **251**: H949–H956.

122. Alyono D, Anderson RW, Parrish DG, Dai XZ, Bache RJ (1986). Alterations of myocardial blood flow associated with experimental canine left ventricular hypertrophy secondary to valvular aortic stenosis. *Circ Res* **58**: 47–57.

123. Tomanek RJ, Gisolfi CV, Bauer CA, Palmer PJ (1988). Coronary vasodilator reserve, capillarity, and mitochondria in trained hypertensive rats. *J Appl Physiol* **64**: 1179–1185.

124. Josko J, Mazurek M (2004). Transcription factors having impact on vascular endothelial growth factor (VEGF) gene expression in angiogenesis. *Med Sci Monit* **10**: RA89–RA98.

125. Conway EM, Collen D, Carmeliet P (2001). Molecular mechanisms of blood vessel growth. *Cardiovasc Res* **49**: 507–521.

126. Hudlicka O, Brown M (2000). *Modulators of Angiogenesis: Hemodynamic Forces* (Marcel Dekker, New York).

127. Legault F, Rouleau JL, Juneau C, Rose C, Rakusan K (1990). Functional and morphological characteristics of compensated and decompensated cardiac hypertrophy in dogs with chronic infrarenal aorto-caval fistulas. *Circ Res* **66**: 846–859.

128. Zheng W, Seftor EA, Meininger CJ, Hendrix MJ, Tomanek RJ (2001). Mechanisms of coronary angiogenesis in response to stretch: role of VEGF and TGF-beta. *Am J Physiol Heart Circ Physiol* **280**: H909–H917.

129. Zheng W, Christensen LP, Tomanek RJ (2004). Stretch induces upregulation of key tyrosine kinase receptors in microvascular endothelial cells. *Am J Physiol* **287**: H2739–H2745.
130. Lehoux S, Tedgui A (1998). Signal transduction of mechanical stresses in the vascular wall. *Hypertension* **32**: 338–345.
131. Sho E, Komatsu M, Sho M, Nanjo H, Singh TM, Xu C, Masuda H, Zarins CK (2003). High flow drives vascular endothelial cell proliferation during flow-induced arterial remodeling associated with the expression of vascular endothelial growth factor. *Exp Mol Pathol* **75**: 1–11.
132. Kassab GS, Imoto K, White FC, Rider CA, Fung YC, Bloor CM (1993). Coronary arterial tree remodeling in right ventricular hypertrophy. *Am J Physiol* **265**: H366–H375.

Regulation of Coronary Vascular Tone and Microvascular Physiology

*by Basel Ramlawi, Munir Boodhwani
and Frank W. Sellke*

1. Introduction

There are many cell types that make up the walls of blood vessels. The innermost layer is made of endothelial cells. This intimal endothelial layer is surrounded by a variable number of layers of smooth muscle cells comprising the medial layer. The adventitial layer surrounds the vascular smooth muscle layers. This last layer is responsible for providing structural integrity to the blood vessel, particularly larger arteries. While initially the endothelium was thought mainly to serve as a barrier to the diffusion of macromolecules, much has recently been learned about the pivotal role it plays in vascular function, regulation of vascular tone and control of local blood flow.[1] Smooth muscle cells also control vascular tone via humoral vasoactive factors, neural mediators or local paracrine factors (Fig. 1).

The classification of microvessels based on structural characteristics is rather arbitrary and there is lack of uniformity in the definitions

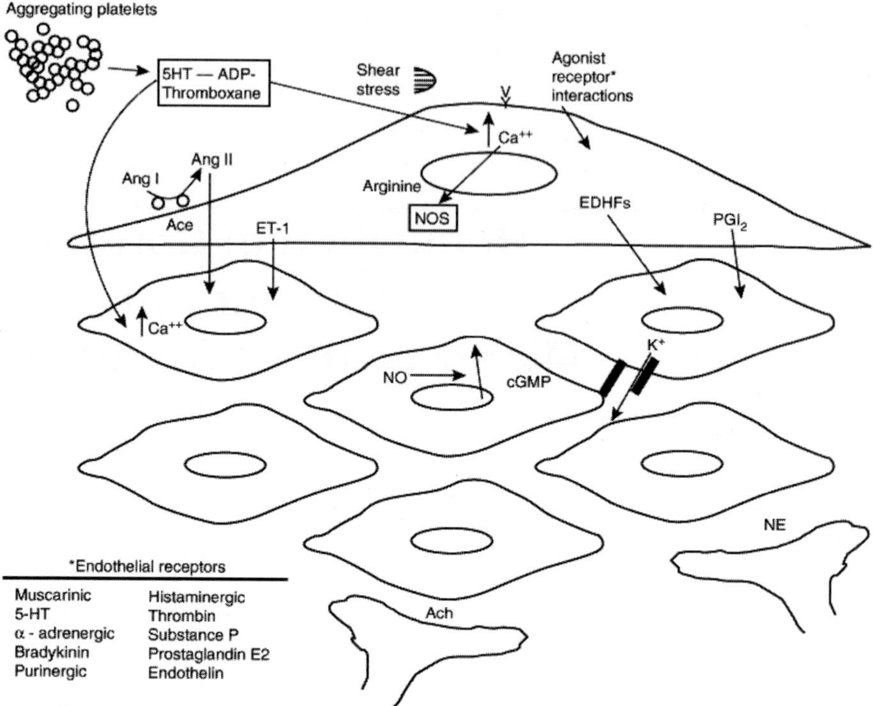

Fig. 1. Regulation of vascular tone by factors released from the endothelium, activated platelets and leukocytes, neuronally released factors and circulating substances. Ang, angiotensin; 5HT, 5-hydroxytrypamine (serotonin); ET, endothelin; ADP, adenosine diphosphate; EDHF, endothelium-derived hyperpolarizing factor; PGI_2, prostaglandin I_2; Ach, acetylcholine; NE, norepinephrine. (*Adapted from Ref. 71.*)

of microvascular segments such as small arteries, arterioles, venules, and so on. The transition between these segments is gradual and there is no clear demarcation between them. In general, "microvessels" are defines as vessels $< 300\,\mu$m in internal diameter. Capillaries are the smallest blood vessels defined as vessels whose walls are composed of only endothelial tubes. The microvessels through which blood flows toward capillaries are defined as "arterial microvessel" and those that drain from capillaries are defined as "venous microvessel."[2] Arterial microvessels usually have three coats, i.e. a thin tunica intima; a relatively thick tunica media, composed of one to several layers of smooth muscle cells disposed circumferentially; and a tunica adventitia, which is made up of fibrous elements and fibroblasts. Venous microvessels collect the blood from capillaries and have thinner vascular walls compared with arterial microvessels. Venules, 50 μm in diameter do not possess smooth muscle cell layers. Smaller venules have only endothelial cells and pericytes, and these venules are the most permeable sites that play an important role in substance exchange.

The various vascular beds within the body possess many similarities and subtle differences. This chapter will particularly focus on the coronary microcirculation. The regulation of myocardial perfusion is dependent on many intrinsic and extrinsic factors that may be affected by atherosclerotic lesions. For this reason, a thorough understanding of coronary flow regulation is critical for optimal care of cardiac patients. It has been shown that vasomotor regulation of coronary vessels, in addition to the actual anatomy, plays an important role in coronary perfusion and operative decision making. Myocardial blood flow is largely also dependent on the resistance generated by the microcirculation. While early coronary vasomotor regulation studies consisted of indirect assessments using measurements of coronary flow and calculations of coronary resistance, more recent investigations yielded much information into the properties of the intact coronary circulation and modern methods of analysis for interpretation of physiological data.[3–6]

The coronary microcirculation possesses unique features that allow it to respond to the dynamic changes in nutrient requirements as well as interact with surrounding contractile tissue. As in other vascular beds, it is composed of arterioles, capillaries and venules. This chapter will focus on issues relating to coronary physiology and pharmacology as well as myocardial perfusion in relation to the microcirculation.

2. Coronary Resistance

An understanding of vascular resistance is important as it is these resistance vessels that cause pressure losses and are responsible for regulation of myocardial perfusion. Initially it was thought that the precapillary arterioles were responsible for vascular resistance, with little resistance involvement by the vessels larger than 25–50 μm in diameter. Subsequent work revealed that over half of total coronary resistance

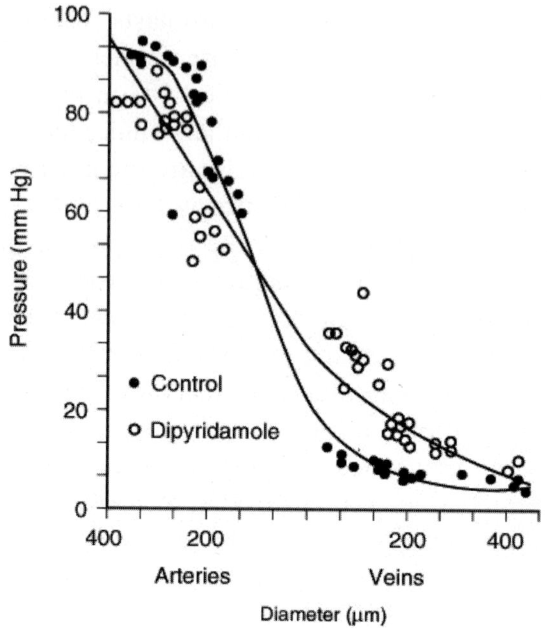

Fig. 2. Intravascular pressures in the coronary microcirculation under basal conditions and during vasodilation with dipyridamole. The distribution of vascular resistance is not static. Rather the size of the vessels regulating vascular tone depends on the tone of the vasculature. (*Adapted from Ref. 9.*)

is caused by vessels larger than 100 μm and can be observed in vessels larger than 300 μm.[7,8] Also, contrary to previous belief, the venous circulation under similar conditions of vasodilation, may account for up to 30% of vascular resistance. Figure 2 shows that under the vasodilatory effects of dipyridamole, larger arteries and veins assume a greater resistance role.[8,9] Similarly, ischemia results in a significant redistribution of vascular resistance.[9] This reveals that the distribution of vascular resistance is dynamic and is dependent on vascular tone among other factors.

The redistribution of microvascular resistance may change the myogenic tone in each microvascular segment because the luminal pressure in a certain vascular segment is determined by the systemic pressure and the relative distribution of vascular resistance. That is, when resistance is shifted upstream by dilation of small arterioles, for instance, the luminal pressure in the upstream microvessels decreases, resulting in myogenic

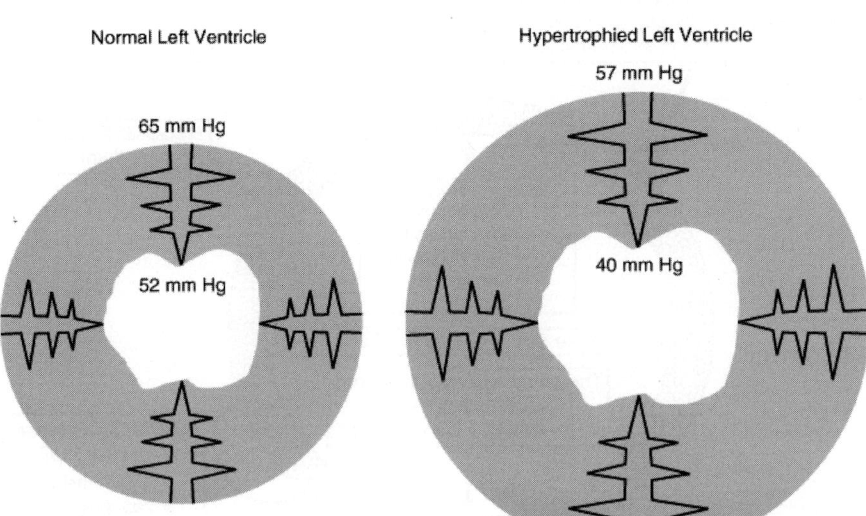

Fig. 3. Transmural losses of coronary perfusion pressure in normal and hypertrophied hearts. Pressures were measured using micropuncture-servo null techniques in hearts perfused via the left main coronary artery at 100 mm Hg. (*Adapted from Ref. 11.*)

dilation. The changes in the venular pressure caused by the resistance redistribution also may critically affect the substance exchange capacity and edema formation.

Pressure losses in the coronary circulation are also caused by vessels as they coarse from the epicardium through the myocardium.[10] This is further accentuated in the setting of cardiac hypertrophy. Such a phenomenon is particularly relevant clinically as it plays a role in explaining the pathophysiology of subendocardial infarcts as is shown in Fig. 3. The hypertrophied pathological state causes a decrease in the perfusion pressure of the subendocardium; predisposing it to ischemia and infarction.[11]

3. Regulation of Coronary Microvascular Tone

3.1. *Intrinsic and extrinsic vasomotor control*

Vessels possess intrinsic control mechanisms for maintaining the homeostasis of the local microenvironment in the face of stress. Vasomotor

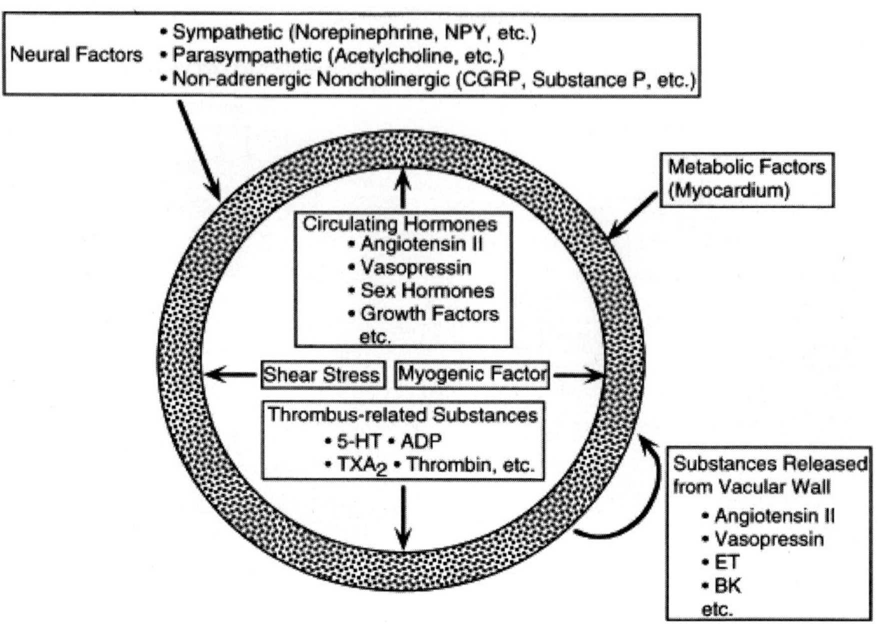

Fig. 4.   Various factors influencing the coronary microvascular tone. (*Adapted from Ref. 2.*)

tone regulation is a complex process that is influenced by the intrinsic properties of the vessel wall, local innervation as well as substances from surrounding parenchymal tissue. Properties intrinsic to the vessel wall and interactions with adjacent tissues work together to promote metabolic regulation and autoregulation. Endothelial regulation of vasomotor tone is also critically involved in regulation of vasomotor tone and hence myocardial perfusion. Microvascular responses to endogenous substances are summarized in Fig. 4. All of these factors play a role in setting the tone of microvessels.[2]

Based on *in vitro* observations, Jones *et al.* found that greater size arterioles are most sensitive to shear stress than myogenic response while small microvessels are most sensitive to metabolic factors.[12] Based on this they proposed three distinct "microdomains" that are governed by distinct forms of regulation. They have divided the coronary arterial microvessels into (1) small arterioles (50 μm), which are most sensitive to metabolic mediators; (2) intermediate arterioles (50–80 μm), where myogenic mechanisms predominate; and (3) large arterioles (80–150 μm), where flow-induced dilation most potently occurs. This model provides an insight into understanding the regulation of coronary microvessels, although there probably is an overlap among these three components.

These authors hypothesized that the longitudinal disposition of these three microdomains may enable the integrated adjustment of coronary flow conductance in the face of various influences, such as increased cardiac metabolism, a reduction in perfusion pressure, and so on, by affecting other microdomains together.[12] For example, the dilation of small arterioles by augmented cardiac metabolism produces a decrease in luminal pressure in upstream microvessels, leading to the dilation of intermediate arterioles by decreasing the myogenic tone. These microvascular dilations can produce an increase in shear stress and can result in enhanced flow-induced dilation in large arterioles. As a result, all sizes of arterial microvessels will dilate in response to the metabolic stimulation. The marked longitudinal heterogeneity of coronary microvascular responses may be at least partly explained by this microdomain hypothesis.

3.2. *Role of the endothelium*

The endothelium plays a pivotal role in vasomotor tone regulation. Many substances can affect coronary tone via endothelial-mediated mechanisms. Coronary endothelial cells also release several substances that affect coronary resistance such as nitric oxide (NO•), prostaglandins, a hyperpolarizing factor, endothelin, and reactive oxygen species (ROS). These are summarized in Fig. 5. As the major regulatory molecule, NO is produced by a constitutively expressed enzyme known as endothelial nitric oxide synthase (eNOS or NOS-3). NO is formed as a result of a series of electron transfer from NADPH to the flavins FAD and FMN on the reductase domain and electron transfer to a prosthetic heme group on the oxygenase domain. When heme reduction occurs, arginine is then catalyzed to citrulline and nitric oxide. The NO• formed diffuses to underlying vascular smooth muscle, where its actions include stimulation of soluble guanylate cyclase,

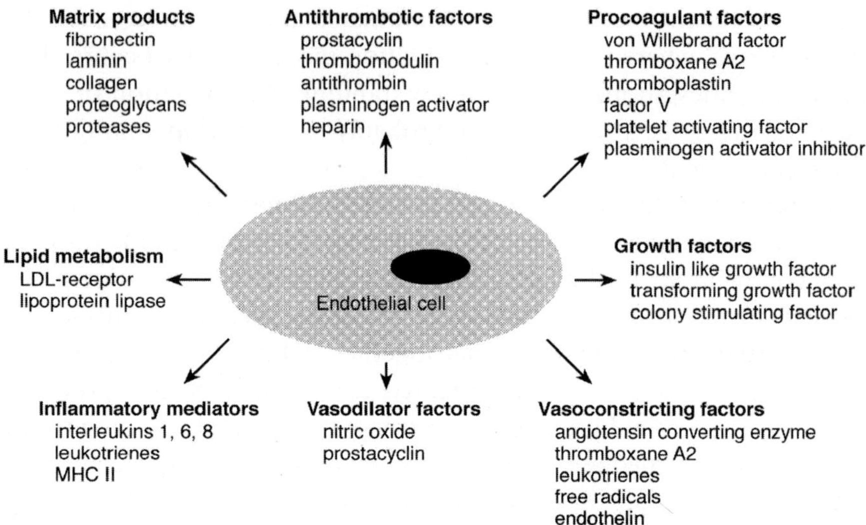

Fig. 5. Endothelial cells have both metabolic and synthetic functions. Through the secretion of a large variety of mediators they are able to influence cellular function throughout the body. LDL, low-density lipoprotein. (*Adapted from Ref. 72.*)

increasing cyclic guanosine monophosphate (cGMP) and prompting vasodilation via activation of cGMP-dependent protein kinase.[13] While binding of calcium/calmodulin is a prerequisite for activity of eNOS, other events such as phosphorylation,[14] membrane binding,[15] binding of eNOS with heat shock protein 90 and association with the integral membrane protein caveolin,[16] can also modulate NOS activity. Also, NO• may undergo reactions with thiol-containing compounds to form biologically active nitroso molecules.[17]

While eNOS is expressed constitutively, it undergoes important gene expression regulation by factors such as shear stress, endothelial cell growth, hypoxia, exposure to oxidized low density lipoprotein, and exposure to cytokines.[18–20] In the coronary circulation, the release of NO• confers a state of basal vasodilation. Hence, administration of NO synthase antagonists produces an increase in resting coronary resistance. On the other hand, when substances such as acetylcholine and bradykinin are administered, coronary microvessels of all sizes dilate. Endothelial NO production is affected by a variety of mechanisms

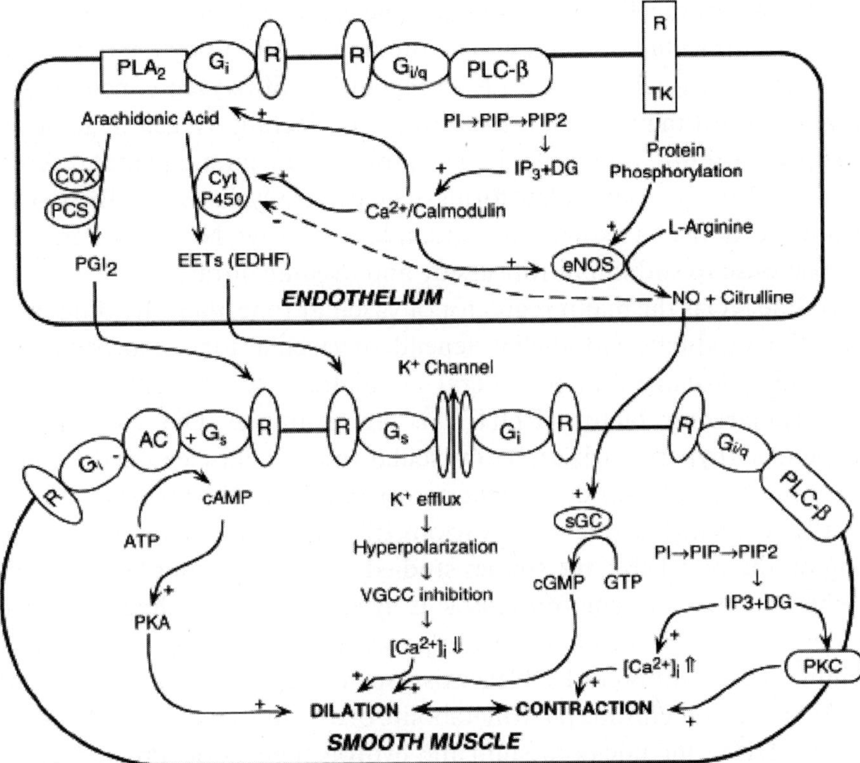

Fig. 6. Signal transduction pathways for the vascular responses to agonists. NO, PGI2 and EDHF are important for the cross-talk between the endothelium and the vascular smooth muscle. AC, adenylyl cyclase; COX, cyclooxygenase; Cyt P450, cytochrome P450; DG, diacylglycerol; Gi, Gi-protein; Gq, Gq protein; PCS, prostacyclin synthase; PI, phosphatidylinositol; PIP, phosphatidylinositol-4-phosphate; PIP2, phosphatidylinositol-4,5-bisphosphate; PKA, protein kinase A; R, receptors; TK, tyrosine kinase; VGCC, voltage-gated Ca^{2+} channel. (*Adapted from Ref. 2.*)

during disease. The signal transduction pathways through which NO acts are summarized in Fig. 6. Likely, the most important pathway involves activation of soluble guanylate cyclase, which catalyzes the formation of cGMP from guanidine triphosphate. cGMP serves as an allosteric regulator of the enzyme cGMP-dependent protein kinase or PKG. PKG phosphorylates contractile proteins and ion channels decreasing intracellular calcium and the sensitivity of contractile proteins to intracellular calcium. The binding of NO to cytochrome oxidase in the mitochondria leads to regulation of oxygen consumption and in return may have a critical role on affecting oxygen demand in the myocardium. Similarly, the receptors of natriuretic peptides ANP and BNP are also particulate forms of guanylate cyclases, and these substances produce vasodilatation via similar pathways. NO is also released in response to sodium nitroprusside and organic nitrates.

While NO is the major regulator of vascular tone, there are other factors that modulate endothelial-dependent vascular tone of microvessels in both coronary and peripheral circulations. Endothelium-derived hyperpolarizing factor (EDHF) is an example. The endothelial-dependent hyperpolarization of vascular smooth muscle is mediated by opening of a calcium-dependent potassium channel or by activating a Na/K ATPase. The role of the various EDHFs probably varies depending on the vessel size, the species studied and the vascular bed studied. When the vascular smooth muscle is hyperpolarized, voltage sensitive calcium channels are closed, leading to a reduction in intracellular calcium. Several different EDHFs exist; such as epoxyeicosatrienoic acid (EET) — a cytochrome p450 metabolite of arachidonic acid. Other possible EDHFs include potassium and hydrogen peroxide. Prostaglandin synthesis by the endothelium also contributes to modulation of tone in the coronary microcirculation. The predominant prostaglandin produced by endothelial cells is prostacyclin or PGI2. There is substantial interaction between nitric oxide, EDHF and prostacyclin. A major stimulus for release of prostacyclin, NO and the EDHF is shear stress, or the tangential force of fluid as it flows over the endothelium, resulting in flow-dependent vasodilatation. Interestingly, the importance of nitric oxide seems to decline and the role of the EDHF increases as blood vessels decrease in size. Consequently, the production of EDHF may increase when nitric oxide is low. The interaction of endothelial cells to vascular smooth muscle cells and the intermediates involved is outlined in Fig. 7.

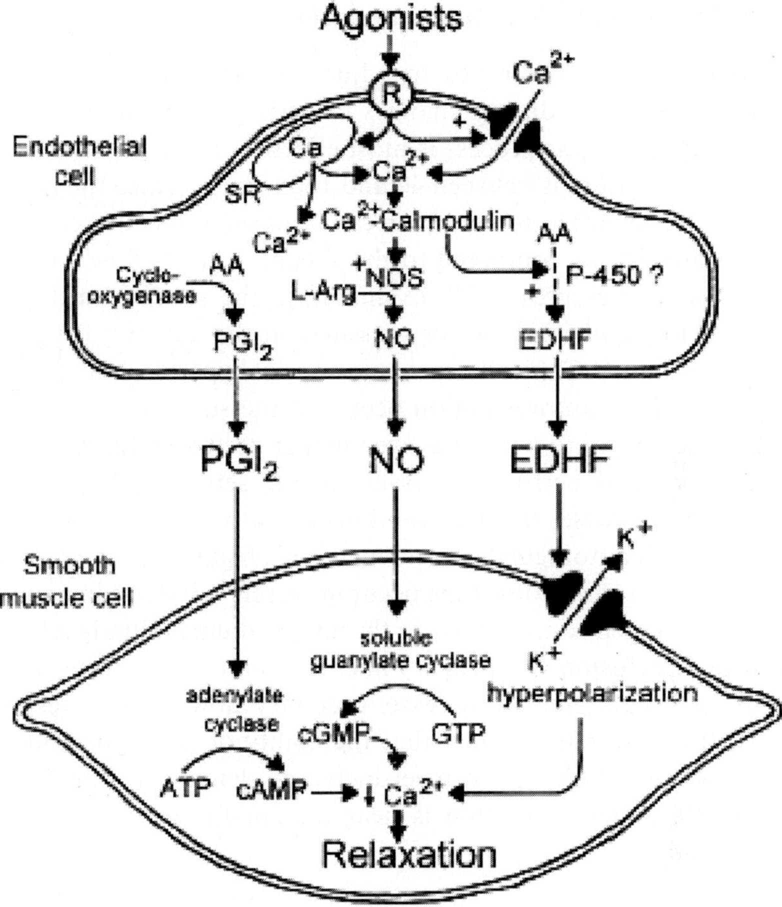

Fig. 7. Role of the increase in cytosolic calcium concentration in the release of endothelium-derived relaxing factors (EDRF). Endothelial receptor activation induces an influx of calcium into the cytoplasm of the endothelial cell; following interaction with calmodulin, this activates NO-synthase and cyclooxygenase, and leads to the release of endothelium-derived hyperpolarizing factor (EDHF). NO causes relaxation by activating the formation of cyclic GMP (cGMP) from GTP. EDHF causes hyperpolarization and relaxation by opening K$^+$ channels. Prostacyclin (PGI2) causes relaxation by activating adenylate cyclase (AC) which leads to the formation of cyclic AMP (cAMP). Any increase in cytosolic calcium (including that induced by the calcium ionophore A23187) causes the release of relaxing factors. When agonists activate the endothelial cells, an increase in inositol phosphate (IP3) may contribute to the increase in cytoplasmic Ca^{2+} by releasing it from the sarcoplasmic reticulum (SR). (*Adapted from Ref. 73.*)

3.3. *Role of metabolism and autoregulation*

The ability of a vascular bed to adjust its tone, in order to maintain a constant flow during changes in perfusion pressure is termed autoregulation.[21] This process is most effective in the coronary circulation when pressure is between 40 and 160 mmHg. Since the range of pressures over which autoregulation can be observed is different for the subendocardium, as compared to the subepicardium, flow will begin to decrease at pressures < 70–75 mmHg in the subendocardium, as compared to significantly lower pressures in the superficial layers of the myocardium.[22] Clinically, systemic arterial hypertension affects the range over which autoregulation occurs in the subendocardium such that flow will begin to decline at even higher pressures. Such a change in subendocardial perfusion pressure in the setting of hypertrophic myocardium increases the likelihood of subendocardial ischemia.

During both autoregulation and metabolic regulation, the predominant changes in vasomotor tone occur in vessels <100 μm in diameter. The rate of oxygen consumption in the myocardium is closely related to myocardial perfusion via coronary microvascular tone. As the myocardial oxygen requirements increase, coronary flow rises in response. Mostly, this is due to the fact that the ability of the myocardium to extract additional oxygen to meet increased demand is limited since myocardial oxygen extraction is near maximum even under resting conditions.

3.4. *Flow-induced dilation*

Flow-induced dilation is a ubiquitous phenomenon of blood vessels in various organs and animals, including humans.[2,23,24] Flow-induced dilation is considered to play important physiological roles as follows: first, to protect the vessel wall against friction-induced injury; second, to prevent the vascular steal phenomenon by dilating upstream vessels in the case of focal hyperemia; third, to reduce the heterogeneity of the coronary flow distribution; and fourth, to buffer the pressure distribution; in the face of rapid pressure changes.

Flow is sensed by endothelial cells through unidentified mechanoreceptors. In contrast to the myogenic response, the endothelium is required for flow-induced dilation, and a study using excised arterioles demonstrated that NO is exclusively responsible for the flow-induced dilation in porcine coronary microvessels, whereas Jimenez *et al.* have shown that prostanoids are also involved in flow mediated microvascular dilation in porcine coronary arterioles.[25,26] Since porcine coronary microvessels were used in both studies, differences in animal age, vessel sizes, or experimental setup can explain the discrepancy. Possible flow-induced arteriolar dilation mechanism is summarized in Fig. 8.

Autoregulation is mediated by the actions of several factors such as NO, EDHF and adenosine. Removal of a particular factor does not prevent autoregulation as the other factors seem to overtake its function. Adenosine and hydrogen peroxide also cause hyperpolarization

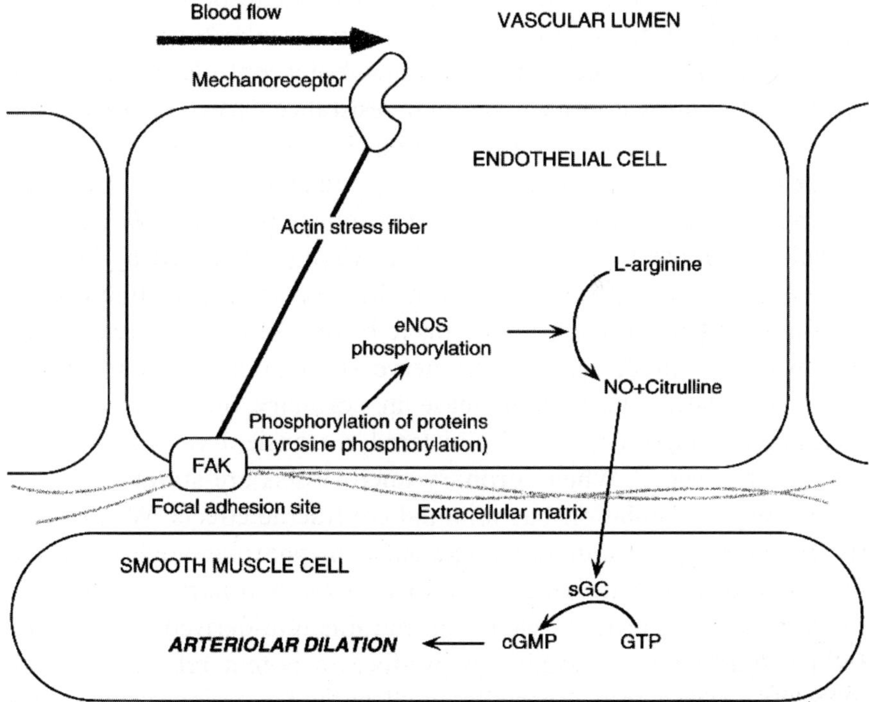

Fig. 8. Possible signal transduction pathway for the flow-induced arteriolar dilation, which is mediated by mechanotransduction via actin stress fibers, and the subsequent activation of the focal adhesion kinase and eNOS phosphorylation. FAK, focal adhesion kinase; sGC, soluble guanylyl cyclase. (*Adapted from Ref. 2.*)

of vascular smooth muscle. Due to the work of Dunker *et al.* as well as others, we now know that several factors work together to influence metabolic regulation and autoregulation leading to adequate regulation of coronary vascular tone despite interruption of any one particular pathway.[21]

3.5. *Neurohumoral influence on microcirculation*

The coronary arterial system is densely innervated with the sympathetic and parasympathetic nervous systems.[27] Neurotransmitters released from nervous tissues and a wide variety of humoral substances significantly affect the microvascular tone. Table 1 summarizes these endogenous substances and their microvascular responses. The effect of neurohumoral factors on coronary microvascular tone, in addition to myogenic, flow-induced, and local metabolic controls all participate in determining the coronary vascular resistance necessary for oxygen and nutrition supply to the myocardium (Fig. 4).

The role of the autonomic sympathetic and parasympathetic nervous systems is important in regulation of coronary perfusion. *In vivo*, the vascular response to sympathetic stimulation is mediated by both α-adrenergic and β-adrenergic receptors. In the coronary circulation, the predominant receptor subtype seems to be the β-adrenergic receptor.[28] For example, direct sympathetic nerve stimulation stimulates coronary vasodilation and an increase in coronary flow occurs. If β-adrenergic antagonists are administered, a transient vasoconstriction can be observed.[28] When coronary microvessels are studied *in vitro*, α-adrenergic stimulation has minimal contractile effects. When selective α_2-adrenergic stimulation is applied using pharmacological stimuli, there is rather potent vasodilation of all sized coronary microvessels, predominantly due to a release of endothelium-derived nitric oxide (NO$^\bullet$). β-adrenergic stimulation produces a potent relaxation of all coronary arteries, but especially small resistance vessels.[28] Also, β_2-adrenergic receptor subtype predominates in vessels less than $10\,\mu$m in diameter in *in vitro* studies, whereas a mixed β_1 or β_2-adrenergic receptor population controls vascular resistance in *in vivo* studies.[28] On the other hand, larger coronary vessels are regulated by a mixed

Table 1. Heterogeneous coronary arterial microvascular responses *in vivo* to endogenous substances.

Agonist	Materials	Small microvesssels*	Large microvessels[†]	Ref. no(s).
Acetylcholine	Dog, Cat	Dilation	Dilation	2, 29
C_1 Stimulation (normal pressure)	Dog	No change	Constriction	10
α_1 Stimulation (low pressure)	Dog	Constriction	Constriction	10
α_2 Stimulation (normal pressure)	Dog	No change	No change	10
α_2 Stimulation (low pressure)	Dog	Constriction	No change	10
NPY	Dog	Constriction	Constriction	74
CGRP	Dog	No change	Dilation	75
Adenosine	Dog	Dilation	No change	76
ET-1 (suffusion)	Dog	Constriction	Constriction	77
ET-1 (intracoronary)	Dog	Dilation	No change	77
5-HT	Cat	Dilation	Constriction	47
Vasopressin	Cat	Constriction	Dilation	47

*Arterial coronary microvessels < 100–150 μm.
[†]Arterial coronary microvessels > 100–150 μm.
Source: Adapted from Ref. 2.

β_1- and β_2-adrenoceptor subtype population. Activation of cholinergic receptors by either vagal stimulation or the infusion of acetylcholine produces uniform vasodilation of coronary vessels.[29] This vasodilation is predominantly mediated by endothelium-derived NO•, although release of EDHF,[30] and the release of prostaglandin substances may also contribute.[31] The coronary flow increase by vagal stimulation can be blunted by a metabolically-mediated flow decrease caused by a decrease in the heart rate and myocardial contractility.[32]

Other neurotransmitters that act on the coronary circulation include neuropeptide Y. NPY is mainly released with norepinephrine as a co-transmitter from sympathetic post-ganglionic nerve terminals upon intense sympathetic activation.[2,33] Intracoronary application of NPY markedly decreases coronary flow, producing myocardial ischemia without large coronary artery constriction.[34,35] These results point to its potent and specific constrictor effects on coronary microvessels. Also, substance P, a potent vasodilator whose effect is dependent on the endothelium, is contained in perivascular nerve fibers and sensory ganglia.[36]

3.6. *Intrinsic myogenic tone*

Myogenic contraction is observed when applying luminal pressure to microvessels causing a development of intrinsic vascular tone as shown by elevated wall tension or a decrease in vessel diameter. The coronary microcirculation, like other vascular beds, possesses this intrinsic myogenic tone response which also contributes to maintaining basal vascular tone and autoregulation.[37] Myogenic microvascular control is endothelium-independent, and is likely to play a critical role in determining the basal tone and maintaining the intraluminal pressure of the downstream exchange vessels within a physiological level.[38,39] Myogenic responses to increases in pressure are greater in subepicardial microvessels than in vessels from the subendocardium. Also, myogenic tone may be reduced during inflammatory states where there is increased expression of iNOS causing altered myocardial perfusion. Increases in myogenic tone, which occur during stretch of vascular smooth muscle, are associated with an increase in inositol-1,4,5-trisphosphate, presumably due to activation of phospholipase C.[37,40,41] Also, the myogenic tone mediator 20-HETE produces constriction of vascular smooth muscle by promoting Ca^{2+}-activated K^+ (BK) channel inhibition. This induces depolarization and increases levels of $[Ca^{2+}]_i$ This effect is likely caused by activation of L-type Ca^{2+} channels and/or the activation of PKC and inhibition of the Na-K-ATPase.[42] Other mediators likely involved in the myogenic tone response include mitogen activated protein (MAP) kinases and Rho protein. The downstream mediator

Rho kinase may modulate myogenic tone by regulation of the actin cytoskeleton. One potential therapeutic agent for treatment of hypertension and coronary spasm is the use of Rho kinase inhibitors. A summary of possible mechanisms of myogenic tone are summarized in Fig. 9.

3.7. *Impact of extravascular and humoral factors on the coronary microcirculation*

In the setting of pathologic processes such as ischemia leading to decreased tissue compliance or increased tissue edema, extravascular forces play an especially important role. For example, collateral perfusion is particularly sensitive to changes in heart rate (more frequent extravascular compression) and ventricular diameter (stretch).[6,43] The coronary circulation is particularly unique in that it is exposed to a large number of extravascular forces produced by contraction of adjacent myocardium and intraventricular pressures. Of relevance to the concept of extravascular forces is the idea that these may collapse coronary vessels under certain circumstances. Of note, flow through the epicardial coronary arteries halted when aortic pressure fell to values ranging from 25 to 50 mmHg raising the possibility that extravascular forces may be sufficiently high to collapse vessels when intraluminal pressure declines to values below this critical value.[44] Flow in the coronary microcirculation continued even when the arterial driving pressure was minimally higher than coronary venous pressure. Based on modeling and various experimental interventions, it was determined that the decrease of antegrade blood flow in larger upstream vessels associated with continued forward flow in microvessels was likely due to capacitance in the coronary circulation.[45] Kanatsuka and colleagues used a floating microscope to visualize epicardial capillaries and were able to show that red cells continued to flow, even after perfusion had stopped in the more proximal vessels.[46] Using this approach, they were able to show that the pressure at which flow stops in the epicardial coronary microvessels was only a few mmHg higher than right atrial pressure.[46] Also, when ventricular diastolic pressure is high, vessels deeper in the subendocardium may be made to collapse by pressure transmitted from

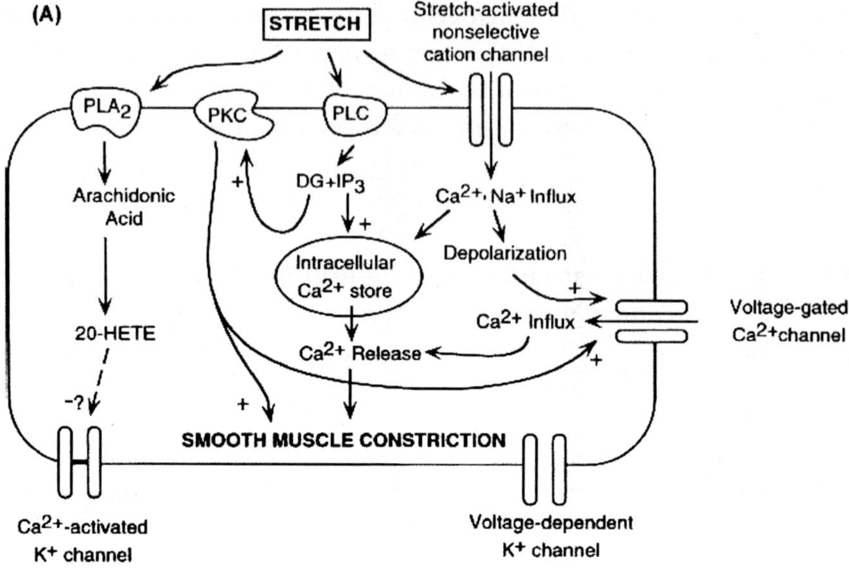

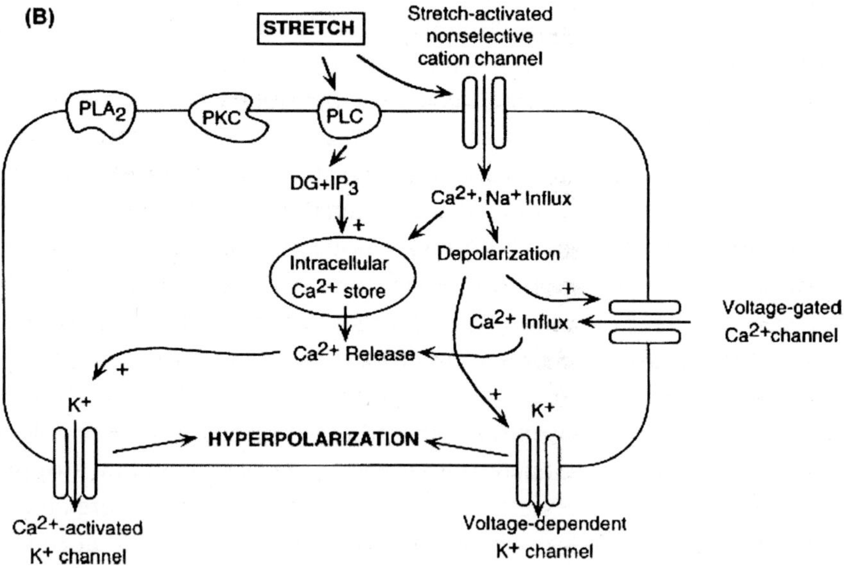

Fig. 9. Schematic illustration for the possible mechanisms of the myogenic constriction (**A**) and its compensatory mechanisms (**B**). DG, Diacylglycerol; 20-HETE, 20-hydroxyeicosatetraenoic acid. (*Adapted from Ref. 2.*)

the ventricular chamber. In contrast, coronary epicardial vessels do not close at any pressure. It therefore seems likely that the concept of "critical closing pressure" is not applicable to all vessels in the coronary circulation.

The coronary microcirculation responds to humoral agents differently depending on vessel size and location. Endothelin-1 produces vasoconstriction when administered to the adventitial surface of coronary microvessels. The degree of constriction produced by endothelin-1 is inversely related to the size of the vessels. In contrast, when endothelin-1 is administered intra-arterially, vasodilation occurs presumably via release of nitric oxide.[29,31] Also, serotonin constricts vessels >100 μm in diameter whereas it causes vasodilation of smaller arteries.[47] Vasopressin, on the other hand, produces greater constriction of microvessels less than 100 μm in diameter than it produces in larger microvessels.[47,48] In the larger epicardial coronary arteries, vasopressin causes predominantly vasodilation.

3.8. *Role of venules in coronary resistance*

Vasomotor regulation is differentially controlled between the venous and arterial microcirculations and certain reactions to pathologic stimuli occur preferentially on one side of the capillary bed. Therefore, consideration of the venous circulation apart from the arterial circulation is needed. Venules have a considerable importance under conditions of vascular dilation such as during exercise, metabolic stress or during reperfusion after myocardial ischemia.[9] The venous circulation may influence myocardial stiffness and relaxation properties of the heart. Veins also respond differently to agonists and neuronal stimulation compared to arteries in the same vascular bed.[31,49] Also, venules are the initiating site of neutrophil adherence and transmigration whereas arterioles seldom manifest these initial changes in the inflammatory response.[50] While ischemia-reperfusion has been determined to cause endothelial dysfunction in veins , under similar conditions, arterioles appear to be more susceptible to a reduction in endothelium-dependent relaxation than are coronary venules, despite the fact that leukocytes preferentially adhere to venular as compared to arterial endothelial cells.[51] In addition, complement fragment C5a causes neutophil adherence in venules but not in arterioles, suggesting that different mechanisms mediate neutrophil-endothelial adherence in the two vessel types.[52]

4. Endothelial Factors in Vascular Growth and Response to Injury

It is important to identify the role of nitric oxide and nitric oxide-related factors in vascular physiology and pathology as summarized in Fig. 10. Nitric oxide inhibits vascular smooth muscle proliferation via apoptosis. Animal models have shown that treatment with L-nitroarginine methyl ester (L-NAME), and inhibitor of NO formation, markedly increases neointimal development following vascular injury.[53] Also,

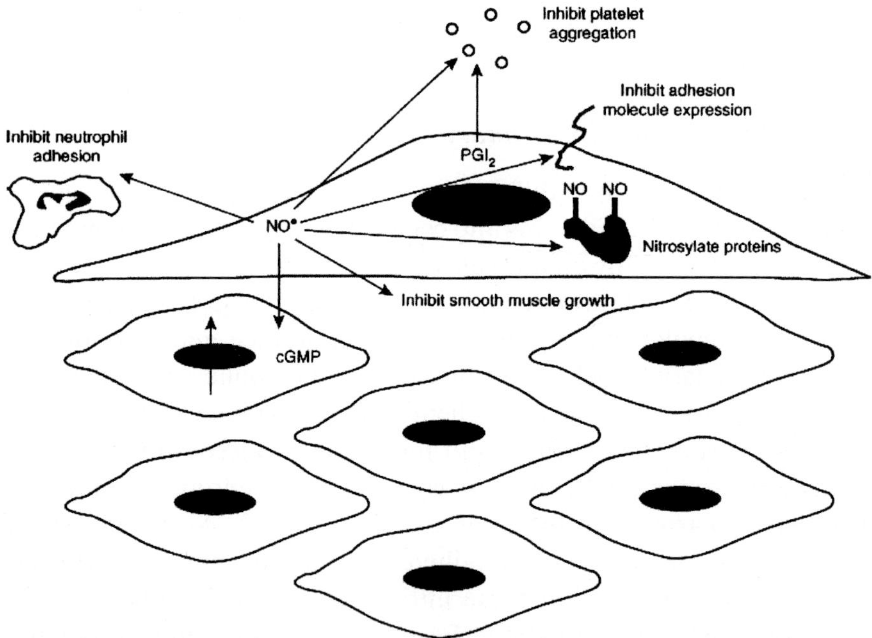

Fig. 10. Schematic representation of endothelium and vascular smooth muscle demonstrating the multifaceted roles of nitric oxide released from the endothelium in the modulation of vascular function, structure, and the response to injury. cGMP, cyclic guanidine monophosphate. (*Adapted from Ref. 71.*)

local transfection with the eNOS cDNA reduces the intimal proliferation which follows balloon injury.[54] The vascular response to injury is enhanced in mice deficient in eNOS.[55,56] Thus, NO• and cyclic GMP elevating agents inhibit the growth of fibroblasts and vascular smooth muscle. This effect of NO on vascular smooth muscle growth is mediated by cGMP and can be mimicked by atrial natriuretic factor.[56,57]

NO plays an important role in supporting the process of angiogenesis; since endothelial cells do not seem to be sensitive to the growth inhibitory effects of nitric oxide. In fact, vascular endothelial growth factor (VEGF) actions during angiogenesis are mediated by NO (see Chapter 14). Endothelial cells in the proliferative phase have a six-fold increase in eNOS expression compared to confluent ones and eNOS knockout mice have little VEGF activity.[19] During the vascular injury response, this feed-forward condition promotes vascular growth — since while endothelial cells are proliferating to form new blood vessels, the high levels of NO promote tube formation. Similarly, in response to the denudation injury, proliferating endothelial cells increase NO production during the growth period to compensate for the lack of endothelial cells in the denuded area while also decreasing platelet adhesion and vascular smooth muscle proliferation in that same area. Moreover, endothelial progenitor cells (EPC) from the bone marrow play a role in repair of denuded vessels as well as angiogenesis. While not completely elucidated, circulating EPCs seem to vary in quantity from one patient to the next depending on the presence of common risk factors such as diabetes (decreased amount) or lipid-lowering drugs such as HMG-Co A reductase inhibitors (increased amount).

5. Impact of Disease States on Coronary Circulation

Coronary microvascular homeostasis may be adversely affected in disease states through variation in their diameter, quantity or responsiveness to humoral factors. Vasomotor tone reliant on endothelial function is particularly vulnerable to pathology such as atherosclerosis, hyperlipidemia, diabetes or the aging process. This mechanism is highlighted in Fig. 11. The mechanisms underlying these abnormal endothelium-dependent responses are likely multifactorial. Factors responsible include abnormalities of G-protein signaling, resulting

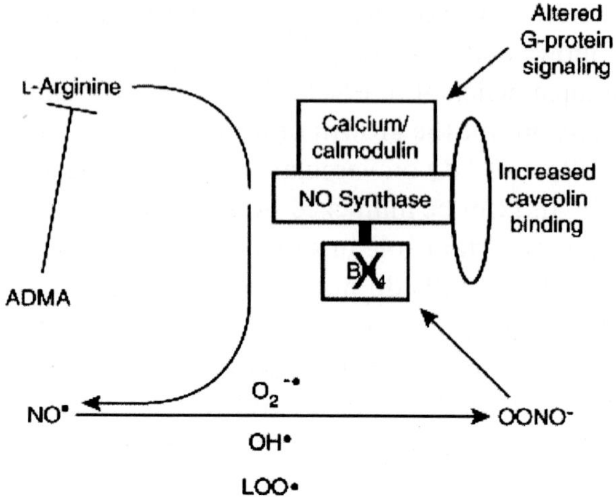

Fig. 11. Reduced production/bioreactivity of endothelium-derived NO in the setting of atherosclerosis, diabetes, and many other pathological conditions. ADMA, asymmetrical dimethylarginine, acts as an antagonist of L-arginine. Superoxide ($O_2^{\bullet-}$) and other oxygen free radicals may interfere with NO availability in conditions of increased oxidant stress. OONO$^{\bullet}$, peroxynitrite radical may inhibit tetrahydrobiopterin (BH$_4$), a cofactor for nitric oxide synthase (NOS). (*Adapted from Ref. 71.*)

in reduced activation of eNOS in response to endothelial cell receptor activation, an alteration of levels of the critical co-factor for eNOS tetrahydrobiopterin (BH$_4$), and an overproduction of the asymmetric dimethylarginine (ADMA) which acts as an antagonist for the eNOS substrate L-arginine. It has been shown that oxidative stress (via increased production of vascular superoxide — $O_2^{\bullet-}$) is particularly increased in the presence of common coronary risk factors. Such an increase in oxidative stress will cause a reduction in endothelium-dependent vasodilatation.

It is currently well proven that diseases that affect endothelial-dependent vascular dilation impact the coronary microcirculation in addition to the larger vessels. Previous experiments have demonstrated that in coronary microvessels from monkeys fed a high cholesterol diet for 18 months, relaxations to acetylcholine, bradykinin, and the calcium ionophore A23187, were dramatically impaired or even produced paradoxical constrictions.[58] Similar findings have been made in other animal models of diet-induced atherosclerosis. Subsequent studies performed using *in vivo* techniques showed that vasoconstriction caused by serotonin and ergonovine (both known to be modulated by the endothelium) was markedly enhanced in the coronary microcirculation of hypercholesterolemic monkeys.[59] These findings are impressive because the coronary microcirculation is spared from the development of overt atherosclerosis. Therefore, in the setting of a risk factor for atherosclerosis, "endothelial dysfunction" occurs leading to an abnormal vascular response. Subsequently, diminished flow responses to acetylcholine have been demonstrated in humans with hypercholesterolemia that were restored with reduction of cholesterol levels.[60] Similar observations have been made in either humans or experimental models of hypertension,[61] ischemia-reperfusion[52,62] and diabetes.[63] It has also been suggested that this endothelial dysfunction plays a role in the development of clinical symptoms despite normal coronary anatomy. Table 2 summarizes studies demonstrating the effects hypercholesterolemia on coronary arterial microvessels.

Impaired endothelial-dependent vasodilation also has been linked to increased cardiovascular events. The loss of NO in cardiovascular disease not only leads to a decrease in vasodilation, but also predisposes to atherosclerotic lesion formation and vascular smooth muscle proliferation. NO also has antioxidant properties and prevents adhesion molecule expression by endothelial cells. An example of relevance to the clinical setting is the endothelial changes in the coronary microcirculation following cardioplegic arrest and cardiopulmonary bypass during cardiac surgery.[64] In this setting, endothelial dysfunction persists for some time after cardiopulmonary bypass, and normalizes thereafter. This has important clinical implications, since it is common for patients undergoing coronary artery bypass grafting, with seemingly complete coronary revascularization, to exhibit signs of myocardial ischemia during the hours following surgery — likely caused by endothelial dysfunction.

Collateral vessels within the coronary circulation are particularly important in coronary disease. These allow for normal resting perfusion to a region of the myocardium that is served by an occluded vessel, albeit at a lower perfusion pressure. However, the coronary arterioles

Table 2. Functional alteration of coronary arterial microvessels by hypercholesterolemia.

Material	Cholesterol feeding	Experiments	Vessel size	Microvascular functional alteration	Ref. no.
Rabbit	0.5% or 2% cholesterol, 10–12 weeks	*In vitro*	290 μm in mean value	Reduced dilation to ACh, ADP	78
Monkey	0.7% cholesterol, 18 months	*In vitro*	12–220 μm	Enhanced contraction to ACh; reduced dilation to BK, A23187	58
Monkey	0.8% cholesterol, 18 months	*In vivo* microvascular pressure	< 190 – 350μm	Increase in microvascular resistance to 5-HT	59
Monkey	0.7% cholesterol, 8–12 weeks or 18–80 weeks	*In vitro*	100–200 μm	Reduced dilation to ADP, 5-HT; paradoxical constriction to thrombin; hyperconstriction to TXA, analogue	79
Pig	3% cholesterol, 16–20 weeks	*In vitro*	30–70 μm	Reduced dilation to flow, ADP, 5-HT histamine, BK	80
Monkey	0.7% cholesterol, 25 months	*In vivo*	~ 300μm	Enhanced constriction to 5-HT in mean value	81
Pig	2% cholesterol, 10–13 weeks	*In vitro*	300–480 μm	Enhanced constriction to HT-1, S6c in mean value	82
Human	CAD patients	*In vitro*	57–183 μm	Reduced dilation to HGF, VEGF	83

CAD, Coronary arterial disease; S6c, semafotoxin, ETB receptor agonist.
Source: Adapted from Ref. 2.

nourished by collaterals develop markedly abnormal vascular reactivity such as by impaired endothelium-dependent vascular relaxations and enhanced constrictions to vasopressin.[48] Possible mechanisms of this impaired microvascular endothelium-dependent relaxation in the collateral-dependent region may include changes in shear stress, pulsatile flow in the collateral-dependent microvasculature or intracellular calcium levels. Such changes may cause disturbance in microvascular tone during a disease state.[18]

It has been also found that treatment of collateral-dependent vessels with angiogenic growth factors may enhance endothelium-dependent relaxation, in addition to improving other aspects of cardiac performance. In patients suffering from disease not amenable to current intervention techniques (e.g. surgical coronary grafting or percutaneous techniques), direct treatment with angiogenic factors can theoretically be the basis for a clinical improvement. Several studies have demonstrated that therapeutic angiogenic interventions, in the setting of chronic ischemia, are associated with a marked improvement in myocardial function, myocardial perfusion and endothelial-dependent vasodilation within the area supplied by collaterals.[65–67] These studies used growth factors such as VEGF, FGF-1 or FGF-2 placement in the perivascular area. Possible mechanisms through which these factors act include FGF-2- and VEGF-induced release of NO, which improves collateral perfusion and decreases tissue ischemia.[68] In addition, it has also been shown that during periods of chronic ischemia there occurs an upregulation of FGF-2 and VEGF receptors. This finding is consistent with results showing that after administration of growth factors, endothelial-dependent relaxation occurs in the collateral-dependent region but not in the myocardium perfused via original vessels.[69] Also, these growth factors may stimulate the release of bone marrow-derived EPCs which promote collateral growth and endothelial function at the treated sites.

6. The Coronary Microcirculation in Hypertophic States

Clinically, patients suffering from ventricular hypertrophy often complain of angina-like symptoms. Animal and human studies have demonstrated that cardiac hypertrophy causes a reduction in the maximal capacity of the coronary circulation to dilate in response to either reactive hyperemia or pharmacological stimuli.[5,61,70] One possible cause for this abnormal response may be caused by a mismatch between the elevated myocardial mass and relatively reduced coronary microcirculation. Peak flow normalized to myocardial mass may be reduced because of this relative paucity of coronary arterioles — since as the myocardium hypertrophies, the coronary resistance circulation may not increase adequately to keep pace with the larger muscle mass (see Chapter 9). Another possible mechanism of impaired vasodilator responses may be explained by endothelial dysfunction since many of the diseases associated with myocardial hypertrophy are also associated with a loss of endothelial NO production.

In normal hearts there is a linear relationship between the diameter of an epicardial coronary artery and the mass of myocardium perfused. Interestingly, epicardial coronary arteries do not enlarge appropriately as the myocardium hypertrophies such that for any diameter coronary artery, the amount of myocardium perfused is increased up to twofold. This phenomenon is particularly relevant in the presence of a coronary stenosis where a small lesion that is otherwise considered minimal, becomes flow limiting in a hypertrophic state.

7. Summary

This chapter provided an overview of some of the newer concepts regarding physiological, pathophysiological, and pharmacological control of the coronary microcirculation. It also emphasized that the properties of peripheral vessels cannot be extrapolated to the coronary circulation. Similarly, properties of one size or class of coronary microvessel cannot be generalized. Clearly, there has been dramatic changes in the technology used during older studies compared to the more recent ones. While we attempted to present studies that have directly examined the coronary microvessels using some of the newer technology (*in vitro* preparations or *in situ* observations), it was also important to present classical studies of the intact coronary circulation performed in intact animals or isolated hearts. Newer research questions have necessitated the use of more basic techniques, including cell culture and molecular biological approaches. Recent development has been the ability to make many *in vivo* measurements of coronary hemodynamics in human subjects in the catheterization laboratory, thus obviating the need for large expenses for large animal flow studies. As the field of vascular biology continues to grow, it will remain necessary for us to validate the observations seen during basic studies in the intact circulation found in our patients so this can be translated to the clinical setting.

References

1. Furchgott RF, Zawadzki JV (1980). The obligatory role of endothelial cells in the relaxation of arterial smooth muscle by acetylcholine. *Nature* **288**: 373–376.
2. Komaru T, Kanatsuka H, Shirato K (2000). Coronary microcirculation: physiology and pharmacology. *Pharmacol Ther* **86**: 217–261.
3. Feigl EO (1983). Coronary physiology. *Physiol Rev* **63**: 1–205.
4. Hoffman JI (1987). Transmural myocardial perfusion. *Prog Cardiovasc Dis* **29**: 429–464.
5. Marcus ML (1983). *The Coronary Circulation in Health and Disease* (McGraw-Hill, New York).
6. Schaper W (1979). *The Pathophysiology of Myocardial Perfusion* (Elsevier/North-Holland Biomedical Press, Amsterdam/New York).
7. Nellis SH, Liedtke AJ, Whitesell L (1981). Small coronary vessel pressure and diameter in an intact beating rabbit heart using fixed-position and free-motion techniques. *Circ Res* **49**: 342–353.
8. Chilian WM, Eastham CL, Marcus ML (1986). Microvascular distribution of coronary vascular resistance in beating left ventricle. *Am J Physiol* **251**: H779–H788.
9. Chilian WM, Layne SM, Klausner EC, Eastham CL, Marcus ML (1989). Redistribution of coronary microvascular resistance produced by dipyridamole. *Am J Physiol* **256**: H383–H390.
10. Chilian WM (1991). Microvascular pressures and resistances in the left ventricular subepicardium and subendocardium. *Circ Res* **69**: 561–570.
11. Fujii M, Nuno DW, Lamping KG, Dellsperger KC, Eastham CL, Harrison DG (1992). Effect of hypertension and hypertrophy on coronary microvascular pressure. *Circ Res* **71**: 120–126.
12. Jones CJ, Kuo L, Davis MJ, Chilian WM (1995). Regulation of coronary blood flow: coordination of heterogeneous control mechanisms in vascular microdomains. *Cardiovasc Res* **29**: 585–596.
13. Murad F (1986). Cyclic guanosine monophosphate as a mediator of vasodilation. *J Clin Invest* **78**: 1–5.

14. Corson MA, James NL, Latta SE, Nerem RM, Berk BC, Harrison DG (1996). Phosphorylation of endothelial nitric oxide synthase in response to fluid shear stress. *Circ Res* **79**: 984–991.

15. Venema RC, Sayegh HS, Arnal JF, Harrison DG (1995). Role of the enzyme calmodulin-binding domain in membrane association and phospholipid inhibition of endothelial nitric oxide synthase. *J Biol Chem* **270**: 14705–14711.

16. Michel JB, Feron O, Sacks D, Michel T (1997). Reciprocal regulation of endothelial nitric-oxide synthase by Ca2+-calmodulin and caveolin. *J Biol Chem* **272**: 15583–15586.

17. Myers PR, Minor RL, Jr, Guerra R, Jr, Bates JN, Harrison DG (1990) Vasorelaxant properties of the endothelium-derived relaxing factor more closely resemble S-nitrosocysteine than nitric oxide. *Nature* **345**: 161–163.

18. Uematsu M, Ohara Y, Navas JP, Nishida K, Murphy TJ, Alexander RW, Nerem RM, Harrison DG (1995). Regulation of endothelial cell nitric oxide synthase mRNA expression by shear stress. *Am J Physiol* **269**: C1371–C1378.

19. Arnal JF, Yamin J, Dockery S, Harrison DG (1994). Regulation of endothelial nitric oxide synthase mRNA, protein, and activity during cell growth. *Am J Physiol* **267**: C1381–C1388.

20. McQuillan LP, Leung GK, Marsden PA, Kostyk SK, Kourembanas S (1994). Hypoxia inhibits expression of eNOS via transcriptional and posttranscriptional mechanisms. *Am J Physiol* **267**: H1921–H1927.

21. Duncker DJ, van Zon NS, Ishibashi Y, Bache RJ (1996). Role of K^+ ATP channels and adenosine in the regulation of coronary blood flow during exercise with normal and restricted coronary blood flow. *J Clin Invest* **97**: 996–1009.

22. Boatwright RB, Downey HF, Bashour FA, Crystal GJ (1980). Transmural variation in autoregulation of coronary blood flow in hyperperfused canine myocardium. *Circ Res* **47**: 599–609.

23. Pohl U, Holtz J, Busse R, Bassenge E (1986). Crucial role of endothelium in the vasodilator response to increased flow *in vivo. Hypertension* **8**: 37–44.

24. Drexler H, Zeiher AM, Wollschlager H, Meinertz T, Just H, Bonzel T. (1989). Flow-dependent coronary artery dilatation in humans. *Circulation* **80**: 466–474.

25. Jimenez AH, Tanner MA, Caldwell WM, Myers PR (1996). Effects of oxygen tension on flow-induced vasodilation in porcine coronary resistance arterioles. *Microvasc Res* **51**: 365–377.

26. Kuo L, Chilian WM, Davis MJ (1991). Interaction of pressure- and flow-induced responses in porcine coronary resistance vessels. *Am J Physiol* **261**: H1706–H1715.

27. Young MA, Knight DR, Vatner SF (1987). Autonomic control of large coronary arteries and resistance vessels. *Prog Cardiovasc Dis* **30**: 211–234.

28. Wang SY, Friedman M, Johnson RG, Weintraub RM, Sellke FW (1994). Adrenergic regulation of coronary microcirculation after extracorporeal circulation and crystalloid cardioplegia. *Am J Physiol* **267**: H2462–H2470.

29. Lamping KG, Chilian WM, Eastham CL, Marcus ML (1992). Coronary microvascular response to exogenously administered and endogenously released acetylcholine. *Microvasc Res* **43**: 294–307.

30. Hammarstrom AK, Parkington HC, Coleman HA (1995). Release of endothelium-derived hyperpolarizing factor (EDHF) by M3 receptor stimulation in guinea-pig coronary artery. *Br J Pharmacol* **115**: 717–722.

31. Sellke FW, Dai HB (1993). Responses of porcine epicardial venules to neurohumoral substances. *Cardiovasc Res* **27**: 1326–1332.

32. Van Winkle DM, Feigl EO (1989). Acetylcholine causes coronary vasodilation in dogs and baboons. *Circ Res* **65**: 1580–1593.

33. Gu J, Polak JM, Adrian TE, Allen JM, Tatemoto K, Bloom SR (1983). Neuropeptide tyrosine (NPY) — a major cardiac neuropeptide. *Lancet* **1**: 1008–1010.

34. Clarke JG, Davies GJ, Kerwin R, Hackett D, Larkin S, Dawbarn D, Lee Y, Bloom SR, Yacoub M, Maseri A (1987). Coronary artery infusion of neuropeptide Y in patients with angina pectoris. *Lancet* **1**: 1057–1059.

35. Maturi MF, Greene R, Speir E, Burrus C, Dorsey LM, Markle DR, Maxwell M, Schmidt W, Goldstein SR, Patterson RE (1989). Neuropeptide-Y. A peptide found in human coronary arteries constricts primarily small coronary arteries to produce myocardial ischemia in dogs. *J Clin Invest* **83**: 1217–1224.

36. Mione MC, Ralevic V, Burnstock G (1990). Peptides and vasomotor mechanisms. *Pharmacol Ther* **46**: 429–468.

37. Kuo L, Davis MJ, Chilian WM (1988). Myogenic activity in isolated subepicardial and subendocardial coronary arterioles. *Am J Physiol* **255**: H1558–H1562.

38. Davis MJ (1988). Microvascular control of capillary pressure during increases in local arterial and venous pressure. *Am J Physiol* **254**: H772–H784.

39. Kuo L, Chilian WM, Davis MJ (1990). Coronary arteriolar myogenic response is independent of endothelium. *Circ Res* **66**: 860–866.

40. Narayanan J, Imig M, Roman RJ, Harder DR (1994). Pressurization of isolated renal arteries increases inositol trisphosphate and diacylglycerol. *Am J Physiol* **266**: H1840–H1845.

41. Osol G, Laher I, Cipolla M (1991). Protein kinase C modulates basal myogenic tone in resistance arteries from the cerebral circulation. *Circ Res* **68**: 359–567.

42. Miller FJ, Jr., Dellsperger KC, Gutterman DD (1997). Myogenic constriction of human coronary arterioles. *Am J Physiol* **273**: H257–H264.

43. Conway RS, Kirk ES, Eng C (1988). Ventricular preload alters intravascular and extravascular resistances of coronary collaterals. *Am J Physiol* **254**: H532–H541.

44. Bellamy RF (1978). Diastolic coronary artery pressure-flow relations in the dog. *Circ Res* **43**: 92–101.

45. Eng C, Jentzer JH, Kirk ES (1982). The effects of the coronary capacitance on the interpretation of diastolic pressure-flow relationships. *Circ Res* **50**: 334–341.

46. Kanatsuka H, Ashikawa K, Komaru T, Suzuki T, Takishima T (1990). Diameter change and pressure-red blood cell velocity relations in coronary microvessels during long diastoles in the canine left ventricle. *Circ Res* **66**: 503–510.

47. Lamping KG, Kanatsuka H, Eastham CL, Chilian WM, Marcus ML (1989). Nonuniform vasomotor responses of the coronary microcirculation to serotonin and vasopressin. *Circ Res* **65**: 343–351.

48. Sellke FW, Quillen JE, Brooks LA, Harrison DG (1990). Endothelial modulation of the coronary vasculature in vessels perfused via mature collaterals. *Circulation* **81**: 1938–1947.

49. Klassen GA, Armour JA (1982). Epicardial coronary venous pressures: autonomic responses. *Can J Physiol Pharmacol* **60**: 698–706.
50. Yuan Y, Mier RA, Chilian WM, Zawieja DC, Granger HJ (1995). Interaction of neutrophils and endothelium in isolated coronary venules and arterioles. *Am J Physiol* **268**: H490–H498.
51. Lefer DJ, Nakanishi K, Vinten-Johansen J, Ma XL, Lefer AM (1992). Cardiac venous endothelial dysfunction after myocardial ischemia and reperfusion in dogs. *Am J Physiol* **263**: H850–H856.
52. Piana RN, Wang SY, Friedman M, Sellke FW (1996). Angiotensin-converting enzyme inhibition preserves endothelium-dependent coronary microvascular responses during short-term ischemia-reperfusion. *Circulation* **93**: 544–551.
53. Cayatte AJ, Palacino JJ, Horten K, Cohen RA (1994). Chronic inhibition of nitric oxide production accelerates neointima formation and impairs endothelial function in hypercholesterolemic rabbits. *Arterioscler Thromb* **14**: 753–759.
54. von der Leyen HE, Gibbons GH, Morishita R, Lewis NP, Zhang L, Nakajima M, Kaneda Y, Cooke JP, Dzau VJ (1995). Gene therapy inhibiting neointimal vascular lesion: in vivo transfer of endothelial cell nitric oxide synthase gene. *Proc Natl Acad Sci U S A* **92**: 1137–1141.
55. Moroi M, Zhang L, Yasuda T, Virmani R, Gold HK, Fishman MC, Huang PL (1998). Interaction of genetic deficiency of endothelial nitric oxide, gender, and pregnancy in vascular response to injury in mice. *J Clin Invest* **101**: 1225–1232.
56. Yu SM, Hung LM, Lin CC (1997). cGMP-elevating agents suppress proliferation of vascular smooth muscle cells by inhibiting the activation of epidermal growth factor signaling pathway. *Circulation* **95**: 1269–1277.
57. Itoh H, Pratt RE, Ohno M, Dzau VJ (1992). Atrial natriuretic polypeptide as a novel antigrowth factor of endothelial cells. *Hypertension* **19**: 758–761.
58. Sellke FW, Armstrong ML, Harrison DG (1990). Endothelium-dependent vascular relaxation is abnormal in the coronary microcirculation of atherosclerotic primates. *Circulation* **81**: 1586–1593.
59. Chilian WM, Dellsperger KC, Layne SM, Eastham CL, Armstrong MA, Marcus ML, Heistad DD (1990). Effects of atherosclerosis on the coronary microcirculation. *Am J Physiol* **258**: H529–H539.
60. Drexler H, Zeiher AM, Meinzer K, Just H (1991). Correction of endothelial dysfunction in coronary microcirculation of hypercholesterolaemic patients by L-arginine. *Lancet* **338**: 1546–1550.
61. Treasure CB, Klein JL, Vita JA, Manoukian SV, Renwick GH, Selwyn AP, Ganz P, Alexander RW (1993). Hypertension and left ventricular hypertrophy are associated with impaired endothelium-mediated relaxation in human coronary resistance vessels. *Circulation* **87**: 86–93.
62. Quillen JE, Sellke FW, Brooks LA, Harrison DG (1990). Ischemia-reperfusion impairs endothelium-dependent relaxation of coronary microvessels but does not affect large arteries. *Circulation* **82**: 586–594.
63. Matsunaga T, Okumura K, Ishizaka H, Tsunoda R, Tayama S, Tabuchi T, Yasue H (1996). Impairment of coronary blood flow regulation by endothelium-derived nitric oxide in dogs with alloxan-induced diabetes. *J Cardiovasc Pharmacol* **28**: 60–67.

64. Sellke FW, Shafique T, Schoen FJ, Weintraub RM (1993). Impaired endothelium-dependent coronary microvascular relaxation after cold potassium cardioplegia and reperfusion. *J Thorac Cardiovasc Surg* **105**: 52–58.

65. Bauters C, Asahara T, Zheng LP, Takeshita S, Bunting S, Ferrara N, Symes JF, Isner JM (1995). Recovery of disturbed endothelium-dependent flow in the collateral-perfused rabbit ischemic hindlimb after administration of vascular endothelial growth factor. *Circulation* **91**: 2802–2809.

66. Harada K, Friedman M, Lopez JJ, Wang SY, Li J, Prasad PV, Pearlman JD, Edelman ER, Sellke FW, Simons M (1996). Vascular endothelial growth factor administration in chronic myocardial ischemia. *Am J Physiol* **270**: H1791–H1802.

67. Sellke FW, Wang SY, Friedman M, Harada K, Edelman ER, Grossman W, Simons M (1994). Basic FGF enhances endothelium-dependent relaxation of the collateral-perfused coronary microcirculation. *Am J Physiol* **267**: H1303–H1311.

68. Sellke FW, Wang SY, Stamler A, Lopez JJ, Li J, Simons M (1996). Enhanced microvascular relaxations to VEGF and bFGF in chronically ischemic porcine myocardium. *Am J Physiol* **271**: H713–H720.

69. Rapps JA, Jones AW, Sturek M, Magliola L, Parker JL (1997). Mechanisms of altered contractile responses to vasopressin and endothelin in canine coronary collateral arteries. *Circulation* **95**: 231–239.

70. Marcus ML, Harrison DG, Chilian WM, Koyanagi S, Inou T, Tomanek RJ, Martins JB, Eastham CL, Hiratzka LF (1987). Alterations in the coronary circulation in hypertrophied ventricles. *Circulation* **75**: I19–I25.

71. Harrison DG, Doughan AR, Sellke FW (2005). Physiology of the coronary circulation. In: Sellke FW, Del Nido PJ, Swanson SJ (eds.), *Sabiston and Spencer Surgery of the Chest*, 7th edn. (Elsevier Inc., Philadelphia), pp. 753–766.

72. Gallery HF, Webster NR (2004). Physiology of the endothelium. *Br J Anaesth* **93**(1): 105–113.

73. Vanhoutte PM, Boulanger CM, Vidal M, Mombouli JV (1993). Endothelium-derived mediators and the renin-angiotensin system. In: Robertson JIS, Nicholls ME (eds.), *The Renin-Angiotensin System* (Gower Medical Publishing, London).

74. Komaru T, Ashikawa K, Kanatsuka H, Sekiguchi N, Suzuki T, Takishima T (1990). Neuropeptide Y modulates vasoconstriction in coronary microvessels in the beating canine heart. *Circ Res* **67**: 1142–1151.

75. Sekiguchi N, Kanatsuka H, Sato K, Wang Y, Akai K, Komaru T, Takishima T (1994). Effect of calcitonin gene-related peptide on coronary microvessels and its role in acute myocardial ischemia. *Circulation* **89**: 366–374.

76. Kanatsuka H, Lamping KG, Eastham CL, Dellsperger KC, Marcus ML (1989). Comparison of the effects of increased myocardial oxygen consumption and adenosine on the coronary microvascular resistance. *Circ Res* **65**: 1296–1305.

77. Lamping KG, Clothier JL, Eastham CL, Marcus ML (1992). Coronary microvascular response to endothelin is dependent on vessel diameter and route of administration. *Am J Physiol* **263**(3 Pt 2): H703–H709.

78. Osborne JA, Siegman MJ, Sedar AW, Mooers SU, Lefer AM (1989). Lack of endothelium-dependent relaxation in coronary resistance arteries of cholesterol-fed rabbits. *Am J Physiol* **256**(3 Pt 1): C591–C597.

79. Quillen JE, Sellke FW, Armstrong ML, Harrison DG (1991). Long-term cholesterol feeding alters the reactivity of primate coronary microvessels to platelet products. *Arterioscler Thromb* **11**: 639–644.
80. Kuo L, Davis MJ, Cannon MS, Chilian WM (1992). Pathophysiological consequences of atherosclerosis extend into the coronary microcirculation. Restoration of endothelium-dependent responses by L-arginine. *Circ Res* **70**: 465–476.
81. Lamping KG, Piegors DJ, Benzuly KH, Armstrong ML, Heistad DD (1994). Enhanced coronary vasoconstrictive response to serotonin subsides after removal of dietary cholesterol in atherosclerotic monkeys. *Arterioscler Thromb* **14**: 951–957.
82. Hasdai D, Mathew V, Schwartz RS, Smith LA, Holmes DR Jr, Katusic ZS, Lerman A (1997). Enhanced endothelin-B-receptor-mediated vasoconstriction of small porcine coronary arteries in diet-induced hypercholesterolemia. *Arterioscler Thromb Vasc Biol* **17**: 2737–2743.
83. Métais C, Li J, Li J, Simons M, Sellke FW (1998). Effects of coronary artery disease on expression and microvascular response to VEGF. *Am J Physiol* **275** (4 Pt 2): H1411–H1418.

11

Kinase Inhibitors: Cancer Drugs Derived from Mechanistic Considerations

*by Karl-Heinz Thierauch
and Andreas Chlistalla*

1. Kinase Inhibition and Tumor Angiogenesis

Angiogenesis describes the process of an expansion of the blood vessel system. It is primarily regulated according to the oxygen needs of growing tissues but is also under hormonal control.[1,2] Tumor angiogenesis is a prerequisite for tumor growth.[3,4] It involves many different cell types of the affected tissue including foremost endothelial cells but also inflammatory cells, fibroblast and other stromal cells which invade the evolving tumors and the tissues around it.[5] Indeed, tumor angiogenesis resembles embryonal angiogenesis in many aspects. The interactive ensemble of these cells is the prerequisite of the angiogenic process. This process is regulated via signaling through mediators which mainly bind to transmembrane receptors, such as cytokines, growth factors and surface ligands, or which permeate directly to the cytoplasm and nucleus, like nuclear hormones.

313

The effect of these factors may be anti-angiogenic or pro-angiogenic. Many of those receptors have a kinase function which are activated by extracellular binding of a ligand and transmit the signal intracellularly via an activation cascade which contains several kinases itself. Therefore, inhibition of any of those kinases may fall under the paradigm of angiogenesis inhibition especially, as proliferating endothelial cells are relatively sensitive to the toxic action of non-specific kinase inhibitors. Many of these kinases have been described as oncogenes involved in tumorigenesis and are potential targets of tumor cell-directed therapy. Some kinases may activate receptor tyrosine phosphatases so their inhibition can be considered as pro-angiogenic.

2. Major Angiogenesis Factors and Receptors

2.1. *VEGF signaling*

In a stricter sense, only kinases which are essential and specific for endothelial cells, or the interaction with them, should be considered as anti-angiogenesis targets. Endothelium-specific kinases, which play a major role in angiogenesis are VEGFR-1, -2 and -3 and Tie1 and -2. VEGFR-2 is considered to play a pivotal role in angiogenesis and is almost exclusively located on the surface of endothelial cells. The inhibition of its kinase function leads to an inhibition of cell proliferation and migration.[6,7] Important exceptions, where other functions are assumed, are found in kidney glomeruli,[8] on hematopoietic stem cell precursors[9] and in a neural stem cell population.[10] Stimulators of VEGFR-2 are the known splice variants of VEGF-A,[11] VEGF-C (processed form[12]) and VEGF-D (processed form[13]). VEGFR-1, which is an indispensable co-receptor of VEGFR-2, also binds to VEGF-B and PlGF.[14] This receptor is found on peripheral blood monocytic cells (PBMC), where it plays a role in monocyte migration and tissue factor expression.[14,15] VEGFR-3, which transduces a signal for lymphangiogenesis, is activated by VEGF-C and VEGF-D.[12]

Other binding proteins for VEGF-A on the surface of endothlial cells are neuropilin-1 and -2.[16,17] At the C-terminus of the neuropilins, there is a short intracellular peptide chain, which makes a direct signaling into the endothelial cell improbable. In addition to VEGF, neuropilins bind semaphorins and hepatocyte growth factor (HGF).[18] This will result in competition between those ligands for the binding site, because some of them are competitive binders. Neuropilins which bind VEGF are co-opted and form heteromers with VEGFRs, plexins and c-met, the receptor of HGF. In addition to their function on endothelial cells, HGF and semaphorins play a role in cancer growth and spread.[19,20] This leads to the present picture, that neuropilins modulate the VEGF signaling in endothelial cells. A bypassing of the kinase activity of VEGFRs in endothelial cells via interaction of VEGF with neuropilin has not been demonstrated unequivocally up till now. Even the activation of VEGFR-2 by phosphorylation through semaphorin 6d and plexin-A1, which is described to occur in the absence of VEGF[21] should be blocked by VEGFR kinase inhibitors. However, VEGF may exert an action on tumor cells competing with sema 3a for the binding to neuropilins, and thus may modulate the migratory response of tumor cells as it was shown for MDA-MB-231 cells.[22]

3. Further Angiogenesis-Related Signaling

PDGFR kinases located mainly on stromal cells are additional important players. They are stimulated by PDGFs secreted by endothelial cells.[23] Also EphB4 receptor kinase shows endothelial specificity but not a completely defined function.[24] EphB4 and its ligand ephrinB2 appear to play a role in the demarcation of boundaries between cells with differential fate.[25] In recent years, several new factors and their role in angiogenesis, especially in the guidance of newly developing vessel sprouts have been found. Amongst them are notch and jagged,[26] wnt and frizzled,[27] slit and robo,[28,29] netrin and Unc5B,[30] and the BMP/TGFβ family with their corresponding receptors.[31,32] From those factors new anti-angiogenic targets will arise, which may capitalize on kinase inhibition.

4. Need for Selectivity of Anti-Angiogenic Kinase Inhibitors

Kinase inhibitors in clinical development as anti-angiogenic agents target VEGFRs with first priority. However, the extent to which the VEGFR inhibitors affect other cellular kinases as well is variable. One method to assess the selectivity of inhibitors for endothelial cells is the study of their effect on cell proliferation. Only VEGF-stimulated endothelial cell proliferation but not tumor cell proliferation should be inhibited. As can be seen in Table 1 many of the angiogenesis kinase inhibitors, which target angiogenesis are not specific for endothelial cells but inhibit tumor cell growth directly. One would like to argue that the mechanism does not matter, since the effect on tumor growth is relevant. However, it has to be taken into consideration that the toxicity of such agents may be paralleling their lack of specificity. Standard tumor therapy generally includes combinations with cytotoxic chemotherapeutics. If a toxic kinase inhibitor is added to the cytotoxic regimen, the quality of life of patients is influenced and a reduction of the dose of chemotherapeutics may be required, thus jeopardizing the overall therapeutic effect.

The more than 500 kinases derived from the human genome have been ordered in a phylogenetic tree and families of kinases are defined.[33] VEGFRs belong to the receptor tyrosine kinase family V with split kinase domain, which is closely related to the receptor tyrosine family III, which also has a split kinase domain but the number of extracellular Ig domains is reduced from seven to five. Fabian *et al.* elegantly showed the breadth of kinases inhibited by a group of inhibitors labeled with the tag "angiogenesis inhibitor".[34] Activity was tested against 113 different kinases. In Fig. 1, the inhibitory spectra are depicted of some of the clinically investigated anti-angiogenic kinase inhibitors taken from this publication, demonstrating the questionable selectivity of presently developed inhibitors, possibly with the exception of PTK787/ZK 222584 (PTK/ZK).

In the following, we discuss several kinase inhibitors in clinical development, that target VEGFR kinases. We will also present basic information of their preclinical properties.

5. Kinase Inhibitors in Clinical Development

The structures of the molecules in clinical development with available information are given in Fig. 2. Data regarding their kinase selectivity and cellular efficacy are summarized in Table 2.

Table 1. Overview of tyrosine kinase inhibitors with anti-angiogenic effects currently in clinical development.

Drug	Phase of dev.	Safe dose	Side effects	Current efficacy data	ASCO 2005 Abstract #
BAY 43-9006 (Sorafenib, Bayer)	Phase III	400 mg bid	Diarrhoea, skin rash, hand-foot syndrome, fatigue, hypertension	Second line RCC positive phase III trial. Phase III in advanced melanoma ongoing. Phase II program ongoing in various indications	3005 3037 3054 3062
PTK787/ZK 222584 (Vatalanib, Novartis, Schering AG)	Phase III	1250 mg QD	Ataxia, vertigo, nausea, hypertension, venous thromboembolism, fatigue	Various phase I/II studies have shown promising activity in RCC, GBM, CRC and mesothelioma. Phase III trial in first line treatment of CRC (CONFIRM1) did not reach primary endpoint	3 5042
SU11248 (Sunitinib, Pfizer)	Phase III	50 mg QD	Fatigue, lethargy, nausea, stomatitis, diarrhea,myelosuppresion	Ongoing phase III evaluation in first and second line RCC and Imatinib-resistant GIST and NSCLC	3006 3040

(*Continued*)

Table 1. (*Continued*).

Drug	Phase of dev.	Safe dose	Side effects	Current efficacy data	ASCO 2005 Abstract #
ZD6474 (Astra-Zeneca)	Phase II	< 300 mg QD	Diarrhoea, hypertension, hepatic toxicity, cutaneous rash, asymptomatic QTc prolongation	Promising data in phase I and phase II especially in NSCLC second line. Phase III in second line NSCLC in combination with docetaxel planned to start in 2005. Phase II in SCLC ongoing	3023 7102
AG-013736 (Pfizer)	Phase II	5 mg bid	Fatigue, nausea, diarrhea, hoarseness, anorexia, weight loss, Grade 3/4 AEs include htn (12%), aggravated htn (6%), diarrhea (6%), fatigue (6%), blister (4%) and limb pain	Promising data in phase II second line RCC; phase II studies now planned in breast cancer, melanoma, NSCLC, thyroid cancer	3003
AEE 788 (Novartis)	Phase I	< 550 mg QD	Diarrhea, skin rash, fatigue, nausea, and anorexia, thrush, and emesis	Phase I trials in GBm and other solid tumors ongoing	3028 3063

Table 1. (*Continued*).

Drug	Phase of dev.	Safe dose	Side effects	Current efficacy data	ASCO 2005 Abstract #
AMG 706 (Amgen)	Phase II	125 QD	Hypertension, fatigue, headache	Evidence of tumor regression in early development. Ongoing phase II studies in imatinib-resistant GIST, NSCLC, breast cancer and CRC	3013
AZD 2171 (Astra-Zeneca)	Phase I	45 mg QD	Diarrhoea, nausea, fatigue, elevated liver markers,vomiting, hypoglycemia, hypertension	Several phase I in solid tumors ongoing	3002 3030 3049
BIB1120 (Boehringer Ingelheim)		300 mg bid	Nausea, vomiting, and diarrhoea. Elevations of hepatic enzymes	Phase I trials in advanced solid malignancies ongoing	3031 3054
Chir-258 (Chiron)	Phase I	100 mg qd	Nausea, vomiting, and diarrhoea, fatgue, anemia, headache, pruritic rush, hypertension (DLT)	Phase I trial in advanced solid malignancies ongoing	3044

(*Continued*)

Table 1. (*Continued*).

Drug	Phase of dev.	Safe dose	Side effects	Current efficacy data	ASCO 2005 Abstract #
GW-786034 (GSK)	Phase II	800 mg qd or 300 mg bid	Nausea, fatigue, hypertension, anorexia, vomiting, hair depigmentation,	Phase I/II in advanced solid tumors ongoing	3012
SU5416 (Semaxinib)	Dev. stopped after phase III	125 mg/ m^2 i.v. twice weekly	Headache, nausea, vomiting, asthenia, phlebitis, dyspnea	Modest signs of efficacy in several phase I/II trials in different cancer populations. Failed to show efficacy in two phase III trials in combination with either 5-FU/LV or Irinotecan/5-FU/LV in first line metastatic colorectal cancer. Compound withdrawn from further development	

Abstracts from the ASCO-Meeting 2005 are obtained from http://www.asco.org/ac/1,1003,_12-002634-00_18-0034,0.asp

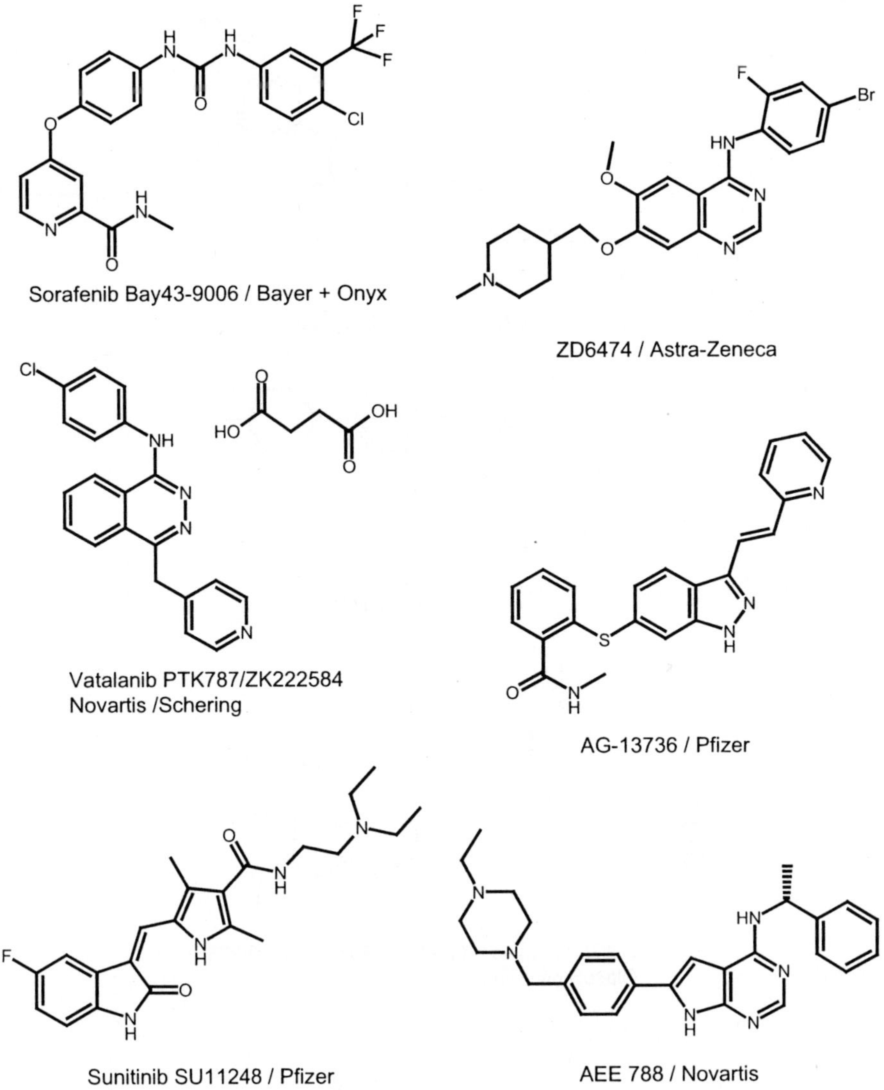

Fig. 1. Available molecular structure of anti-angiogenic kinase inhibitors in development.

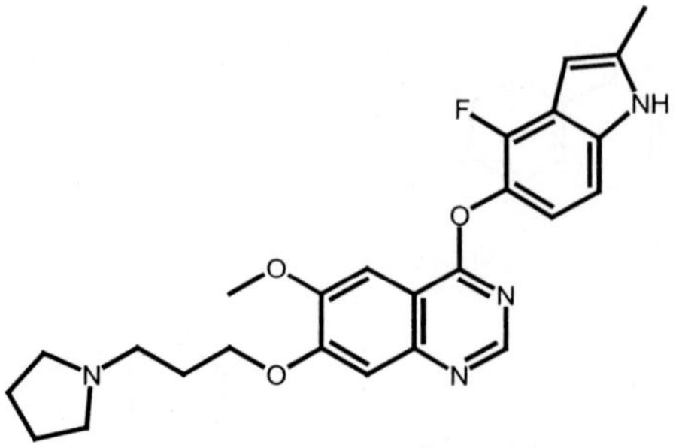

AZD2171 / Astra-Zeneca

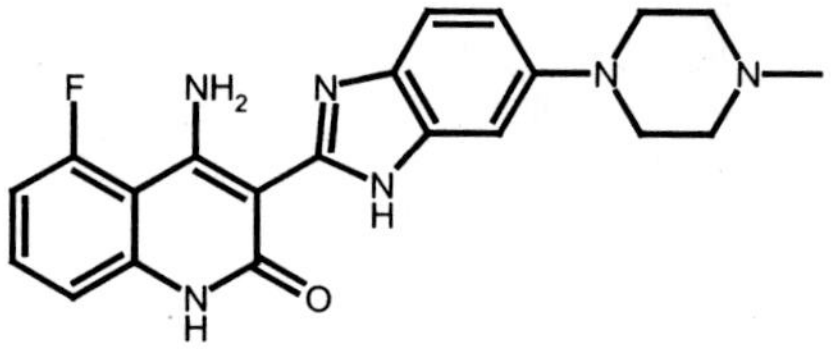

CHIR258 / Chiron

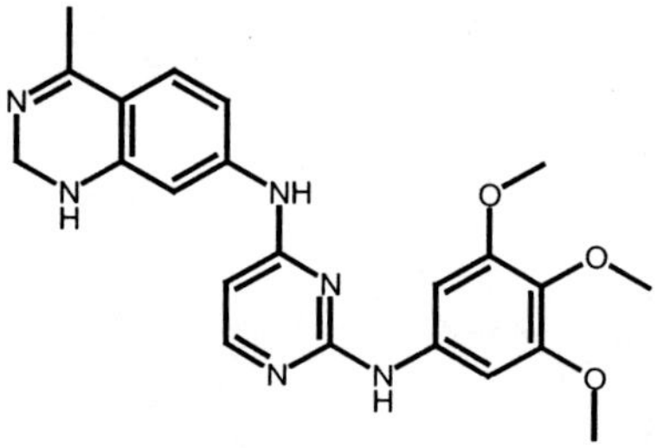

GW78034 (also named SB78034 and GW2286)
GlaxoSmithKline

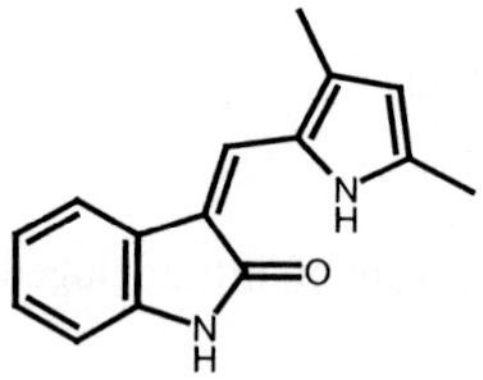

SU5416 SUGEN / Pfizer

Fig. 1. (*Continued*).

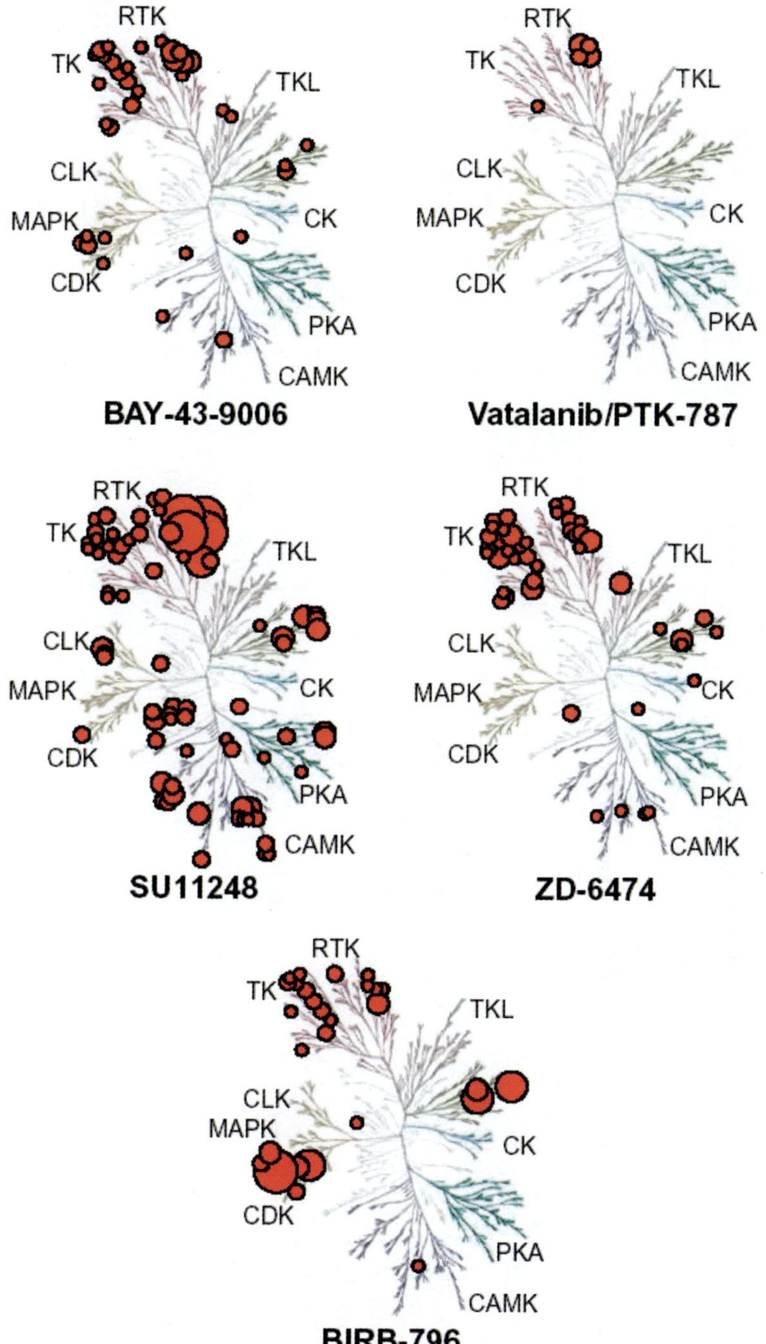

Fig. 2. Selectivity pattern of anti-angiogenic kinase inhibitors according to Fabian *et al.* (*Ref. 34*, with permission of the editors).

Table 2. Selectivity and cellular activity of anti-angiogenic kinase inhibitors in clinical development (IC50 [nM]).

Agent	VEGFR2	PDGFR	c-kit	FGFR	b-raf	Others	Inhib. endoth. cell prolif.	Inh. tumor cell prolif.	Ref(s).
BAY 43-9006 (Sorafenib, Bayer)	90	60	70	600	20–40		1500	2600	80
PTK787/ZK 222584 (Vatalanib, Novartis, Schering AG)	40	600	400	>10000	>10000	c-fms ~600	60	> 100000	69, 81
SU11248 (Sunitinib, Pfizer)	10	10	10–100	800		FLT-3 ~30 IGFR 2400 Src 600	4	1000	82, 83
ZD6474 (Astra-Zeneca)	40				>10000	EGFR~100 Fyn ~300	60		84
AG 13736 (Pfizer)	10	n.a.	50	n.a.	>10000	Fyn ~1000	1	3000	unpub. KHT

Table 2. (*Continued*).

Agent	VEGFR2	PDGFR	c-kit	FGFR	b-raf	Others	Inhib. endoth. cell prolif.	Inh. tumor cell prolif.	Ref.
AEE 788 (Novartis)	80	300	800			EGFR ~2 c-src ~60		3000	85
AMG 706 (Amgen)	5	200	10			Ret ~100			86
AZD 2171 (Astra-Zeneca)	1	5	2	30		EGFR~1600 Src ~130 InsR ~700#	4	1000	87
BIBF1120	20	60		70		Lck ~20 Src ~150	10		88
Chir-258	10	27	2	8		FLt-3 ~1 CSF-1R ~36 Insr ~2100			89
GW786034 No data available									
SU5416	300	300	200			Insr ~ 7000 Flt-3 100 nM	400		90 unpub. KHT

5.1. *BAY 43-9006 (Sorafenib)*

BAY 43-9006 is an orally available inhibitor of Raf-1 with additional activity against VEGFR-2, -3 and PDGFR-β kinases and others.

Clinical phase I trials in several oncologic indications showed promising signs of efficacy with regards to tumor regression.[35] A dosing regimen of 400 mg bid was established to be safe. Encouraging results of the phase II program led to the launch of a randomized phase III trial in advanced renal cell carcinoma (RCC). The preliminary analysis of progression free survival (PFS) showed an increase from 12 weeks in the best supportive care group to 24 weeks in the BAY 43-9006 treatment group (p < 0.00001, HR = 0.44). The trial is ongoing for final analyses of overall survival (OS). Predominant side effects are rash, diarrhea, hand-foot skin reaction, fatigue and hypertension.[36] Rash, as an adverse event, may pinpoint to a residual inhibition of EGFR kinase in dermal tissue, whereas hypertension seems to occur as adverse event in all therapies effectively targeting the VEGF pathway.

BAY 43-9006 is currently in clinical testing in randomized phase III trials in advanced melanoma and hepatocellular carcinoma, as well as in phase II studies in thyroid cancer and prostate cancer. Filing of a New Drug Application for Sorafenib for patients with advanced RCC based on the PFS data has been completed in mid-2005.

5.2. *PTK/ZK (Vatalanib)*

PTK/ZK is an orally available inhibitor of VEGFR-1, -2 and -3 tyrosine kinases which, less potently, also targets PDGFR-β and c-kit. Phase I/II studies in colorectal cancer, glioblastoma multiforme (GBM), RCC, prostate cancer, ovarian cancer and mesothelioma showed a safe side-effect profile and promising activity.[37–41] A continuous oral dosing regimen of 1250 mg qd without the need for treatment interruption has been established to be safe.

Two randomized, placebo-controlled, double-blinded phase III trials in colorectal cancer (CONFIRM 1 and CONFIRM 2) have been launched, investigating the safety and efficacy of PTK/ZK added to the FOLFOX4 regimen in the first (CONFIRM 1) and second (CONFIRM 2) line treatment of metastatic colorectal cancer. In the analysis of PFS based on a central assessment of CONFIRM 1, PTK/ZK did not demonstrate a statistically significant improvement in the overall patient population, although a statistically significant improvement was reached in the investigator assessment. A strong treatment benefit can be observed in the patient population with high LDH levels at baseline. The trial is ongoing for evaluation of OS, which is the subsequent primary endpoint. Predominant side effects in the PTK/ZK group were hypertension, nausea, vomiting, fatigue, dizziness and thrombembolic events.[42]

Efficacy of PTK/ZK is also currently investigated in a phase II study in advanced non-small cell lung cancer (NSCLC) and in a randomized phase II study in newly diagnosed GBM.

For the phase I studies, dynamic contrast-enhanced magnetic resonance imaging (DCE-MRI) was used as a pharmacodynamic marker of activity. A bidirectional transfer constant K_{trans}, which is a measure of the extravasation of the MRI contrast medium Magnevist$^{®}$ into the tissue was found to be reduced up to 50% after dosing higher than 750 mg qd. Changes in K_{trans} represent alterations in blood vessel permeability, vascularity, blood flow and extracellular space.[43]

5.3. *SU11248 (Sunitinib)*

SU11248 is an orally available broad activity kinase inhibitor with high activity against VEGF receptor tyrosine kinases, c-Kit and FLT-3. The phase I program in different patient populations showed efficacy with regard to tumor regressions and established a safe dosing schedule of 50 mg qd for four weeks followed by a two-week interruption. Two consecutively conducted phase II trials showed that SU11248 has substantial efficacy as second line therapy in RCC. Side-effect profile comprises mainly fatigue, nausea, diarrhea and stomatitis.[44] The need for a treatment interruption may be a consequence of the broader kinase inhibition activity as shown in Fig. 1 and does not permit to completely exploit the potential of anti-angiogenic treatment. SU11248 also showed clinical activity in imatinib refractory gastrointestinal stromal tumors (GIST) and metastatic neuroendocrine tumors (NET).

SU11248 is currently investigated in a phase III trial comparing its monotherapy efficacy to interferon-α in first line treatment of metastatic RCC.

5.4. *ZD6474*

ZD6474 is an oral inhibitor of VEGFR-2 and the EGFR-tyrosine kinases. Safety and tolerability of ZD6474 have been evaluated in two phase I studies with refractory tumors including a large proportion of NSCLC cancer patients.[45,46] An oral administration of doses of 300 mg or less was generally safe and well tolerated. Predominant side effects are diarrhea, rash and asymptomatic corrected QT interval prolongation. The characteristics of this side-effect profile (rash) is attributable to the EGFR-inhibiting properties, whereas the dominant side-effects caused by VEGF inhibition known from other compounds such as hypertension are not reported in high frequency. This leads to the hypothesis that the clinical activity of ZD6474 has a predominant component of EGFR inhibition.

ZD6474 had been investigated in a phase II trial in patients with previously treated metastatic breast cancer without objective responses at doses of 100 and 300 mg qd.[47] Interestingly the K_{trans} values obtained from MRI measurements of extravasating contrast media did not change with treatment outside the normal variability of the technique. Changes in K_{trans} values are described for other compounds targeting the VEGF pathway.[43] The lack of these changes again supports the hypothesis that ZD6474 exhibits its clinical effects primarily by targeting the EGFR pathway.

ZD6474 is currently investigated in two different dosing schedules in two randomized double-blinded phase II trials comparing the safety and efficacy of ZD6474 once added to docetaxel in second line NSCLC and in combination with paclitaxel/carboplatin in first line NSCLC. Preliminary PFS data of the second line docetaxel combination trial were reported in a corporate symposium at the World Congress of Lung Cancer 2005, which showed a statistically significant prolongation of PFS once 100 mg qd ZD6474 was combined with docetaxel. Interestingly in the combination the safe dose of 300 mg qd was less effective than the dose of 100 mg qd. These results led to the launch of a phase III trial in second line non-small cell lung cancer evaluating the same regimen. An additional randomized phase II trial comparing the efficacy of ZD6474 with Gefitinib in the treatment of second or third line non-small cell lung cancer is also in progress.[48] In another still ongoing randomized phase II trial, the efficacy of ZD6474 is compared to placebo in patients with small cell lung cancer who responded to previous treatments.[49]

5.5. *AG-013736*

AG-013736 is an orally available compound with strong inhibitory activity against VEGFR-1 and -2 and PDGFR-β. Phase I trials established an oral dosing regimen of 5 mg bid to be safe.[50] In a phase II trial in cytokine refractory metastatic RCC, AG-013736 showed an impressive 46% response rate and 40% disease stabilization rate. Side-effect profile was characterized by hypertension, fatigue, nausea, diarrhea, hoarseness, anorexia and weight loss.[51] The kinase inhibitory activity of this drug appears to be relatively broad.

The drug is currently in phase II development in thyroid cancer, melanoma and NSCLC.

5.6. *AEE788*

AEE788 is an orally active inhibitor of multiple tyrosine kinases, including EGFR, HER2 and VEGFR-2. EGFR kinase inhibition occurs at 100-fold lower concentration than VEGFR kinase inhibition, explaining a side-effect profile that is characteristic for EGFR blockade.

The compound is currently in several phase I studies in GBM, NSCLC and other solid tumors to establish a safe dosing schedule. Dose limiting toxicity seems to occur at dosing levels of 500 to 550 mg qd. The side-effect profile is characterized by diarrhea, fatigue, asthenia, anorexia, rash, nausea, vomiting and liver enzyme elevation.[52] Interestingly, AEE788 does not show many side effects attributable to VEGFR inhibition paralleling the observations with ZD6474.

5.7. *AMG 706*

AMG 706 is an orally available inhibitor targeting the VEGF and PDGF receptor tyrosine kinases, as well as c-kit and Ret. The molecular structure has not been published. In a phase I study, a continuous oral dosing of 125 mg qd was established to be safe. Adverse event profile was characterized by hypertension, fatigue and headache.[53] AMG 706 is currently explored in phase II studies in NSCLC, thyroid cancer and gastrointestinal cancer.

5.8. *AZD 2171*

AZD 2171 is an orally available kinase inhibitor of VEGFR-2, PDGFR, c-kit and FGFR, with an inhibitory activity for these kinases at a low nanomolar range. Preliminary results from phase I studies in different tumor types show that AZD 2171 is safe and generally well tolerated in a once daily dosing regimen. The terminal half-life is estimated to be in the range of 40 hours. Oral administration of doses below or equal to 45 mg qd seems to be safe and the adverse event profile consists of fatigue, nausea, diarrhea, vomiting and hypertension.[54]

5.9. *BIBF 1120*

BIBF 1120 is a potent orally available inhibitor of VEGF, PDGF and FGF receptor kinases. The molecular structure has not been published. In an ongoing phase I clinical trial, doses of 400 mg qd, or 250 mg bid have been established to be safe. Side-effect profile is characterized by nausea, vomiting, diarrhea, fatigue and reversible elevations of liver enzymes in serum.[55,56]

5.10. *Chir-258*

Chir-258 is a broader tyrosine kinase inhibitor of VEGFR, FLT-3, FGFR, PDGFR, c-kit and CSF-R. The molecular structure has not been published. Preliminary results of an ongoing phase I study in patients with advanced solid tumors demonstrate that doses up to 100 mg qd are tolerable. Major side-effects are hypertension, nausea, vomiting, diarrhoea, fatigue, anemia, headache and transient pruritic rash. The latter points to EGFR-dependent side-effects.[57]

5.11. *GW786034*

GW786034 is a tyrosine kinase inhibitor of VEGFR-1, -2 and -3, PDGFR and c-kit. Preliminary results from an ongoing phase I clinical trial suggest that doses of 800 mg qd and 300 mg bid can be safely administered. Side-effect profile is characterized by hypertension, nausea, diarrhea, fatigue, anorexia, vomiting and hair depigmentation.[58,59]

5.12. *SU5416 (Semaxinib)*

SU5416 is a lipophilic, highly protein-bound non-selective tyrosine kinase inhibitor of VEGFR-2, c-Kit and FLT3. This compound was the first VEGF receptor tyrosine kinase inhibitor to be tested clinically. In several phase I/II trials in various patient populations, the compound showed signs of clinical activity with regards to complete remissions and partial responses in monotherapy treatment.[60-63] SU5416 had to be dissolved in a cremophor plus ethanol vehicle for an i.v. administration route, with co-administration of steroids to prevent hypersensitivity reactions. Dosing schedule was 145 mg/m^2 twice weekly for eight weeks by central catheter. In two large randomized trials, SU5416 was tested in combination with 5-FU/leucovorin and with irinotecan/5-FU/leucovorin in first line treatment of patients with metastatic colorectal cancer.[64] The results of these completed trials have not been published. The development program has been stopped after the two negative phase III trials in metastatic colorectal cancer.

Toxicities as known from the phase I/II development include headache, nausea, vomiting, asthenia, pain at the infusion site, infections and phlebitis.

The formulation and administration mode may partly explain the high rate of infusion site pain and also a high rate of immunosuppression-related infections. A prolonged binding of the compound to VEGFR kinase had been postulated[65] for an explanation of the efficacy even though the $t_{1/2}$ in humans is below 40 minutes.[66]

6. Challenges and Future Directions

The development of VEGFR kinase inhibitors as anti-cancer drugs is a consequence of the search for compounds, that interfere specifically with the VEGF signal transduction pathway. The search is based on inhibitors competing for the ATP binding site in kinases which are very similar to each other. Such compounds were available from earlier drug finding programs investigating kinase inhibition of EGFR and PKC. These searches led to compounds which inhibit VEGFR kinases and also other kinases such as EGFR.

The efficacy of bevacizumab in the clinical setting demonstrates, that interfering with VEGF signaling is a valid strategy to slow down tumor growth. It is logical that VEGF signal blockade is supposed to be a non-curative treatment. It should prolong the life of advanced cancer patients by several months without eradicating tumor cells directly. The consistent finding in several phase III studies showing that bevacizumab increases the response rate to standard chemotherapy cannot be explained easily by mechanistic considerations. Various explanatory hypotheses have been put forward including the idea that the reduction of interstitial pressure in tumor tissues by specific signal transduction antagonists may lead to an improved blood flow, which in consequence may increase the uptake of cytotoxic drugs.[67] Others, however, did not find an increased uptake of cytotoxic drugs combining PTK/ZK with cytotoxic agents (Hess-Stumpp, Haberey and Thierauch, unpublished). Daldrup *et al.* even found a *reduced* tumor uptake of the strongly albumin-bound chemotherapeutic cisplatin but no change for 5-FU.[68] It still remains to be proven, whether small molecule tyrosine kinase inhibitors will succeed in improving the efficacy of chemotherapy.

As mentioned before, the clinically investigated VEGFR kinase inhibitors are based on the competitive inhibition of ATP binding in the intracellular kinase portion of the receptor. They can be distinguished from each other with regard to potency against the target and kinetic parameters. However, the most important distinction appears to be the selectivity of the agents towards different kinases. Practically, all of them inhibit the VEGFR family of tyrosine kinase receptors. In addition, the PDGFR family is affected in varying potency. No further kinases outside of the VEGFR and PDGFR family are inhibited by the most selective compounds (e.g. PTK/ZK).[34,69] Other compounds such as ZD6474 strongly affect the EGF receptors amongst others (Fig. 1). Only when it was recognized that this property is supportive in NSCLC therapy the development of the compound was resumed. AEE788 was explicitly selected as a compound with the dual efficacy combining VEGFR blockade with a very potent EGFR blockade. Such compounds most likely have a maximally tolerated dose defined by EGFR inhibition and produce the side-effect profile attributable to EGFR signal blockade.[70,71]

Some of the agents discussed have a rather broad kinase inhibitory action such as Su11248.[34] This has its repercussions in the side-effect profile and consequently treatment schedules. SU11248 has to be given intermittently due to severe fatigue affecting the patients and requiring several weeks of recovery before continuation of treatment allowing recovery of the vessels in the meantime. Preclinical investigations point to a rapid regrowth of blood vessels along the preformed fabric of extracellular vessel fibres forming basement membrane ghosts, when treatment is interrupted.[72]

Interestingly, all VEGFR kinase inhibitors show the phenomenon of fatigue. The penetration of the blood-brain barrier by the compound may be an explanation for this side-effect.

Non-selective kinase inhibitory drugs are conceivably more effective compared to selective agents with regard to the observation of tumor regressions in monotherapy. Their disadvantage is a diminished versatility in their combination with cytotoxic drugs as the toxicities are cumulative and may require dose reductions, which compromises the efficacy of each drug. They compromise the quality of life for patients, which are exposed to long treatment periods, more than selective compounds. These multi-target kinase inhibitors may find their special applications such as Su11248 in RCC and relapsing gastrointestinal stromal tumors or like ZD 6474 in NSCLC.

For several of the drugs, changes in hair pigmentation have been described (e.g. SU11248 and AG-013736). Apparently inhibition of c-kit is the underlying cause. It is known that mutations either in c-kit or in c-kit ligand lead to color changes of the hair.[73,74]

The hypertensive reaction observed in some patients treated with VEGFR kinase inhibitors may be a consequence of two effects: antagonism to VEGF-induced increase in vascular permeability[75] and the blockade of VEGF-induced release of the vasodilator NO.[76] This side-effect has been manageable by standard care treatment.

Anti-angiogenic therapy is intended for a long treatment period in order to assure a prolonged control of disease progression (maintenance therapy). In many cases it will remain necessary to reduce tumor size with surgery, cytotoxic therapy or radiotherapy. Thus for anti-angiogenic therapy, side-effects and the drug-drug interaction potential are of utmost importance. Such an agent can be combined with independent treatment modalities like anti-hormonal therapy in breast or prostate cancer and cytotoxic treatments as indicated. If an anti-angiogenic drug has a mild side-effect profile, a long-term treatment should not have an untolerable impact on the quality of life of patients. Bevacizumab was a first step in this direction. Further generations of anti-angiogenic drugs with largely improved pharmacokinetic parameters and higher efficacy will fulfill such requirements.

To optimize the anti-angiogenic treatment even more, several questions should be answered:

- For how long during a day does the VEGFR function need to be blocked by kinase inhibitors to achieve the maximum anti-angiogenic effect? The VEGF signal is tightly controlled: the heterozygous deletion of the VEGF gene results in a lethal phenotype in mice.[77]
- A circadian rhythm of VEGF secretion is postulated.[78] Does this have an impact on the dosing schedule?
- Angiogenesis inhibition leads to endothelial cell apoptosis. How fast is the reappearance of proliferating and migrating endothelial cells forming new blood vessels and what is their origin? Original reports of a strong bone marrow-derived endothelial cell participation have been disproved.[79]
- Why is VEGFR kinase inhibition not effective in all tumor models and in all patients? In such cases, are other pathways which organize tumor angiogenesis, bypassing the VEGF signal, involved? Are existing vessels co-opted for tumor growth?

- What is the best schedule to combine anti-angiogenic treatment with cytotoxic agents?
- How is efficacy and safety of kinase inhibitors combined with cyto-toxic drugs or radiotherapy impacted by specific kinase selectivities?
- Are there markers that describe the inhibition of angiogenesis or efficacy of angiogenesis inhibitors as tumor growth inhibitors?
- Are such markers tumor-, angiogenesis- or compound-specific?

The answers to these questions may have consequences for treatment schedules and may facilitate the development of kinase inhibitory drugs. Only in the next couple of years will the potential of anti-angiogenic kinase inhibitors in cancer therapy and beyond surface for the benefit of various patient populations.

Acknowledgments

We thank David Lockhard, Ambit Biosciences, San Diego, and *Nature Biotechnology* for the permission to reproduce Fig. 2 taken from Fabian *et al.*[34] Furthermore, we thank Frank Hilberg, Boehringer Ingelheim, Vienna, for providing three posters about BibF1120. We also wish to acknowledge the critical reading of the manuscript by Jia Li and Hans Menssen, Schering AG.

References

1. Shweiki D, Itin A, Soffer D, Keshet E (1992). Vascular endothelial growth factor induced by hypoxia may mediate hypoxia-initiated angiogenesis. *Nature* **359**(6398): 843–845.
2. Shweiki D, Itin A, Neufeld G, Gitay-Goren H, Keshet E (1993). Patterns of expression of vascular endothelial growth factor (VEGF) and VEGF receptors in mice suggest a role in hormonally regulated angiogenesis. *J Clin Invest* **91**(5): 2235–2243.
3. Greenblatt M, Shubi P (1968). Tumor angiogenesis: transfilter diffusion studies in the hamster by the transparent chamber technique. *J Natl Cancer Inst* **41**(1): 111–124.
4. Folkman J (1971). Tumor angiogenesis: therapeutic implications. *N Engl J Med* **285**(21): 1182–1186.
5. Ryu S, Albert DM (1979). Evaluation of tumor angiogenesis factor with the rabbit cornea model. *Invest Ophthalmol Vis Sci* **18**(8): 831–841.
6. de Vries C, Escobedo JA, Ueno H, Houck K, Ferrara N, Williams LT (1992). The fms- like tyrosine kinase, a receptor for vascular endothelial growth factor. *Science* **255**(5047): 989–991.

7. Grosskreutz CL, Anand-Apte B, Duplaa C, Quinn TP, Terman BI, Zetter B, D'Amore PA (1999). Vascular endothelial growth factor-induced migration of vascular smooth muscle cells *in vitro*. *Microvasc Res* **58**(2): 128–136.
8. Simon M, Grone HJ, Johren O, Kullmer J, Plate KH, Risau W, Fuchs E (1995). Expression of vascular endothelial growth factor and its receptors in human renal ontogenesis and in adult kidney. *Am J Physiol* **268**(2 Pt 2): F240–F250.
9. Kabrun N, Buhring HJ, Choi K, Ullrich A, Risau W, Keller G (1997). Flk-1 expression defines a population of early embryonic hematopoietic precursors. *Development* **124**(10): 2039–2048.
10. Schanzer A, Wachs FP, Wilhelm D, Acker T, Cooper-Kuhn C, Beck H, Winkler J, Aigner L, Plate KH, Kuhn HG (2004). Direct stimulation of adult neural stem cells *in vitro* and neurogenesis *in vivo* by vascular endothelial growth factor. *Brain Pathol* **14**(3): 237–248.
11. Cohen T, Gitay-Goren H, Sharon R, Shibuya M, Halaban R, Levi BZ, Neufeld G (1995). VEGF121, a vascular endothelial growth factor (VEGF) isoform lacking heparin binding ability, requires cell-surface heparan sulfates for efficient binding to the VEGF receptors of human melanoma cells. *J Biol Chem* **270**(19): 11322–11326.
12. Joukov V, Sorsa T, Kumar V, Jeltsch M, Claesson-Welsh L, Cao Y, Saksela O, Kalkkinen N, Alitalo K (1997). Proteolytic processing regulates receptor specificity and activity of VEGF-C. *EMBO J* **16**(13): 3898–3911.
13. Stacker SA, Stenvers K, Caesar C, Vitali A, Domagala T, Nice E, Roufail S, Simpson RJ, Moritz R, Karpanen T, Alitalo K, Achen mg (1999). Biosynthesis of vascular endothelial growth factor-D involves proteolytic processing which generates non-covalent homodimers. *J Biol Chem* **274**(45): 32127–32136.
14. Clauss M, Weich H, Breier G, Knies U, Rockl W, Waltenberger J, Risau W (1996). The vascular endothelial growth factor receptor Flt-1 mediates biological activities. Implications for a functional role of placenta growth factor in monocyte activation and chemotaxis. *J Biol Chem* **271**(30): 17629–17634.
15. Shen H, Clauss M, Ryan J, Schmidt AM, Tijburg P, Borden L, Connolly D, Stern D, Kao J (1993). Characterization of vascular permeability factor/vascular endothelial growth factor receptors on mononuclear phagocytes. *Blood* **81**(10): 2767–2773.
16. Soker S, Takashima S, Miao HQ, Neufeld G, Klagsbrun M (1998). Neuropilin-1 is expressed by endothelial and tumor cells as an isoform-specific receptor for vascular endothelial growth factor. *Cell* **92**(6): 735–745.
17. Klagsbrun M, Eichmann A (2005). A role for axon guidance receptors and ligands in blood vessel development and tumor angiogenesis. *Cytokine Growth Factor Rev* **16**(4–5): 535–548.
18. West DC, Rees CG, Duchesne L, Patey SJ, Terry CJ, Turnbull JE, Delehedde M, Heegaard CW, Allain F, Vanpouille C, Ron D, Fernig DG (2005). Interactions of multiple heparin binding growth factors with neuropilin-1 and potentiation of the activity of fibroblast growth factor-2. *J Biol Chem* **280**(14): 13457–13464.
19. Neufeld G, Shraga-Heled N, Lange T, Guttmann-Raviv N, Herzog Y, Kessler O (2005). Semaphorins in cancer. *Front Biosci* **10**: 751–760.
20. Comoglio PM, Tamagnone L, Giordano S (2004). Invasive growth: a two-way street for semaphorin signalling. *Nat Cell Biol* **6**(12): 1155–1157.

21. Toyofuku T, Zhang H, Kumanogoh A, Takegahara N, Yabuki M, Harada K, Hori M, Kikutani H (2004). Guidance of myocardial patterning in cardiac development by Sema6D reverse signalling. *Nat Cell Biol* **6**(12): 1204–1211.

22. Bachelder RE, Lipscomb EA, Lin X, Wendt MA, Chadborn NH, Eickholt BJ, Mercurio AM (2003). Competing autocrine pathways involving alternative neuropilin-1 ligands regulate chemotaxis of carcinoma cells. *Cancer Res* **63**(17): 5230–5233.

23. Fox PL, DiCorleto PE (1984). Regulation of production of a platelet-derived growth factor-like protein by cultured bovine aortic endothelial cells. *J Cell Physiol* **121**(2): 298–308.

24. Gerety SS, Wang HU, Chen ZF, Anderson DJ (1999). Symmetrical mutant phenotypes of the receptor EphB4 and its specific transmembrane ligand ephrin-B2 in cardiovascular development. *Mol Cell* **4**(3): 403–414.

25. Augustin HG, Reiss Y (2003). EphB receptors and ephrinB ligands: regulators of vascular assembly and homeostasis. *Cell Tissue Res* **314**(1): 25–31.

26. Zimrin AB, Pepper MS, McMahon GA, Nguyen F, Montesano R, Maciag T (1996). An antisense oligonucleotide to the notch ligand jagged enhances fibroblast growth factor-induced angiogenesis *in vitro*. *J Biol Chem* **271**(51): 32499–32502.

27. Goodwin AM, D'Amore PA (2002). Wnt signaling in the vasculature. *Angiogenesis* **5**(1–2): 1–9.

28. Huminiecki L, Gorn M, Suchting S, Poulsom R, Bicknell R (2002). Magic roundabout is a new member of the roundabout receptor family that is endothelial specific and expressed at sites of active angiogenesis. *Genomics* **79**(4): 547–552.

29. Park KW, Morrison CM, Sorensen LK, Jones CA, Rao Y, Chien CB, Wu JY, Urness LD, Li DY (2003). Robo4 is a vascular-specific receptor that inhibits endothelial migration. *Dev Biol* **261**(1): 251–267.

30. Lu X, Le Noble F, Yuan L, Jiang Q, De Lafarge B, Sugiyama D, Breant C, Claes F, De Smet F, Thomas JL, Autiero M, Carmeliet P, Tessier-Lavigne M, Eichmann A (2004). The netrin receptor UNC5B mediates guidance events controlling morphogenesis of the vascular system. *Nature* **432**(7014): 179–186.

31. RayChaudhury A, D'Amore PA (1991). Endothelial cell regulation by transforming growth factor-beta. *J Cell Biochem* **47**(3): 224–229 (Review).

32. Kozian DH, Ziche M, Augustin HG (1997). The activin-binding protein follistatin regulates autocrine endothelial cell activity and induces angiogenesis. *Lab Invest* **76**(2): 267–276.

33. Manning G, Whyte DB, Martinez R, Hunter T, Sudarsanam S (2002). The protein kinase complement of the human genome. *Science* **298**(5600): 1912–1934.

34. Fabian MA, Biggs WH 3rd, Treiber DK, Atteridge CE, Azimioara MD, Benedetti mg, Carter TA, Ciceri P, Edeen PT, Floyd M, Ford JM, Galvin M, Gerlach JL, Grotzfeld RM, Herrgard S, Insko DE, Insko MA, Lai AG, Lelias JM, Mehta SA, Milanov ZV, Velasco AM, Wodicka LM, Patel HK, Zarrinkar PP, Lockhart DJ (2005). A small molecule-kinase interaction map for clinical kinase inhibitors. *Nat Biotechnol* **23**(3): 329–336.

35. Ratain MJ, Flaherty KT, Stadler WM (2004). Preliminary antitumor activity of BAY 43-9006 in metastatic renal cell carcinoma and other advanced refractory solid tumors in phase II randomized discontinuation trial (RDT). *J Clin Oncol* **22** (Suppl): 4501 (Abstract).

36. Escoudier B, *et al.* (2005). *ASCO* (Abstract #4510).
37. Drevs J, Mross K, Medinger M *et al.* (2003) Phase I dose-escalation and phar-macokinetic (PK) study of he VEGF inhibitor (PTK/ZK) in patients with liver metastases. *Proc Am Soc Clin Oncol* **22**: 284 (Abstract #1142).
38. Reardon, *et al.* (2003). *ASCO* (Abstract #412).
39. George D, *et al.* (2003). *ASCO* (Abstract #1548).
40. Schroeder W., *et al.* (2005). *ASCO* (Abstract #5042).
41. Jahan T, *et al.* (2005). *IASCLC* (Poster #402).
42. Hecht R, *et al.* (2005). *ASCO* (Abstract #3).
43. Morgan B, Thomas AL, Drevs J, Hennig J, Buchert M, Jivan A, Horsfield MA, Mross K, Ball HA, Lee L, Mietlowski W, Fuxuis S, Unger C, O'Byrne K, Henry A, Cherryman GR, Laurent D, Dugan M, Marme D, Steward WP (2003). Dynamic contrast-enhanced magnetic resonance imaging as a biomarker for the pharmaco-logical response of PTK787/ZK 222584, an inhibitor of the vascular endothelial growth factor receptor tyrosine kinases, in patients with advanced colorectal can-cer and liver metastases: results from two phase I studies. *J Clin Oncol* **21**(21): 3955–3964 (Epub 29 Sep 2003).
44. Motzer RJ, *et al.* (2005). *ASCO* (Abstract #4508).
45. Hurwitz H, *et al.* (2002). *ASCO* (Abstract #325).
46. Minami H, *et al.* (2003) *ASCO* (Abstract #778).
47. Miller KD, Trigo JM, Wheeler C, Barge A, Rowbottom J, Sledge G, Baselga J (2005). A multicenter phase II trial of ZD6474, a vascular endothelial growth factor receptor-2 and epidermal growth factor receptor tyrosine kinase inhibitor, in patients with previously treated metastatic breast cancer. *Clin Cancer Res* **11**(9): 3369–3376.
48. Natale R, *et al.* (2005). *IASCLC*.
49. Bates D (2003). ZD-6474. AstraZeneca. *Curr Opin Investig Drugs* **4**(12): 1468–1472 (Review); Bielenberg DR, Hida Y, Shimizu A, Kaipainen A, Kreuter M, Kim CC, Klagsbrun M (2004). Semaphorin 3F, a chemorepulsant for endothelial cells, induces a poorly vascularized, encapsulated, nonmetastatic tumor pheno-type. *J Clin Invest* **114**(9): 1260–1271.
50. Liu G, Rugo HS, Wilding G, McShane TM, Evelhoch JL, Ng C, Jackson E, Kelcz F, Yeh BM, Lee FT Jr, Charnsangavej C, Park JW, Ashton EA, Steinfeldt HM, Pithavala YK, Reich SD, Herbst RS (2005). Dynamic contrast-enhanced magnetic resonance imaging as a pharmacodynamic measure of response after acute dosing of AG-013736, an oral angiogenesis inhibitor, in patients with advanced solid tumors: results from a phase I study. *J Clin Oncol* **23**(24): 5464–5473.
51. Rini B, *et al.* (2005). *ASCO* (Abstract #4509).
52. Martinelli G, *et al.* (2005). *ASCO* (Abstract #3039).
53. Rosen L, *et al.* (2005). *ASCO* (Abstract #3013).
54. Drevs J, *et al.* (2005). *ASCO* (Abstract #3002).
55. Mross KB, *et al.* (2005). *ASCO* (Abstract #3031).
56. Lee CP, *et al.* (2005). *ASCO* (Abstract #3054).
57. Sarker D, *et al.* (2005). *ASCO* (Abstract #3044).
58. Hurwitz H, *et al.* (2005). *ASCO* (Abstract #3012).

59. Suttle AB, *et al.* (2005). *ASCO* (Abstract #3054).
60. Stopeck A, Sheldon M, Vahedian M, Cropp G, Gosalia R, Hannah A (2002). Results of a Phase I dose-escalating study of the anti-angiogenic agent, SU5416, in patients with advanced malignancies. *Clin Cancer Res* **8**(9): 2798–2805.
61. Giles FJ, Stopeck AT, Silverman LR, Lancet JE, Cooper MA, Hannah AL, Cherrington JM, O'Farrell AM, Yuen HA, Louie SG, Hong W, Cortes JE, Verstovsek S, Albitar M, O'Brien SM, Kantarjian HM, Karp JE (2003). SU5416, a small molecule tyrosine kinase receptor inhibitor, has biologic activity in patients with refractory acute myeloid leukemia or myelodysplastic syndromes. *Blood* **102**(3): 795–801.
62. Fiedler W, Mesters R, Tinnefeld H, Loges S, Staib P, Duhrsen U, Flasshove M, Ottmann OG, Jung W, Cavalli F, Kuse R, Thomalla J, Serve H, O'Farrell AM, Jacobs M, Brega NM, Scigalla P, Hossfeld DK, Berdel WE (2003). A phase 2 clinical study of SU5416 in patients with refractory acute myeloid leukemia. *Blood* **102**(8): 2763–2767.
63. Peterson AC, Swiger S, Stadler WM, Medved M, Karczmar G, Gajewski TF (2004). Phase II study of the Flk-1 tyrosine kinase inhibitor SU5416 in advanced melanoma. *Clin Cancer Res* **10**(12 Pt 1): 4048–4054.
64. Tabernero J, Salazar R, Casado E, Martinelli E, Gomez P, Baselga J (2004). Targeted therapy in advanced colon cancer: the role of new therapies. *Ann Oncol* **15** (Suppl 4): iv48–iv58 (Review).
65. Mendel DB, Schreck RE, West DC, Li G, Strawn LM, Tanciongco SS, Vasile S, Shawver LK, Cherrington JM (2000). The angiogenesis inhibitor SU5416 has long-lasting effects on vascular endothelial growth factor receptor phosphorylation and function. *Clin Cancer Res* **6**(12): 4848–4858.
66. Cooney MM, Tserng KY, Makar V, McPeak RJ, Ingalls ST, Dowlati A, Overmoyer B, McCrae K, Ksenich P, Lavertu P, Ivy P, Hoppel CL, Remick S (2005). A phase IB clinical and pharmacokinetic study of the angiogenesis inhibitor SU5416 and paclitaxel in recurrent or metastatic carcinoma of the head and neck. *Cancer Chemother Pharmacol* **55**(3): 295–300.
67. Heldin CH, Rubin K, Pietras K, Ostman A (2004). High interstitial fluid pressure — an obstacle in cancer therapy. *Nat Rev Cancer* **4**(10): 806–813 (Review).
68. Daldrup-Link HE, Okuhata Y, Wolfe A, Srivastav S, Oie S, Ferrara N, Cohen RL, Shames DM, Brasch RC (2004). Decrease in tumor apparent permeability-surface area product to a MRI macromolecular contrast medium following angiogenesis inhibition with correlations to cytotoxic drug accumulation. *Microcirculation* **11**(5): 387–396.
69. Hess-Stumpp H, Haberey M, Thierauch KH (2005). PTK **787/ZK** 222584, a tyrosine kinase inhibitor of all known VEGF receptors, represses tumor growth with high efficacy. *Chembiochem* **6**(3): 550–557.
70. Robert C, Spatz A, Faivre S, Armand JP, Raymond E (2003). Tyrosine kinase inhibition and grey hair. *Lancet* **361**(9362): 1056.
71. Robert C, Soria JC, Spatz A, Le Cesne A, Malka D, Pautier P, Wechsler J, Lhomme C, Escudier B, Boige V, Armand JP, Le Chevalier T (2005). Cutaneous side-effects of kinase inhibitors and blocking antibodies. *Lancet Oncol* **6**(7): 491–500.
72. Inai T, Mancuso M, Hashizume H, Baffert F, Haskell A, Baluk P, Hu-Lowe DD, Shalinsky DR, Thurston G, Yancopoulos GD, McDonald DM (2004). Inhibition

of vascular endothelial growth factor (VEGF) signaling in cancer causes loss of endothelial fenestrations, regression of tumor vessels, and appearance of basement membrane ghosts. *Am J Pathol* **165**(1): 35–52.

73. Geissler EN, Ryan MA, Housman DE (1988). The dominant-white spotting (W) locus of the mouse encodes the c-kit proto-oncogene. *Cell* **55**(1): 185–192.

74. Moss KG, Toner GC, Cherrington JM, Mendel DB, Laird AD (2003). Hair depigmentation is a biological readout for pharmacological inhibition of KIT in mice and humans. *J Pharmacol Exp Ther* **307**(2): 476–480.

75. Senger DR, Van de Water L, Brown LF, Nagy JA, Yeo KT, Yeo TK, Berse B, Jackman RW, Dvorak AM, Dvorak HF (1993). Vascular permeability factor (VPF, VEGF) in tumor biology. *Cancer Metastasis Rev* **12**(3–4): 303–324.

76. Murohara T, Horowitz JR, Silver M, Tsurumi Y, Chen D, Sullivan A, Isner JM (1998). Vascular endothelial growth factor/vascular permeability factor enhances vascular permeability via nitric oxide and prostacyclin. *Circulation* **97**(1): 99–107.

77. Ferrara N, Carver-Moore K, Chen H, Dowd M, Lu L, O'Shea KS, Powell-Braxton L, Hillan KJ, Moore MW (1996). Heterozygous embryonic lethality induced by targeted inactivation of the VEGF gene. *Nature* **380**(6573): 439–442.

78. Koyanagi S, Kuramoto Y, Nakagawa H, Aramaki H, Ohdo S, Soeda S, Shimeno H (2003). A molecular mechanism regulating circadian expression of vascular endothelial growth factor in tumor cells. *Cancer Res* **63**(21): 7277–7283.

79. Rajantie I, Ilmonen M, Alminaite A, Ozerdem U, Alitalo K, Salven P (2004). Adult bone marrow-derived cells recruited during angiogenesis comprise precursors for periendothelial vascular mural cells. *Blood* **104**(7): 2084–2086.

80. Wilhelm SM, Carter C, Tang L, Wilkie D, McNabola A, Rong H, Chen C, Zhang X, Vincent P, McHugh M, Cao Y, Shujath J, Gawlak S, Eveleigh D, Rowley B, Liu L, Adnane L, Lynch M, Auclair D, Taylor I, Gedrich R, Voznesensky A, Riedl B, Post LE, Bollag G, Trail PA (2004). Bay 43-9006 exhibits broad spectrum oral antitumor activity and targets the RAF/MEK/ERK pathway and receptor tyrosine kinases involved in tumor progression and angiogenesis. *Cancer Res* **64**(19): 7099–7109.

81. Wood JM, Bold G, Buchdunger E, Cozens R, Ferrari S, Frei J, Hofmann F, Mestan J, Mett H, O'Reilly T, Persohn E, Rosel J, Schnell C, Stover D, Theuer A, Towbin H, Wenger F, Woods-Cook K, Menrad A, Siemeister G, Schirner M, Thierauch KH, Schneider MR, Drevs J, Martiny- Baron G, Totzke F (2000). PTK787/ZK 222584, a novel and potent inhibitor of vascular endothelial growth factor receptor tyrosine kinases, impairs vascular endothelial growth factor-induced responses and tumor growth after oral administration. *Cancer Res* **60**(8): 2178–2189.

82. Mendel DB, Laird AD, Xin X, Louie SG, Christensen JG, Li G, Schreck RE, Abrams TJ, Ngai TJ, Lee LB, Murray LJ, Carver J, Chan E, Moss KG, Haznedar JO, Sukbuntherng J, Blake RA, Sun L, Tang C, Miller T, Shirazian S, McMahon G, Cherrington JM (2003). *In vivo* antitumor activity of SU11248, a novel tyrosine kinase inhibitor targeting vascular endothelial growth factor and platelet-derived growth factor receptors: determination of a pharmacokinetic/pharmacodynamic relationship. *Clin Cancer Res* **9**(1): 327–337.

83. O'Farrell AM, Abrams TJ, Yuen HA, Ngai TJ, Louie SG, Yee KW, Wong LM, Hong W, Lee LB, Town A, Smolich BD, Manning WC, Murray LJ, Heinrich MC,

Cherrington JM (2003). SU11248 is a novel FLT3 tyrosine kinase inhibitor with potent activity *in vitro* and *in vivo*. *Blood* **101**(9): 3597–3605.

84. Wedge SR, Ogilvie DJ, Dukes M, Kendrew J, Chester R, Jackson JA, Boffey SJ, Valentine PJ, Curwen JO, Musgrove HL, Graham GA, Hughes GD, Thomas AP, Stokes ES, Curry B, Richmond GH, Wadsworth PF, Bigley AL, Hennequin LF (2002). ZD6474 inhibits vascular endothelial growth factor signaling, angiogenesis, and tumor growth following oral administration. *Cancer Res* **62**(16): 4645–4655.

85. Traxler P, Allegrini PR, Brandt R, Brueggen J, Cozens R, Fabbro D, Grosios K, Lane HA, McSheehy P, Mestan J, Meyer T, Tang C, Wartmann M, Wood J, Caravatti G (2004). AEE788: a dual family epidermal growth factor receptor/ErbB2 and vascular endothelial growth factor receptor tyrosine kinase inhibitor with antitumor and anti-angiogenic activity. *Cancer Res* **64**(14): 4931–4941.

86. Herbst R, Kurzrock R, Parson M, Benjamin R, Chen L, Ng C, Ingram M, Wong S, Chang D, Rosen L (2004). AMG706 first in human, open-label, dose-finding study evaluating the safety and pharmacokinetic (PK) in subjects with advanced solid tumors. *Eur J Cancer Supp* **2**(8), (Abstract #151).

87. Wedge SR, Kendrew J, Hennequin LF, Valentine PJ, Barry ST, Brave SR, Smith NR, James NH, Dukes M, Curwen JO, Chester R, Jackson JA, Boffey SJ, Kilburn LL, Barnett S, Richmond GH, Wadsworth PF, Walker M, Bigley AL, Taylor ST, Cooper L, Beck S, Jurgensmeier JM, Ogilvie DJ (2005). AZD2171: a highly potent, orally bioavailable, vascular endothelial growth factor receptor-2 tyrosine kinase inhibitor for the treatment of cancer. *Cancer Res* **65**(10): 4389–4400.

88. Hilberg F, Tontsch-Grunt U, Colbatzky F, Heckel A, Lotz R, van Mee, JCA, Roth GJ (2004). BIBF1120 a novel, small molecule triple angiokinase inhibitor: profiling as a clinical candidate for cancer therapy. *Eur J Cancer Supp* **2**(8), (Abstract #158).

89. Lee SH, Lopes de Menezes D, Vora J, Harris A, Ye H, Nordahl L, Garrett E, Samara E, Aukerman SL, Gelb AB, Heise C (2005). *In vivo* target modulation and biological activity of CHIR-258, a multitargeted growth factor receptor kinase inhibitor, in colon cancer models. *Clin Cancer Res* **11**(10): 3633–3641.

90. Fong TA, Shawver LK, Sun L, Tang C, App H, Powell TJ, Kim YH, Schreck R, Wang X, Risau W, Ullrich A, Hirth KP, McMahon G (1999). SU5416 is a potent and selective inhibitor of the vascular endothelial growth factor receptor (Flk-1/KDR) that inhibits tyrosine kinase catalysis, tumor vascularization, and growth of multiple tumor types. *Cancer Res* **59**(1): 99–106.

12

Therapeutic Angiogenesis — An Overview

by Masahiro Murakami and Michael Simons

1. Introduction

Recent advance of our understanding in biological processes underlying blood vessel growth has laid the foundation for new possibilities in the treatment of ischemic diseases over the conventional drug-based therapy and invasive procedures such as coronary bypass surgery and percutaneous catheter-based angioplasty. These new approaches to facilitation of the natural revascularization process have been termed therapeutic angiogenesis. The potential impact of therapeutic angiogenesis in clinical medicine is considerable, enabling us to control tissue perfusion by manipulating endogenous blood vessel growth. However, we still face formidable challenges in applying angiogenic therapies to clinical settings. In the last couple of decades since the identification and purification of angiogenic growth factors, extensive research efforts have been focused on the basic and clinical angiogenesis research. As a result, we have accumulated an enormous amount of knowledge in the field. Based on our understanding of this subject, in this chapter, we discuss current concept, strategy, and future prospective of therapeutic angiogenesis.

2. Concepts and Rationales

The concept of therapeutic angiogenesis is to facilitate blood vessel growth to restore the perfusion to and function of the ischemic tissue. Tissue ischemia refers to a situation when the oxygen supply does not meet the demand necessary to maintain normal tissue function and homeostasis, resulting in impaired organ function and endangered viability. In clinical situations, acute or chronic occlusion of the main feeding artery is largely responsible for development of ischemia although diffuse small vessel arterial disease is also a fairly frequent cause.

Natural biological responses to ischemia include *in situ* upregulation of angiogenic factors in conjunction with mobilization and recruitment of various cellular components, promoting new vessel growth and arterial remodeling. However, in most circumstances, this endogenous response does not achieve a full compensation of original blood supply, resulting in compromised tissue function and clinical symptoms. Therefore, currently the rationale of therapeutic angiogenesis resides in augmentation and manipulation of this revascularization process to gain a maximum restoration of tissue function by administrating exogenous angiogenic growth factors or cellular products.

The latest progress of vascular biology research has expanded the notion of therapeutic angiogenesis to encompass other types of vascular growth, namely, arteriogenesis and vasculogenesis, defining the term as general enhancement of blood vessel growth.[1] For this reason, although the term "therapeutic angiogenesis" is still used and will continue to be used here, "therapeutic neovascularization" is, perhaps, a more appropriate term. While we use therapeutic angiogenesis in a broad sense, blood vessel growth in general is referred to as neovascularization in this chapter. As currently understood, adult neovascularization occurs as a result of several processes, including angiogenesis, arteriogenesis, and potentially vasculogenesis (Table 1).

In its strictest sense angiogenesis, defined as growth of new capillaries, takes place at the site of ischemia, by promoting formation of new capillaries from post-capillary venules. Therefore, it does not augment arterial inflow into the region. In contrast, arteriogenesis, referred to as positive remodeling of pre-existing collaterals or *de novo* growth of

Table 1. Types of neovascularization.

	Definition	Mechanism	Driver	Cell	Effect
Angiogenesis	*De novo* capillary formation from post-capillary venules	Ischemia-driven Regulated by local HIF-1α expression	VEGF FGF Ang1 HGF	EC	Small increase in blood flow
Arteriogenesis	Remodeling of pre-existing arteries or *de novo* formation of arteries	Shear stress-induced.	MCP-1 FGF PDGF PlGF	MNC EC SMC	Large increase in blood flow
Vasculogenesis	*De novo* formation or re-modeling of pre-existing vessels by vascular progenitors	Local ischemia- or injury-driven	VEGF SDF-1 TGF-β	EPC	Unclear

conduit arteries, typically occurs in the upstream area of ischemia in response to increased shear stress and endothelial activation coupled with the subsequent influx of blood derived-mononuclear cells. The physiological significance of arteriogenesis is well recognized clinically by the development of collateral vessels that bypass the occluded artery and supply arterial inflow to various degrees.

From the point of view of therapy in most cardiovascular settings such as coronary or peripheral arterial disease, arteriogenesis is more appealing as it can increase tissue perfusion to a greater magnitude in comparison to angiogenesis.[2] Moreover, increased arterial inflow can trigger tissue regeneration efficiently coupled with concomitant angiogenesis in the ischemic area. However, the precise mechanism underlying arterial growth is less well understood with its complex nature involving many cell types and driving factors. Angiogenic growth factors are, in general, believed to be positive regulators of arteriogenesis; however, it appears that monocyte chemoattractant factors, such as MCP-1, GM-CSF and PlGF, are another entity of a potent driving force of arteriogenesis.

Vasculogenesis refers to the process of an *in situ* formation of blood vessels from circulating or tissue-resident endothelial progenitor cells (EPC) and vascular progenitor cells. While probably real, the frequency, feasibility and physiological significance of adult vasculogenesis in the setting of ischemic diseases have not been established conclusively.

3. Strategy

The basic strategy of therapeutic angiogenesis thus far has been reduced to administration in or recruitment to an ischemic area of an angiogenic agent or cellular products. However, the choice of a therapeutic angiogenic agent has been extensively revised as the knowledge of vascular biology grew in the last decade. The biological agents used in therapeutic angiogenesis primarily include angiogenic growth factors in the form of peptide, plasmid DNA and viral vector encoding a cognate sequence, and lately cellular components such as fractionated or unfractionated mononuclear cells. Combination of angiogenic factors and utilization of a master gene that can transcriptionally upregulate multiple angiogenic factors and their receptors have also been explored recently.

At this point, target diseases for therapeutic angiogenesis approaches are limited to peripheral artery disease (critical limb ischemia and claudication), ischemic cardiomyopathy and chronic coronary artery disease, including acute myocardial infarction in case of certain cell therapies. Stroke and its less severe form such as brain hypoperfusion due to carotid occlusion can theoretically be proximal candidates for therapeutic angiogenesis as well. Cell therapy for cardiac repair, in which stem cells are used with the intention of the functional recovery of infracted heart, has lately drawn considerable attention and shown some evidence of improvement of cardiac function.[3,4] Apart from the authenticity of the original concept where transplanted cells transd-ifferentiate to functioning cardiomyocytes, the alternative mechanism also suggests that these cells stimulate angiogenesis by secreting various growth factors, thus facilitating vascularization in the hibernated myocardium and improving functionality of the heart.

In the last couple of decades, we have experienced three strategic phases of therapeutic angiogenesis.[5] With the discovery of angiogenic growth factors in the late 70s and 80s, the early attempts to perform therapeutic angiogenesis have been initiated with the assumption that these growth factors are capable of enhancing vascular growth in the ischemic area. The first phase of therapeutic angiogenesis, the angiogenic approach, involved administration of a single angiogenic growth factor such as VEGF and FGF in the form of protein therapy and gene therapy. The methods of delivery included direct injection in the ischemic or periischemic area or a catheter-based infusion. To achieve sustained local levels of a growth factor, a heparin-alginate formulation had also been tested. Although results of initial animal experiments and open label clinical trials were encouraging, double-blind, placebo-controlled, randomized trials failed to show definitive functional improvement in the patients with coronary heart disease and peripheral arterial disease. A number of issues need to be considered in reaching a conclusion regarding the failure of these early approaches. However, it appears that while such an approach was valid in healthy young animals, it is probably not applicable in older end-stage ischemic disease patients.

The second strategic phase of therapeutic angiogenesis began shortly after the identification of circulating bone marrow-derived endothelial progenitor cells (EPC) with their possible contribution to adult vasculogenesis. The understanding and clarification of these progenitors which express markers of both hematopoietic (CD133, CD34, c-kit) and endothelial (VEGF-R2) lineage have accelerated the shift in the strategy of therapeutic angiogenesis: the angiogenic approach to the vasculogenic approach. In this approach, bone marrow-derived or circulating endothelial progenitor cells were administrated or recruited to the site of ischemia in expectation of not only efficient vascular growth, but also transdifferentiation of these cells into other tissue-specific cell types, facilitating functional improvement and regeneration of the compromised tissue. Attempts to apply these cells in experimental and preclinical settings confirmed beneficial effects in many cases. It is, nonetheless, still controversial whether the effects are truly attributed to the "progenitorship" of these cells or a bystander effect that can be achieved by non-progenitor cells.

That is, although it has been observed that these bone marrow-derived cells, under some circumstance, indeed contribute to functional recovery of ischemic tissue, this contribution may not be due to cell incorporated into the vasculature or tissue as structural or functional components. Instead, such cells may reside in the perivascular region as paracrine cells, facilitating vascular growth by secreting angiogenic and chemoattracting factors. Furthermore, the entity of these cells may not necessarily be "endothelial" precursors. Therefore, the third phase of therapeutic angiogenesis explored the possibility of non-progenitor cells. This paracrine approach involves organizer cells, for instance, bone marrow-derived or peripheral blood-derived mononuclear cells (BM-MNC or PB-MNC), capable of regulating various aspects of vascular growth.

4. Clinical Trials

4.1. *Growth factor-based, angiogenic approach*

Phase II/III clinical trials that address the efficacy of therapeutic angiogenesis using the growth factor-based, angiogenic approach are summarized in Table 2. This approach has been focused mainly on $VEGF_{165}$, $VEGF_{121}$ and FGF2 with limited data available on HGF. As delivery strategies, protein therapy including heparin-alginate formulation and gene therapy have been tested in these trials. Among them, VIVA, FIRST and TRAFFIC trials are well designed and adequately powered, suggesting several important lessons despite overall disappointing results. One of them is the difficulty in translating animal models in which most of the growth factors work efficiently in clinical settings where the patient population is more heterogeneous and refractory to angiogenesis. This will require careful patient selection and ideal development of biomarkers that enable us to predict neovascularization

Table 2. Growth factor-based therapy clinical trials.

Growth factor	Type of study	Indication	Delivery	Patient (n)	Study phase	Results	Ref(s).
VEGF	rVEGF *VIVA trial*	IHD	IC + IV	178	Phase II/III, DBR, placebo-controlled	Tolerated, no improvement vs. placebo	21
	Plasmid VEGF-2	IHD	IM, NOGA-guided	29	Phase I/II, DBR, dose-ranging, placebo-controlled	Safe, reduction in angina class	38–40
	Plasmid VEGF$_{121}$ *RAVE Trial*	CLI	IM	105	Phase II, DBR, placebo-controlled	No difference between groups	41
	Adenovirus VEGF$_{121}$ *REVASC trial*	IHD	IM	67	Phase II Controlled vs. maximum medical therapy	Objective improvement in exercise-induced ischemia	42
	Adenovirus VEGF$_{165}$ or plasmid VEGF$_{165}$ Liposome *KAT trial*	IHD PCI	IC	103	Phase II, DBR, placebo-controlled	Safe, no differences in clinical restenosis rate, better myocardial perfusion in Ad-VEGF-group	43

Table 2. (*Continued*).

Growth factor	Type of study	Indication	Delivery	Patient (n)	Study phase	Results	Refs.
FGF	rFGF2 (heparin-alginate microcapsule)	IHD CABG	IM	24	Phase I/II, DBR, placebo-controlled	Safe, improved symptom and myocardial perfusion in high dose FGF2 group for 3 years.	9, 44
	rFGF2 *FIRST Trial*	IHD	IC	337	Phase II, DBR, placebo-controlled	No improvement in ETT or myocardial perfusion vs. placebo	20
	rFGF2 *TRAFFIC Trial*	CLI	IA	190	Phase II, DBR, placebo-controlled	Transient benefit only at 3 months	45
	Adenovirus FGF4 *AGENT 2 Trial*	IHD	Sole therapy IC l	52	Phase II, DBR, placebo-controlled	Safe, no adverse effects Trends to improve ETT, perfusion	46, 47

IHD, ischemic heart disease; CLI, critical limb ischemia; PCI, percutaneous coronary intervention; IC, intracoronary; IV, intravenous; IM, intramyocardial; CABG, coronary artery bypass grafting; DBR, double-blind randomized; ETT, exercise treadmill tests.

responsiveness. Furthermore, we have learned the importance of randomization. Most open label Phase I trials claimed significant improvement in patient's symptoms as well as objective measurement of cardiac function. However, the same agents were not effective in the Phase II trials partially because of the unexpected placebo effect. Spontaneous improvement in the end-stage patients is highly significant, which can overshadow the agent's true effect. It is also noted that fluctuation of the end-point is often observed in the patient population. This is not only manifested in subjective, soft end-points, but also in hard end-points such as MR and PET perfusion.

Gene therapy is an alternative to protein therapy with its ability to provide more sustained presence of the desired agent in the target tissue. While some of gene therapy trials including Phase I open label trials suggested potential beneficial effects, none of double-blind, adequately powered trials demonstrated definitive benefit as we experienced in the protein therapy trials. Conclusively, we have learned from growth factor-based studies that one- or two-time administration of a single angiogenic growth factor, regardless of the delivery strategy, is not sufficient to provide therapeutic effect.

4.2. *Cell therapy-based, vasculogenic and paracrine approach*

The next generation of therapeutic angiogenesis is cell-based therapy which includes the vasculogenic and paracrine approaches, although the difference of these two is practically not discernable in many cases. The vasculogenic approach utilizes progenitor cells which are, in principal, fractionated before administration by using progenitor cell markers.[6] Although there is no consensus with regard to the surface marker representing genuine endothelial progenitor cells, many studies use CD34 and/or CD133 as "EPC markers." However, in reality most large clinical trials thus far use the crude bone marrow mononuclear cell population that is heterogeneous and presumably contain a significant number of progenitor or stem cells in the preparation.

In contrast, the paracrine approach uses unfractionated mononuclear cells either from bone marrow or more recently from peripheral blood. This approach does not rely on the progenitor ability of the cell population; instead, expectation here is paracrine function of mononuclear cells that can regulate neovascularization processes in many steps. Some of relatively large, randomized, controlled cell-based trials are shown in Table 3. Most of the trials initially intended cardiac repair in MI patients; however, it has been suggested that beneficial effects are possibly derived from an angiogenic effect which revascularizes the ischemic tissue. At this point, it is premature to conclusively evaluate the beneficial effect of cell therapy with a limited number of large, double-blind, randomized trials.

5. Issues Regarding Current Strategy

5.1. *Choice of biological agent*

Blood vessel growth is a complex event which involves the regulation of a number of genes with multiple growth factors, cytokines and modulators acting at different phases. Therefore, the supplementation of a single growth factor that can only trigger the neovascularization process is most likely insufficient, although the rest of the process may be carried out endogenously. The rationale for the multiple growth factor approach is the utilization of different types of biological reagents that can augment neovascularization in a complementary or synergistic fashion. Hence, one agent initiates the growth of new vascular structures while another induces their maturation, thereby ensuring the stability of new blood vessels. This coordinated therapeutic strategy, however, requires a detailed understanding of the kinetics of vessel growth and the ability to assess the state of neovascular response.

The other side of approach is that because there is still a long way to a full understanding of the blood vessel growth process at this point, we should use a "magic bullet" — a mixed bag of biologic agents without a defined composition that demonstrates functional effectiveness. This is, in essence, the premise inherent in the cell therapy approach where the ability to induce a functional benefit has significantly outstripped a thorough understanding of biology. The expectation here is that such biological materials will spontaneously regulate the neovascularization process by releasing all of the factors, and these factors will figure out by themselves how to grow vessels effectively.

Table 3. Cell-based therapy clinical trials (randomized, controlled trials).

Study	Cell type/ treatment	Indication	Delivery	Patient (n)	Study design	Results
BOOST Trial[48,49]	BM-MNC	AMI, PCI with stent	IC	60	Randomized, controlled vs. standard therapy for MI	Safe, improved LV function at 6 months, not at 18 months
Janssens[50]	BM-MNC	AMI, PCI	IC	67	DBR, placebo-controlled	No difference in LV function
ASTAMI trial[51]	BM-MNC	AMI, PCI	IC	100	Randomized, controlled vs. standard therapy for MI	Safe
REPAIR-AMI Trial[52]	BM-MNC	AMI, PCI	IC	204	DBR, placebo-controlled	Increased LVEF
Erbs[53]	CPC	OMI, PCI	IC	26	DBR, placebo-controlled	Increased LVEF, reduced infarct size
MAGIC Cell-3-DES Trial[54]	PB-MNC mobilized by G-CSF	AMI+OMI, PCI + DES	IC	96	Randomized, controlled vs. standard therapy for MI	Increased LVEF in AMI group, not in OMI group
Zohlnhofer[55]	G-CSF	AMI, PCI	SC	114	DBR, placebo-controlled	No difference
STEMMI[56]	G-CSF	AMI, PCI + stent	SC	62	DBR, placebo-controlled	No difference
START Trial[12]	GM-CSF	CLI	SC	40	DBR, placebo-controlled	No difference

BM, bone marrow; PB, peripheral blood; MNC, mononuclear cells; CPC, circulating progenitor cells; G-CSF, granulocyte colony-stimulating factor; AMI, acute myocardial infarction; OMI, old myocardial infarction; PCI, percutaneous coronary intervention; DES, drug eluting stent; IC, intracoronary; SC, subcutaneous; DBR, double-blind randomized; LV, left ventricular; LVEF, left ventricular ejection fraction.

With regard to the selection of an agent, as we begin to realize different forms of vascular growth, it may be better to consider the arteriogenic approach rather than the angiogenic approach as arteriogenesis has more impact on the flow recovery in the ischemic vascular bed. An interesting observation implies the weak contribution of angiogenesis in the cardiovascular mortality. Down's syndrome patients carrying an extra copy of collagen XVIII gene have increased endostatin levels and appear to be protected from the development of cancer. This is attributed to impaired angiogenesis necessary for tumor growth. However, the same study indicates the same mortality rate of ischemic heart disease in the Down's and non-Down's population, suggesting the limited role of angiogenesis *per se* as a determinant of prognosis in ischemic heart disease.[7]

Shear response of the endothelium is of prime importance in the process of arteriogenesis, which is followed by mononuclear cell influx. Therefore, one means of inducing arteriogenesis may be the restoration of effective endothelial signaling or alteration of mononuclear cell function rather than administration of a growth factor. The other appealing option is the enhancement of an endogenous vasculogenic process. If vasculogenesis plays an important role in adult tissue neovascularization, agents that mobilize progenitor cells and promote this process may prove effective. We need to be, however, cautious that the effectiveness of these agents may be limited in the end-stage patient population because stem cell functionality declines with age and possibly with disease.[8]

5.2. *Pharmacokinetics and delivery mode*

Understanding pharmacokinetics of an angiogenic agent is prerequisite for developing a delivery strategy that effectively promotes therapeutic angiogenesis. However, in animal experiments, the point has been mainly focused on the angiogenic outcome, rather than the pharmacokinetics of an agent. Taking into consideration pharmacokinetics, the effective delivery strategy needs to fulfill several criteria: a necessary concentration of the agent that can initiate neovascularization, specific delivery of an agent at the desired site or cell population such as endothelial cells or monocytes, and sustained effectiveness of an agent for a duration of time (weeks or even longer) sufficient to allow maturation of newly formed blood vessels. Most likely this cannot be achieved with single-dose administration of proteins or peptides with their short half-lives. Instead, a slow-release gel formulation can be an alternative despite the inconvenient administration with invasive means. A sustained benefit of heparin-alginate-based FGF2 delivery provides a strong endorsement of this strategy.[9]

Gene therapy is one of the promising modalities to deliver a therapeutic agent for a long period of time; however, neither of two-vector systems currently in use, plasmid- or adenovirus-based, provide more than a few weeks of high-level expression. More long-lived gene transfer vectors such as adeno-associated virus (AAV) or lentiviruses have not yet been tested. In any vector system, we can only control the amount and duration of expression without fine tuning of expression levels. This will bring about other concerns such as a long, unregulated expression of an angiogenic agent, possibly leading to substantial side effects like development of atherosclerosis and cancer.

The remaining options include the systemic administration of an agent that specifically acts only in the desired tissue or provide therapeutic effect only at the desired site. For example, PlGF appears to induce vessel growth only in the setting of ischemia,[10] and endocrine tissue-specific VEGF raises the possibility that tissue-specific growth factors exist that can be used in a systemic fashion.[11]

In the premise of augmenting arteriogenesis, the delivery modality is a more complex issue because of the limited accessibility to the endothelium of relatively large vessels in comparison to the endothelium of capillaries. GM-CSF has been shown to enhance collateral growth in experimental models by promoting mobilization of bone marrow cells. However, the START trial, in which patients with critical limb ischemia were treated with subcutaneous injection of GM-CSF, failed to demonstrate efficacy.[12]

Thus, given our understanding of the biology of neovascular development, especially arteriogenesis, a prolonged treatment modality that can influence endothelial cells, monocyte and/or vascular smooth muscle cells appears to be necessary. This will most likely be achieved by an organizer-type growth factor such as FGF family, rather than an endothelial-specific growth factor such as VEGF and angiopoietin that only stimulates endothelial function. At present, delivery strategies that can accomplish prolonged and targeted exposure include either sustained-release delivery of growth factor proteins or systemic administration of organ- or tissue-specific agents.

5.3. *Monitoring of neovascularization*

Monitoring the effect of angiogenic therapy has been a long-standing challenge. In principle, this can be accomplished by either directly monitoring blood vessel growth or evaluating the functional effects of the therapy. In any case, it needs to be sensitive enough to detect small changes of vascularization that can account for the symptomatic improvement of the patient. The development of monitoring modality capable of non-invasively assessing neovascularization is essential for the success of therapeutic angiogenesis.

Recently molecular imaging of angiogenesis has received considerable attention, with a number of reports demonstrating the feasibility of detecting blood vessel growth in tissues by targeting "angiogenic" endothelial cell-specific antigens.[13] Among them, $\alpha v \beta 3$ integrin, a member of the family of adhesion molecules, has been studied as a marker of tumor angiogenesis. This integrin is known to bind to matrix proteins with an exposed RGD (arginine-glycine-aspartate) sequence. Recently, a clinical study has shown that Galacto-RGD with positron-emission tomography (PET) enables non-invasive quantitative assessment of the $\alpha v \beta 3$ expression pattern on tumor and endothelial cells in cancer patients.[14]

Direct visualization of new vasculature is a convincing way to detect collateral vessel growth, although it does not necessarily address the functional consequences of the new vessel growth. Large collaterals ($>130 \, \mu$m diameter) can be observed and perhaps quantified with standard angiographic techniques; however, a number of artifacts influence the angiographic appearance of vessels, including vascular tone, amount of the injected contrast, force of injection, and medication. A particularly interesting technology is the 3D reconstruction of micro-CT angiographic images, which enables depiction of patterns of arterial growth and quantification of blood vessel density and volume.[15] However, micro-CT is unsuitable for repeated non-invasive measurements of vessels, as it requires long scan periods and high X-ray doses. Volumetric computed tomography (VCT) which combines advantages of micro-CT and clinical CT scanners is tested in experimental and preclinical applications.[16] Finally, magnetic resonance angiography can also provide visualization of collaterals.[17] Although relatively effective in the limb, the sensitivity of MRI coronary reconstruction is not yet sufficient.

The alternative to direct visualization of the vasculature is the assessment of the physiological consequences of vessel growth, such as improvement in tissue perfusion, oxygenation, or function. In coronary artery disease trials, nuclear perfusion scan imaging has been used as a surrogate end-point. Remarkably little effect was observed with this imaging modality even when clinical symptoms of patients appeared to be improved. This raised questions about the spatial resolution and sensitivity of single-photon emission computed tomography (SPECT) imaging. Positron-emission tomography (PET) and MRI are the main alternatives to SPECT imaging. PET boasts somewhat higher spatial resolution, elimination of attenuation, and quantitative assessment of perfusion, providing comprehensive insights into vascular development. A variety of PET imaging probes for imaging neovascular development have been synthesized, including perfusion tracers such as ^{15}O-labeled water, monoclonal anti-VEGF antibody, and ligands for $\alpha v\beta 3$ integrin.[13] MRI can determine perfusion and blood volume changes in response to treatment; however, no agreement has been reached with regard to the methodology of the measurement.[18] One approach relies on assessing relative differences in perfusion between normal and ischemic zones, thereby providing an assessment of the ischemic zone size. Alternative approaches attempt to measure perfusion by the first-pass technique. The advantages of MRI are its high spatial resolution and sensitivity to small changes in flow. Similar to PET, however, experience with the use of MRI perfusion in large clinical trials is limited.

5.4. *Study design*

Appropriate study design is essential for translating findings of basic research into clinical setting. An important part of therapeutic angiogenesis trials is the selection of an appropriate patient population.

As with all new therapies, there is a tendency to initially restrict the therapy to the no-option population. Indeed, most therapeutic angiogenesis trials have been carried out in symptomatic patients who have exhausted standard therapy modalities.[19] These patients tend to be older, with more extensive disease and clinical evidence of not being responsive to standard therapies, thus suggesting defects in intrinsic neovascularization response. These characteristics can make these patients especially poor candidates for neovascularization.

Another issue that has been arduously learned from early clinical trials is the occurrence of a significant placebo effect.[20,21] Although placebo effects are well described in many fields of medicine, the sheer magnitude of the effect observed in these trials was surprising. In the placebo group of the end-stage, no-option patients, their exercise capacity increased by 45 to 60 seconds while scores on the Seattle Angina Questionnaire and Short Form-36 as well as pill counts also showed surprising changes. Regardless of the reason why the placebo response is so prominent and significant in this patient population, the importance of this phenomenon clearly indicates that small open label studies can only be used for the assessment of safety and tolerability; assessment of efficacy should be evaluated by a double-blind randomized manner.

6. Emerging Concepts of Therapeutic Angiogenesis

Despite frustrating results of initial large clinical trials, the underlying premise of therapeutic angiogenesis seems still valid: augmentation of blood vessel growth to compensate for insufficient blood supply to the compromised tissue. The currently tested concept of therapeutic angiogenesis is to induce the increased presence of an angiogenic factor or cellular components in the target area. However, such logic may be flawed. The underlying assumption is that the endogenous biological response to ischemia is impaired because of the lack of angiogenic stimuli, thus justifying exogenous supplementation of growth factors. It appears to be more likely that a defective angiogenic response is due to a defective endothelial signaling especially in cases of diseases such as diabetes.[22] Furthermore, there is no evidence demonstrating that levels of growth factors are indeed decreased in the ischemic tissue, resulting in the obstruction of the neovascularization process. Even if this is the case, we need to be more deliberate with regard to which growth factor is missing and responsible for the impaired angiogenic response in individual patients. Moreover, in the current concept we also assume supplementation of the superphysiological amount of growth factor simply augments normal angiogenic response in old, end-stage patients with ischemic diseases.

6.1. *Neovascularization responsiveness*

Due to long-term, sustained endogenous angiogenic stimuli or defective endothelial function, the angiogenic adaptation process of these patients may have blunted significantly, resulting in the loss of tissue responsiveness to angiogenesis — a situation similar to other conditions such as insulin resistance in type 2 diabetes and tolerance developed by repeated drug usage. It is known that general sensitivity of the endothelium to angiogenic growth factors is an important determinant of angiogenic response. This may explain the discrepancy in the neovascularization response between healthy young animals and diseased patients with systemic illnesses. To overcome this issue, the focus may need to be more on increasing tissue responsiveness rather than increasing angiogenic stimuli.

To address the neovascularization responsiveness issue, we need to clarify the differences between healthy endothelium and dysfunctional endothelium. It is widely accepted that endothelial function progressively declines with age,[23] moreover, dysfunction of the endothelium is well described in patients with atherosclerosis, diabetes, and other risk factors for vascular diseases. With accumulating evidence showing a loss of endothelial homeostasis is a prime event leading to cardiovascular diseases, clinical and basic research have focused on elucidating the role of endothelial dysfunction in influencing vascular disease progression.

Endothelial dysfunction, clinically assessed by endothelium-dependent vasodilator responses, is a broad term that implies diminished production or bioavailability of nitric oxide (NO). It is also affected by an imbalance of endothelium-derived relaxing and contracting factors such as endothelin-1, angiotensin-II and antioxidants. In addition to the vasodilatory effect, NO is a versatile biomediator involved in protection against vascular injury, inflammation, thrombosis, angiogenesis, and EPC mobilization.[24] Although mixed reports exist with regard to NO production in aged arteries, several studies have shown expression of endothelial NO synthase (eNOS) is attenuated in aging, rendering endothelial cells more susceptible to apoptotic death. This may explain why physical activity prevents age-related impairment in NO availability in elderly people; shear stress is one of the strongest stimuli for the expression of eNOS.[25] Moreover, secretion of many growth factors and hormones such as growth hormone and steroid hormones decline with age. Although the relevance to angiogenic responsiveness has not been established, decreased estrogen levels have been implicated as a risk factor of atherosclerotic disease for post-menopausal women.

Furthermore, it is widely recognized that aging is associated with oxidative stress and related cellular damage, which may well result in refractory neovascular development in old patients.[26] Endothelium is an important source of reactive oxygen species (ROS) that is required for normal endothelial functions. However, continuous production of ROS and impaired ROS scavenging system may cause mitochondrial dysfunction in the long run by damaging mitochondrial DNA and contribute to age-dependent cellular dysfunction.

Endothelial dysfunction eventually leads to endothelial senescence, a condition in which cells lose the capacity to divide and enter a state of irreversible growth arrest. Impaired wound healing and angiogenesis observed in older people are attributed to endothelial senescence. One of the indices to evaluate endothelial senescence is the length of telomeres which are essential for maintaining genome stability and integrity, contributing to extended proliferative life span both in cultured cells and in organisms. It is suggested that age-dependent telomere shortening occurs in human endothelium, which results in impaired angiogenesis.[27] The activity of telomerase reverse transcriptase (TERT), one of the components of telomerase which serves for preserving telomeric DNA length, is attenuated by oxidative stress, facilitating telomerase ablation in aged cells. Constitutive hTRET expression enhances the regenerative capacity of endothelial progenitor cells. The idea of telomerase rescue may provide with a new approach to therapeutic angiogenesis as an animal experiment using hTERT-transduced endothelial progenitors improved neovascularization in ischemic limbs.[28]

6.2. *Genetic determination of neovascularization*

Recent advances in molecular medicine led us to take into account the genetic aspect of the neovascularization response. Increasing evidence shows that genetic background and related predisposed conditions can affect neovascular development. Genetic loci which can influence angiogenic responses induced by VEGF or FGF2 were mapped in the mouse genome, suggesting that variation of the genetic background plays a role in determining the angiogenic responsiveness. Moreover, a number of sporadic mutations and polymorphisms which can alter angiogenic response have also been identified.[29] In contrast to mutations which in general affect health to a greater extent, smaller genetic changes often referred to as polymorphism do not manifest overt phenotypes and are much more common in the human population. Among them VEGF polymorphisms in the promoter region which are implicated in a wide variety of angiogenesis-dependent diseases can predispose to a person the altered neovascularization response. Making an allowance for anatomic severity and duration of disease, clinical observations have long noted a variable presence or absence of collateral circulation on coronary angiograms. Of particular interest is the ability of monocytes from different individuals to respond to hypoxia by increasing HIF-1α expression correlated with the extent of collateral development.[30] Although the degree of collaterarization is independent of plasma levels of pro- or anti-agniogenic factors, monocytes functionality discriminates patients with different degree of collateral development, suggesting genetic differences in the patient population.[31] In an extension of this idea, a further study has identified, using monocyte transcriptomes from CAD patients with and without well developed collaterals, a set of molecular markers characteristic of a "non-collateralgenic" phenotype.[32]

7. Future Prospective

One of the fundamentally new approaches to therapeutic angiogenesis is to improve tissue responsiveness to angiogenic stimuli. To achieve this goal, more research efforts need to be focused on basic research to investigate what determines angiogenic responsiveness. Endothelial function is probably of prime importance to determine the course of neovascularization as the initial response of endothelial cells plays a key role in the angiogenic and arteriogenic processes. The other key cell type is a monocyte or an EPC. Recently, circulating EPCs have been shown to correlate inversely with the risk of cardiovascular events.[33,34] Hypercholesterolemia has been shown to delay arteriogenic process by affecting monocyte function.[35] A similar observation has been made in diabetic patients.[36] By analogy with coronary risk factors, we may be able to identify neovascularization risk factors as these two conditions often co-exist in the patient population.

We also need to explore the contribution of the genetic background in the neovascularization process since it will enable us to estimate neovascularization potential and predict the efficacy of the therapy. While this custom-made approach requires careful dissection of patient population according to the various genetic backgrounds underlying the ischemic disease, it undoubtedly shows great promise for the future. For example, in the case of a polymorphism that reduces the VEGF promoter activity, the supplement of VEGF may be beneficial. Furthermore, we may be able to utilize genetically modified monocytes and EPCs depending on the abnormality in the patients' genetic program.

Whatever strategy we use in therapeutic angiogenesis, it is essential to expand our knowledge on arterial development as an arteriogenesis approach will be the mainstream of therapeutic angiogenesis in the future. With regard to the mechanism of arteriogenesis, we still do not know what the most important driving force is and whether larger arteries can be formed *de novo* in adult humans. In the embryonic development, circulatory dynamics were previously thought to play a major role in establishing arterial-venous determination; however, recent molecular analyses have demonstrated the underlying process of this specification.[37] It has been clear that arterial and venous endothelial cells are molecularly distinct even before the initiation of the first embryonic heartbeat, revealing the existence of genetic programs coordinating arterial-venous differentiation. Recent studies have identified VEGF, Notch, and Ephrin/Eph signaling as critical factors determining arterial and venous cell fate.

8. Summary

Therapeutic modulation of angiogenesis is still at its infant stages, but it is showing considerable potential. The ultimate success of this therapeutic modality will depend on careful translation of research programs that will incorporate the ever-evolving basic understanding of the biology of neovascular development into effective, adequately powered clinical trial programs. The likely keys include the careful choice of a biological agent that may be different for different indications, effective delivery modality with pharmacokinetic properties matching biological needs, and the meticulous choice of patient population and end-points for the study. While still controversial, cell therapy has an enormous potential that is yet to be explored. In addition, with the advance of basic research, different strategies for therapeutic angiogenesis have come into view that can fundamentally change our approach to this issue. These include increasing tissue neovascularization responsiveness and genetic analysis of neovascularization potentials. Therefore, new developments will be expected in this area in the next decade enabling us to treat better end-stage ischemic diseases.

References

1. Simons M (2005). Angiogenesis: where do we stand now? *Circulation* **111**(12): 1556–1566.
2. Heil M, Schaper W (2004). Influence of mechanical, cellular, and molecular factors on collateral artery growth (arteriogenesis). *Circ Res* **95**(5): 449–458.
3. Boyle AJ, *et al.* (2006). Controversies in cardiovascular medicine: ready for the next step. *Circulation* **114**(4): 339–352.
4. Oettgen P, *et al.* (2006). Controversies in cardiovascular medicine. *Circulation* **114**(4): 353–358.
5. Semenza GL (2006). Therapeutic angiogenesis: another passing phase? *Circ Res* **98**(9): 1115–1116.
6. Stamm C, *et al.* (2003). Autologous bone-marrow stem-cell transplantation for myocardial regeneration. *Lancet* **361**(9351): 45–46.
7. Yang Q, Rasmussen SA, Friedman JM (2002). Mortality associated with Down's syndrome in the USA from 1983 to 1997: a population-based study. *Lancet* **359**(9311): 1019–1025.
8. Rando TA (2006). Stem cells, ageing and the quest for immortality. *Nature* **441**(7097): 1080–1086.
9. Laham RJ, *et al.* (1999). Local perivascular delivery of basic fibroblast growth factor in patients undergoing coronary bypass surgery: results of a phase I randomized, double-blind, placebo-controlled trial. *Circulation* **100**(18): 1865–1871.

10. Luttun A, *et al.* (2002). Revascularization of ischemic tissues by PlGF treatment, and inhibition of tumor angiogenesis, arthritis and atherosclerosis by anti-Flt1. *Nat Med* **8**(8): 831–840.

11. LeCouter J, *et al.* (2001). Identification of an angiogenic mitogen selective for endocrine gland endothelium. *Nature* **412**(6850): 877–884.

12. van Royen N, *et al.* (2005). START trial: a pilot study on STimulation of ARTeriogenesis using subcutaneous application of granulocyte-macrophage colony-stimulating factor as a new treatment for peripheral vascular disease. *Circulation* **112**(7): 1040–1046.

13. Czernin J, Weber WA, Herschman HR (2006). Molecular imaging in the development of cancer therapeutics. *Annu Rev Med* **57**: 99–118.

14. Haubner R, *et al.* (2005). Noninvasive visualization of the activated alphavbeta3 integrin in cancer patients by positron emission tomography and [18F]Galacto-RGD. *PLoS Med* **2**(3): e70.

15. Li W, *et al.* (2006). High-resolution quantitative computed tomography demonstrating selective enhancement of medium-size collaterals by placental growth factor-1 in the mouse ischemic hindlimb. *Circulation* **113**(20): 2445–2453.

16. Kiessling F, *et al.* (2004). Volumetric computed tomography (VCT): a new technology for noninvasive, high-resolution monitoring of tumor angiogenesis. *Nat Med* **10**(10): 1133–1138.

17. Pearlman JD, *et al.* (2002). Medical imaging techniques in the evaluation of strategies for therapeutic angiogenesis. *Curr Pharm Des* **8**(16): 1467–1496.

18. Viswamitra S, *et al.* (2004). Magnetic resonance imaging in myocardial ischemia. *Curr Opin Cardiol* **19**(5): 510–516.

19. Simons M, *et al.* (2000). Clinical trials in coronary angiogenesis: issues, problems, consensus: An expert panel summary. *Circulation* **102**(11): E73–E86.

20. Simons M, *et al.* (2002). Pharmacological treatment of coronary artery disease with recombinant fibroblast growth factor-2. Double-blind, randomized, controlled clinical trial. *Circulation* **105**: 788–793.

21. Henry T, *et al.* (2003). The VIVA trial: Vascular endothelial growth factor in ischemia for vascular angiogenesis. *Circulation* **107**: 1359–1365.

22. Waltenberger J, Lange J, Kranz A (2000). Vascular endothelial growth factor-A-induced chemotaxis of monocytes is attenuated in patients with diabetes mellitus: A potential predictor for the individual capacity to develop collaterals. *Circulation* **102**(2): 185–190.

23. Brandes RP, Fleming I, Busse R (2005). Endothelial aging. *Cardiovasc Res* **66**(2): 286–294.

24. Forstermann U, Munzel T (2006). Endothelial nitric oxide synthase in vascular disease: from marvel to menace. *Circulation* **113**(13): 1708–1714.

25. Taddei S, *et al.* (2000). Physical activity prevents age-related impairment in nitric oxide availability in elderly athletes. *Circulation* **101**(25): 2896–2901.

26. Storz P (2006). Reactive oxygen species-mediated mitochondria-to-nucleus signaling: a key to aging and radical-caused diseases. *Sci STKE* **2006**(332): re3.

27. Serrano AL, Andres V (2004). Telomeres and cardiovascular disease: does size matter? *Circ Res* **94**(5): 575–584.

28. Murasawa S, *et al.* (2002). Constitutive human telomerase reverse transcriptase expression enhances regenerative properties of endothelial progenitor cells. *Circulation* **106**(9): 1133–1139.

29. Rogers MS, D'Amato RJ (2006). The effect of genetic diversity on angiogenesis. *Exp Cell Res* **312**(5): 561–574.

30. Schultz A, *et al.* (1999). Interindividual heterogeneity in the hypoxic regulation of VEGF: significance for the development of the coronary artery collateral circulation. *Circulation* **100**(5): 547–552.

31. Sherman JA, *et al.* (2006). Humoral and cellular factors responsible for coronary collateral formation. *Am J Cardiol* **98**: 1194–1197.

32. Chittenden TW, *et al.* (2006). Transcriptional profiling in coronary artery disease: Indications for novel markers of coronary collateralization *Circulation* **114**: 1811–1820.

33. Schmidt-Lucke C, *et al.* (2005). Reduced number of circulating endothelial progenitor cells predicts future cardiovascular events: proof of concept for the clinical importance of endogenous vascular repair. *Circulation* **111**(22): 2981–2987.

34. Werner N, *et al.* (2005). Circulating endothelial progenitor cells and cardiovascular outcomes. *N Engl J Med* **353**(10): 999–1007.

35. Tirziu D, *et al.* (2005). Delayed arteriogenesis in hypercholesterolemic mice. *Circulation* **112**(16): 2501–2509.

36. Waltenberger J (2001). Impaired collateral vessel development in diabetes: potential cellular mechanisms and therapeutic implications. *Cardiovasc Res* **49**(3): 554–560.

37. Torres-Vazquez J, Kamei M, Weinstein BM (2003). Molecular distinction between arteries and veins. *Cell Tissue Res* **314**(1): 43–59.

38. Losordo D, *et al.* (2002). Phase 1/2 placebo-controlled, double-blind, dose-escalating trial of myocardial vascular endothelial growth factor 2 gene transfer by catheter delivery in patients with chronic myocardial ischemia. *Circulation* **105**: 2012–2018.

39. Fortuin F, *et al.* (2003). One-year follow-up of direct myocardial gene transfer of vascular endohelial growth factor-2 using naked plasmid deoxyribonucleic acid by way of thoractomy in no-option patients. *Am J Cardiol* **92**: 436–439.

40. Reilly JP, *et al.* (2005). Long-term (2-year) clinical events following transthoracic intramyocardial gene transfer of VEGF-2 in no-option patients. *J Interv Cardiol* **18**(1): 27–31.

41. Rajagopalan S, *et al.* (2003). Regional angiogenesis with vascular endothelial growth factor in peripheral arterial disease: a phase II randomized, double-blind, controlled study of adenoviral delivery of vascular endothelial growth factor 121 in patients with disabling intermittent claudication. *Circulation* **108**(16): 1933–1938.

42. Stewart DJ, *et al.* (2006). Angiogenic gene therapy in patients with nonrevascularizable ischemic heart disease: a phase 2 randomized, controlled trial of AdVEGF(121). (AdVEGF121) versus maximum medical treatment. *Gene Ther* **13**: 1503–1511.

43. Hedman M, *et al.* (2003). Safety and feasibility of catheter-based local intracoronary vascular endothelial growth factor gene transfer in the prevention of postangioplasty and in-stent restenosis and in the treatment of chronic myocardial

ischemia: phase II results of the Kuopio Angiogenesis Trial (KAT). *Circulation* **107**(21): 2677–2683.

44. Ruel M, *et al.* (2002). Long-term effects of surgical aniogenic therapy with fibroblast growth factor 2 protein. *J Thorac Cardiovasc Surg* **124**: 28–34.

45. Lederman RJ, *et al.* (2002). Therapeutic angiogenesis with recombinant fibroblast growth factor-2 for intermittent claudication (the TRAFFIC study): a randomised trial. *Lancet* **359**(9323): 2053–2058.

46. Grines CL, *et al.* (2003). A randomized, double-blind, placebo-controlled trial of Ad5FGF-4 gene therapy and its effect on myocardial perfusion in patients with stable angina. *J Am Coll Cardiol* **42**(8): 1339–1347.

47. Grines C, *et al.* (2003). Angiogenic gene therapy with adenovirus 5 fibroblast growth factor-4 (Ad5FGF-4): a new option for the treatment of coronary artery disease. *Am J Cardiol* **92**(9B): 24N–31N.

48. Wollert KC, *et al.* (2004). Intracoronary autologous bone-marrow cell transfer after myocardial infarction: the BOOST randomised controlled clinical trial. *Lancet* **364**(9429): 141–148.

49. Meyer GP, *et al.* (2006). Intracoronary bone marrow cell transfer after myocardial infarction: eighteen months' follow-up data from the randomized, controlled BOOST (BOne marrOw transfer to enhance ST-elevation infarct regeneration) trial. *Circulation* **113**(10): 1287–1294.

50. Janssens S, *et al.* (2006). Autologous bone marrow-derived stem-cell transfer in patients with ST-segment elevation myocardial infarction: double-blind, randomised controlled trial. *Lancet* **367**(9505): 113–121.

51. Lunde K, *et al.* (2005). Autologous stem cell transplantation in acute myocardial infarction: The ASTAMI randomized controlled trial. Intracoronary transplantation of autologous mononuclear bone marrow cells, study design and safety aspects. *Scand Cardiovasc J* **39**(3): 150–158.

52. Cleland JG, *et al.* (2006). Clinical trials update from the American Heart Association: REPAIR-AMI, ASTAMI, JELIS, MEGA, REVIVE-II, SURVIVE, and PROACTIVE. *Eur J Heart Fail* **8**(1): 105–110.

53. Erbs S, *et al.* (2005). Transplantation of blood-derived progenitor cells after recanalization of chronic coronary artery occlusion: first randomized and placebo-controlled study. *Circ Res* **97**(8): 756–762.

54. Kang HJ, *et al.* (2006). Differential effect of intracoronary infusion of mobilized peripheral blood stem cells by granulocyte colony-stimulating factor on left ventricular function and remodeling in patients with acute myocardial infarction versus old myocardial infarction: the MAGIC Cell-3-DES randomized, controlled trial. *Circulation* **114**(Suppl 1): I145-I151.

55. Zohlnhofer D, *et al.* (2006). Stem cell mobilization by granulocyte colony-stimulating factor in patients with acute myocardial infarction: a randomized controlled trial. *JAMA* **295**(9): 1003–1010.

56. Ripa RS, *et al.* (2006). Stem cell mobilization induced by subcutaneous granulocyte-colony stimulating factor to improve cardiac regeneration after acute ST-elevation myocardial infarction: result of the double-blind, randomized, placebo-controlled stem cells in myocardial infarction (STEMMI) trial. *Circulation* **113**(16): 1983–1992.

13

Hepatocyte Growth Factor

by Ryuichi Morishita and Toshio Ogihara

1. Hepatocyte Growth Factor in Cardiovascular System

Hepatocyte growth factor (HGF), a mesenchyme-derived pleiotropic growth factor, is considered a humoral mediator of the epithelial-mesenchymal interactions responsible for morphogenic tissue interactions during embryonic development and organogenesis (Fig. 1).[1] Although HGF was originally identified as a potent hepatocyte mitogen, it is also a very potent endothelial cell mitogen.[2,3] Moreover, both HGF and its receptor, c-met, are expressed in vascular cells and cardiac myocytes *in vitro* as well as *in vivo*.[4] Production of local HGF in vascular cells is regulated by various cytokines including transforming growth factor (TGF)-β and angiopoietin (Ang) II,[5] as well as by HGF itself via induction of Ets activity, which plays important roles in regulating gene expression in response to multiple developmental and mitogenic signals. The promoter region of HGF contains a number of putative regulatory elements, such as a B cell- and a macrophage-specific transcription factor binding site (PU.1/ETS), as well as an interleukin-6 response element (IL-6 RE), a TGF-β inhibitory element (TIE), and a camp response element (CRE).[6] Interestingly, serum HGF concentration is significantly correlated with blood pressure. Thus HGF secretion

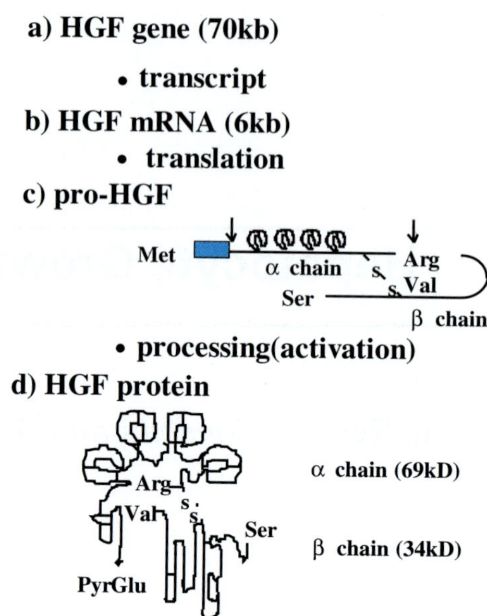

Fig. 1. Structure of hepatocyte growth factor. HGF belongs to the kringle family. The pro-HGF polypeptide comprises an N-terminal secretory signal that is cleaved during the maturation process of the protein. Arrows point to the cleavage sites of pro-HGF. For gene therapy experiments/trials, the cDNA encoding full-length pro-HGF is inserted in an expression cassette; HGF gene transfer is performed with this HGF expression construct (HGF minigene).

may be elevated in response to high blood pressure as a counter-system against endothelial dysfunction, and may be viewed as an indicator of severity of hypertension.[7]

2. HGF Signaling in Endothelial Cells

HGF acts as a mitogen, dissociation factor, and motility factor for many epithelial cells in culture through its tyrosine kinase receptor, c-met.[2,8] Various intracellular signaling pathways have been shown to be activated by tyrosine kinases linked to c-met. As shown in Fig. 2, the biological responses mediated by c-met are triggered by the tyrosine phosphorylation of a single multifunctional docking site located in the receptor's carboxy terminal.[9] This sequence, containing two phosphotyrosines, interacts with several cytoplasmic signal transducers either directly or indirectly through molecular adapters such as Grb2,

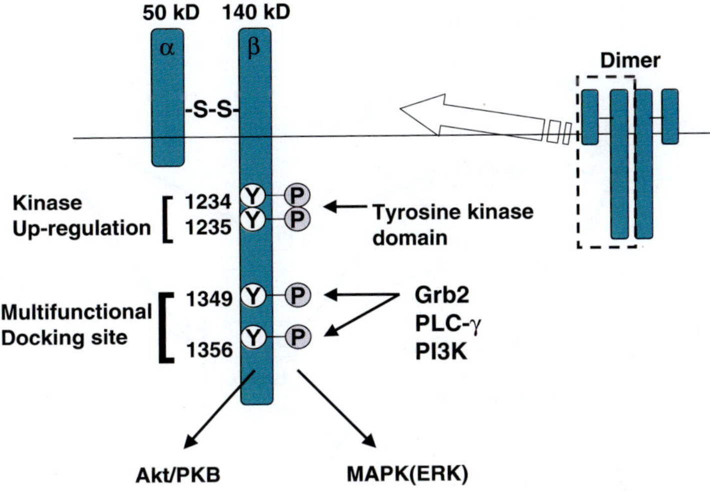

Fig. 2. Scheme of c-met (HGF receptor) structure. HGF receptor (c-met) consists of an α (50 kDa) and a β (140 kDa) chain, which make a heterodimer of each other. The biological responses mediated by c-met are triggered by the tyrosine phosphorylation of a single multifunctional docking site located in the carboxy terminal tail of the α chain. This sequence, containing two phosphotyrosines, interacts with several cytoplasmic signal transducers either directly or indirectly through molecular adapters such as Grb2. After HGF stimulation, c-met binds and activates phosphatidylinositol-3-OH kinase (PI3K), which then activates Akt/PKB (protein kinase B), and recruits the Grb-SOS complex, stimulating the Ras-MAP kinase cascade.

Shc and Gab1.[10,11] After HGF stimulation, c-met binds and activates phosphatidylinositol-3-OH kinase (PI3K) and recruits the Grb-SOS complex, stimulating the Ras-MAP kinase cascade.[12,13] In addition, the induction of epithelial tubules by HGF is dependent on activation of the STAT pathway and, importantly, c-met/the HGF tyrosine receptor can bind and directly phosphorylate STAT3.[14] HGF also stimulates cell proliferation through the ERK-STAT3 pathway and has an anti-apoptotic activity through the PI3K-Akt pathway in human aortic endothelial cells.[15] Interestingly, HGF also increases expression of the anti-apoptotic gene bcl-2 and inhibits translocation of a trigger of apoptosis, bax protein, from cytosol to the mitochondrial membrane.[16] It has also been reported that HGF can protect against cell death via inhibition of bad translocation, which is regulated by phosphorylation,

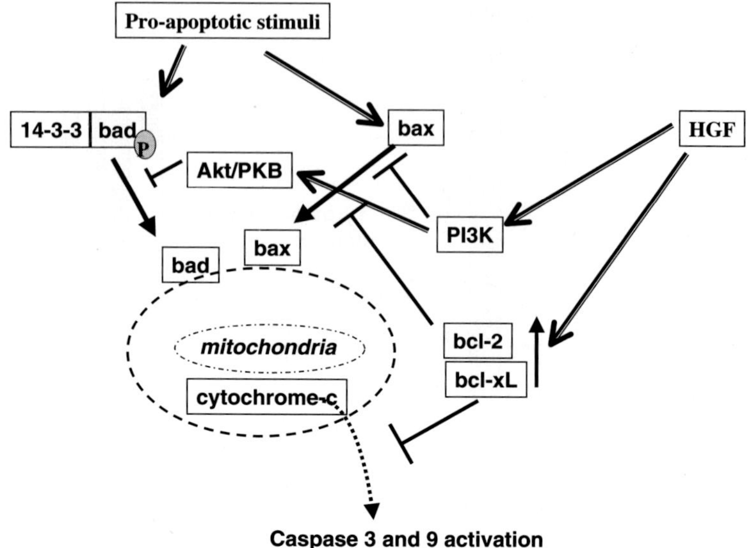

Fig. 3. Potential mechanisms of anti-apoptotic action of HGF. Pro-apoptotic stimuli increase pro-apoptotic genes, such as bax, and also stimulate the translocation of bax and/or bad to the mitochondrial heavy membrane. This bad translocation was regulated by binding to 14-3-3 protein through phosphorylation of bad by PI3K-Akt/PKB pathway. Since HGF can activate PI3K-Akt/PKB pathway and significantly increase bcl-2 and/or bcl-x/L protein, it can block the translocation of bax and/or bad. These changes in bax and/or bad protein release cytochrome c from mitochondria, resulting in activation of the caspase cascade. Therefore, HGF can block the release of cytochrome c through both direct action on mitochondria and blockade of bax and/or bad translocation.

and bax translocation, regulated by the conformational change resulting in the exposure of its BH3 domain via PI3K (Fig. 3).[17]

3. Angiogenic Therapy for Ischemic Peripheral Arterial Diseases

Critical limb ischemia is estimated to develop in 500 to 1000 individuals per million people in the general population per year. In a large proportion of these patients, the anatomic extent and the distribution of arterial occlusive disease make these individuals unsuitable for operative or percutaneous revascularization. Thus the disease frequently follows an inexorable downhill course and there is no optimal medical therapy for critical limb ischemia, as the Consensus Document of the European Working Group on Critical Limb Ischemia concluded. Therefore novel therapeutics are required to treat these patients. In the presence of obstruction of a major artery, blood flow to the ischemic tissue is often dependent on collateral vessels. When spontaneous development of collateral vessels is insufficient to allow normal perfusion of the tissue at risk, residual ischemia occurs. Preclinical and clinical studies have demonstrated that angiogenic growth factors can stimulate the development of collateral arteries,[18–22] a concept called therapeutic angiogenesis.

As intra-arterial administration of recombinant HGF protein induced angiogenesis in a rabbit hindlimb ischemia model,[23] the feasibility of gene therapy using HGF rather than recombinant protein therapy was examined to treat peripheral arterial disease. Intramuscular injection of a "naked" human HGF plasmid resulted in a significant increase in blood flow and capillary density in mice, rat and rabbit models of hind limb ischemia. Importantly, the degree of angiogenesis induced by transfection of HGF plasmid was significantly greater than that caused by a single injection of recombinant HGF protein. Angiogenic property of HGF was also demonstrated in high risk conditions such as diabetes mellitus and high Lp(a) concentration models.[24–28] One may assume that overexpression of an angiogenic growth factor can enhance tumor growth. To resolve this issue, we examined the overexpression of HGF in tumor-bearing mice. Tumor growth was initiated with an intradermal inoculation of A431, human epidermoid cancer cells expressing c-met. The mice were then intramuscularly injected with a human HGF or control plasmid into the femoral muscle. Analysis of human HGF expression noted increased concentration only in the injected femoral muscle, but not in blood. Although recombinant HGF stimulated the growth of A431 cells *in vitro*, no effect on tumor growth was detected in these mice.[29]

4. Clinical Trial in PAD

With this preclinical HGF data in hand, we investigated the safety and efficiency of HGF plasmid DNA transfection in patients with critical limb ischemia in an open-labeled phase I clinical trial.[30] Patients could be enrolled if they (1) had chronic critical limb ischemia, including rest pain or non-healing ischemic ulcers, for a minimum of four weeks; (2) were resistant to conventional drug therapy at least for more than four weeks after hospitalization; (3) were not candidates for surgical or percutaneous revascularization based on usual practice standards; (4) did not have cancer or a history of cancer; and (5) did not have severe unstable retinopathy. Objective documentation of ischemia, included a resting ankle brachial index (ABI) of less that 0.6 in the affected limb on two consecutive examinations performed one week apart. Patients were observed for four weeks under conventional therapy to confirm that their clinical symptoms and objective parameters were not improved.

Intramuscular injection of the naked HGF plasmid DNA was performed in ischemic limbs of 22 patients with arteriosclerosis obliterans or Buerger disease graded as Fontaine III or IV. The primary endpoints were safety and improvement of ischemic symptoms at 12 weeks after transfection. Throughout the gene therapy periods, there were no signs of systemic or local inflammatory reactions. No serious side-effects related to gene therapy were seen. To date, development of tumors or progression of diabetic retinopathy has not been observed in any patient transfected with HGF plasmid DNA during the trial. Two-month follow-up studies showed no evidence of the development of neoplasm or hemangioma. In addition, no significant increase in serum HGF concentration was observed throughout the gene therapy periods.

The preliminary evaluation of initial six patients demonstrated beneficial effects of HGF gene therapy.[30] Although ABI could not be measured in one patient because of uncompressible severely calcified vessels, ABI was significantly increased from 0.426 ± 0.046 (n = 5) at baseline (before administration) to 0.626 ± 0.071 ($P = 0.0155$; n = 5) at 4 weeks after the second injection, and to 0.596 ± 0.046 ($P = 0.0360$; n = 5) at eight weeks after the second injection. Increase in ankle pressure index more than 0.1 was observed in five of five patients. In addition, the change in transcutaneous PO_2 ($TcPO_2$) after O_2 stimulation was significantly increased at eight weeks compared with baseline ($P < 0.05$). To evaluate the effects of HGF gene therapy on clinical symptoms, we used the change in the ischemic ulcer and visual analogue scale. In this trial, a total of 11 ischemic ulcers were found in four patients. Two of 11 ulcers completely disappeared. Considering an improvement of ischemic ulcers of more than 25% to be evaluated as positive, eight of 11 ulcers (72%) improved. Three of four patients demonstrated an improvement of the maximum ischemic ulcer diameter of > 25% (efficacy rate = 75%). Also, we evaluated resting pain using a visual analog scale, a standard method for the evaluation of pain, where 0.0 cm means "pain free" or no pain, and 10 cm means more severe pain. Pain was significantly improved in a time-dependent manner. Thus, on the basis of this small trial, it appeared that intramuscular injection of naked HGF plasmid is safe, feasible, and can achieve successful improvement of ischemic limbs. It is noteworthy that no edema has been observed in this trial, although transient lower-extremity edema was reported with clinical gene therapy using the VEGF gene because of an increase in vascular permeability.

One of the distinguishing features of HGF is that it stimulates the migration of VSMC without stimulating their replication.[31] As shown in Fig. 4a, the initial event in angiogenesis induced by VEGF is the migration of endothelial cells, leading to the sprouting of blood vessels. Later, the migration of VSMC occurs due to the release of PDGF. However, a delay in the maturation of blood vessels may exist in the case of angiogenesis induced by VEGF. In contrast, HGF simultaneously stimulated the migration of both endothelial cells and VSMC (Fig. 4b). Thus, the blood vessels may mature at an earlier time point, thereby avoiding the release of blood-derived cells into the extracellular space, although further studies may be necessary to examine the angiogenic properties among various angiogenic growth factors including HGF, VEGF and FGF. Although these trials have not been finished, the feasibility of gene therapy using angiogenic growth factors to treat peripheral arterial disease seems to be the realm in the near future.

Clinical studies of alternative dosing regiments of gene therapy with randomized placebo-controlled trials were designed. Currently, phase III double-blinded randomized placebo-controlled studies in Japan and phase II studies in USA[32] are ongoing.

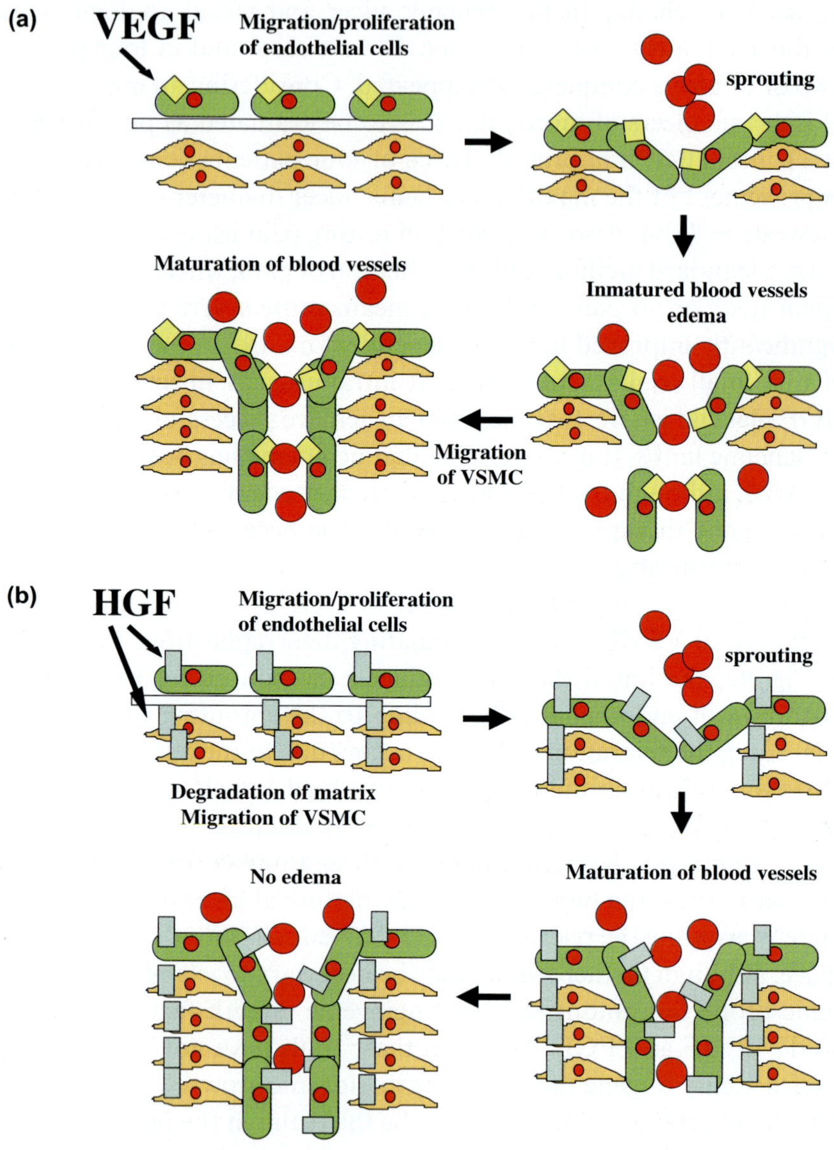

Fig. 4. Model of collateral formation induced by VEGF (**a**) and HGF (**b**). HGF stimulated the growth and migration of endothelial cells together with the migration, but not proliferation, of VSMC through c-met. In contrast, VEGF only stimulated the growth and migration of endothelial cells without the migration or proliferation of VSMC, due to lack of VEGF receptors in VSMC.

5. HGF Gene Therapy for Myocardial Ischemia

In addition to stimulating peripheral angiogenesis, overexpression of
HGF in the myocardium was also reported to stimulate angiogenesis
and collateral formation in a rat myocardial infarction model.[33] Moreover, intramuscular injection of the HGF plasmid into the ischemic
myocardium resulted in a significant increase in blood flow and prevention of cardiac dysfunction in a canine model.[34] We also injected human
HGF plasmid DNA at doses of 0.4 or 4 mg into ischemic myocardium
of pigs induced by ameroid constrictor using the NOGA™ system. At
one month after injection, the ischemic area was significantly reduced
in the 4 mg HGF group, accompanied by a significant increase in the
capillary density and regional myocardial perfusion in the ischemic area
as compared to the control group. These favorable outcomes suggest
potential utility of HGF gene transfer for treatment of patients with
ischemic heart disease.

The molecular mechanisms of the angiogenic activity of HGF seem
to be largely dependent on the Ets pathway. Members of Ets family
of transcription factors share a DNA-binding domain that binds to
a core GGA(A/T) DNA sequence. *In situ* hybridization studies have
revealed that the proto-oncogene c-Ets-1 is expressed in endothelial cells
at the start of blood vessel formation, under both normal and pathological conditions. Thus, the Ets family may activate the transcription
of genes encoding collagenase-1, stromelysine-1 and urokinase plasminogen activator, which are proteases involved in extracellular matrix
degradation. It is believed that activation of these proteases is a major
role played by Ets transcription factors in the regulation of angiogenesis. Our previous study demonstrated that HGF upregulated Ets activity and Ets-1 protein expression in a myocardial infarction model.[33]
In addition, exogenously expressed HGF also stimulated endogenous
HGF expression through induction of Ets activity (Fig. 5), since the
promoter region of the HGF gene contains an Ets binding site.[6]

More recently, an anti-fibrotic action of HGF has been identified,
as HGF inhibits collagen synthesis via downregulation of TGF-β and
stimulates collagen degradation via upregulation of MMP-1 and uPA.[35]
Although the mechanisms of HGF inhibition of TGF-β synthesis are

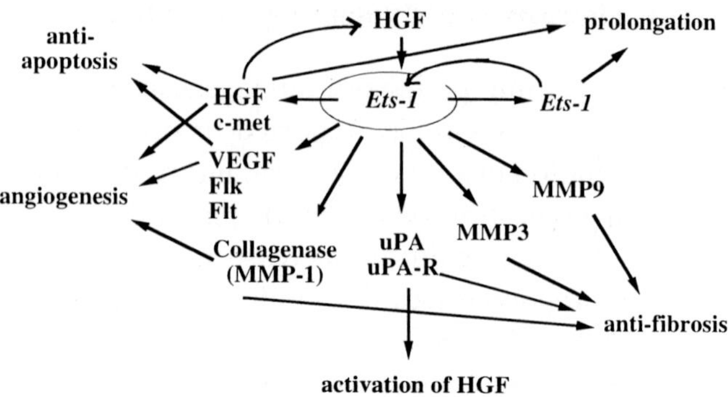

Fig. 5. Molecular mechanisms of angiogenesis induced by HGF through Ets-1. HGF stimulated various actions on collateral formation through Ets-1, revealing that HGF plays a pivotal role as a master gene in the cascade of angiogenesis.

not clear, HGF stimulates, as already mentioned, various metalloproteases such as MMP-1 through induction of Ets-1 activity.[36] Prevention of fibrosis by HGF was confirmed by studies in which administration of human rHGF protein or gene transfer of human HGF prevented and/or induced regression of fibrosis in liver and pulmonary injury models.[37,38] Similar findings were obtained in the myocardium, where overexpression of HGF reduced fibrosis in a cardiomyopathic hamster model.[36] In addition, our recent study demonstrated a significant decrease in a fibrotic area by HGF in porcine chronic angina model, associated with a significant decrease in collagen I, III and TGF-β synthesis as compared to the control (unpublished data). Thus, HGF may also provide a new therapeutic opportunity to treat fibrotic cardiovascular diseases such as cardiomyopathy. Of importance, decrease in the ventricular fibrillation (VF) frequency and increased VF threshold were reported after HGF gene therapy.[39] These findings suggest that HGF gene therapy may have an anti-arrhythmic effect after myocardial ischemia. Overall, HGF is a very unique angiogenic growth factor which has a number of cardioprotective effects including inhibition of fibrosis and apoptosis and suppression of arrhythmias. Currently, a phase I study using human HGF plasmid DNA in USA is ongoing to test the validity of this concept. Overall, coronary artery disease may also be curable using therapeutic angiogenesis by gene therapy.

6. HGF Gene Therapy for Restenosis After Angioplasty

Another important cardiovascular disease potentially amenable to gene therapy is post-angioplasty restenosis. Intimal hyperplasia develops in a large part as a result of vascular smooth muscle cell (VSMC) proliferation and migration induced by a complex interaction of multiple growth factors that are activated by vascular "injury". It has been hypothesized that rapid regeneration of endothelial cells without replication of VSMC may also modulate vascular growth, because multiple anti-proliferative endothelium-derived substances (PGI_2, NO, CNP) are secreted from endothelial cells. This concept was first tested by over-expression of the VEGF-165 gene.[40] Using a similar strategy, we also reported preclinical experiments in which overexpression of the HGF gene in balloon-injured arteries can accelerate re-endothelialization, thereby attenuating intimal hyperplasia.[41] In this study, we found that re-endothelialized balloon-injured arteries showed improvement in endothelial dysfunction induced by balloon angioplasty.[41] HGF also has a strong anti-apoptotic effect in endothelial cells.[42–47] Of interest, HGF can abrogate the decrease in DNA synthesis and cell death of endothelial cells mediated by serum-free treatment.[42] An additive effect of HGF and FGF2 was observed in the prevention of endothelial cell death, equivalent to the effect of serum treatment.[47] Thus HGF can be classified as a new member of a class of growth factors, such as FGFs and VEGFs, with anti-cell death activity. Interestingly, high D-glucose-induced aortic endothelial cell death was also attenuated by recombinant HGF.[43] The mechanisms by which HGF prevented endothelial cell death mediated by these conditions are still unclear. One possibility is HGF upregulation of an anti-apoptotic factor, bcl-2, and inhibition of the translocation of bax from the cytosol to the mitochondria membrane in human endothelial cells.[44,46] HGF is known to stimulate phosphatidylinositol-3′-kinase (PI3K), protein tyrosine phosphatase-2, phospholipase C-r, pp60[c-src] and grb2/hSos1.[48–50] Moreover, HGF also stimulates the Rho- and Ras-mediated signal transduction pathways[51] as well as ERK and Akt signaling that play pivotal roles in the mitogenic and anti-apoptosis actions of HGF in endothelial cells.[52] A unique feature of the HGF signal transduction system is re-phosphorylation of ERK by HGF.[52] This re-phosphorylation of ERK may be due to auto-induction of endogenous HGF, given previously demonstrated ability of exogenous HGF to induce endogenous HGF expression.[53] Indeed, addition of neutralizing anti-HGF antibody after HGF stimulation attenuated the re-phosphorylation of ERK.[52] This unique property of the HGF signal transduction system is involved in the potent mitogenic activity and anti-apoptotic action of HGF. Further studies are needed to clarify the utility of gene therapy to treat restenosis after angioplasty.

7. Next Five Years Perspective — Future Direction of HGF Therapy

To potentially improve the efficiency of angiogenic therapy with HGF we tested a new strategy, combining transfection of HGF and prostacyclin synthase genes.[54] Prostacyclin synthase was chosen because of the utility of vasodilator agents such as prostaglandins and phosphodiesterase type III inhibitors to treat patients with peripheral artery disease. A combination of angiogenesis induced by HGF and vasodilation of newly generated blood vessels induced by prostacyclin may enhance blood flow recovery and maintain new vessel formation. The combined therapy resulted in an improvement in peripheral neuropathy, characterized by significant slowing of nerve conduction velocity and was more effective than a single-gene transfection.

Peripheral neuropathy is common and ultimately accounts for significant morbidity in diabetes. However, there are currently no therapeutic options for patients with diabetic neuropathy. Earlier work using animal models of hindlimb ischemia also documented favorable effects of VEGF gene transfer on ischemic peripheral neuropathy.[55] It is intriguing to note that the neurological and neurophysiological findings in a prospective study of patients undergoing phVEGF$_{165}$ gene transfer for critical limb ischemia showed clinical improvement in electrophysiological measurements in diabetic patients. Although the model used in this study was more severe compared with the previous work, co-transfection of HGF and prostacyclin synthase genes was able to improve the electrophysiological measures. As HGF has been reported to have direct effects on nerve cells, the results of these experiments suggest a possible contribution of direct effects of HGF on nerve integrity.

HGF also acts as a neurotrophic factor.[56,57] We examined the therapeutic effects of HGF on brain injury in a rat permanent middle cerebral artery occlusion model because an ideal therapeutic approach to treat ischemia may have both aspects of enhancement of collateral formation and prevention of neuronal death.[58–61] Gene transfer of the HGF plasmid into the brain by injection into the cerebrospinal fluid via the cisterna magna resulted in a significant decrease in the infarcted brain area after 24 hours of ischemia.[61] Consistently, the neurological deficit was significantly attenuated in rats transfected with the HGF gene at 24 hours after the ischemic event. Stimulation of angiogenesis was also detected in rats transfected with the HGF gene compared with controls. Of importance, no cerebral edema or destruction of the blood-brain barrier was observed in rats transfected with the HGF gene. In particular, the reduction of brain injury by HGF may provide a new therapeutic option to treat cerebrovascular disease.

Gene therapy will be useful for treatment of many diseases, including peripheral arterial disease, myocardial infarction, and heart transplantation rejection among others. Since the first federally approved human gene therapy protocol started on 14 September 1990 for adenosine deaminase deficiency, more than 4000 patients have been treated using gene therapy approaches. Although there are still many unsolved issues in the clinical application of gene therapy, it now appears to be not far from reality and it is time to take a hard look at practical issues that will determine the real clinical potential, including further innovations in gene transfer methods, well-defined disease targets, cell-specific targeting strategies, and effective and safe delivery systems.

Acknowledgments

This work was partially supported by a Grant-in-Aid from the Program for Promotion of Fundamental Studies in Health Sciences of the National Institute of Biomedical Innovation (NIBIO), the Organization for Pharmaceutical Safety and Research, a Grant-in-Aid from the Ministry of Public Health and Welfare, a Grant-in-Aid from the Japan

Promotion of Science, and through Special Coordination Funds of the Ministry of Education, Culture, Sports, Science and Technology, the Japanese Government.

References

1. Nakamura T, Nishizawa T, Hagiya M, *et al*. (1989). Molecular cloning and expression of human hepatocyte growth factor. *Nature* **342**: 440–443.
2. Nakamura Y, Morishita R, Higaki J, *et al*. (1996). Hepatocyte growth factor is a novel member of the endothelium-specific growth factors: additive stimulatory effect of hepatocyte growth factor with basic fibroblast growth factor but not with vascular endothelial growth factor. *J Hypertens* **14**: 1067–1072.
3. Van Belle E, Witzenbichler B, Chen D, *et al*. (1998). Potentiated angiogenic effect of scatter factor/hepatocyte growth factor via induction of vascular endothelial growth factor: the case for paracrine amplification of angiogenesis. *Circulation* **97**: 381–390.
4. Nakamura Y, Morishita R, Higaki J, *et al*. (1995). Expression of local hepatocyte growth factor system in vascular tissues. *Biochem Biophys Res Commun* **215**: 483–488.
5. Nakano N, Moriguchi A, Morishita R, *et al*. (1997). Role of angiotensin II in the regulation of a novel vascular modulator, hepatocyte growth factor (HGF), in experimental hypertensive rats. *Hypertension* **30**: 1448–1454.
6. Liu Y, Michalopoulos GK, Zarnegar R(1994). Structural and functional characterization of the mouse hepatocyte growth factor gene promoter. *J Biol Chem* **269**: 4152–4160.
7. Nakamura Y, Morishita R, Nakamura S, *et al*. (1996). A vascular modulator, hepatocyte growth factor, is associated with systolic pressure. *Hypertension* **28**: 409–413.
8. Bussolino F, Di Renzo MF, Ziche M, *et al*. (1992). Hepatocyte growth factor is a potent angiogenic factor which stimulates endothelial cell motility and growth. *J Cell Biol* **119**: 629–641.
9. Ponzetto C, Bardelli A, Zhen Z, *et al*. (1994). A multifunctional docking site mediates signaling and transformation by the hepatocyte growth factor/scatter factor receptor family. *Cell* **77**: 261–271.
10. Pelicci G, Giordano S, Zhen Z, *et al*. (1995). The motogenic and mitogenic responses to HGF are amplified by the Shc adaptor protein. *Oncogene* **10**: 1631–1638.
11. Weidner KM, Di Cesare S, Sachs M, *et al*. (1996). Interaction between Gab1 and the c-Met receptor tyrosine kinase is responsible for epithelial morphogenesis. *Nature* **384**: 173–176.
12. Graziani A, Gramaglia D, Cantley LC, *et al*. (1991). The tyrosine-phosphorylated hepatocyte growth factor/scatter factor receptor associates with phosphatidylinositol 3-kinase. *J Biol Chem* **266**: 22087–22090.
13. Graziani A, Gramaglia D, dalla Zonca P, *et al*. (1993). Hepatocyte growth factor/scatter factor stimulates the Ras-guanine nucleotide exchanger. *J Biol Chem* **268**: 9165–9168.

14. Boccaccio C, Ando M, Tamagnone L, *et al*. (1998). Induction of epithelial tubules by growth factor HGF depends on the STAT pathway. *Nature* **391**: 285–288.

15. Nakagami H, Morishita R, Yamamoto K, *et al.*(2001). Mitogenic and antiapoptotic actions of hepatocyte growth factor through ERK, STAT3, and AKT in endothelial cells. *Hypertension* **37**: 581–586.

16. Nakagami H, Morishita R, Yamamoto K, *et al*. (2002). Hepatocyte growth factor prevents endothelial cell death through inhibition of bax translocation from cytosol to mitochondrial membrane. *Diabetes* **51**: 2604–2611.

17. Gilmore AP, Metcalfe AD, Romer LH, *et al*. (2000). Integrin-mediated survival signals regulate the apoptotic function of Bax through its conformation and subcellular localization. *J Cell Biol* **149**: 431–446.

18. Bauters C, Asahara T, Zheng LP, *et al*. (1994). Physiological assessment of augmented vascularity induced by VEGF in ischemic rabbit hindlimb. *Am J Physiol* **267**: H1263–1271.

19. Isner JM, Pieczek A, Schainfeld R, Blair R, Haley L, Asahara T, Rosenfield K, Razvi S, Walsh K, Symes JF (1996). Clinical evidence of angiogenesis after arterial gene transfer of phVEGF165 in patient with ischaemic limb. *Lancet* **348**: 370–374

20. Isner JM, Baumgartner I, Rauh G, Schainfeld R, Blair R, Manor O, Razvi S, Symes JF (1998). Treatment of thromboangiitis obliterans (Buerger's disease) by intramuscular gene transfer of vascular endothelial growth factor: preliminary clinical results. *J Vasc Surg* **28**: 964–973.

21. Baumgartner I, Pieczek A, Manor O, Blair R, Kearney M, Walsh K, Isner JM (1998). Constitutive expression of phVEGF165 after intramuscular gene transfer promotes collateral vessel development in patients with critical limb ischemia. *Circulation* **97**: 1114–1123.

22. Baumgartner I, Rauh G, Pieczek A, Wuensch D, Magner M, Kearney M, Schainfeld R, Isner JM (2000). Lower-extremity edema associated with gene transfer of naked DNA encoding vascular endothelial growth factor. *Ann Intern Med* **132**: 880–884.

23. Morishita R, Nakamura S, Hayashi S, *et al*. (1999). Therapeutic angiogenesis induced by human recombinant hepatocyte growth factor in rabbit hind limb ischemia model as cytokine supplement therapy. *Hypertension* **33**: 1379–1384.

24. Belle EV, Witzenbichler B, Chen D, Silver M, Chang L, Schwall R, Isner JM (1998). Potentiated angiogenic effect of scatter factor/hepatocyte growth factor via induction of vascular endothelial growth factor: the case for paracrine amplification of angiogenesis. *Circulation* **97**: 381–390.

25. Hayashi S, Morishita R, Nakamura S, Yamamoto K, Moriguchi A, Nagano T, Taizi M, Noguchi H, Matsumoto K, Nakamura T, Higaki J, Ogihara T (1999). Potential role of hepatocyte growth factor, a novel angiogenic growth factor, in peripheral arterial disease: down-regulation of HGF in response to hypoxia in vascular cells. *Circulation* **100**: II301–II308.

26. Taniyama Y, Morishita R, Aoki M, *et al*. (2001). Therapeutic angiogenesis induced by human hepatocyte growth factor gene in rat and rabbit hindlimb ischemia models: preclinical study for treatment of peripheral arterial disease. *Gene Ther* **8**: 181–189.

27. Taniyama Y, Morishita R, Hiraoka K, Aoki M, Nakagami H, Yamasaki K, Matsumoto K, Nakamura T, Kaneda Y, Ogihara T (2001). Therapeutic

angiogenesis induced by human hepatocyte growth factor gene in rat diabetic hind limb ischemia model: molecular mechanisms of delayed angiogenesis in diabetes. *Circulation* **104**: 2344–2350.

28. Morishita R, Sakaki M, Yamamoto K, Iguchi S, Aoki M, Yamasaki K, Matsumoto K, Nakamura T, Lawn R, Ogihara T, Kaneda Y (2002). Impairment of collateral formation in Lp(a) transgenic mice: therapeutic angiogenesis induced by human hepatocyte growth factor gene. *Circulation* **105**: 1491–1496.

29. Matsuki A, Yamamoto S, Nakagami H, *et al.* (2004). No influence of tumor growth by intramuscular injection of hepatocyte growth factor plasmid DNA: safety evaluation of therapeutic angiogenesis gene therapy in mice. *Biochem Biophys Res Commun* **315**: 59–65.

30. Morishita R, Aoki M, Hashiya N, Makino H, Yamasaki K, Azuma J, Sawa Y, Matsuda H, Kaneda Y, Ogihara T (2004). Safety evaluation of clinical gene therapy using hepatocyte growth factor to treat peripheral arterial disease by therapeutic angiogenesis. *Hypertension* **44**: 203–209.

31. Morishita R, Aoki M, Hashiya N, Yamasaki K, Makino H, Wakayama K, Azuma J, Ogihara T (2002). Hepatocyte growth factor (HGF) angiogenic gene therapy: promises for cardiovascular diseases. *Gene Ther Regul* **1**: 343–359.

32. Powell RJ, Dormandy J, Simons M, Morishita R, Annex BH (2004). Therapeutic angiogenesis for critical LIMB ischemia: design of the haptocyte growth factor therapeutic angiogenesis clinical trial. *Vasc Med* **9**: 1–6.

33. Aoki M, Morishita R, Taniyama Y, Kida I, Moriguchi A, Matsumoto K, Nakamura T, Kaneda Y, Higaki J, Ogihara T (2000). Angiogenesis induced by hepatocyte growth factor in non-infarcted myocardium and infarcted myocardium: up-regulation of essential transcription factor for angiogenesis, ets. *Gene Ther* **7**: 417–427.

34. Ueda H, Sawa Y, Matsumoto K, Kitagawa-Sakakida S, Kawahira Y, Nakamura T, Kaneda Y, Matsuda H (1999). Gene transfection of hepatocyte growth factor attenuates reperfusion injury in the heart. *Ann Thorac Surg* **67**: 1726–1731.

35. Taniyama Y, Morishita R, Nakagami H, Moriguchi A, Sakonjo H, Shokei-Kim, Matsumoto K, Nakamura T, Higaki J, Ogihara T (2000). Potential contribution of a novel antifibrotic factor, hepatocyte growth factor, to prevention of myocardial fibrosis by angiotensin II blockade in cardiomyopathic hamsters. *Circulation* **102**: 246–252.

36. Taniyama Y, Morishita R, Aoki M, Hiraoka K, Yamasaki K, Hashiya N, Matasumoto K, Nakamura T, Kaneda Y, Ogihara T (2002). Angiogensis and antifibrotic action by hepatocyte growth factor in cardiomyopathic hamster. *Hypertension* **40**: 47–53.

37. Ueki T, Kaneda Y, Tsutsui H, Nakanishi K, Sawa Y, Morishita R, Matsumoto K, Nakamura T, Takahashi H, Okamoto E, Fujimoto J (1999). Hepatocyte growth factor gene therapy of liver cirrhosis in rats. *Nat Med* **5**: 226–230.

38. Nakamura T, Sakata R, Ueno T, Sata M, Ueno H (2000). Inhibition of transforming growth factor beta prevents progression of liver fibrosis and enhances hepatocyte regeneration in dimethylnitrosamine-treated rats. *Hepatology* **32**: 247–255.

39. Yumoto A, Fukushima Kusano K, Nakamura K, Hashimoto K, Aoki M, Morishita R, Kaneda Y, Ohe T (2005). Hepatocyte growth factor gene therapy reduces ventricular arrhythmia in animal models of myocardial ischemia. *Acta Med Okayama* **59**: 73–78.

40. Asahara T, Bauters C, Pastore C, Kearney M, Rossow S, Bunting S, Ferrara N, Symes JF, Isner JM (1995). Local delivery of vascular endothelial growth factor accelerates reendothelialization and attenuates intimal hyperplasia in balloon-injured rat carotid artery. *Circulation* **91**: 2793–2801.

41. Hayashi K, Nakamura S, Morishita R, Moriguchi A, Aoki M, Matsumoto K, Nakamura T, Kaneda Y, Sakai N, Ogihara T (2000). *In vivo* transfer of human hepatocyte growth factor gene accelerates re-endothelialization and inhibits neointimal formation after balloon injury in rat model. *Gene Ther* **7**: 1664–1671.

42. Yo Y, Morishita R, Nakamura S, Tomita N, Yamamoto K, Moriguchi A, Matsumoto K, Nakamura T, Higaki J, Ogihara T (1998). Potential role of hepatocyte growth factor in the maintenance of renal structure: anti-apoptotic action of HGF on epithelial cells. *Kidney Int* **54**: 1128–1138.

43. Morishita R, Nakamura S, Nakamura Y, Aoki M, Moriguchi A, Kida I, Yo Y, Matsumoto K, Nakamura T, Higaki J, Ogihara T (1997). Potential role of endothelium-specific growth factor, hepatocyte growth factor, on endothelial damage in diabetes mellitus. *Diabetes* **46**: 138–142.

44. Yamamoto K, Morishita R, Hayashi S, Matsushita H, Nakagami H, Moriguchi A, Matsumoto K, Nakamura T, Kaneda Y, Ogihara T (2001). Contribution of Bcl-2, but not Bcl-xL and Bax, to anti-apoptotic actions of hepatocyte growth factor in hypoxic conditioned human endothelial cells. *Hypertension* **37**: 1341–1348.

45. Nakagami H, Morishita R, Yamamoto K, Taniyama Y, Aoki M, Matsumoto K, Nakamura T, Kaneda Y, Horiuchi M, Ogihara T (2001). Mitogenic and anti-apoptotic actions of HGF through ERK, STAT3 and Akt in endothelial cells. *Hypertension* **37**: 581–586.

46. Nakagami H, Morishita R, Yamamoto K, Taniyama Y, Aoki M, Yamasaki K, Matasumoto K, Nakamura T, Kaneda Y, Ogihara T (2002). Hepatocyte growth factor (HGF) prevents endothelial cell death through inhibition of bax translocation from cytosol to mitochondrial membrane. *Diabetes* **51**: 2604–2611.

47. Nakamura Y, Morishita R, Higaki J, Kida I, Aoki M, Moriguchi A, Yamada K, Hayashi S, Yo Y, Nakano H, Matsumoto K, Nakamura T, Ogihara T (1996). Hepatocyte growth factor (HGF) ia a novel member of endothelium-specific growth factors: additive stimulatory effect of HGF with basic fibroblast growth factor, but not vascular endothelial growth factor. *J Hypertens* **14**: 1067–1072.

48. Ponzetto C, Bardelli A, Zhen Z, Maina F, Zonca P, Giordano S, Graziani A, Panayoyou G, Comoglio PM (1994). A multifunctional docking site mediates signaling and transformation by the HGF/SF receptor family. *Cell* **77**: 261–271.

49. Rosen EM, Nigam SK, Goldberg ID (1994). Scatter factor and the c-met receptor: a paradigm for mesenchymal/epithelial interaction. *J Cell Biol* **127**: 1783–1787.

50. Canley LG, Cantley LC (1995). Signal transduction by the hepatocyte growth factor receptor, c-met. Activation of the phosphatidylinositol 3-kinase. *J Am Soc Nephrol* **5**: 1872–1881.

51. Ridley AJ, Comoglio PM, Hall A (1995). Regulation of scatter factor/hepatocyte growth factor responses by ras, rac and rho in MDCK cells. *Mol Cell Biol* **15**: 1111–1122.

52. Nakagami H, Morishita R, Yamamoto K, Taniyama Y, Aoki M, Kim S, Matsumoto K, Nakamura T, Higaki J, Ogihara T (2000). Anti-apoptotic action of hepatocyte growth factor (HGF) through mitogen-activated protein kinase on human aortic endothelial cells. *J Hypertens* **18**: 1411–1420.

53. Hayashi S, Morishita R, Higaki J, Aoki M, Moriguchi A, Kida I, Sawa Y, Matsumoto K, Nakamura T, Kaneda Y, Ogihara T (1996). Autocrine-paracrine effects of over-expression of hepatocyte growth factor gene on growth of endothelial cells. *Biochem Biophys Res Commun* **220**: 539–545.

54. Koike H, Morishita R, Iguchi S, *et al.* (2003). Enhanced angiogenesis and improvement of neuropathy by cotransfection of human hepatocyte growth factor and prostacyclin synthase gene. *FASEB J* **17**: 779–781.

55. Simovic D, Isner JM, Ropper AH, *et al.* (2001). Improvement in chronic ischemic neuropathy after intramuscular phVEGF165 gene transfer in patients with critical limb ischemia. *Arch Neurol* **58**: 761–768.

56. Korhonen L, Sjoholm U, Takei N, *et al.* (2000). Expression of c-Met in developing rat hippocampus: evidence for HGF as a neurotrophic factor for calbindin D-expressing neurons. *Eur J Neurosci* **12**: 3453–3461.

57. Ebens A, Brose K, Leonardo ED, *et al.* (1996). Hepatocyte growth factor/scatter factor is an axonal chemoattractant and a neurotrophic factor for spinal motor neurons. *Neuron* **17**: 1157–1172.

58. Hayashi K, Morishita R, Nakagami H, Yoshimura S, Hara A, Matsumoto K, Nakamura T, Kaneda Y, Ogihara T, Sakai N (2001). Gene therapy for preventing neuronal death using hepatocyte growth factor: *in vivo* gene transfer of HGF to subarachnoid space prevents delayed neuronal death in gerbil hippocampal CA1 neurons. *Gene Ther* **8**: 1167–1173.

59. Yoshimura S, Morishita R, Hayashi K, Kokuzawa J, Aoki M, Matsumoto K, Nakamura T, Ogihara T, Kaneda Y, Sakai N (2002). Gene transfer of hepatocyte growth factor to subarachnoid space in cerebral hypoperfusion model. *Hypertension* **39**: 1028–1034.

60. Kokuzawa J, Yoshimura S, Kitajima H, Shinoda J, Kaku Y, Iwama T, Morishita R, Shimazaki T, Okano H, Kunisada T, Sakai N (2003). Hepatocyte growth factor promotes proliferation and neuronal differentiation of neural stem cells from mouse embryos. *Mol Cell Neurosci* **24**: 190–197.

61. Shimamura M, Sato N, Oshima K, *et al.* (2004). Novel therapeutic strategy to treat brain ischemia: overexpression of hepatocyte growth factor gene reduced ischemic injury without cerebral edema in rat model. *Circulation* **109**: 424–431.

14

Role of Nitric Oxide in Adult Angiogenesis: Therapeutic Potential of Endothelial Nitric Oxide Synthase Gene Transfer

by Gabor M. Rubanyi

1. Endothelial Nitric Oxide in Health and Disease

Cardiovascular homeostasis under physiological conditions is maintained by a complex system of regulatory mediators. One such mediator is endothelium-derived relaxing factor ("EDRF") originally described in 1980 by Furchgott and Zawadzki[1] and eventually identified as nitric oxide (NO).[2,3]

1.1. *Nitric oxide synthases*

Nitric oxide is synthesized from the guanidino nitrogens of L-arginine through a process that consumes five electrons and results in the formation of the co-product L-citrulline by a family of nitric oxide synthase (NOS) enzymes.[4] The process involves the transfer of electrons between five co-factors including flavin adenine dinucleotide (FAD), flavin mononucleotide (FMN), tetrahydrobiopterin (BH_4), heme and

385

calmodulin (CaM), and requires three co-substrates including L-arginine, nicotinamide adenine dinucleotide phosphate (NADPH) and molecular oxygen.[4]

Three isoforms of NOS, encoded by three distinct genes on different chromosomes, have been isolated and purified. Both the neuronal (nNOSor NOS-I) and endothelial (eNOS or NOS-III) isoforms are constitutively activated and expressed upon calcium-calmodulin binding following an increase in intracellular calcium. The inducible isoform (iNOS or NOS-II) is activated by cytokines independent of calcium (calmodulin is tightly bound to NOS-II in contrast to the constitutive isoforms, probably due to the lack of an auto-inhibitory loop on NOS-II).[5] All NOS isoenzymes form homodimers, and contain a heme oxygenase domain and a cytochrome P-450 reductase domain.

1.2. *Physiological role of endothelial NO ("EDNO")*

Under physiological conditions endothelial NOS-III-derived NO, released by receptor activation or shear stress, freely diffuses from the endothelium towards the lumen and abluminally towards the underlying vessel wall, and plays a key role in the maintenance of vascular homeostasis.[6]

Endothelium-derived NO (EDNO) is a potent vasodilator, which led to its discovery as "EDRF"[1] and later to its identification as NO[2,3] using bioassay systems allowing the assessment of its biological half-life,[7,8] and describing the characteristics of "EDRF" including its interaction with superoxide anion radical $(.O^{2-})$[9] and its release by increased flow/shear stress.[10]

The vasodilating activity of EDNO is mediated by activation of soluble guanylate cyclase (sGC) and elevation of cGMP.[3,11] NO inhibits platelet adhesion and aggregation,[12] also through a cGMP-mediated pathway. NO inhibits vascular SMC proliferation[13] while promoting endothelial cell growth.[14,15] NO have been shown to reduce leukocyte infiltration of the endothelial barrier.[16]

Oxidatively modified LDL (oxLDL) is a major contributor to the pathogenesis of atherosclerosis. NO have been shown to inhibit oxidative modification of LDL.[17] NO attenuates smooth muscle proliferation and inhibits neointima formation.[18] On the other hand NO protects endothelial cells from apoptotic stimuli,[19,20] which play an important role in its participation in the angiogenic process.

Endothelial NOS is localized to the caveolae of the endothelial plasma membrane[21] in close proximity to key membrane receptors, ion channels and signaling molecules, positioning the NOS-III/EDNO system upstream of many regulatory pathways determining cell function and phenotype.

1.3. *Endothelial NO-deficiency in cardiovascular diseases*

Availability and biological activity of endothelial NO are regulated by the expression of its generating enzyme, NOS-III, as well as by the activity of the NOS-III enzyme, which is tightly controlled by co-factor and substrate availability, post-translational modifications (myristoylation, palmitoylation and phosphorylation), protein-protein interactions (e.g. caveolin and Hsp90) and subcellular localization. In addition, accumulation of endogenous NOS inhibitors (e.g. ADMA) and increased oxidative degradation of NO can also lead to diminished availability/bioactivity of endothelial NO.

"Endothelial NO-deficiency" is an early phenomenon in the progression of various cardiovascular diseases.[22,23] Impaired continuous basal NO synthesis may be the first detectable evidence of endothelial dysfunction.[24–26] Early signs of endothelial function are easily assessable by measuring endothelium-dependent vasoconstriction to NOS inhibitors or endothelium-dependent vasodilation in response to increased flow or receptor agonists, such as acetylcholine. Diagnostic approaches, like quantitative coronary angiography and new ultrasound/Doppler devices, are becoming mainstream tools for early detection of EDNO-deficiency in high risk patients.[27]

Impaired EDNO activity is associated with several cardiovascular diseases,[28] including atherosclerosis, systemic and pulmonary hypertension, congestive heart failure, peripheral arterial occlusive disease as well as cardiovascular complications of diabetes. The apparent "NO-deficiency" is the net result of several different pathological processes interfering with one or more of the components regulating the availability and/or bioactivity of NO in the vascular wall. These processes can decrease the amount of endothelial NO at different levels, which include (a) reduced NOS-III gene expression (both at the transcription and mRNA stability level), (b) reduced activity of the NOS-III enzyme via diminished co-factor and substrate availability, or by modifications in post-translational processes, cellular localization or protein-protein interactions, and (c) reduced biological activity of NO (e.g. through oxidative inactivation).

Proof for the numerous physiological (mostly vasculoprotective) roles of endothelial NO in the cardiovascular system was provided by the development of the NOS-III-deficient (NOS-III-KO) mouse,in which NOS-III expression was genetically disrupted.[29] Homozygous NOS-III-KO mice have elevated mean arterial blood pressure, consistent with the role of endothelial NO in the regulation of blood pressure and vascular tone.[30] Isolated aortic rings with intact endothelium from NOS-III-KO mice do not relax to acetylcholine, which provides genetic evidence that the NOS-III gene is required for the "EDRF" activity. These mice show markedly decreased bleeding times,[31] exhibit enhanced leukocyte adhesion associated with elevated surface expression of P-selection in the microcirculation[32] and impaired angiogenic response.[33] Myocardial ischemia and reperfusion injury were significantly exacerbated in NOS-III-KO mice.[34] NOS-III deficiency also resulted in enlarged cerebral infarcts following permanent middle cerebral artery occlusion.[35]

1.4. *Therapeutic restoration of endothelial NO production in cardiovascular diseases*

Contribution of diminished endothelial NO production to the pathomechanism and the progression of different cardiovascular diseases have been demonstrated under numerous experimental and clinical conditions. Therefore drugs that can improve endothelial NO production may have significant therapeutic benefits in these pathological conditions.

Endothelial function can be restored by different classes of compounds directly or indirectly targeting regulatory mechanisms of

NO production by NOS-III. These include the female sexual steroid hormone, 17β-estradiol;[36–38] lipid lowering agents, represented by HMG-CoA reductase inhibitors ("statins");[39] blood pressure lowering drugs, represented by ACE inhibitors;[40] and antioxidants, represented by antioxidant vitamins (e.g. vitamin E and C).[41] Some of these treatments lead to clinical efficacy in cardiovascular diseases and evidence have been accumulating for the role of NO as an important contributor to these therapeutic effects.

In a randomized, double-blind, placebo-controlled clinical trial (TREND = Trial on Reversing ENdothelial Dysfunction), the ACE inhibitor quinapril have been shown to improve endothelium-dependent relaxation after six months of treatment.[42] In another study the effect of the HMG-CoA reductase inhibitor, lovastatin, was examined in patients with hyperlipidemia and coronary artery disease. Benefit on endothelium-dependent relaxation was shown after five and a half months of treatment.[43]

2. Nitric Oxide and Angiogenesis

2.1. *VEGF causes endothelium-dependent vasodilation mediated by EDNO*

The endothelial cell-specific vascular endothelial growth factor (VEGF) increases cytosolic free calcium in cultured endothelial cells.[44] In isolated canine coronary artery segments, VEGF causes endothelium-dependent relaxation that can be attenuated by pretreatment with N^G-monomethyl-L-arginine (L-NMMA).[45] Isner's group demonstrated the release of NO from arteries after treatment with VEGF[46] and extended these studies to demonstrate that VEGF promoted recovery of endothelium-dependent relaxation in ischemic rabbit hindlimb. Subsequent work by the same group showed that VEGF infusion lead to EDNO-mediated hypotension in animals[47] as well as in humans.[48] The signaling link between VEGF and EDNO release include VEGF/KDR interaction-induced Akt phosphorylation (activation), which in turn phosphorylates NOS-III at Serine 1177, leading to increased NO production (Fig. 1).

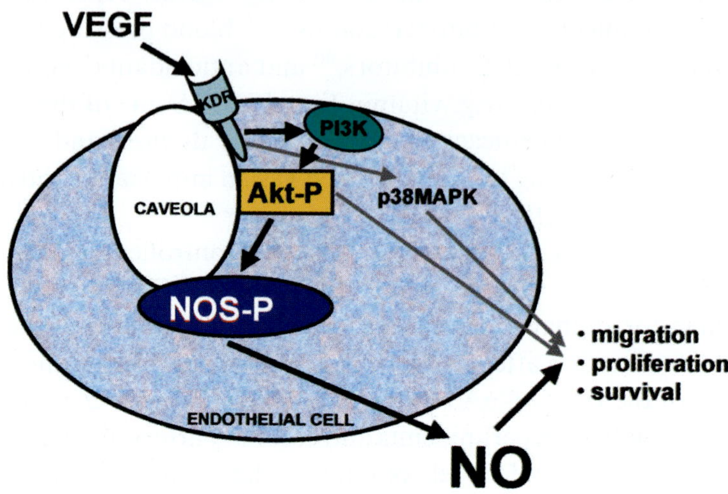

Fig. 1. Current concept of the role of the NOS-III/NO system in angiogenesis. The endothelial cell-specific angiogenic growth factor, VEGF, interacts with its receptor (KDR) on endothelial cells and via several mechanisms stimulates the proliferation, migration and survival of these cells, which ultimately leads to new vessel formation (angiogenesis). An intact NOS-III/NO system is required for these actions of VEGF (and of other growth factors, including FGF and HGF, not shown). NOS-III, localized to the caveolae in endothelial cell plasma membrane is activated by VEGF/KDR interaction via P13K-induced activation (phosphorylation, P) of Akt, which in turn phosphorylates serine 1177 in NOS-III enzyme (NOS-P), which augments EDNO production several-fold (for further details see text).

2.2. *Tumor angiogenesis and NO*

Indirect evidence that NO may be involved in the angiogenic process was provided by studies evaluating tumor angiogenesis. Increased levels of NO have been reported in human tumors,[49,50] and it has been demonstrated that transfection of the inducible NOS (NOS-II) gene into an adenocarcinoma cell line gave rise to more vascularized tumors than the wild-type cells when injected into animals.[51] Other experiments using inhibitors of NO synthase also strongly support the importance of NO in xenografted tumor neovascularization.[52,53] Ziche and colleagues have demonstrated that vascularization of tumors that have been initiated with breast cancer cells overexpressing VEGF can be attenuated by treatment with NO synthase inhibitors.[54]

2.3. *Evidence in cultured endothelial cells and in rabbit cornea*

A more direct link between the production of NO induced by VEGF and angiogenesis have been demonstrated by Papapetropoulos *et al.*[55,56] and Morbidelli *et al.*[57] describing that many of the angiogenic responses to VEGF *in vitro* were associated with increases in cGMP in cultured endothelial cells. Inhibitors of NO synthesis attenuated VEGF-induced angiogenic responses, including cultured endothelial cell proliferation, migration and tube formation.

In vivo evidence for the link between NO and angiogenesis in adult animals was first described by Ziche and co-workers, who demonstrated that, in rabbits, corneal angiogenesis induced by VEGF can be inhibited by treatment of the animals with inhibitors of NO production.[15,54]

2.4. *Role of NO in post-ischemic revascularization*

In the rabbit ischemic hindlimb model, dietary L-arginine supplementation significantly decreased blood pressure, reduced arterial resistance, increased flow at rest, and increased flow reserve in the ischemic limb.[58] In hypercholesterolemic mice, elevated level of plasma asymmetric dimethylarginine (ADMA), an endogenous inhibitor of NO synthesis,[59] is associated with impaired angiogenic response to ischemia.[60] This effect of ADMA can be reversed by administration of the NO precursor L-arginine or mimicked in normal animals by administration of an NO synthase inhibitor.[61]

Isner's group was the first to investigate the effects of ischemia on angiogenesis in NOS-III-KO mice.[58] The degree of angiogenesis (revascularization) was determined using both laser Doppler imaging *in situ* and by measuring capillary density. In the NOS-III-KO mice the spontaneous angiogenic response to ischemia was severely attenuated as compared to normal mice. The levels of VEGF were comparable in the ischemic limbs of both the control and NOS-III-KO mice, suggesting that the reason for the attenuated response to ischemia was not the lack of ischemia-induced upregulation of VEGF production. Although

VEGF gene transfer has already been shown to stimulate the angiogenic response in Apo E-KO mice,[62] treatment of NOS-III-KO mice with either a vector carrying the VEGF gene, or injection of recombinant VEGF protein, failed to improve post-ischemic revascularization.[58]

The ability to restore the effects of ischemia on angiogenesis in the NOS-III-KO mice with a systemically administered NO donor was not successful,[58] so a direct proof for the therapeutic benefits of exogenously applied NO on angiogenesis still remained to be demonstrated.

2.5. *Role of NO in exogenous VEGF and FGF-induced revascularization*

Therapeutic angiogenesis in several animal models of myocardial and hindlimb ischemia have been demonstrated using VEGF and FGF protein or gene delivery. The essential role of NO in exogenous VEGF-induced post-ischemic revascularization have been demonstrated in NOS-III-KO mice.[58] In a rat hindlimb ischemia model, treatment with both VEGF and FGF significantly augmented post-ischemic revascularization. Pretreatment with L-NAME prevented therapeutic arteriogenesis in this rat model by exogenous VEGF-121 and FGF-2 protein treatment[63] or by exercise training,[64] indicating that impaired EDNO production can reduce angiogenesis and arteriogenesis in response to both exogenously administered VEGF and FGF.

2.6. *Molecular mechanisms*

A number of studies indicate that NO is an endothelial cell survival factor, inhibiting apoptosis.[65–67] Under certain experimental conditions NO enhances endothelial cell proliferation.[57] Another prerequisite for the formation of new vessels is migration of endothelial cells. NO is known to enhance endothelial migration, by stimulating endothelial cell podokinesis,[68] and/or by enhancing the expression of $av\beta3$, an endothelial integrin involved in attachment and migration.[69] Migration of endothelial cells also requires the dissolution of the surrounding extracellular matrix. NO may contribute to this process by increasing the production of urokinase-type plasminogen activator (uPA).[70] Finally, the hemodynamic effects of this potent vasodilator may play a role in its angiogenic effects. Increases in blood flow, induced by a vasodilator, may stimulate the proliferation of endothelial cells. Increased shear stress in skeletal muscle microcirculation is associated with increased uptake of bromodeoxyuridine by capillary endothelial cells.[71] Increased shear stress also plays a role in the remodeling of existing capillary structures, leading to the development of arterialized collateral vessels.[72]

3. NOS Gene Transfer

The lack of effect of systemic NO donor treatment on post-ischemic revascularization in NOS-III-KO mice[58] may be due (at least in part) to insufficient local NO concentration in the ischemic hindlimb. In order to overcome this limitation, the effect of local delivery of NOS-III gene was tested in several animal models of limb ischemia.

3.1. *Gene delivery vectors*

Local and systemic delivery of NOS-III gene has been achieved using both viral and non-viral gene delivery vectors.[73–77] Viral vectors currently used for cardiovascular gene transfer include adenovirus, adeno-associated virus, and most recently, lentivirus.[78–81] These recombinant viruses are genetically modified to be replication-incompetent, and they contain an inserted cDNA sequence of interest ("transgene") and an appropriate promoter (i.e. CMV). Non-viral vectors that have been studied in vascular gene transfer include naked plasmids and liposomes. Viral vectors generally have higher transfection efficiency but are also more immunogenic than non-viral vectors. A detailed comparison of these vectors with major advantages and disadvantages for cardiovascular gene transfer and the various ways to deliver them can be found in a recent review article.[82]

Naked plasmid and recombinant adenovirus have become the most frequently used vectors for cardiovascular gene transfer studies.[83] The human adenovirus is a non-enveloped linear double-stranded DNA virus with a 36 kb viral genome.[84] The advantage of adenoviral vectors for cardiovascular gene transfer includes their ability to transduce both dividing and non-dividing cells with high efficiency and the ability to generate high titer stock vector (i.e. up to 10^{12} pfu/ml). The viruses enter the host cells via an endocytosis process through the interaction between viral fiber and penton proteins and their cell surface receptors (CAR and integrins, respectively).[85,86] Upon entry into the cell, the virus is taken up by endosomes, which are disrupted by the virus, resulting in viral DNA release into the cytoplasm. The viral DNA then enters the nucleus, where it is not incorporated into the host chromosome but remains episomal. The major drawbacks of first generation adenoviral vectors include the capsid protein-induced, cell-mediated (i.e. CD8[+] T-cell) immune response which may limit the duration of transgene expression and may prevent repeated administration of the vector.

3.2. *NOS-III gene transfer*

Potential advantages of targeted local delivery of the NOS enzyme gene include sufficient (local) increase in NO production in target tissues without systemic hypotensive side effects, as observed after systemic treatment with small molecule NO donors. Attempts so far included plasmid/liposome gene transfer of NOS-III gene resulting in restored NO production and concomitant inhibition of intimal hyperplasia in balloon-injured rat carotid arteries.[76] *Ex vivo*, adventitial delivery of NOS-III gene using adenoviral vectors demonstrated improved NO-dependent vasorelaxation.[77,87] NOS-III gene transfer was also shown to reverse hyperlipidemia-induced endothelial dysfunction.[73]

Although these studies demonstrated favorable outcome of NOS-III gene transfer under well-defined experimental conditions, utility of NOS overexpression will depend on the disease and the mechanism of endothelial dysfunction. In conditions when substrate or co-factor availability may be limited or NOS-III activation is prohibited by risk factors (such as ox LDL or hyperglycemia), wild-type NOS-III over-expression may not result in efficacious restoration of NO-mediated functions.

3.3. *NOS-II gene transfer*

The inducible isoform of NOS (NOS-II) is associated with disease pathophysiology such as systemic vasodilation and hypotension in sepsis[88] as well as in the pathogenesis of autoimmune diseases.[89–92]

Nonetheless, there may be advantages of using NOS-II over NOS-III for certain cardiovascular disease indications. An example is the prevention of re-stenosis after balloon angioplasty. With even the most efficient delivery system available (adenoviral vectors), gene transfer efficiency may be low during clinical applications. The potential advantage of NOS-II gene transfer is that NOS-II synthesizes much larger quantities of NO[93] which can diffuse to a large number of neighboring cells.[94] In comparison, many more cells would have to express NOS-III to synthesize a similar amount of NO. In addition, NOS-II enzymatic activity will be maximally activated in the absence of agonist stimulation.

4. Therapeutic Angiogenesis with NOS-III Gene Transfer for Critical Limb Ischemia

4.1. *Impaired angiogenesis and arteriogenesis in patients with critical limb ischemia*

In healthy young individuals, myocardial ischemia induces collateral vessel development, which provides certain protection from subsequent coronary events.[95] Ischemia-induced arteriogenesis and subsequent angiogenesis are compensatory mechanisms to restore adequate blood supply to ischemic tissues. In the coronary circulation and peripheral vasculature, ischemia-initiated opening of pre-existent collaterals and arterialization of these immature vascular channels preserves blood flow and contributes to the extent of ischemic reserve capacity in the heart and leg.[72] However, in older patients with chronic diseases and multiple risk factors, arterio- and angiogenesis in response to ischemia is severely limited or absent. Indeed, it has been shown that common risk factors, such as diabetes, hypercholesterolemia and advanced age, impair angiogenesis and arteriogenesis[96–100] which may contribute to the increased severity of cardiovascular diseases in this patient population.

Peripheral arterial occlusive disease (PAOD) is a highly prevalent disease lacking adequate treatment especially for its most advanced stage, critical limb ischemia (CLI).[101] Local delivery of genes encoding angiogenic growth factors such as vascular endothelial growth factors

(VEGF), fibroblast growth factors (FGF) or hepatocyte growth factor (HGF) emerged as a promising new therapeutic approach (termed "therapeutic angiogenesis") for ischemic cardiovascular disease.[102] However, clinical results so far lack unequivocal proof for efficacy by therapeutic angiogenesis. It may be due to inefficient gene delivery or because some patients are refractory to exogenously administered growth factors.[102]

4.2. *Animal models of impaired ischemia-induced angiogenesis and arteriogenesis*

Impaired ischemic limb arteriogenesis and angiogenesis was described in mice deficient in placental-derived growth factor,[103] interleukin-1,[104] the angiotensin II type-1 receptor,[105] matrix metalloproteinase-9,[106] adiponectin,[107] and caveolin-1.[108] In addition, studies in experimental mouse models of hindlimb ischemia demonstrated impaired revascularization and CLI-like symptoms (i.e. tissue necrosis) in old animals[98,109] and in animals with compromised immune system.[110] Diminished angiogenesis in some of these animal models corresponded with decreased expression of VEGF.[98,99,111] Interestingly, substituting endogenous VEGF by exogenously delivered protein or gene has not shown therapeutic benefit in either aged mice[109] or in the immune-compromised balb/c mice,[110] unlike in other mouse models of hindlimb ischemia.[99]

4.2.1. *NOS-III-KO mice*

Mice deficient in NOS-III gene (NOS-III-KO) also show a severe form of critical limb ischemia after femoral artery ligation.[58,112–114] Moreover, the ability of statin-based drugs, angiotensin II, neuropeptide Y, and stromal cell-derived factor-1α (SDF-1α) to improve limb angiogenesis are absent in mice deficient in NOS-III.[115–118] The molecular mechanisms for how NOS-III regulates ischemia-triggered arteriogenesis and angiogenesis are related, in part, to the inability of NOS-III-KO mice to respond to vascular endothelial growth factor (VEGF). Indeed, VEGF-mediated increased permeability, angiogenesis, and endothelial cell precursor mobilization is markedly impaired in mice lacking NOS-III.[58,112,119,120]

Post-ischemic recovery of hindlimb blood flow is impaired in NOS-III-KO mice,[113,114] which is associated with severe tissue injury (necrosis and auto-amputation of the affected foot and lower limb).[113,114] These symptoms, which resemble those observed in patients with CLI, are the consequences of impaired post-ischemic angiogenesis and collateral formation in the absence of NOS-III.

Four weeks after femoral artery ligation, there was an increase in the number of PECAM-1-positive endothelial cells (a measure of angiogenesis) surrounding the skeletal muscle myocytes in wild-type mice (Fig. 2A), an effect markedly diminished in NOS-III-KO (Fig. 2B).[113]

Ischemia-initiated arteriogenesis (collateral formation) was examined by quantitative angiography.[114] After ten days of ischemia in wild-type mice, there was an increase in the number of vessels on the ischemic side compared with the contralateral leg, reflecting an increase in the density of collateral vessels. This result was quantified as increased vessel area in the adductor muscle region (Fig. 3). In contrast, ischemia did not increase collateral formation in NOS-III-KO mice (Fig. 3) similar to that described in the same model using a different technique.[113]

4.2.2. *NOS-III transgenic mice*

Cardiovascular diseases and diabetes are associated with reduced EDNO production/bioactivity, but the "EDNO-deficiency" never reaches the level observed in NOS-III-KO animals. To assess the potential usefulness of EDNO elevation in patients with normal or moderately reduced EDNO levels, a study was performed by Amano *et al.*[121] using NOS-III transgenic (NOS-Tg) mice. Aortic NOS-III expression was increased 3.3-fold in NOS-Tg mice compared to wild-type mice. Blood flow recovery in ischemic hindlimb, following femoral artery resection, was > 40% higher in NOS-Tg mice compared to wild-type mice. Angiography revealed significant increase in basal and ischemia-induced collateral vessel formation, while histochemistry showed increase in basal capillary density and ischemia-induced neocapillary formation in NOS-Tg mice versus wild-type mice. These results indicate that increase of NOS-III expression above the "normal" levels enhances ischemia-induced angiogenesis and arteriogenesis. It

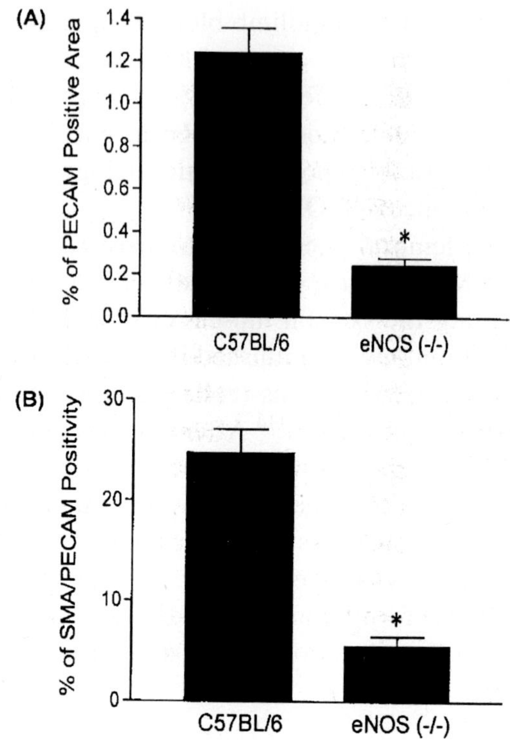

Fig. 2.   Impaired ischemia–induced angiogenesis in NOS-III-KO mice. Gastrocnemius muscles from the ischemic hindlimb of C57B1/J6 and NOS-III-KO mice (eNOS(-/-)) were stained with the endothelial cell marker, PECAM-1 alone **(A)** or PECAM-1 and the smooth muscle/pericyte marker SMA **(B)** four weeks after arteriectomy of the left femoral artery. After harvest, the muscles were methanol-fixed and paraffin-embedded. Tissue sections (5 mm thick) were stained using anti-PECAM-1 antibody (Pharmingen) and anti-smooth muscle α-actin (SMA) antibody (DAKO). Bound primary antibodies were detected by using avidin-biotin-peroxidase and quantified. "Arteriogenesis" was estimated by the ratio of SMA/PECAM-1 staining. Data represent mean $\pm$ SEM of $n = 8$ mice per group. (Modified from Ref. 113 and reproduced with permission from the National Academy of Sciences, USA.)

is therefore plausible that "normal" EDNO production is still not sufficient to maximally support ischemic revascularization, and that NOS-III overexpression may have a therapeutic benefit in patients with "normal" or moderately reduced EDNO levels.

LACK OF COLLATERAL FORMATION IN NOS-III-KO MICE

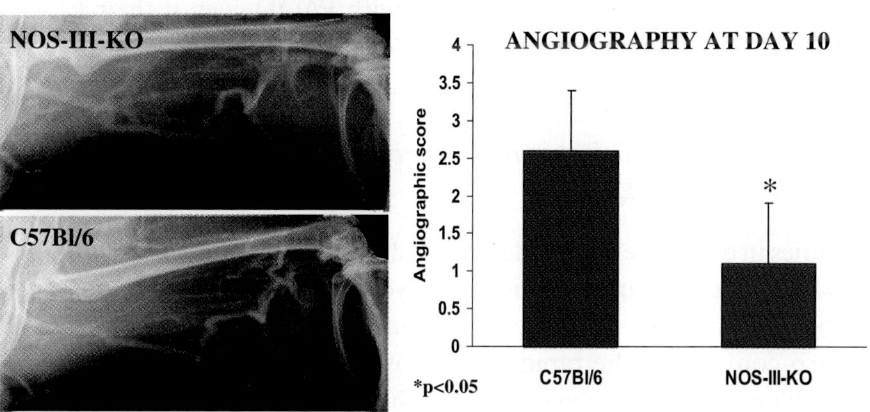

Fig. 3. Angiographic evidence of impaired ischemia-induced collateral vessel formation in the hindlimbs of NOS-III-KO mice. C57Bl/J6 and NOS-III-KO mice underwent surgical arteriectomy of the left femoral artery. Ten days after surgery, the mice were anesthetized, heparinized and perfused with PBS, containing vasodilators, for five minutes at physiological pressure through the descending aorta followed by contrast agent infusion. Microangiography was performed with a cineangiography device and serial pictures were analyzed during the arterial filling stage. The number of visible vessels crossing the midpoint of the femur in the upper hindlimb were counted on both the right (intact) and left (arteriectomy) limb of each animal. Angiographic score values were expressed as the ratio of ischemic to non-ischemic vessel number (a score of one indicates no ischemia-induced collateral formation). Data are shown as mean ± SEM of $n = 5$ animals in each group.

4.2.3. *Wild-type NOS-III gene transfer in normal rats*

Experimental proof for the suggestion that NOS-III overexpression can have therapeutic benefits when EDNO levels are "normal" was provided by a study of Smith *et al.*[122] Injection of human NOS-III gene (Ad5NOS-III) into the adductor muscle of "normal" rats immediately after femoral artery ligation lead to significant elevation of NOS-III protein (persisting for at least four weeks), post-ischemic blood flow recovery (assessed by LDPI) and capillary density (CD31-immunostaining) compared to animals injected with Ad5Luc. These results with local overexpression of NOS-III in the ischemic hindlimb of rats confirms the results from the study on NOS-Tg mice (systemic overexpression of NOS-III), and combined they provide strong rationale for the use of NOS-III gene transfer in patients with PAOD even if their EDNO levels are not significantly decreased.

4.3. *Impaired post-ischemic flow recovery is age dependent in NOS-III-KO mice*

Studies in experimental models of hindlimb ischemia reported impaired revascularization in old mice (> 2 years of age) and demonstrated progression of the ischemic injury into skin ulcers and limb loss, in sharp contrast to the rapid flow recovery in younger animals.[96,98,123] These symptoms were associated with reduced collateral development and decreased expression of VEGF in hindlimb muscle samples in the old animals.[98] Exogenously delivered VEGF was ineffective in old animals,[109] which was shown to be effective in other mouse hindlimb ischemia models using younger animals.[99,123]

The levels of ischemia-induced angiogenesis, the severity of the ischemic damage and extent of limb necrosis were all dependent on the age of NOS-III-KO mice.[124] In three-month-old NOS-III-KO mice, unilateral femoral artery ligation led to increased angiogenesis in the gastrocnemius muscle, whereas in six-month-old animals the severe ischemic disease phenotype corresponded with significantly impaired angiogenesis (Fig. 4). Three-month-old NOS-III-KO mice showed no defect in post-ischemic blood flow recovery (Fig. 5), but at age of six months flow recovery was already severely limited in NOS-III-KO mice, which further deteriorated by the age of 12 months (Fig. 5). In 21-month-old control mice, the flow recovery was somewhat reduced, but not to the extent observed in six-month-old NOS-III-KO mice (Fig. 5).

The similarity between the ischemic phenotype in six-month-old NOS-III-KO mice and that seen in old (> 2 years of age) wild-type mice provides (at least circumstantial) evidence that reduced EDNO production is indeed an important factor in the etiology of "cardio-vascular aging." The accelerated "aging" process in NOS-III-KO mice (where progressive EDNO decrease does not occur with aging) suggests that factors other then EDNO-deficiency should also contribute to age-induced acceleration of ischemic limb damage.

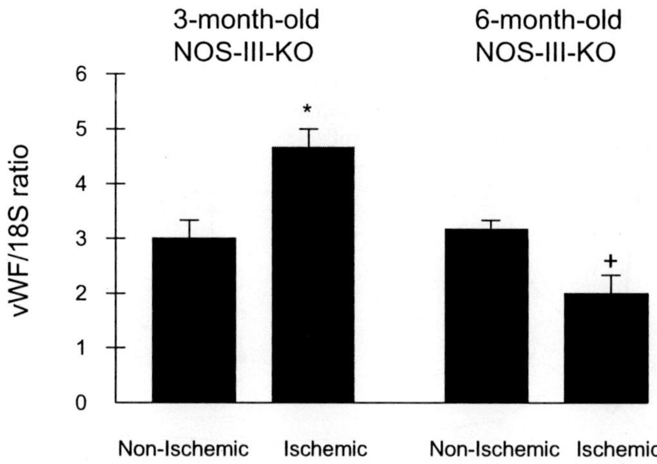

Fig. 4. Age-dependent decrease of post-ischemic angiogenesis in NOS-III-KO mice. Quantitation of angiogenesis was performed by measuring the endothelial cell-specific marker, von Willebrand factor (vWF) expression. Total RNA, isolated from skeletal muscle homogenate at the end of the 21-day recovery period, was used to quantitate angiogenesis by determining the expression level of the endothelial-specific vWF gene. Significant (*p < 0.05) increase in vWF expression was measured in samples from the ischemic limb (crosshatched bars) of the three-month-old NOS-III-KO mice compared to contralateral non-ischemic (filled bars) samples ($n = 8$). In contrast, impaired angiogenesis was found in samples from the hindlimbs of six-month-old NOS-III-KO mice exhibiting severe ischemic injury ($n = 8$). The difference in vWF expression between three-month-old versus six-month-old post-ischemic samples was statistically significant ($+p < 0.05$) indicating impaired age-dependent angiogenesis.

A recent study demonstrated significantly attenuated KDR expression in six-month-old compared to three-month-old NOS-III-KO,[124] suggesting that VEGF receptor expression deficiency (in combination with EDNO deterioration) may contribute to the observed phenomenon.

4.4. *Advantages of a constitutively more active and "risk factor-resistant" mutant NOS-III gene*

In order to avoid potential shortcomings of the wild-type NOS-III enzyme (i.e. low NO production, requirement for agonist stimulation and impairment by common risk factors), a mutation of a

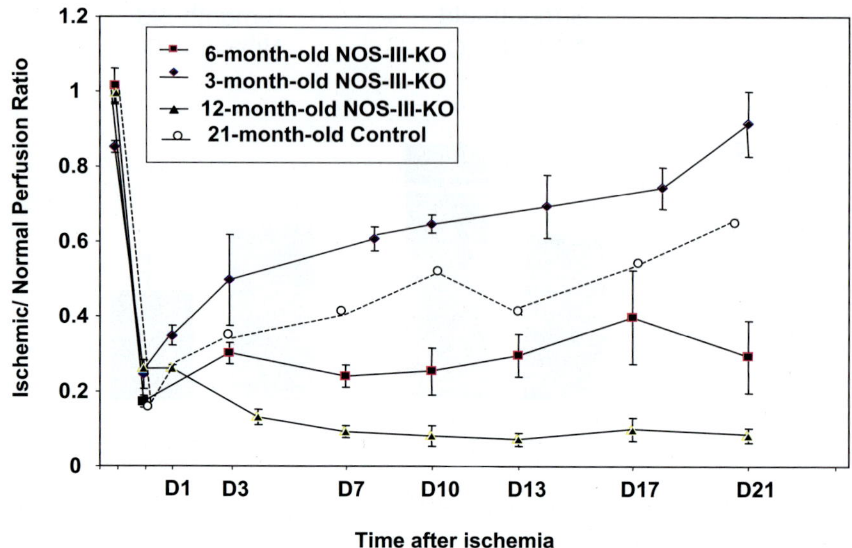

Fig. 5. NOS-III-KO mice exhibit progressive age-dependent impairment in post-ischemic blood flow recovery. Twenty-one-month-old C57Bl/J6 mice ($n = 5$) exhibited flow recovery (measured by LDPI, at days [D] 1, 3, 7, 10, 14, 17 and 21 following surgery) somewhat lower than that observed in three-month-old NOS-III-KO mice ($n = 8$). Six- and 12-month-old NOS-III-KO mice ($n = 6$ each) showed significantly and progressively impaired flow recovery compared to three-month-old NOS-III-KO mice or the 21-month-old C57Bl/J6 mice. Data are expressed as the ratio of ischemic/non-ischemic limb flow and shown as mean $\pm$ SEM. (Reproduced with modifications by permission of the *Journal of Cardiovascular Pharmacology*.)

key phosphorylation site (Serine 1179 in the bovine enzyme or Serine 1177 in the human enzyme) to an aspartate residue have been accomplished.[125,126] Serine 1177 is phosphorylated by several kinases including AMPK, Akt, PKG and PKA, and phosphorylation of this residue increases the basal synthesis of NO.[125,126] Moreover, substitution of Aspartate (D) for Serine (S) mimics the negative charge of the phosphate group and renders NOS-III constitutively more active by increasing the rate of electron flux through the protein and increasing basal NO production several fold.[127] Indeed, overexpression of this mutant NOS-III protected aged cells from apoptosis without additional stimulation of NO release, such as shear stress, unlike transfection with the wild-type NOS-III enzyme.[128] The mutant NOS-III (NOS1177D) enzyme substitutes for Akt-phosphorylated wild-type NOS-III enzyme, which results from KDR-mediated VEGF activation. Therefore the mutant gene will not require intact agonist pathway for its full activity. Furthermore, common risk factors for cardiovascular disease (oxLDL or hyperglycemia)[129–131] and angiogenesis inhibitors, such as endo-statin,[132] interrupt Akt-mediated Serine 1177 phosphorylation in wild-type NOS-III which can all be avoided with the mutant enzyme.[129]

4.5. *NOS-III-S1177D gene transfer in animal models of hindlimb ischemia*

Two recent publications showed that transfer of this constitutively active mutant form of NOS-III gene using either adenovirus[113] or naked plas-mid (combined with electroporation)[114] effectively rescued the struc-tural and functional defects in angiogenesis and arteriogenesis in the ischemic hindlimbs of NOS-III-KO mice. In addition, improvement in blood flow recovery and angiogenesis in response to NOS-III gene transfer have been also demonstrated in several animal models of hind-limb ischemia without genetic deficiency in NOS-III (Akt-KO miceand non-obese diabetic (NOD) mice).

4.5.1. *Plasmid delivery of the NOS1177D gene*

Skeletal muscle expression of NOS1177D (the human form of mutant NOS-III) was tested in NOS-III-KO mice using intramuscular plas-mid injection in combination with electroporation.[114] There was no detectable NOS-III protein (measured by specific ELISA) in hindlimb musculature of NOS-III-KO mice (Fig. 6). In wild-type mice, the aver-age level of NOS-III protein was approximately 30 pg/mg skeletal mus-cle wet weight. Quantitation of NOS-III protein expression from skele-tal muscle homogenates by NOS-III-specific ELISA showed that the transgene expression after NOS1177D gene delivery can reach or exceed levels seen in wild-type mice (Fig. 6).

To evaluate the therapeutic potential of local mutant human NOS-III (NOS1177D) gene delivery, two groups of six-month-old NOS-III-KO mice were injected intramuscularly either with NOS1177D plasmid (pNOS1177D) or an empty vector (pNull).[114] Treatment with the

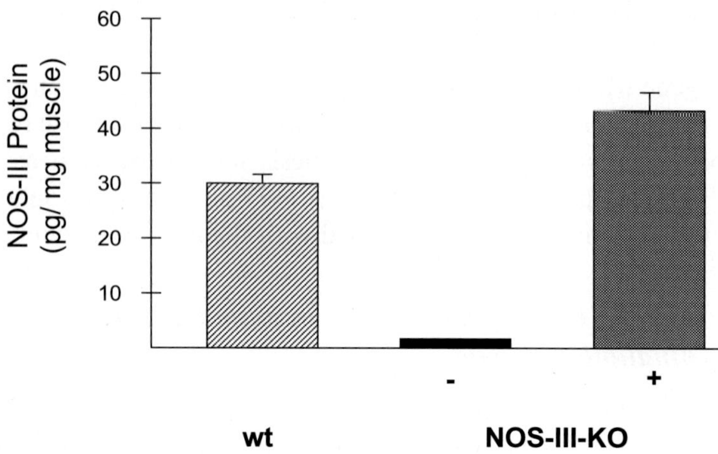

Fig. 6. NOS-III (ecNOS) protein expression in skeletal muscle following intramuscular pNOS1177D delivery to NOS-III-KO mice. NOS-III protein level was determined by an NOS-III-specific ELISA. There was no detectable NOS-III protein in untreated hindlimb of NOS-III-KO mice (−). Treatment with pNOS1177D (+) resulted in measurable levels of NOS-III protein in NOS-III-KO mice, not different from levels measured in control (wt) mice. (Reproduced with modification by permission from the Nature Publishing Group.)

mutant NOS-III gene resulted in a significantly improved blood flow recovery compared to the pNull-treated animals (Fig. 7A). Without improvement in hindlimb perfusion, four out of eight pNull-treated mice lost the ischemic limb by day 28 (Fig. 7B). In contrast, treatment with pNOS1177D prevented limb loss in all treated animals (Fig. 7B).

4.5.2. *Adenoviral delivery of the NOS1179D gene*

Intramuscular injection of AdNOS1179D (the bovine form of mutant NOS-III) into the adductor muscle group of ischemic NOS-III-KO mice resulted in expression of NOS-III protein (detected by immunofluoresence microscopy).[113] Intramuscular administration of AdNOS-III1179D, but not Ad-GFP, at the time of femoral arterectomy markedly improved blood flow recovery at two and four weeks after ischemia (Fig. 8), which was associated with increased angiogenesis (Fig. 8A) and arteriogenesis (Fig. 8B) in the treated limb musculature.

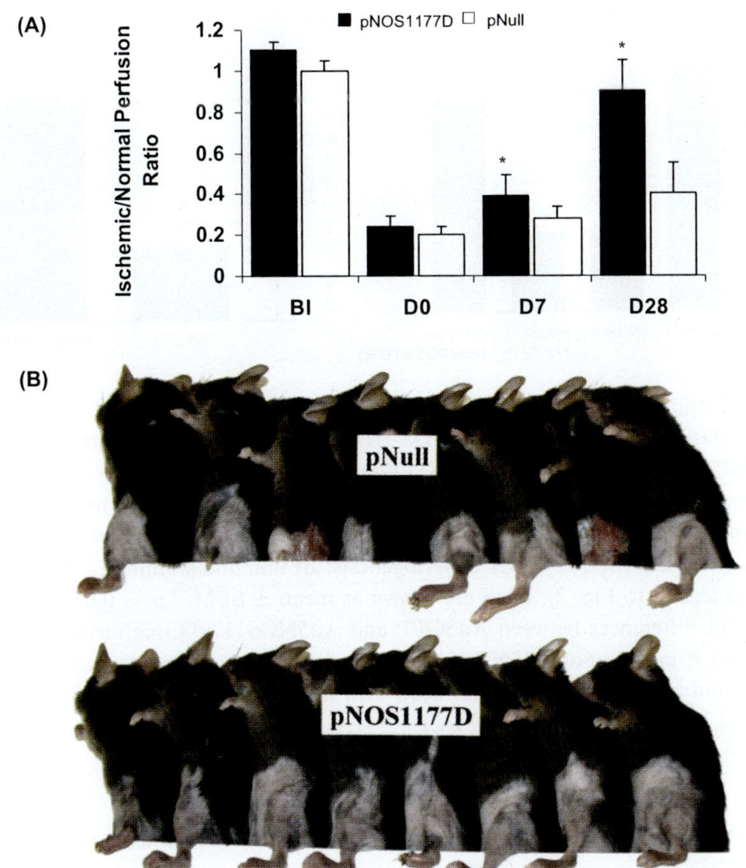

Fig. 7. Intramuscular injection of plasmid NOS1177D augments post-ischemic flow recovery (A) and prevents limb loss (necrosis and auto-amputation) (B) in NOS-III-KO mice. NOS-III-KO mice underwent unilateral femoral artery resection and three days later were injected with an empty plasmid ("pNull") ($n = 8$) or a plasmid carrying the mutant NOS-III-S1177D gene (pNOS1177D) ($n = 8$) followed by electroporation. (A) Limb blood flow was measured by laser Doppler perfusion imaging (LDPI) at various time points after surgery. Data are shown as mean $\pm$ SEM and expressed as the ratio of perfusion in the ischemic versus non-ischemic hindlimb. NOS-III S1177D gene transfer (filled columns) significantly improved LDPI flow on days 7 and 28 (D7, D28) compared to pNull treatment (open columns). $*p < 0.05$; BI = before ischemia). (B) Ischemic tissue damage of the hindlimb was evaluated by taking photographs of the limbs on the same days when LDPI measurements were made. By day 28, the evidence of limb loss was significantly greater in the pNull–treated (four out of eight animals lost their limb) than in the pNOS1177D–treated group (none of the animals lost their limb). (Reproduced with modifications by permission from the Nature Publishing Group.)

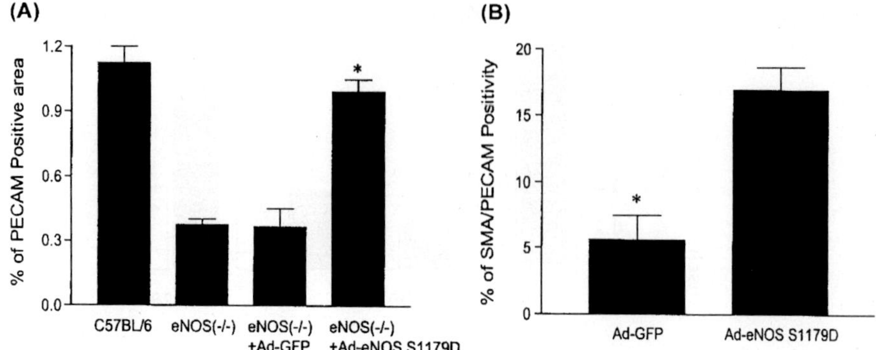

Fig. 8. Intramuscular injection of Ad5NOS1179D improves ischemia–induced angiogenesis (**A**) and arteriogenesis (**B**) in NOS-III-KO mice. NOS-III-KO mice were injected with Ad5GFP ($n = 5$) or Ad5NOS1179D ($n = 5$) in the adductor muscle of the ischemic left hindlimb. Four weeks after surgery and gene injection, mice were sacrificed and the gastrocnemius muscle harvested and tested for PECAM-1 (angiogenesis; A) and PECAM-1 + SMA (arteriogenesis; B) immuno-staining (as described in detail in legend to Fig. 2). Data are shown as mean ± SEM. *$p < 0.05$: statistically significant differences between Ad5GFP and Ad5NOS1179D treatment. In contrast to Ad5GFP, injection of Ad5NOS1179D significantly increased post-ischemic angiogenesis and arteriogenesis. (Reproduced with modifications by permission from the National Academy of Sciences, USA.)

4.5.3. *Effect of NOS1177D gene transfer in mouse CLI models without genetic deficiency of NOS-III*

In a rat model of hindlimb ischemia, intramuscular injection of adenovirus carrying the wild-type human NOS-III gene increased post-ischemic flow recovery and angiogenesis.[122] Similar to NOS-III-KO mice, balb/c mice also respond to unilateral surgical femoral artery occlusion with severe, VEGF refractory hindlimb ischemia and auto-amputation.[110,111] In this mouse strain, which does not have a genetic deficiency of NOS-III expression, delivery of pNOS1177D resulted in significantly improved post-ischemic blood flow recovery (K. Kauser and G. M. Rubanyi — unpublished observation). Non-obese diabetic (NOD) mice develop severe impairment in hindlimb blood flow recovery after femoral artery ligation, potentially due to diabetes-induced endothelial dysfunction resulting from lost nitric oxide activity.[133] This

model is relevant since there is a high incidence of diabetic patients in
the CLI population.

The phosphatidylinositol-3-OH kinase (PI3K)/Akt pathway have
been implicated in the shear stress-induced phosphorylation of NOS-III
leading to an increase in nitric oxide production. Although Akt-1-KO
mice have an intact NOS-III gene present, the NOS-III activation pathway may be inhibited, resulting in a decrease in nitric oxide levels (W.
Sessa — unpublished observations).

Hindlimb ischemia models were developed in both the Akt-1 KO
mice (eight to 12 weeks old) and NOD mice (12 weeks old) by ligation
and dissection of the proximal end of the femoral artery, proximal site of
the popliteal artery, and distal portion of the saphenous artery. Immediately after surgery, AdNOS1177D (2.5×10^9 pfu for Akt-1-KO mice;
1×10^9 pfu for NOD mice), AdLacZ (1×10^9 pfu for NOD mice only),
or saline was injected into three different sites of the adductor magnus
and adductor longus muscle. Blood flow in the left (ischemic) and right
(non-ischemic) hindlimbs was measured from the gastrocnemius muscle prior to, and immediately after surgery, and at one, two, and four
weeks after surgery by using the PeriFlux system with Laser Doppler
Perfusion Module (LDPM). Akt-1-KO mice develop CLI-like disease
phenotype in response to surgical hindlimb ischemia, similar to what
has been observed in NOS-III-KO mice. Treatment with AdNOS1177D
resulted in a significant improvement in blood flow at two and four
weeks following ischemia (Fig. 9A). Treatment with AdNOS1177D in
NOD mice resulted in a significant improvement in blood flow compared to AdLacZ or saline-treated animals (Fig. 9B).

5. Potential Therapeutic Utility of NOS-III Gene Transfer in the Heart

5.1. *Facilitation of coronary angiogenesis and ischemia-induced collateral growth*

Although it have been less studied than in the peripheral (limb) vasculature, nitric oxide was shown to play a key role in angiogenesis and
collateral growth also in the heart.

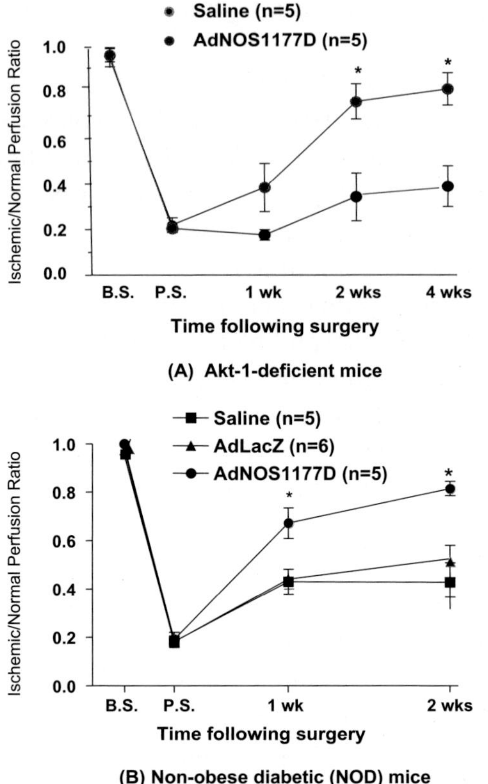

Fig. 9. Intramuscular injection of Ad5NOS1177D improves blood flow recovery in ischemic hindlimb of mice without genetic deficiency of NOS-III. The effect of Ad5NOS1177D treatment was tested in two animal models of hindlimb ischemia, where endothelial NO production (Akt-1-KO mice; A) or NO availability (NOD mice; B) are reduced without genetic deficiency in NOS-III. (**A**) Akt-1-KO mice underwent arteriectomy of the left femoral artery followed by injection of saline ($n = 5$) or Ad5NOS1177D ($n = 5$) in the adductor muscle of the ischemic limb. Post-ischemic flow recovery was significantly (*$p < 0.05$) higher in the Ad5NOS1177D–treated than in the saline-treated group, two and four weeks after surgery. Data are shown as mean ± SEM and expressed as the ratio of ischemic versus non-ischemic hindlimb perfusion measured by LDPI. (**B**) Non-obese diabetic (NOD) mice (12 weeks old, blood glucose > 500 mg / dl) underwent arteriectomy of the left femoral artery followed by injection of saline ($n = 5$), Ad5LacZ ($n = 6$) or Ad5NOS1177D ($n = 5$) in the adductor muscle of the ischemic hindlimb. Ad5NOS1177D treatment significantly (*$p < 0.05$) augmented post-ischemic flow recovery when compared to Ad5LacZ or saline treatment, one and two weeks after surgery (B.S. = before surgery; P.S. = post-surgery). Data are shown as mean ± SEM and expressed as the ratio of ischemic versus non-ischemic hindlimb perfusion.

In coronary post-capillary (venular) endothelial cells, nitric oxide mediates the angiogenic effect of VEGF[57] by activation of ERK $^1/_2$.[134] In a canine model of repetitive myocardial ischemia, collateral blood flow (measured by microspheres) progressively increased during the 21-day experimental period, which was prevented by treatment of dogs with the NOS inhibitor L-NAME.[135] Similarly, in a rat model of chronic myocardial ischemia, treatment with L-NAME significantly reduced basal and maximum left ventricular blood flow (measured by MRI) and angiogenesis.[136] Ischemia-induced upregulation of VEGF production in the myocardium was not prevented by L-NAME, indicating, that similar to the peripheral vasculature, NO mediates VEGF-induced angiogenesis and collateral growth in the coronary circulation as well. An additional mechanism of NO-induced facilitation of collateral growth in the ischemic heart is suppression of anti-angiogenic molecules such as angiostatin, via downregulation of tissue matrix metalloproteinases (MMPs), MMP-2 and MMP-9, which generate angiostatin.[135,137]

5.2. *NOS-III-derived nitric oxide facilitate myocardial gene transfer by adenoviral vectors*

In pre-immunized pigs (by intravenous injection of control adenovirus causing elevation of anti-adenoviral neutralizing antibody titer similar to that found in the majority of coronary artery bypass graft patients), co-injection of Ad5Luc and Ad5NOS-III via the great cardiac vein (retrograde) resulted in > 200-fold higher luciferase expression than after retrograde injection of Ad5Luc alone.[138] Ad5NOS-III co-injection also reduced Ad5Luc injection-induced T-cell-mediated inflammation and cardiac myocyte apoptosis.[138] These results suggested that intracardiac NOS-III gene transfer may reduce some of the known barriers to adenovirus-mediated myocardial gene transfer. Another limitation of effective myocardial gene transfer by adenoviral vectors is poor penetration of the microvascular (endothelial) barrier. It has been reported that increasing vascular permeability by VEGF pre-administration significantly augments adenoviral transfection efficiency in the isolated perfused rabbit heart, which can be inhibited by L-NAME and mimicked by the NO donor nitroglycerin.[139]

5.3. *Nitric oxide protects against acute ischemia-induced myocyte death (acute myocardial infarction) and pathological ventricular remodeling (chronic heart failure)*

The two constitutive NOS isoforms (NOS-I and NOS-III) are present in the myocardium and are compartmentalized: the neuronal NOS (NOS-I) is localized to the ryanodine receptor of the sacroplasmic reticulum and the endothelial NOS (NOS-III) localizing to the caveolae of the sarcrolemma in close proximity to β-adrenergic receptors and the L-type calcium channel.[140]

The effect of NO (by the various NOS isoforms) on the heart is complex and controversial.[140–142] Recent reports showed that local overexpression of NOS-III in the heart of transgenic mice provided protection against the acute consequences of ischemia-reperfusion[143] and the longer-term ventricular remodeling after myocardial infarction.[144,145]

In a rat model of coronary artery ligation-induced left ventricular remodeling,[146] adenoviral delivery of NOS-III gene resulted in elevated NOS-III and cGMP levels in the heart and significantly reduced myocardial infarction-induced left ventricular remodeling, fibrosis and myocyte apoptosis.[147]

6. Conclusions

Based on experimental evidence accumulated over the past decade and briefly reviewed in this chapter, the key role of endothelial NO in angiogenesis is well established. Nitric oxide is essential for the angiogenic effect of VEGF and other growth factors (e.g. FGF and HGF) in the adult organism. It has also been well documented that EDNO plays an essential role in post-ischemic revascularization (neoangiogenesis) in the limb musculature and in the heart, in part by mediating the actions of angiogenic growth factors. EDNO overexpression (in transgenic mice and by local delivery of the NOS-III gene) facilitates ischemic flow recovery and neoangiogenesis in the hindlimb of animals with physiological EDNO levels, indicating that NOS-III gene transfer may be therapeutically beneficial in patients with chronic peripheral or myocardial ischemia even if their EDNO level is not significantly decreased. Common cardiovascular risk factors (such as oxidized LDL or hyperglycemia) and endogenous anti-angiogenic molecules (such as endostatin) can lead to "EDNO-deficiency" by disruption of NOS-III activation via Akt-induced phosphorylation at Serine 1177. The development of a constitutively active phosphomimetic NOS-III mutant (S1177D) provided an "improved" NOS-III enzyme, which is "resistant" to the above risk factors and endostatin. The potential therapeutic utility of this mutant enzyme has already been shown in animal models of critical limb ischemia.

Acknowledgments

The valuable contribution of Dr. Katalin Kauser (Boehringer-Ingelheim), Dr. William Sessa (Yale University) and Dr. Husheng Qian (Berlex Biosciences) to some of the experimental work presented in this chapter is highly appreciated.

References

1. Furchgott RF, Zawadzki JV (1980). The obligatory role of endothelial cells in the relaxation of arterial smooth muscle by acetylcholine. *Nature* **228**: 373–376.
2. Furchgott RF (1998). Studies on relaxation of rabbit aorta by sodium nitrite: basis for the proposal that the acid-activatable component of the inhibitory factor from retractor penis is inorganic nitrate and the endothelium-derived relaxing factor is nitric oxide. In: Vanhoutte PM (ed)., *Mechanism of Vasodilatation* (Raven Press, New York), pp. 401–414.
3. Ignarro LJ, Buga GM, Wood KS, Byrns RE, Chaudhuri G (1987). Endothelium-derived relaxing factor produced and released from artery and vein is nitric oxide. *Proc Natl Acad Sci USA* **84**: 9265–9269.
4. Knowles RG, Moncada S (1994). Nitric oxide synthase in mammals. *Biochem J* **298**: 249–258.
5. Salerno JC, Harris DE, Irizarry K, Patel B, Morales AJ, Smith SME, *et al.* (1997). An auto-inhibitory control element defines calcium-regulated isoforms of nitric oxide synthase. *J Biol Chem* **272**: 29769–29777.
6. Rubanyi GM (1993). The role of endothelium in cardiovascular homeostasis and diseases. *J Cardiovasc Pharmacol* **22**(Suppl): S1–S14.
7. Griffith TM, Edwards DH, Lewis MJ, Newby AC, Henderson AH (1984). The Nature of endothelium-derived relaxant factor. *Nature* **308**: 645–647.
8. Rubanyi GM, Lorenz RR, Vanhoutte PM (1985). Bioassay of endothelium-derived relaxing factors(s): inactivation by catecholamines. *Am J Physiol* **249**: 95–101.
9. Rubanyi GM, Vanhoutte PM (1986). Superoxide anions and hyperoxia inactivate endothelium-derived relaxing factor. *Am J Physiol* **250**: 822–827.

10. Rubanyi GM, Romero JC, Vanhoutte PM (1986). Flow-induced release of endothelium-derived relaxing factor. *Am J Physiol* **250**: 1145–1149.

11. Palmer R, Ferrige MJ, and Moncada S (1987). Nitric oxide release accounts for the biological activity of endothelium-derived relaxing factor. *Nature* **327**: 524–526.

12. Radomski MW, Palmer RMJ, Moncada S (1987). The anti-aggregating properties of vascular endothelium: interactions between prostacyclin and nitric oxide. *Br J Pharmacol* **92**: 639–646.

13. Garg UC, Hassid A (1989). Nitric oxide-generating vasodilators and 8-bromo-cyclic guanosine monophosphate inhibit mitogenesis and proliferation of cultared rat vascular smooth muscle cells. *J Clin Invest* **83**: 1774–1771.

14. Guo JP, Panday MM, Consigny PM, Lefer AM (1995). Mechanisms of vascular preservation by a novel NO donor following rat carotid artery intimal injury. *Am J Physiol* **269**: 1122–1131.

15. Ziche M, Morbidelli L, Masini E, Amerini S, Granger HJ, Maggi CA, Geppetti P, Ledda F (1994). Nitric oxide mediates angiogenesis *in vivo* and endothelial cell growth and migration *in vitro* promoted by substance P. *J Clin Invest* **94**: 2036–2044.

16. Kubes PM, Suzuki M, Granger DN (1991). Nitric oxide: an endogenous modulator of leukocyte adhesion. *Proc Natl Acad Sci USA* **88**: 4651–4655.

17. Wang JM, Chow SN, Lin JK (1994). Oxidation of LDL by nitric oxide and its modification by superoxides in-macrophage and cell-free systems. *FEBS Lett* **342**: 171–175.

18. Tarry WC, Makhoul RG (1994). L-arginine improves endothelium-dependent vasorelaxation and reduces intimal hyperplasia after balloon angioplasty. *Arterioscler Thromb* **14**: 938–943.

19. Dimmeler S, Fissthaler B, Fleming I, Assmus B, Hermann C, Zeiher AM (1998). Shear stress stimulates the protein kinase Akt: involvement in regulation of the endothelial nitric oxide synthase. *Circulation* **17**: I–312.

20. Dimmeler S, Fleming I, Fisslthaler B, Hermann C, Busse R, Zeiher AM (1999) .Activation of nitric oxide synthase in endothelial cells by Akt-dependent phosphorylation. *Nature* **399**: 601–605.

21. Garcia-Cardena G, Martasek P, Masters BS, Skidd PM, Couet J, Li S, *et al.* (1997). Dissecting the interaction between nitric oxide synthase (NOS) and caveolin. Functional significance of the NOS caveolin binding domain *in vivo*. *J Biol Chem* **272**: 25437–25440.

22. Zeiher AM, Drexler H, Wollschlager, H, Just H (1991). Endothelial dysfunction of the coronary microvasculature is associated with impaired coronary blood flow regulation in patients with early atherosclerosis. *Circulation* **84**: 1984–1992.

23. McAllister A-S, Atkinson A-B, Johnston G-D, Hadden D-R, Bell P-M, McCance D-R (1999). Basal nitric oxide production is impaired in offsprings of patients with essential hypertension. *Clin Sci* **97**: 141–147.

24. Rubanyi GM, Freay AD, Kauser K, Sukovich D, Burfton G, Lubahn D, Couse JF, Curtis SN, Korach KS (1997). Vascular estrogen receptors and endothelium-derived nitric oxide production in the mouse aorta: gender difference and effect of estrogen receptor gene distruption. *J Clin Invest* **99**: 2429–2437.

25. Sudhir K, Chou TM, Chatterjee K, Smith EP, Williams TC, Kane JP, Malloy MJ, Korach KS, Rubanyi GM (1997). Premature coronary artery disease associated with a disruptive mutation in the estrogen receptor gene in a man. *Circulation* **96**: 3774–3777.

26. Kauser K, da Cunha V, Fitch RM, Mallari C, Rubanyi GM (2000). Role of endogenous nitric oxide in the progression of atherosclerosis in the apolipoprotein E-deficient mouse. *Am J Physiol* **278**: 1679–1685.

27. Edmundowicz D (2002). Noninvasive studies of coronary and peripheral arterial blood-flow. *Curr Atheroscler Rep* **4**: 381–385.

28. Gokce N, Keaney JF Jr, Hunter LM, Watkins MT, Nedelljkovic ZS, Menzoian JO, Vita JA (2003). Predictive value of noninvasively determined endothelial dysfunction for long-term cardiovascular events in patients with peripheral vascular disease. *J Am Coll Cardiol* **41**: 1769–1775.

29. Huang PL, Huang Z, Mashimo H, bloch KD, Moskowitz MA, Bevan JA, *et al.* (1995). Hypertension in mice lacking the gene for endothelial nitric oxide synthase. *Nature* **377**: 239–242.

30. Shesely E-G, Maeda N, Kim H-S, Desai K-M, Krege J-H, Laubach V-E, Sherman P-A, *et al.* (1996). Elevated blood pressures in mice lacking endothelial nitric oxide synthase. *Proc Natl Acad Sci* **93**: 13176–13181.

31. Freedman JE, Sauter R, Battinelli EM, Adult K, Knowles C, Huang PL, Loscalzo J (1999). Deficient platelet-derived nitric oxide and enhanced hemostasis in mice lacking the NOS-III gene. *Circ Res* **84**: 1416–1421.

32. Lefer DJ, Jones SP, Girod WG, Baines A, Grisham MB, Cockrell AS, *et al.* (1999). Leukocyte-endothelial cell interactions nitric oxide synthase-deficient mice. *Am J Physiol (Heart Circ Physiol* 45) **276**: 1943–1950.

33. Lee PC, Salyapongse AN, Bragdon GA, Shears LL 2nd, Watkins SC, Edington HD, Billiar TR (1999). Impaired wound healing and angiogenesis in ecNOS-deficient mice. *Am J Physiol* **277**: 1600–1608.

34. Jones SP, girod WG, Palazzo AJ, Granger DN, Grisham MB, Jourd'heuil D, *et al.* (1999). Myocardial ischemia-reperfusion injury is exacerbated in absence of endothelial cell nitric oxide synthase. *Am J Physiol (Heart Circ Physiol)* **276**: 1576–1573.

35. Huang Z, Huang PL, Ma J, Meng W, Ayata C, Fishman MC, Moskowitz MA (1996). Enlarged infarcts in endothelial nitric oxide synthase knockout mice are attenuated by nitro-L-arginine. *J Cereb Blood Flow Metab* **16**: 9891–9987.

36. Kauser K, Rubanyi GM (1994). Effect of 17b-estradiol on endothelial dysfunction in the aorta of spontaneously hypertensive rats. *Endothelium* **25**: 517–523.

37. Kauser K, Rubanyi GM (1997). Potential cellular signaling mechanisms mediating upregulation of endothelial nitric oxide production by estrogen. *J Vasc Res* **34**: 229–236.

38. Sudhir K, Jennings GL, Funder JW, Komesaroff PA (1996). Estrogen enhances basal nitric oxide release in the forearm vasculature in perimenopausal women. *Hypertension* **28**: 330–334

39. Laufs U, Fata VL, Plutzky J, Liao, J-K (1998). Upregulation of endothelial nitric oxide synthase by HMG-CoA reductase inhibitors. *Circulation* **97**: 1129–1135.

40. Sudhir K, Chou TM, Hutchison SJ, Chatterjee SJ (1996). Coronary vasodilation induced by angiotensin-converting enzyme inhibition *in vivo*; differential contribution of nitric oxide and bradykinin in conductance and resistance arteries. *Circulation* **93**: 1734–1739.

41. Kauser K, Himmerich S, Blasko E, Glaser CB, Paulo E, Humm R, Rubanyi GM (1999). Increase in endothelial nitric oxide synthase (NOS-3) activity by antioxidant vitamins is mediated by tetrahydrobiopterin (BH$_4$). *Acta Physiol Scand* **167**: 60.

42. Mancini GBJ, Henry GC, Macaya C, O'Neill BJ, Pucillo AL, Carere RG, *et al.* (1996). Angiotensin-converting enzyme inhibition with quinapril improves endothelial vasomotor dysfunction in patients with coronary artery disease: the TREND (trial on reversing endothelial) study. *Circulation* **94**: 258–265.

43. Treasure CB, Klein JL, Weintraub WS, Talley JD, Stillabower ME, Kosinski AS, *et al.* (1995). Beneficial effects of cholesterol-lowering therapy on the coronary endothelium in patients with coronary artery disease. *N Engl J Med* **332**: 481–487.

44. Brock TA, Dvorak HF, Senger DR (1991). Tumor-secreted vascular permeability factor increases cytosolic Ca^{2+} and von Willebrand factor release in human endothelial cells. *Am J Pathol* **138**: 213–221.

45. Ku DD, Zaleski JK, Liu S, Brock TA (1993). Vascular endothelial growth factor induces EDRF-dependent relaxation in coronary arteries. *Am J Physiol* **265**: 586–592.

46. Van der Zee R, Murohara T, Luo Z, Zollman F, Passeri J, Lekutat C, Isner JM (1997). Vascular endothelial growth factor (VEGF)/vascular permeability factor (VPF) augments nitric oxide release from quiescent rabbit and human vascular endothelium. *Circulation* **95**: 1030–1037.

47. Horowitz JR, Rivard A, Van der Zee R, Hariawala MD, Sheriff DD, Esakof DD, Chaudry GM, Symes JF, Isner JM (1997). Vascular endothelial growth factor/vascular permeability factor produces nitric oxide-dependent hypotension. *Arterioscler Thromb Vasc Biol* **17**: 2793–2800.

48. Isner JM (1998). Angiogenesis. In: Topol EJ (ed.), *Textbook of Cardiovascular Medicine* (Lippincott-Raven, Philadelphia), pp. 2491–2518.

49. Thomsen LL, Lawton FG, Knowles RG, Beesley JE, RiverosMoreno V, Moncada S (1994). Nitric oxide synthase activity in human gynecological cancer. *Cancer Res* **54**: 1352–1354.

50. Cobbs CS, Brenman JE, Alpade KD, Bredt DS, Israe MA (1995). Expression of nitric oxide synthase in human central nervous system tumors. *Cancer Res* **55**: 727–730.

51. Jenkins DC, Charles IG, Thomsen LL, Moss DW, Holmes LS, Baylis SA, Rhodes P, Westmore K, Emson PC, Moncada S (1995). Role of nitric oxide in tumor angiogenesis. *Proc Natl Acad Sci USA* **92**: 4392–4396.

52. Kennovin GD, Hirst DG, Stratford MRL, Flitney FW (1994). Inducible nitric oxide synthase is expressed in tumor associated vasculature; inhibition retards tumor growth *in vivo*. In: Moncada S, Feelish M, Busse R, Higgs EA (eds.), *Biology of Nitric Oxide: Enzymology, Biochemistry and Immunology* (Portland Press, London), Vol. 4, pp. 473–479.

53. Orucevic A, Lala PK (1996). N^G-nitro-L-arginine methyl ester, an inhibitor of nitric oxide synthesis, ameliorates interleukin 2-induced capillary leakage and reduces tumor growth in adenocarcinoma-bearing mice. *Br J Cancer* **73**: 189–196.

54. Ziche M, Morbidelli L, Choudhuri R, Zhang H-T, Donnini S, Granger HJ, Bicknell R (1997). Nitric oxide synthase lies downstream from vascular endothelial growth factor-induced but not fibroblast growth factor-induced angiogenesis. *J Clin Invest* **99**: 2625–2634.

55. Papapetropoulos A, Desai KM, Rudic RD, Mayer B, Zhang R, Ruiz-Torres' MP, Garcia-Cardena G, Madri JA, Sessa WC (1997). Nitric oxide synthase inhibitors attenuate transforming-growth-factor-b1-stimulated capillary organization *in vitro*. *Am J Pathol* **150**: 1835–1844.

56. Papapetropoulos A, Garcia-Cardena G, Madri JA, Sessa WC (1997). Nitric oxide production contributes to the angiogenic properties of vascular endothelial growth factor in human endothelial cells. *J Clin Invest* **100**: 3131–3139.

57. Morbidelli L, Chang C-H, Douglas JG, Granger HJ, Ledda F, Ziche M (1995). Nitric oxide mediates mitogenic effect of VEGF on coronary venular endothelium. *Am J Physiol* **270**: 411–415.

58. Murohara T, Asahara T, Silver M, Bauters C, Masuda H, Kalka C, Kearney M, Chen D, Symes JF, Fishman MC, Huang PL, Isner JM (1998). Nitric oxide synthase modulates angiogenesis in response to tissue ischemia. *J Clin Invest* **101**: 2567–2578.

59. Vallance P, Leone A, Calver A, Collier J, Moncada S (1992). Endogenous dimethylarginine as an inhibitor of nitric oxide synthesis. *J Cardiovasc Pharmacol* **20**: S60–S62.

60. Heeschen C, Ho H-KV, Kaji S, Jang SJ, Stuehlinger M, Yang X, Yang P, Hu BS, Cooke JP (1999). Hypercholesterolemia impairs angiogenic response to hindlimb ischemia: Role of ADMA. [Abstract] *Circulation* (Suppl I): I–285.

61. Boger RH, Bode-Boger SM, Szuba A, Tsao PS, Chan JR, Tangphao O, Blaschke TF, Cooke JP (1998). Asymmetric Dimethylarginine (ADMA): a novel risk factor for endothelial dysfunction. Its role in hypercholesterolemia. *Circulation* **98**: 1842–1847.

62. Couffinhal T, Silver M, Zheng LM, Kearney M, Witzenbichler B, Isner JM (1996). Angiogenesis is impaired in ApoE Knock out mice due to a reduced expression of vascular endothelial growth factor (abstr). *Circulation* **94**: 1–102.

63. Yang HT, Yan Z, Abraham JA, Terjung RL (2001). VEGF (121) and bFGF-induced increase in collateral blood flow requires normal nitric oxide production. *Am J Physiol Heart Circ Physiol* **280**: H1097–H1104.

64. Prior BM, Lloyd PG, Ren J, Li Z, Yang HT, Laughlin MH, Terjung RL (2003). Arteriogenesis: role of nitric oxide. *Endothelium* **10**: 207–216.

65. Dimmeler S, Hermann C, Galle J, Zeiher AM (1999). Upregulation of superoxide dismutase and nitric oxide synthase mediates the apoptosis-suppressive effects of shear stress on endothelial cells. *Arterioscler Thromb Vasc Biol* **19**: 656–664.

66. Rossig L, Fichtlscherer B, Breitschopf K, Haendeler J, Zeiher AM, Mulsch A, Dimmeler S (1999). Nitric oxide inhibits caspase-3 by S-nitrosation *in vivo*. *J Biol Chem* **274**: 6823–6826.

67. Ceneviva GD, Tzeng E, Hoyt DG, Yee E, Gallagher A, Engelhardt JF, Kim YM, Billiar TR, Watkins SA, Pitt BR (1998). Nitric oxide inhibits lipopolysaccharide-induced apoptosis in pulmonary artery endothelial cells. *Am J Physiol* **275**: L717–728.

68. Noiri E, Lee E, Testa J, Quigley J, Colflesh D, Keese CR, Giaever I, Goligorsky MS (1998). Podokinesis in endothelial cell migration: role of nitric oxide. *Am J Physiol* **274**: C236–244.

69. Murohara T, Asahara T, Takahashi T (1998). Role of endothelial nitric oxide synthase in endothelial cell migration: endogenous nitric oxide maintains integrin $\alpha\upsilon\beta3$ expression in endothelial cells. [Abstract] *Circulation* (Suppl) **98**: I-730.

70. Ziche M, Parenti A, Ledda F, Dell'Era P, Granger HJ, Maggi CA, Presta M (1997). Nitric oxide promotes proliferation and plasminogen activator production by coronary venular endothelium through endogenous bFGF. *Circ Res* **80**: 845–852.

71. Hudlicka O (1998). Is physiological angiogenesis in skeletal muscle regulated by changes in microcirculation? *Microcirculation* **5**: 7–23

72. Heil M, Schaper W (2004). Influence of mechanical, cellular, and molecular factors on collateral artery growth (arteriogenesis). *Circ Res* **95**: 449–459.

73. Channon K-M, Qian H-S, Neplioueva V, Blazing M-A, Olmez E, Shetty G, *et al.* (1998). *In vivo* gene transfer of nitric oxide synthase enhances vasomotor function in carotid arteries from normal and cholesterol-fed rabbits. *Circulation* **98**: 1905–1911.

74. Janssens S, Flaherty D, Nong Z, Varenne O, van Pelt N, Haustermans C, Zoldhelyi P, Gerard R, Collen D (1998). Human endothelial nitric oxide synthase gene transfer inhibits vascular smooth muscle cell proliferation and neointima formation after balloon injury in rats. *Circulation* **97**: 1274–1281.

75. Gaballa M-A, Goldman S (1999). Overexpression of endothelium nitric oxide synthase reverses the diminished vasorelaxation in the hindlimb vasculature in ischemic heart failure *in vivo*. *J Mol Cell Cardiol* **31**: 1243–1252.

76. Von der Leyen HE, Gibbons GH, Morishita R, Lewis NP, Zhang L, Nakajima M, *et al.* (1995). Gene therapy inhibiting neointimal vascular lesion: *in vivo* transfer of endothelial cell nitric oxide synthase gene. *Proc Natl Acad Sci USA* **92**: 1137–1141.

77. Kullo U, Mozes G, Schwartz RS, Gloviczki P, Crotty TB, Barber DA, Katusic ZS, O'Brien T (1997). Adventitial gene transfer of recombinant endothelial nitric oxide synthase to rabbit carotid arteries alters vascular reactivity. *Circulation* **96**: 2254–2261.

78. Nabel EG (1995). Gene therapy for cardiovascular disease. *Circulation* **91**: 541–548.

79. Reifers F, Kreuzer J (1995). Current aspects of gene therapy: implications for vascular interventions. *J Mol Med* **73**: 595–602.

80. Heistad DD, Faraci FM (1996). Gene therapy for cerebral vascular disease. *Stroke* **27**: 1688–1693.

81. O'Brien T (1996). Gene therapy and inherited dyslipidemia. *Endocrinol Pract* **2**: 37–43.

82. Yla-Herttuala S, Martin JF (2000). Cardiovascular Gene Therapy. *The Lancet* **355**: 213–222.

83. Schneider MD, French BA (1993). The advent of adenovirus: gene therapy for cardiovascular disease. *Circulation* **88**: 1937–1942.

84. Wilson JM (1996). Adenoviruses as gene-delivery vehicles. *Mol Med* **334**: 1185–1187.

85. Wickham TJ, Mathias P, Cheresh DA, Nemerow GR (1993). Integrins alpha v beta 3 and alpha v beta 5 promote adenovirus internalization but not virus attachment. *Cell* **73**: 309–319.

86. Bergelson JM, Cunningham JA, Droguett G, Kurt-Jones EA, Krithivas A, Hong JS, *et al.* (1997). Isolation of a common receptor for coxsackie B virus and adenoviruses 2 and 5. *Science* **275**: 1320–1323.

87. Ooboshi H, Chu Y, Rios CD, Faraci FM, Davidson BL, Heistad DD (1997). Altered vascular function following adenovirus-mediated overexpression of endothelial nitric oxide synthase. *Am J Physiol* **273**: H265–H270.

88. Kilbourn RG, Jubran A, Groaa SS, Griffith OW, Levi R, Adams J, Lodato RF (1990). Reversal of endotoxin-mediated shock by N^G-monomethyl-L-arginine, an inhibitor of nitric oxide synthesis. *Biochem Biophys Res Commun* **72**: 1132–1138.

89. Weinberg JB, Granger DL, Pisetsky DS, Seldin MF, Misukonis MA, Mason SN, Pippen AM, Ruiz P, Wood ER, Gilkeson GS (1994). The role of nitric oxide in the pathogenesis of spontaneous murine autoimmune disease: increased nitric oxide production and nitric oxide synthase expression in MRL-lpr/lpr mice, and reduction of spontaneous glomerulonephritis and arthritis by orally administered N^G-monomethyl-L-arganine. *J Exp Med* **79**: 651–660.

90. Schmidt HHHW, Warner TD, Ishii K, Sheng H, Murad F (1992). Insulin-secretion from pancreatic B-cells caused by L-arginine-derived nitrogen oxides. *Science* **255**: 721–723.

91. McCarthy-Francis N, Allen JB, Mizel DE, Albina JE, Xie Q-W, Nathan CF, Wahl SM (1993). Suppression of arthritis by an inhibitor of nitric oxide synthase. *J Exp Med* **78**: 749–754.

92. Corbett JA, Lancaster JR Jr, Sweetland MA, McDaniel ML (1991). Interleukin-1-β-induced formation of EPR-detectable iron-nitrosyl complexes in Islets of Langerhans: role of nitric oxide in interleukin-1-β-induced inhibition of insulin secretion. *J Biol Chem* **266**: 21351–21354.

93. Nathan C, Xie Q (1994). Nitric oxide synthases: roles, tolls, and controls. *Cell* **78**: 915–918.

94. Lancaster JR Jr (1994). Simulation of the diffusion and reaction of endogenously produced nitric oxide. *Proc Natl Acad Sci USA* **91**: 8137–8141.

95. Kodama K, Kusuoka H, Sakai A, Adachi T, Hasegawa S, Ueda Y, *et al.* (1993). Collateral channels that develop after an acute myocardial infarction prevent subsequent left ventricular dilation. *J Am Coll Cardiol* **27**: 1133–1139.

96. Anversa P, Li P, Sonnenblick EH, Olivetti G (1994). Effects of aging on quantitative structural properties of coronary vasculature and microvasculature in rats. *Am J Physiol* **267**: H1062–H1073.

97. Gerhard M, Roddy MA, Creager SJ, Creager MA (1996). Aging progressively impairs endothelium-dependent vasodilation in forearm resistance vessels of humans. *Hypertension* **27**: 849–853.

98. Rivard A, Fabre JE, Silver M, Chen D, Murohara T, Kearney M, *et al.* (1999). Age-dependent impairment of angiogenesis. *Circulation* **99**: 111–120.

99. Couffinhal T, Silver M, Kearney M, Sullivan A, Witzenbichler B, Magner M, *et al.* (1999). Impaired collateral vessel development associated with reduced expression of vascular endothelial growth factor in ApoE $-/-$ mice. *Circulation* **99**: 3188–3198.

100. Panchal VR, Rehman J, Nguyen AT, Brown JW, Turrentine MW, Mahomed Y, *et al.* (2004). Reduced pericardial levels of endostatin correlate with collateral development in patients with ischemic heart disease. *J Am Coll Cardiol* **43**: 1383–1387.

101. Rissanen TT, Vajanto I, Yla-Herttuala S (2001). Gene therapy for therapeutic angiogenesis in critically ischaemic lower limb on the way to the clinic. *Eur J Clin Invest* **31**: 651–666.

102. Khan TA, Sellke FW, Laham RJ (2003). Gene therapy progress and prospects: therapeutic angiogenesis for limb and myocardial ischemia. *Gene Therapy* **10**: 285–291.

103. Carmeliet P, Moons L, Luttun A, Vincenti V, Compernolle V, De Mol M, Wu Y, Bono F, Devy L, Beck H, *et al.* (2001). Synergism between vascular endothelial growth factor and placental growth factor contributes to angiogenesis and plasma extravasation in pathological conditions. *Nat Med* **7**: 575–583.

104. Silvestre JS, Mallat Z, Duriez M, Tamarat R, Bureau MF, Scherman D, Duverger N, Branellec D, Tedgui A, Levy BI (2000). Antiangiogenic effect of interleukin-10 in ischemia-induced angiogenesis in mice hindlimb. *Circ Res* **87**: 448–452.

105. Sasaki K, Murohara T, Ikeda H, Sugaya T, Shimada T, Shintani S, Imaizumi T (2002). Evidence for the importance of angiotensin II type 1 receptor in ischemia-induced angiogenesis. *J Clin Invest* **109**: 603–611.

106. Johnson C, Sung HJ, Lessner SM, Fini ME, Galis ZS (2004). Matrix metalloproteinase-9 is required for adequate angiogenic revascularization of ischemic tissues: potential role in capillary branching. *Circ Res* **94**: 262–268.

107. Shibata R, Ouchi N, Kihara S, Sato K, Funahashi T, Walsh K (2004). Adiponectin stimulates angiogenesis in response to tissue ischemia through stimulation of amp-activated protein kinase signaling. *J Biol Chem* **279**: 28670–28674.

108. Sonveaux P, Marinive P, DeWever J, Batova Z, Daneau G, Pelat M, Ghisdal P, Gregoire V, Dessy C, Balligand JL, Feron O (2004). Caveolin-1 expression is critical for vascular endothelial growth factor-induced ischemic hindlimb collateralization and nitric oxide-mediated angiogenesis. *Circ Res* **95**: 154–161.

109. Pola R, Ling LE, Aprahamian TR, Barban E, Bosch-Marce M, Curry C, *et al.* (2003). Postnatal recapitulation of embryonic hedgehog pathway in response to skeletal muscle ischemia. *Circulation* **108**: 479–485.

110. Fukino K, Sata M, Seko Y, Hirata Y, Nagai R (2003). Genetic background influences therapeutic effectiveness of VEGF. *Biochem Biophys Res Commun* **310**: 143–147.

111. Masaki I, Yonemitsu Y, Yamashita A, Sata S, Tanii M, Komori K, *et al.* (2002). Angiogenic gene therapy for experimental critical limb ischemia: acceleration of limb loss by overexpression of vascular endothelial growth factor 165 but not of fibroblast growth factor-2. *Circ Res* **90**: 966–973.

112. Aicher A, Heeschen C, Mildner-Rihm C, Urbich C, Ihling C, Technau-Ihling K, Zeiher AM, Dimmeler S (2003). Essential role of endothelial nitric oxide synthase for mobilization of stem and progenitor cells. *Nat Med* **9**: 1370–1376.

113. Yu J, deMuinck ED, Zhuang Z, Drinane M, Kauser K, Rubanyi GM, *et al.* (2005). Endothelial nitric oxide synthase is critical for ischemic remodeling, mural cell recruitment, and blood flow reserve. *Proc Natl Acad Sci USA* **102**: 10999–11004.

114. Qian HS, Liu P, Huw L-Y, *et al.* (2006). Effective treatment of vascular endothelial growth factor refractory hindlimb ischemia by a mutant endothelial nitric oxide synthase gene. *Gene Therapy* **13**: 1342–1350.

115. Tamarat R, Silvestre JS, Kubis N, Benessiano J, Duriez M, deGasparo M, Henrion D, Levy BI (2002). Endothelial nitric oxide synthase lies downstream from angiotensin II-induced angiogenesis in ischemic hindlimb. *Hypertension* **39**: 830–835.

116. Sata M, Nishimatsu H, Suzuki E, Sugiara S, Yoshizumi M, Ouchi Y, Hirata Y, Nagai R (2001). Endothelial nitric oxide synthase is essential for HMG-CoA reductase inhibitor cerivastin to promote collateral growth in response to ischemia. *FASEB J* **15**: 2530–2532.

117. Lee EW, Michalkiewicz M, Kitlinska J, Kalezic I, Switalska H, Yoo P, Sangkharat A, Ji H, Li L, Michalkiewicz T, *et al.* (2003). Neuropeptide Y induces ischemic angiogenesis and restores function of ischemic skeletal muscles. *J Clin Invest* **111**: 1853–1862.

118. Hiasa K, Ishibashi M, Ohtani K, Inoue S, Zhao Q, Kitamoto S, Sata M, Ichiki T, Takeshita A, Egashira K (2004). Gene transfer of stromal cell-derived factor–1alpha enhances ischemic vasculogenesis and angiogenesis via vascular endothelial growth factor/endothelial nitric oxide synthase-related pathway: next-generation chemokine therapy for therapeutic neovascularization. *Circulation* **109**: 2454–2461.

119. Fukumura D, Gohongi T, Kadambi A, Izumi Y, Ang J, Yun CO, Buerk DG, Huang PL, Jain RK (2001). Predominant role of endothelial nitric oxide synthase in vascular endothelial growth factor-induced angiogenesis and vascular permeability. *Proc Natl Acad Sci USA* **98**: 2604–2609.

120. Gratton JP, Lin MI, Yu J, Weiss ED, Jiang ZL, Fairchild TA, Iwakiri Y, Groszmann R, Claffey KP, Cheng YC, Sessa WC (2003). Selective inhibition of tumor microvascular permeability by cautratin blocks tumor progression in mice. *Cancer Cell* **4**: 31–39.

121. Amano K, Mastubara H, Iba O, Okigaki M, Fujiyama S, Imada T, Kojima H, Nozawa Y, Kawashima S, Yokoyama M, Iwasaka T (2003). Enhancement of ischemia-induced angiogensis by eNOS overexpression. *Hypertension* **41**: 156.

122. Smith RS, Lin K-F, Agata J, Chao L, Chao J (2002). Human endothelial nitric oxide synthase gene delivery promotes angiogenesis in a rat model of hindlimb ischemia. *Arterioscler Thromb Vasc Biol* **22**: 1279–1285.

123. Shimada T, Takeshita Y, Murohara T, *et al.* (2004). Angiogenesis and vasculogenesis are impaired in the precocious-aging klotho mouse. *Circulation* **110**: 1148–1155.

124. Qian HS, Monterio de Resende M, Beausejour C, Huw LY, Liu P, Rubanyi GM, Kauser K (2006). Age-dependent acdeleration of ischemic injury in endothelial nitric oxide synthase deficient mice: potential role of impaired VEGF receptor 2 expression. *J Cardiovasc Pharmacol* **47**: 587–593.

125. Dimmeler S, Fleming I, Fisslthaler B, Hermann C, Busse R, Zeiher AM (1999). Activation of nitric oxide synthase in endothelial cells by Akt-dependent phosphorylation. *Nature* **399**: 601–605.

126. Fulton D, Gratton JP, McCabe TJ, Fontana J, Fujio Y, Walsh K, Franke TF, Papapetropoulos A, Sessa WC (1999). Regulation of endothelium-derived nitric oxide production by the protein kinase Akt. *Nature* **399**: 597–601.

127. McCabe TJ, Fulton D, Roman LJ, Sessa WC (2000). Enhanced electron flux and reduced calmodulin dissociation may explain "calcium-independent" eNOS activation by phosphorylation. *J Biol Chem* **275**: 6123–6128.

128. Hoffman J, Haendeler J, Aicher A, Rossig L, Vasa M, Zeiher AM, *et al.* (2001). Aging enhances the sensitivity of endothelial cells toward apoptotic stimuli: important role of nitric oxide. *Circ Res* **89**: 709–715.

129. Chavakis E, Dernbach E, Hermann C, Mondorf UF, Zeiher AM, Dimmeler S (2001). Oxidized LDL inhibits vascular endothelial growth factor-induced endothelial cell migration by an inhibitory effect on the Akt/endothlial nitric oxide synthase pathway. *Circulation* **103**: 2102–2107.

130. Sasso FC, Torella D, Carbonara O, *et al.* (2005). Increased vascular endothlelial growth factor expression but impaired vascular endothelial growth factor receptor signaling in the myocardium of type 2 diabetic patients with chronic coronary heart disease. *J Am Coll Cardiol* **45**: 827–834.

131. Schnyder B, Pittet M, Durand J, Schnyder-Candrian S (2002). Rapid effects of glucose on the insulin signaling of endothelial NO generation and epithelial Na transport. *Am J Physiol Endocrinol Metab* **282**: 87–94.

132. Urbich C, Reissner A, Chavakis E, Dernbach E, Haendeler J, Fleming I, Zeiher A, Kaszkin M, Dimmeler S(2002). Dephosphorylation of endothelial nitric oxide synthase contributes to the antiangiongenic effects of endostatin. *FASEB J* **16**: 706–708.

133. Rivard A, Silver M, Chen D, *et al.* (1999). Rescue of diabetes-related impairment of angiogenesis by intramuscular gene therapy with adeno-VEGF. *Am J Pathol* **154**: 355–363.

134. Parenti A, Morbidelli L, Cui XL, *et al.* (1998). Nitrix oxide is an upstream signal of vascular endothelial growth factor-induced extracellular signal-regulated kinase1/2 activation in postcapillary endothelium. *J Biol Chem* **273**: 4220–4226.

135. Matsunaga T, Warltier DC, Weihrauch DW, *et al.* (2000). Ischemia-induced coronary collateral growth is dependent on vascular endothelial growth factor and nitric oxide. *Circulation* **102**: 3098–3103.

136. Waller C, Hiller KH, Rudiger T, *et al.* (2005). Nonivasive imaging of angiogenesis inhibition following nitric oxide synthase blockade in the ischemic rat heart *in vivo. Microcirculation* **12**: 339–347.

137. Matsunaga T, Weihrauch DW, Moniz MC, *et al.* (2002). Angiostation inhibits coronary angiogenesis during impaired production of nitrix oxide. *Circulation* **105**: 2185–2191.

138. Szelid Z, Sinnaeve P, Vermeersch P, *et al.* (2002). Preexisting antiadenoviral immunity and regional myocardial gene transfer: modulation by nitric oxide. *Hum Gene Ther* **13**: 2185–2195.

139. Nagata K, Marban E, Lawrence J, Donahue JK, *et al.* (2001). Phosphodiesterase inhibitor-mediated potentiation of adenovirus delivery to myocardium. *J Mol Cell Cardiol* **33**: 575–580.

140. Barouch LA, Harrison RW, Skaf MW, *et al.* (2002). Nitric oxide regulates the heart by spatial confinement of nitric oxide synthase isoforms. *Nature* **416**: 337–339.

141. Massion PB, Feron O, Dessy C, Balligand JL (2003). Nitric oxide and cardiac function: ten years after, and continuing. *Circ Res* **93**: 388–398.

142. Hare JM (2003). Nitric oxide and excitation-contraction coupling. *J Mol Cell Cardiol* **35**: 719–729.

143. Elrod JW, Greer JJ, Bryan NS, *et al.* (2006). Cardiomyocyte-specific overexpression of NO synthase–3 protects against myocardial ischemia-reperfusion injury. *Arterioscler Thromb Vasc Biol* **26**: 1517–1523.

144. Janssens S, Pokreisz P, Schoonjans L, *et al.* (2004). Cardiomyocyte-specific overexpression of nitric oxide synthase 3 improves left ventricular performance and reduces compensatory hypertrophy after myocardial infarction. *Circ Res* **94**: 1256–1262.

145. Massion PB, Dessy C, Desjardins F, *et al.* (2004). Cardiomyocyte-restricted overexpression of endothelial nitric oxide synthase (NOS3) attenuates beta-adrenergic stimulation and reinforces vagal inhibition of cardiac contraction. *Circulation* **110**: 2666–2672.

146. Pfeffer MA, Pfeffer JM, Fishbein MC, Fletcher PJ, Spadaro J, Kloner RA, Braunwald E, (1979). Myocardial infarct size and ventricular function in rats. *Circ Res* **44**: 503–512.

147. Smith RS, Agata J, Xia C, *et al.* (2005).Human endothelial nitric oxide synthase gene delivery protects against cardiac remodeling and reduces oxidative stress after myocardial infarction. *Life Sci* **76**: 2457–2471.

Index

$av\beta3$, 392
  integrin, 356
α-adrenergic receptor, 294
β-adrenergic receptor, 294
β_2-adrenoceptor subtype, 295
3-O-sulfotransferae (3OST), 127
17β-estradiol, 389
20-hydroxyeicosatetraenoic acid
  (20-HETE), 296
26S proteasome, 182

aberrant sprouting of intersomitic vessel,
  49
acetylcholine, 295, 387, 388
acute myocardial infarction, 264, 410
Ad5NOS-III, 399
AdCA5, 185, 187
adeno-associated virus, 393
adenosine, 293
adenovirus, 393
  adenoviral delivery, 404
AdLacZ, 407
AdNOS1177D, 407
AEE788, 329
AG-013736, 329
age-dependent impairment, 400, 402
aging process, 301, 400
Akt, 183, 229, 389
  Akt-KO mice, 403
  Akt-1-KO mice, 407, 408
  Akt-phosphorylated wild-type
    NOS-III enzyme, 403
AMG 706, 330
angiogenesis, 109, 110, 175, 184, 186, 198,
  217, 219, 220, 225, 229, 253, 260, 385,
  389, 391, 397

angiogenesis-dependent
  transcription factor, 233
  corneal, 391
  coronary, 407
  impaired, 400
  ischemia–induced, 406
  post-ischemic, 401
  postnatal, 51, 226
  therapeutic, 343, 395
  tumor, 67, 75, 105, 107, 198, 313, 390
  repair associated, 78
angiogenic
  growth factor, 305
  therapy, 370
  response, 217
angiography, 198
  evidence, 399
angiopoietin-1 (Ang-1), 218, 223, 230,
  232
angiopoietin-2 (Ang-2), 185, 367
angiostatin, 409
angiotensin II (Ang II), 218, 223, 227
  type 1 receptor, 266
angiotensin-converting enzyme (ACE)
  inhibition, 256, 389
animal model, 396
ankle brachial index (ABI), 372
anti-angiogenic
  effect, 200
  kinase inhibitor, selectivity of, 315
anti-cancer agent, 200
anti-CD31 immunohistochemistry, 198
antioxidant, 220, 221
  vitamin, 389
aortocaval fistula, 258
Apo E-KO mice, 392

apolipoprotein-E deficient (ApoE$^{-/-}$) mice, 234
apoptosis, 217, 300, 402
   cardiac myocyte, 409
arterial occlusion rabbit model, 197
arterial-venous capillary boundary formation, 45
arteriogenesis, 101, 175, 186, 198, 253, 260, 344, 397, 406
arteriolar length density, 258
asymmetric dimethylarginine (ADMA), 302, 391
atherosclerosis, 95, 218, 228, 301, 386
atherosclerotic plaque, 105
ATP, 176
atrial natriuretic peptide (ANP), 223, 290
autoimmune disease, 394
autoregulation, 286, 292
axon guidance factor, 4
AZD 2171, 330

balb/c mice, 406
BAY 43-9006 (Sorafenib), 326
bcl-2, 369
BIBF 1120, 330
blood pressure (BP), 198, 388
blood vessel permeability, 327
bone marrow-derived mononuclear cell (BM-MNC), 348
brain natriuretic peptide (BNP), 290
Buerger disease, 372

c-kit, 330
c-met, 367, 368
Ca^{2+}-activated K^+ (BK) channel, 296
Caenorhabditis elegans, 176
cancer drug, 313
capillary, 282
   angiogenesis, 100
   capillary: myocyte ratio, 198
capillary growth, 263
   inadequate, 256
cellular adenoviral receptor (CAR), 394
cardiac hypertrophy, 253, 255, 306
   model, 254
cardiomyocyte, 226

cardiovascular
   disease, 387
   gene transfer, 393
   system, 367
catalase, 221
CBP co-activator, 183
cell migration, 4
cerebral artery occlusion, 388
chemokine, 120
Chir-258, 330
cholinergic receptor, 295
chorioallantoic membrane (CAM) model, 221
chronic heart failure, 410
clinical trial, 348
collapsin response mediator protein-2 (CRMP-2), 12
collateral formation, 375
collateral growth, 409
   ischemia-induced, 407
collateral vessel, 303, 399
   arterialized, 393
   development, 395
colorectal cancer, 326
congestive heart failure, 387
converting enzyme, 97
copper/zinc SOD (Cu/ZnSOD), 222
   cytosolic, 219
coronary flow, 253
   regulation, 283
coronary reactive hyperemia, 260
coronary reserve, 255, 263
   decline, 255
coronary resistance, 283
coronary vascular tone, 281
critical closing pressure, 299
critical limb ischemia (CLI), 370, 395
   mouse model, 406
crosstalk, endostatin-VEGF, 40
cyclic guanidine monophosphate (cGMP), 288, 386, 391, 410
   c-GMP-dependent protein kinase, 288
cyclooxygenase, 219
cytochrome p450, 219

diabetes, 218, 228, 301, 387
diastolic stress, 258
dipeptidyl peptidase IV (DPPIV), 97
docetaxel, 328
Drosophila melanogaster, 176

electron transfer, 177
embryonic development, 367
endocardial/epicardial flow ratio, 258
endostatin, 403
endothelial
    capillary morphogenesis, 33
    factors in vascular growth, 300
    function, 387
    NO production, 288
    senescence, 360
endothelial cell (EC), 185, 217, 256, 281,
    368, 397
    adhesion, 98
    arterial, 187
    capillary, 393
    coronary post-capillary (venular),
        409
    cultured, 389
    differentiation, 98
    dysfunction, 228, 303, 359, 387, 406
    human microvascular, 220
    migration, 98, 222, 392
    precursor, 396
    proliferation, 95, 98, 222, 316, 391,
        392
endothelial nitric oxide synthase (eNOS),
    98, 219, 229, 287, 301, 386, 410
    eNOS/NO pathway, 232
    gene transfer, 385
    protein expression, 404
endothelial progenitor cell (EPC), 228,
    301, 347
endothelin, 287
    endothelin-1, 184, 299
endothelium, 287
    endothelium-dependent relaxation,
        305, 389
    endothelium-dependent
        vasodilation, 387, 389
    impaired, 303

endothelium-derived hyperpolarizing
    factor (EDHF), 290, 293
endothelium-derived NO (EDNO), 386
    overexpression, 410
    deficiency, 387, 411
endothelium-derived relaxing factor
    (EDRF), 385
endothelium-specific kinase, 314
endurance training, 263
EPAS1 gene, 181
Eph, 156
    family, 27
    receptor, 28
EphA receptor, 27
EphA2, 59
    attractive target for selective anti-
        angiogenic therapy, 59
    signaling in endothelial cell migration
        and vascular assembly, 34
EphB, 35
    receptor, 27
EphB4, 44, 315
    venous endothelial cell marker, 44
ephrin, 156
ephrin-A1, 28, 33
ephrin-B, 35
ephrin-B reverse signaling, 37
    angiogenic effect, 38
ephrin-B2, 44
    arterial endothelial cell molecular
        marker, 44
epicardial coronary artery, 297
epidermal growth factor (EGF), 230
epoxyeicosatrienoic acid (EET), 290
ERK, 184
    ERK-STAT3 pathway, 369
erythropoietin (EPO), 178
Ets
    Ets-1, 233, 376
    family of transcription factors, 375
    pathway, 375
extracellular matrix, 217
extracellular SOD (ecSOD), 219
extravascular force, 297

fetal heart, 269

fibroblast growth factor (FGF), 67, 77,
  390, 392, 396, 410
    administration, 79
    FGF-1, 305
    FGF-2, 98, 266, 305, 377, 392
fibroblast growth factor receptor
  (FGFR), 133
    soluble, 76
fibrosis, 266
fibstatin, 77
flow-induced dilation, 287, 292
Flt-1, 229
forward signaling, 158

Gab1, 369
gain-of-function study, 186
gastrointestinal stromal tumor (GIST),
  327
Gefitinib, 329
gene delivery vector, 393
gene therapy, 355
glycosylphosphatidylinositol (GPI)
  linkage, 29
glypican, 68
granulocyte macrophage colony-
  stimulating factor (GM-CSF), 355
Grb-SOS complex, 369
green tea catechin, 229
growth factor, 201
GTPase, 155
    GTPase activating protein (GAP)-
      like domain, 5
guidance cue, 147
GW786034, 331

heart, 407
    failure, 218
hemodynamic force, 271
heparan sulfate proteoglycan (HSPG),
  68, 119
heparin, 121
    heparin-binding growth factor, 121
hepatocyte growth factor (HGF), 315,
  367, 390, 396, 410
    gene therapy for myocardial
      ischemia, 375

plasmid DNA, 371
signaling, 368
herparan sulfate (HS)
  development, 131
Hif1a$^{flox/flox}$; Tie2-Cre mice, 186
HIF-α, 262
HIF-1α, 99, 179, 198, 233
    O_2-dependent degradation, 182
    transactivation domain, 183
HIF-1β, 179
HIF-2αs, 181
HIF-3α, 182
hindlimb
    blood flow, 397
    ischemia, 403
histone acetyltransferase, 181
    humoral substance, 294
Homo sapiens, 178
hydrogen peroxide (H_2O_2), 217, 219, 220,
  232, 293
hydroxyl radical ($^\bullet OH$), 219
hyperglycemia, 228, 394, 403
hyperlipidemia, 301
hyperoxia, 177
hyperpolarizing factor, 287
hypertension, 218, 254, 255, 303, 330, 368
    hypertensive reaction, 334
hypertrophy
    exercise induced, 263
    hypoxia induced, 261
    myocardial infarction induced, 264
    pressure overload induced, 255
    right ventricular, 256
    thyroxine induced, 260
    volume overload induced cardiac,
      258
hypochlorous acid (HOCl), 219
hypoxia, 137, 177, 181, 184, 187, 201, 220,
  288
hypoxia-inducible factor-1 (HIF-1), 175,
  178, 185, 186, 233
hypoxia response element (HRE), 178
    binding site, 178
    target gene, 204

IGF-R1, 201

immune response, 394
   regulation of, 92
inducible nitric oxide synthase (iNOS), 221, 386
inhibitory PAS (IPAS), 182
insulin-like growth factor (IGF)-2, 202
intrinsic
   control mechanism, 285
   myogenic tone, 296
ischemia, 297
   ischemia-induced myocyte death, 410
   ischemia-induced retinal neovascularization, 192
   ischemia-reperfusion, 303
   ischemic heart disease, 375
   ischemic ulcer, 373

jagged1, 70

KDR expression, 401
KDR/Flk1, 229
kinase inhibitor, 313

L-arginine, 391
left ventricular end diastolic volume, 261
lentivirus, 393
limb loss, 400, 405
liposome, 393
L-nitroarginine methyl ester (L-NAME), 300, 392, 409
loss-of-function study, 186
lovastatin, 389
low density lipoprotein (LDL), 394

macrophage, 218, 220
magnetic resonance imaging (MRI), 357
maximal vasodilation, 254
metabolic regulation, 286
matrix metalloproteinase (MMP), 234, 409
   MMP-1, 376
   MMP-2, 409
   MMP-9, 409
metastatic
   breast cancer, 328
   neuroendocrine tumor (NET), 327

micro-CT, 356
microangiography, 399
microcirculation, neurohumoral influence on, 294
microvascular
   physiology, 281
   resistance, redistribution of, 284
microvessel
   arterial, 282
   classification, 281
   venous, 283
minimal coronary vascular resistance, 254
mitochondria, 176
mitochondrial-
   restricted manganese SOD (MnSOD), 219
mitogenic effect, 97
monocyte, 218
monocyte chemoattractant protein (MCP)-1, 233
mouse, 388
mTOR, 183
myeloperoxidase (MPO), 219
myocardial blood flow, 258
   redistribution, 258
myocardial
   gene transfer, 409
   ischemia, 388
   perfusion, 254

N-acetylcysteine (NAC), 221
N^G-monomethyl-arginine (L-NMMA), 389
Na/K ATPase, 290
NAD(P)H oxidase, 218, 222, 224, 225
   phagocytic, 222
   vascular, 222
naked plasmid, 393
Nck-NIK-JNK pathway, 36
neointima, 105, 387
neovascularization, 95, 192, 344
nerve conduction velocity, 378
neural crest origin, 109
neuroblastoma, 106
   growth factor, 107

neurogenesis, 109, 110
neurogenic trophic factor, 92
neuronal
    precursor, 107
    sprouting, 110
neuropeptide Y, 91, 296
neuropilin, 314
    neuropilin-1, 148
neuronal nitric oxide synthase (nNOS), 386
neurotransmitter, 294
    release, inhibition of, 92
nitric oxide (NO), 98, 232, 287, 293, 300, 334, 359, 377, 385, 395, 410
    donor, 393
    nitroglycerin, 409
    NO•, 219, 288
NOGA™ system, 375
non-insulin diabetes mellitus (NIDDM) rat model, 227
non-obese diabetic (NOD) mice, 403, 406, 408
non-small cell lung cancer (NSCLC), 327
non-viral vector, 393
NOS gene transfer, 393
NOS-II gene transfer, 394
NOS-III, 387, 410
    adventitial delivery, 394
    gene transfer, 394
    genetic deficiency, 406
    mutant gene, 401, 402
    NOS-III-deficient mice, 388
    NOS-III-KO mice, 391, 396, 402
    NOS-III-S1177D gene transfer, 403
    phosphomimetic mutant, 411
    serine 1177 in enzyme, 390
    transgenic mice, 397
    wild-type enzyme, 402
    wild-type gene transfer, 401
NOS1177D, 402
notch, 70, 132
Nox2, 226
    knockout mice, 226

oxygen (O$_2$)
    atmospheric concentration, 176
    consumption, 177
    delivery, 176
    diffusion, 176, 254
    homeostasis, 175, 177
    utilization, 253
old animals, 400
organogenesis, 367
oxidative
    phosphorylation, 176
    stress, 218, 302, 360
oxidatively modified LDL (oxLDL), 386, 403

p22phox, 222
    transgenic mice overexpressing, 226
p300, 183
p40phox, 222
p47phox, 222
p67phox, 222
paclitaxel/carboplatin, 328
parasympathetic nervous system, 294
PDGFR-β kinase, 326
PECAM-1, 397
percutaneous revascularization, 370
pericyte, 204
peripheral arterial disease (PAD), 370
    clinical trial, 371
peripheral arterial occlusive disease, 387, 395
peripheral blood-derived mononuclear cell (PB-MNC), 348
peripheral neuropathy, 378
perlecan, 68
peroxynitrite (ONOO$^-$), 219
phagocyte, 218
pharmacokinetics, 354
phase II study, 326
phase III trial, 326
pheochromocytoma, 108
phosphatidylinositol-3-OH kinase (PI3K), 369
photosynthesis, 176
PI3K-Akt pathway, 369
pigment epithelium-derived factor (PEDF), 221
placebo effect, 358

placental growth factor (PLGF), 185, 192
plasmid, 403
plasminogen activator inhibitor-1 (PAI-1), 234
platelet-derived growth factor (PDGF), 315
   PDGF-B, 185
   PDGF-BB, 270
plexin D1, 151
positron-emission tomography (PET), 357
post-ischemic
   flow recovery, 405
   revascularization, 391
postsynaptic density 95, disk large, zona occludens-1 (PDZ), binding motif, 6, 154
pre-immunized pig, 409
progenitor cell, 42
prostacyclin synthase, 378
prostaglandin, 287
prostaglandin I_2 (PGI_2), 290, 377
prostate cancer, 228
protein tyrosine phosphatase, 230
PTK/ZK (Vatalanib), 326
pulmonary artery coarctation, 254

QT interval prolongation, 328
quinapril, 389

Rac1, 222
Raf-1, 326
rapamycin, 183
Ras-MAP kinase cascade, 369
reactive oxygen species (ROS), 217, 218, 222, 287, 360
   molecular target, 230
receptor tyrosine kinase family V, 316
red wine, 221
redox
   signaling, 229
   state, 217
renal cell carcinoma (RCC), 326
renin-angiotensin system, 227
reoxygenation, 181
reperfusion injury, 388

restenosis after angioplasty, 377
resting pain, 373
retinal neovascularization, 221
retinopathy, 103, 104
reverse signaling, 160
Rho guanine exchange factor (GEF), 155

Sef, 72
sema domain, 4
   sema3C, 151
   sema3E, 151
semaphorin signaling, 148
sepsis, 394
serotonin, 299
Shc, 369
shear stress, 286, 288, 393, 402
single-photon emission computed tomography (SPECT), 357
skin ulcer, 400
smooth muscle cell, 281
soluble guanylate cyclase (sGC), 386
split kinase domain, 316
spontaneously hypertensive rat, 256
Src family kinases, 38
STAT pathway, 369
statin, 227, 389
stem cell, 42
stromal-derived growth factor-1 (SDF-1), 185
SU11248 (Sunitinib), 327
SU5416 (Semaxinib), 331
subendocardial infarct, 285
substance P, 296
Sulf1, 138
Sulf2, 138
superoxide anion radical ($O_2^{\bullet-}$), 217, 219, 302, 386
superoxide dismutase (SOD), 221
sympathetic nervous system, 294
syndecan, 68
   syndecan-1, 41
synergism, FGF and VEGF, 77

telomere, 360
tempol, 221
therapeutic trial

for coronary disease, 79
for intermittent claudication, 79
Tie-1, 314
Tie-2, 262, 266
Tie-2 receptor, 230
tissue edema, 297
transcutaneous PO_2 (TcPO$_2$), 372
transforming growth factor (TGF)-α, 202
TGF-β, 367
tumor cell proliferation, 316
tumor vasculature, 54
tyrosine phosphorylation, 368
tyrosine-kinase Fes/Fps, 5

ubiquitination, 182
urokinase plasminogen activator (uPA), 234, 392

vagal stimulation, 295
vascular
    cell, 367
    growth, 254
    injury, 300
    permeability, 198
    remodeling, 95
    resistance, 254
    tone, 388
vascular cell-adhesion molecule-1 (VCAM-1), 233
vascular endothelial growth factor (VEGF), 1, 40, 77, 179, 192, 200, 204, 217–219, 223, 225, 227, 256, 262, 266, 301, 305, 328, 373, 389, 391, 392, 396, 400, 409, 410
    gene, 184
    polymorphism, 361
    signaling, 134
    VEGF-121, 392
    VEGF-165 gene, 377
    VEGF-A, 314

VEGF-C, 314
VEGF-D, 314
VEGF/KDR, 389
vascular smooth muscle cell (VSMC), 47
    migration, 373
    proliferation, 97
vasculature, 218
vasculogenesis, 267, 346
vasoconstriction, 92
    effect, 97
vasomotor tone, endothelial regulation of, 286
vasopressin, 299
VEGF receptor (VEGFR), 201, 270, 314, 401
    tyrosine kinase, 327, 328
    VEGFR-1, 229
    VEGFR-2, 98, 184, 229
ventricular remodeling, 410
viral vector, 393
volumetric computed tomography (VCT), 357
von Hippel-Lindau (VHL)
    null renal carcinoma cell, 179
    tumor suppressor protein, 182
von Willebrand factor (vWF), 401

white adipose tissue (WAT), 102
wound
    healing, 101
    repair, 101

xanthine oxidase, 218

Y1 receptor, 94, 95
Y1/Y5 receptor, 97
Y2 receptor, 95
Y2/Y5 receptor, 97

ZD6474, 328